THE LANDAU THEORY OF PHASE TRANSITIONS

THE LANDAU THEORY OF PHASE TRANSITIONS

Application to Structural, Incommensurate, Magnetic, and Liquid Crystal Systems.

Jean-Claude
TOLÉDANO
Centre National d'Etudes des
Télécommunications
FRANCE

Pierre
TOLÉDANO
University of Amiens
FRANCE

Published by

World Scientific Publishing Co. Pte. Ltd.
P.O. Box 128, Farrer Road, Singapore 9128

U.S.A. office: World Scientific Publishing Co., Inc.
687 Hartwell Street, Teaneck NJ 07666, USA

Library of Congress Cataloging-in-Publication data is available.

THE LANDAU THEORY OF PHASE TRANSITIONS

ISBN 9971-50-025-6
9971-50-026-4 (pbk)

Printed in Singapore by Kim Hup Lee Printing Co. Pte. Ltd.

A NOS PARENTS

FOREWORD

Why a book on Landau's theory of phase transitions ? To many physicists working in the field of phase transitions, this question will appear as doubly relevant. Indeed, why describe in detail the foundations and consequences of a theory whose basic hypothesis (the absence of singularity in the transition free-energy), and whose essential physical result (the specification of a critical behaviour) have been known for 40 years to be questionable ? Why, on the other hand, restrict to this phenomenological approach, at a time when microscopic models can be handled by various theoretical and numerical methods, and provide a "royal way" to the investigation of phase transitions ?

The existence of a satisfactory answer to these questions is attested by the fact that, ever since its formulation, Landau's theory has been used without interruption as the theoretical background of many studies of systems undergoing phase transitions. More strikingly, it is in the last 15 years, after the advent of the modern statistical theory of critical phenomena, that the utility of Landau's theory has been demonstrated most clearly, when it has been applied to the intricate patterns of transitions observed in the structural, magnetic, and liquid-crystalline systems, and more recently, to the investigation of the stabilities and properties of the incommensurate phases, and of the icosahedral quasi-crystalline phases.

There are several reasons to this persistent use of Landau's theory. A first set of reasons is related to the fact that the objected lack of validity of the basic assumptions and results of Landau's theory is not of a clearcut nature. Thus, the symmetry aspects which constitute an important part of the theory, are rigorous. Besides, there are classes of systems, governed by long range interactions (e.g. elastic interactions) for which the critical behaviour is expected to be correctly described by Landau's theory. More significantly, the temperature range adjacent to the transitions, in which the behaviour of a system is dominated by the fluctuations (and in which, accordingly Landau's theory fails), usually constitutes a small fraction of the temperature range of experimental interest. In most of the latter range, the Landau theory is an adequate tool to investigate the physical behaviour of the system.

Another important set of reasons pertains to the possibility of manipulating, through a mathematically simple and flexible theory, the complicated degrees of freedom which describe the states of real systems (e.g. sets of collective atomic displacements, or intricate spin configurations). Likewise, one has the possibility of relating simply to eachother a variety of physical properties (mechanical, optical, lattice-dynamical, structural,...) whose microscopic description would be very difficult.

The mathematical simplicity of the theory is a consequence of the clever manner by which Landau defines the order-parameter through the substitution of a small set of scalar (spatially uniform) quantities, to a set of functions having rapid variations at the atomic level. This substitution which appears as a "trick" of mere mathematical convenience, has the important consequence of permitting the description of the evolution of a complex spatial configuration of particles by means of an ordinary polynomial expansion. This trick also gives to the order-parameter a duality of meanings. As a spatially uniform quantity (or a smoothly varying one, in certain cases) it can be considered as of macroscopic nature. On the other hand, the functions it substitutes are clearly of microscopic nature. At choice, one can put the accent on one interpretation or the other (e.g. on the dielectric polarization of a ferroelectric crystal, or on the structural changes and lattice dynamical mode related to the polarization).

The flexibility of the theory resides in its modular character. Aside from the primary order-parameter, additional degrees of freedom can be incorporated in the theory as measure as the acquisition of the experimental data requires interpreting a larger set of results. For instance, in the study of a crystalline transition, once the primary order-parameter is given a sense in terms of atomic displacements, one can focus successively on the anomalies induced by the considered transition in the thermal expansion, the optical properties, the vibrational atomic spectrum, etc... In this view one will add terms in the Landau free-energy respectively corresponding to the mechanical deformations, to the dielectric polarization, to other collective atomic displacements, etc...

The only rigid feature of the theory is its symmetry framework which imposes the form of the interactions between the various degrees of freedom, and the number of adjustable phenomenological coefficients. The form of the interactions determines the relationship between the laws governing the behaviour of the various physical quantities. The explanatory power of Landau's theory resides in the checking of the overall consistency of the observed laws.

Finally, an additional reason of consideration of Landau's theory is its specific status in respect to the statistical theory of critical phenomena (Wilson's theory). From this standpoint, Landau's theory appears as a necessary point of passage, and also as a tool. On the one hand, it is the order-parameter defined by Landau's theory whose fluctuations give rise to a singularity at the transition point. Accordingly, it is on this set of degrees of freedom that the statistical theory operates. On the other hand, Landau's theory provides the rules for constructing the effective Hamiltonian density, function of the order-parameter and of the secondary degrees of freedom, on which the renormalization-group transformations act.

The contents of this book stems from three different objectives. In the first place, it is an introduction to the basic principles and techniques of Landau's theory, which is intended for teaching purposes. In this spirit, it includes an introduction to the peripheric group-theoretical and crystallographic concepts required to work out the theory. This part of the book is an expanded version of courses taught by the authors in various circumstances. Chapter I is a self-contained, simplified, introduction to the basic aspects of Landau's theory, which is well adapted to a teaching at the undergraduate level. The first paragraphs of chapter II constitute a complete presentation of the theory. They involve a thorough discussion of the starting assumptions and an explicit decomposition of the steps of the argumentation. The same pedagogical purpose has presided over the writing of chapter III, of the two first paragraphs of chapter IV, and of the four first paragraphs of chapters VI and VII. These chapters are respectively devoted to the applications of Landau's theory to structural transitions, to first-order transitions, and to magnetic and liquid-crystalline systems. Their contents is rooted in courses given at the graduate level.

A second purpose of the book is to provide the practical "recipes" for applying Landau's theory to complex systems. In this view, each element of the method is illustrated by several examples and the intermediate steps of many calculations are explicitely reproduced. Thus, one can find the constructions of the matrices of the order-parameter representation, or corepresentation (chaps. II, III, V and VI), the construction of the Landau free-energy (chaps. II and V), the description of the procedures of its minimization, and the method of identification of the low symmetry group for structural (chap. III) magnetic (chap. VI) and liquid crystal (chap. VI) systems. The procedure of application of the Landau criterion (chap. II) and of the Lifschitz one (chap. III) are exposed in details. For incommensurate systems (chap. V), an extensive description is given of the construction of the Lifschitz-invariant and other spatially dispersive terms, as well as the resolution of the equations relative to the standard situation of a two-component order-parameter. In the chapter devoted to liquid crystals, we have adopted a unified description of the various types of transitions occuring in these systems while existing theories often derive from a variety of approaches (chap. VII).

The last objective of the book is to incorporate the developments which have arisen in the last 15 years from the extensive application of the theory to a variety of physical systems. These developments involve several aspects. On the one hand, certain bases of the theory itself have been discussed. The meaning of the Lifschitz criterion has been analyzed by a number of authors and substantially clarified through the study of incommensurate systems. Its initially derivation by Lifschitz (chap. V) has been replaced by other derivations, physically more transparent, and mathematically more correct, though more complex (chap. III). Conversely, it has been understood that Lifschitz's derivation provided a method for the study of incommensurate systems (chap. V).

The second aspect of progress concerns the specification of the essential symmetries underlying the Landau theory, through the replacement of groups acting in the physical space, by groups acting in the order parameter space. One has been able, by this means, to express in a more efficient way the intrinsic symmetry of the order-parameter, the symmetry of the truncated free-energy expansion, and the characteristics of the symmetry breaking across the transition. New procedures of minimization of the free-energy have been based on the consideration of these essential symmetries. These methods have clarified the nature of the mathematical problem set by the Landau theory : find the absolute minimum of a mth degree polynomial in several variables, i) having a local maximum at the origin, ii) positive and infinite at infinity in any direction, iii) whose extrema have a symmetry-specified degeneracy, and iiii) which possesses obligatory extrema along symmetry directions in the order-parameter space (chap. II, paragraph 4).

Finally, a large part of the book is devoted to the systems, already mentioned above, which have been studied, in recent years, by means of Landau's theory : continuous or discontinuous structural transitions involving coupling between several relevant degrees of freedom (chaps III and IV) magnetic transitions (chap. VI), transitions in liquid-crystals (chap. VII), incommensurate phases (chap. V). We have also outline the principles of the application of the theory to icosahedral phases and to defects (chap. VIII).

In certain chapters, we had the choice between various distinct approaches. It is worth pointing out that the theory of first-order transitions is essentially inspired by the works of Gufan and co-workers (chap. IV). The theories of magnetic and liquid crystal systems have respectively their roots in methods elaborated by Dzyaloshinski and by Indenbom and coworkers (chaps VI and VII).

The multiplicity of objectives pursued has the consequence that the different chapters or paragraphs are not treated evenly. In certain parts of the book, each statement is justified, while in others, dealing with more recently developped fields, the reader is directed, for complete justification to appropriate reference works. The latter situation will be found, in particular, in large fractions of chapters IV-VIII.

In writing this book, we feel indebted to a number of colleagues. We are especially greatful to Louis Michel. From discussions with him we have learned most of the considerations pertaining to the essential symmetries of the Landau theory, which are included in the book. We have also benefitted from meeting several times Yu M. Gufan and V.P. Dimitriev who have shared with us their deep understanding of the physical implications of the Landau theory. The enlightening explanations of E. Brézin have been very helpful to clarify our view of the situation of Landau's theory from the standpoint of statistical physics. We had stimulating discussions with several experts in the handling of Landau's theory, namely N. Boccara, V. Dvorak, and A.P. Levanyuk.

We are also indebted to F.W. Ainger,R. Chaves, A. Janner,T. Janssen J.Mozrzymas, P.M. Raccah, H. Schmid, and D. Weigel,who have invited us to teach the courses which have eventually resulted in this book. We acknowledge stimulating collaborations in various fields of application of Landau's theory with M. Clin, G. Errandonéa, J. Schneck, H. Schmid, P. Schobinger-Papamantelos, and R. Tekaia.

It is a pleasure to thank M. Coiret, P. Durand, and M.C. Mourier for their competent realization of the huge task of typing the manuscript.

THE LANDAU THEORY OF PHASE TRANSITIONS

TABLE OF CONTENTS

Chapter V : LANDAU THEORY OF INCOMMENSURATE PHASES

Chapter VI : THE LANDAU THEORY OF MAGNETIC TRANSITIONS

CHAPTER I

INTUITIVE APPROACH TO THE BASIC IDEAS OF LANDAU'S THEORY

1. INTRODUCTION

The occurence of a phase transition at a certain temperature and pressure can be easily perceived in certain physical systems. This is, for instance, the case of the evaporation, or the melting of a substance, for which the two phases differ greatly in some of their physical properties (e.g. mechanical, optical,...) and, in consequence, can be straightforwardly distinguished by visual inspection. In many other systems, the existence of a phase transition can only be infered from observation of subtle changes in physical quantities whose measurement require sophisticated equipment (e.g. neutron scattering). Generally speaking, the detection of a phase transition implies that, at least one specific physical quantity differs in the two phases. The identification of this property is an important step in the description and understanding of the considered transition.

In the case of the liquid-vapour transition, a relevant property is the fluid's density (or the specific volume) the value of which changes by several orders of magnitude between the two phases. However, it is to be noted that this quantity does not only change between the two phases, but also in each phase when the temperature or pressure is modified. The transition point is singled out by the fact that a discontinuous (finite) change of the density will occur as opposed to its continuous variations in either phases. In fluids, the presence of a discontinuity appears as a requirement for revealing the transition through the variations of the relevant physical quantity. Such a requirement does not hold necessarily for other types of systems. In the liquid-vapour transition, it derives from the fact that the two phases adjacent to the transition can be considered as "quantitatively distinct" but "qualitatively identical". Indeed ,beyond the critical point (which is the end-point of the line of discontinuous transitions in the pressure-temperature plane) the two phases are essentially undistinguishable. The concept of a "qualitative difference" between phases can be precisely defined as their difference of symmetry.

We can provisionally define the symmetry of a given phase (this point will be further discussed in chapter II) as the set of geometrical transformations which leave unchanged the equilibrium spatial configuration of the particles (atoms) constituting the system, in the considered phase. In both the liquid, and the vapour phase, the distribution of atoms in space is "isotropic" and "homogeneous". The former property ensures that these phases are invariant by all rotations and reflexions about any point of the medium, while the latter implies the invariance by any translation. The two phases have the same symmetry, defined by the set of the enumerated geometrical transformations.

A different situation arises when the two phases have different symmetries and thus differ "qualitatively". In this case, whatever the temperature and the pressure, the two phases can be distinguished by their symmetries, and, accordingly, there can be no "end-point" as in fluids. Besides, the transition point can be defined, independently from the discontinuous jump of a physical parameter, by the point of occurence of a symmetry change.

The Landau theory of phase transitions pertains to the latter category of transitions. It is a phenomenological theory : it assumes the existence of a phase transition in the considered system, as well as the occurence of a symmetry change across the transition point. Its aim is to establish the mutual compatibility of the symmetry characteristics, and of the physical characteristics of the transition : relationship between the symmetries of the two phases, consistency between the nature of the symmetry change and the nature of the physical quantities behaving anomalously across the transition. It achieves this aim by means of introducing two basic concepts, the <u>order-parameter</u> (OP) and the <u>Landau-free-energy</u>.

The working out of the theory will be performed in chapter II, and its application to various types of systems will be described in chapters III-VII. In the present chapter, we rely on an elementary example , taken in the class of "crystalline" phase transitions, in order to provide an inductive introduction to the basic arguments used in the theory.

2. MODEL EXAMPLE OF PHASE TRANSITION IN A CRYSTAL.

2.1. The system and its degrees of freedom

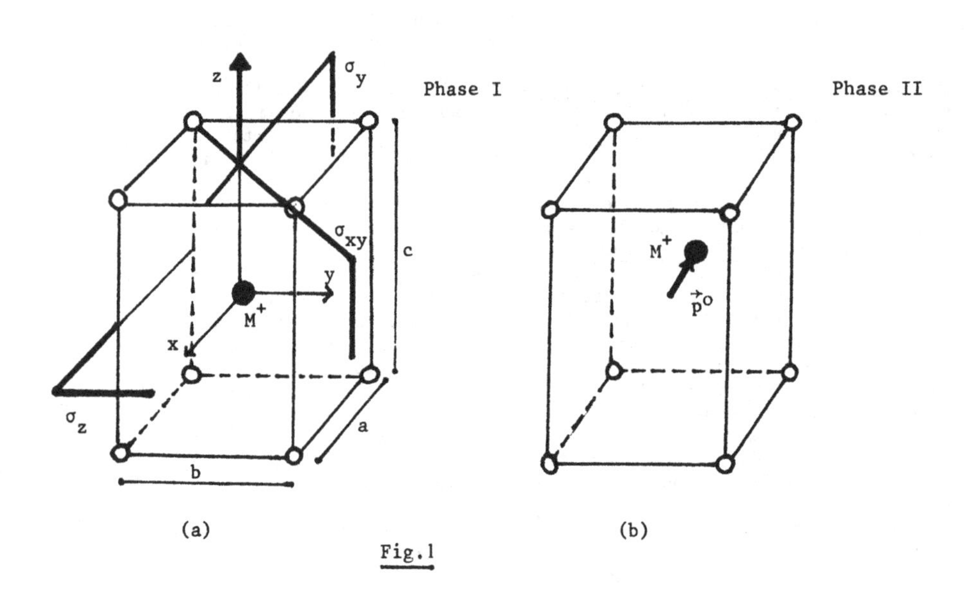

Fig.1

We consider a crystalline substance in which a phase transition is assumed to take place at a given temperature and pressure (T_c, P_c). In one of the phases (denoted phase I, and assumed to be stable for $T > T_c$), the substance consists in the juxtaposition of unit-cells all identical to the one represented on figure 1a). This "tetragonal" cell is a regular prism with a square basis ($a = b \neq c$). Its vertices and its center are respectively occupied by negative ions and by a positive ion M^+. We assume, on the other hand, that the other phase (denoted II, and stable for $T < T_c$) is also constituted by the juxtaposition of identical tetragonal unit-cells, only differing from the ionic configuration of phase I by the fact that the M^+ ion is displaced out of the center of the cell, in an unspecified direction. The displacement is assumed to be the same in all the unit-cells of the crystal. The relevant degree of freedom allowing to distinguish one phase from the other is the set of coordinates of the off-center displacement of the M^+ ion. As this displacement generates, in each cell, an electric dipole $\vec{p}_o = (p_x^o, p_y^o, p_z^o)$, we can also characterize phases I and II, respectively by $\vec{p}_o = 0$ and $\vec{p}_o \neq 0$ [in an equivalent way we can also use the resultant of the dipoles per macroscopic unit volume, i.e. the dielectric polarisation $\vec{P}_o$].

2.2. Symmetry of the two phases

Given the assumed identity of the different unit-cells of the crystal, in the two phases, let us consider that the structure of the crystal is entirely represented by the ionic configuration of one unit-cell. In each phase, this configuration is left unchanged by a set of rotations and reflexions, refered to the center of the cell, whose elements are easily enumerated.

As apparent on figure 1a), phase I is left invariant by fourfold ($\pi/2$) rotations around the z-axis, by reflexions about planes σ_y, σ_z, or σ_{xy} perpendicular to the coordinate axes, or to the diagonals of the square basis of the unit-cell. It is also unchanged by the inversion I about the center of the cell. Clearly the product of any two of the preceding geometrical transformations will also leave the structure unchanged : the set of "symmetry" transformations is a group G_o which contains 16 elements (Its crystallographic label in either of the two currently used conventions is 4/mmm or D_{4h} .Cf. chapter III §2).

The set of transformations leaving invariant phase II is a group G which depends of the direction of the displacement of M^+. If the dipole $\vec{p}_o$ associated to this displacement is along the z-direction, G contains the fourfold rotations around z as well as the various reflexions σ_x, σ_{xy}, in planes containing the z-axis. It does not include other elements of G_o such as the inversion I or the reflexion σ_z since these transformations reverse the sense of $\vec{p}_o$, and accordingly do not preserve the displaced position of M^+. In this case G is a group of 8 elements (whose standard label is 4 mm or C_{4v}) all belonging to G_o : G is a subgroup of G_o.

It is easy to check (figure 1b), in the cases where $\vec{p}_o$ belongs to the (x,y) plane, that, depending on the direction of $\vec{p}_o$

in this plane, the structure is invariant by sets of transformations G forming subgroups of G_o of various orders. For instance if $\vec{p}_o$ is along the x-axis, G is a group of 4 elements generated by the 2 reflexions in the planes σ_z and σ_y (labelled $mm2_x$ or C_{2v}). Table 1 lists the symmetry groups of phase II for the other possible directions of $\vec{p}_o$, within the (x,y) plane or inclined on this plane.

Table 1

Direction	Symmetry-group	Direction	Symmetry-group
(0,0, p)	4mm	$(p_x,0,p_z)$	(E,σ_y)
(p,0,0)	$mm2_{\pm x}$	$(0,p_y,p_z)$	(E,σ_x)
(0, p,0)	$mm2_{\pm y}$	(p,p,p_z)	(E,σ_{xy})
(p,p,0)	$mm2_{\pm\bar{x}y}$	$(-p,p,p_z)$	(E,σ_{xy})
(-p,p,0)	$mm2_{\pm xy}$	$(p_x,p_y\neq p_x,p_z)$	E
$(p_x,p_y\neq p_x,0)$	(E,σ_z)		

2.3. Variational free-energy associated to the system

A basic idea of Landau's theory is to consider an arbitrary displacement of M^+, specified by the dipole $\vec{p}$, as a variational degree of freedom of the system, and to note that the equilibrium value $\vec{p}_o(T,P)$, at any temperature and pressure, can be determined by minimizing with respect to the components of $\vec{p}$, a variational "free-energy" $F(T,P,\ p_x,\ p_y,\ p_z)$, referred, for instance ,to one unit-cell of the system. (F is not necessarily a free-energy in the strict sense. It is rather the thermodynamic potential whose minimum, in the given external conditions imposed to the system, determines the equilibrium state of the system. We will use, nevertheless the term "free-energy" for it).

The next step is to show that, given certain general assumptions, a form of F, valid in the neighborhood of T_c, can be determined. We assume, first, that the transition is continuous, i.e. that the components of $\vec{p}_o(P,T)$ vary continuously across T_c. On the other hand, we assume that the regularity of the function F is such that a Taylor expansion of F can be performed nearby $(T_c,\ P_c,\ \vec{p}_o)$.

Invoking the continuity, we note that in the vicinity of the transition, $|\vec{p}_o|$ is zero, or small. Accordingly, the determination of the functional form of F can be restricted to small $|\vec{p}|$, $|T-T_c|$ and $|P-P_c|$. F is then equal to the sum of the first relevant terms of its Taylor expansion as a function of $\vec{p}$, $(T-T_c)$ and $(P - P_c)$.

2.4. Symmetry property of F and form of its 2nd degree Taylor expansion

The important symmetry argument which allows to specify the form of F is that F is a function of $\vec{p}$, invariant by all the geometrical transformations constituting the group G_o of phase I. Indeed, an arbitrary displacement ($\vec{p}$) and the displacement ($\vec{p}'$) transformed from ($\vec{p}$) by any element of G_o (e.g. a fourfold rotation) correspond to identical relative positions of the negative and positive ions in the system. Only the global orientation of this configuration is changed. As the free-energy F only depends of the "internal" state of the system, and not of its absolute orientation (or, equivalently, of the coordinate framework adopted), F has to be invariant when the transformations $\vec{p} \to \vec{p}'$ are performed.

This invariance with respect to G_o also holds for the terms of each degree belonging to the Taylor expansion of F.

Indeed, any element of G_o will transform the components (p_x, p_y, p_z) into linear combinations of the same components. The degree of each term of the Taylor expansion will be preserved, and accordingly terms of different degrees will not "mix" in the transformation (A less general argument could also be invoked relying on the fact that, for small $|\vec{p}|$, the terms of different degrees have different orders of magnitude, and have to be separately invariant by G_o).

Up to second degree terms in the p_i, the expansion of F can be written as :

$$F(T,P,\vec{p}) = F_o(T,P) + \Sigma a_i(T,P).p_i + \Sigma b_{ij}(T,P).p_i.p_j \qquad (2.1)$$

As T and P are scalar quantities the coefficients a_i and b_{ij} do not change under the application of a geometrical transformation.

a) Absence of a linear term

Consider the 3 reflexions σ_x, σ_y and σ_z belonging to G_o. Each σ_i reverses the corresponding component p_i and leaves unchanged the two other components. The action of these transformations shows that the linear term in (2.1) cannot be invariant by all the transformations of G_o, as required, unless the coefficients $a_i(T,P)$ are identically zero. The invariance of F with respect to G_o therefore implies that its Taylor expansion does not contain a linear term in the p_i.

b) Form of the second-degree term

The action of the σ_i also precludes the existence in (2.1) of any "cross"-term $p_i.p_j$ ($i \neq j$). On the other hand, the action of a fourfold rotation around z exchanges p_x^2 and p_y^2. This implies $b_{11}=b_{22}$. The latter condition completes the restrictions set by the requirement of invariance of the second-degree term in (2.1) with respect to G_o. Indeed, any element of G_o is a product of the elements already considered. Hence a term invariant by these elements will be invariant by their successive, or repeated application : it is sufficient to check the invariance by the "generators" of the group (in the present case G_o can be generated by a fourfold rotation and by the two reflexions σ_x and σ_z). Relabelling the coefficients b_{ii}, we can write the form of (2.1), up to the second degree terms (which are the relevant terms of lowest degree of the expansion):

$$F(T,P,\vec{p}) = F_o(T,P) + \frac{\alpha_2(T,P)}{2}(p_x^2 + p_y^2) + \frac{\alpha_1(T,P)}{2} p_z^2 \qquad (2.2)$$

2.5. Decoupling of the (p_x,p_y) and p_z degrees of freedom. Order-parameter

Let us now describe an important step of the method, and show that either p_z, or (p_x,p_y) can take non-zero values below T_c, but not both sets of components. The meaning of this result will be clarified by the symmetry considerations developed in §.2.8. Its derivation relies on the proof that one only of the two coefficients α_i in eq.(2.2) vanishes at (T_c,P_c) while the other one remains strictly positive in the neighbourhood of the phase transition. In turn, this proof is deduced from the "reasonable" assumption that these two coefficients are different (independent) functions of T and P. Indeed, the symmetry of the considered structure, which is the starting information of the model, does not impose any relation between α_1 and α_2. Consequently, in the phenomenological framework adopted here, α_1 and α_2 must be considered as unrelated.

a) positiveness of the α_i for $T > T_c$

For $T > T_c$, the equilibrium state has been assumed to be $\vec{p}_o(T,P) = 0$. This state corresponds to the minimum of F with respect to p_x, p_y, p_z. For vanishingly small $\vec{p}$, the expression of F is provided by (2.2) which will have $\vec{p}_o = 0$ as its minimum if, and only if, $\alpha_1 \geqslant 0$, $\alpha_2 \geqslant 0$.

b) vanishing of at least one α_i for $T = T_c$

The positiveness of the α_i (in a broad sense) holds for $T = T_c$. However, for the latter temperature this positiveness cannot be strict for all the α_i. Indeed, if $\alpha_i > 0$ $(i = 1,2)$ for $T = T_c$, then, due to the regularity of F and of its coefficients, the same strict inequality would be fulfilled slightly below T_c. In the range $T < T_c$, the minimum of the free-energy (2.2) would again correspond to $\vec{p}_o = 0$, in contradiction with the basic assumption (§.2.1) that for $T < T_c$, $\vec{p}_o(T,P_c) \neq 0$. Hence, one at least of the $\alpha_i(T,P)$ vanishes for (T_c,P_c).

c) vanishing of a single α_i, and exclusion of "singular" transitions

The different temperature and pressure dependence of α_1 and α_2, discussed above, implies that their simultaneous vanishing can only occur at certain isolated points of the (T,P) plane. Consider, for instance, the situation depicted on figure 2. For P above P_1, on lowering the temperature, α_1 vanishes at T'_c and α_2 remains positive in the neighbourhood of T'_c. Hence the equilibrium value of the set (p_x,p_y) remains equal to zero on either sides

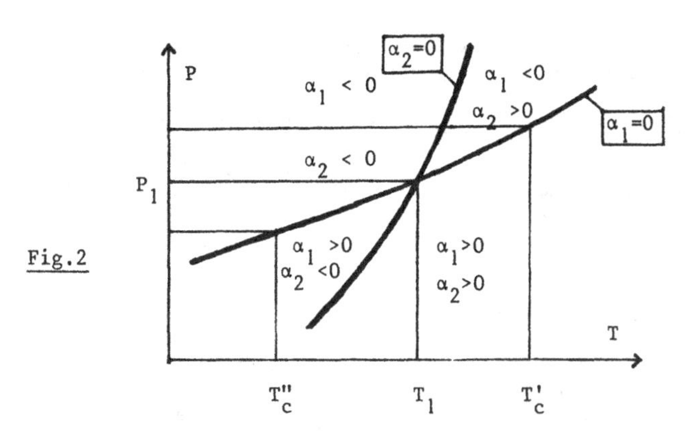

Fig.2

of T'_c. A transition at T'_c only concerns a possible change in p_z^o. Likewise, for P below P_1, a transition at T''_c only concerns a possible change in the set (p_x^o, p_y^o), the third component p_z^o remaining equal to zero on either sides of T''_c. Finally, for $P = P_1$, the transition at T_1 [where $\alpha_1 = \alpha_2 = 0$] can concern a change in the equilibrium value of all 3 components.

Hence, a small change of the pressure above or below P_1 modifies the direction of $\vec{p}_o$ in such a way (table 1) that the symmetry of phase II will change. We can qualify this situation by saying that, at the point (T_1,P_1) a slight change of pressure modifies "qualitatively" the nature of the transition (i.e. its symmetry characteristics). By contrast, away from (T_1,P_1) a slight shift of the pressure does not change the symmetry characteristics of the transition.

The Landau theory considers those "non-singular" transitions which exist with preserved symmetry characteristics in a portion of line in the pressure-temperature plane, i.e. it excludes the transition points of the type (T_1,P_1).

We will assume that the transition at (T_c, P_c) discussed here is not singular from the symmetry point of view. This implies that one only of the two α_i coefficients vanishes at T_c for the pressure P_c imposed to the system.

d) reduction of the number of relevant degrees of freedom. Order parameter

As emphasized above, if $\alpha_1(T_c,P_c) = 0$, and $\alpha_2(T_c,P_c) > 0$, the transition concerns a possible change in p_z^o, while in the converse case the change will only concern (p_x^o,p_y^o). The two sets of components (p_z) and (p_x,p_y) pertain to two disjoint types of transitions. For each type, one set of p-components remains equal to zero across T_c, and can be neglected in the description of the transition (at least in a first approximation : Cf.§ 2.9).

The remaining set, whose equilibrium value is modified by the transition, is the order-parameter of the transition.

2.6. Need for an expansion of degree higher than 2, below T_c

The vanishing at T_c of one α_i necessarily occurs with a change of sign across the transition. A positive sign above and below T_c would imply $\vec{p}_o = 0$ below T_c. Thus α_i becomes negative below T_c. The first term of the Taylor expansion of $\alpha_i(T,P_c)$ as a function of $(T-T_c)$ is :

$$\alpha_i = a(P_c).(T-T_c) \quad \text{with} \quad a(P_c) > 0 \tag{2.3}$$

(A first term of any odd-degree would also satisfy the stated requirement of sign-change of α_i ; however as no characteristic of the system imposes the vanishing of the linear term, assuming the absence of this term would be an unjustified complication of the theory).

The condition $\alpha_i < 0$ below T_c ensures that $\vec{p}_o = 0$ is not a minimum of F below T_c. Reduced to the second degree terms in the order-parameter components (either p_z, or (p_x,p_y)), the free-energy becomes infinitely negative for $|\vec{p}| \to \infty$. In order to determine the possible existence of a minimum of F for finite OP values, and locate the position of this minimum, it is necessary to expand F as a function of the OP components up to degrees higher than two.

Let us examine the situations respectively encountered when $p_z(\alpha_1(T_c,P_c) = 0)$, or $(p_x,p_y)(\alpha_2(T_c,P_c) = 0)$ are the OP of the system.

2.7. Simple physical consequences of the 4th degree order-parameter expansion

2.7.1. Order-parameter coinciding with p_z

The form of the successive powers of p_z in the Taylor expansion of F are straightforwardly obtained by using the property of invariance of the nth degree term of the expansion by the group G_o. Due to the occurence in G_o of transformations reversing p_z (e.g. σ_z or I), no odd power of p_z is invariant by G_o : F only contains even powers of p_z. Up to the 4th degree, the expression of F is, omitting the explicit dependence of the coefficients on pressure ,

$$F = F_o + \frac{a}{2}\,(T - T_c).p_z^2 + \frac{\beta(T)}{4}\,p_z^4 \tag{2.4}$$

Clearly, to ensure the existence of a minimum for finite values of p_z we must have $\beta(T) > 0$ for $T < T_c$. As no condition is imposed to $\beta(T)$ for $T > T_c$, the simplest function satisfying the preceding condition, in the vicinity of T_c, is a positive constant β .

If $\beta < 0$ for $T < T_c$, one should search for higher degree terms

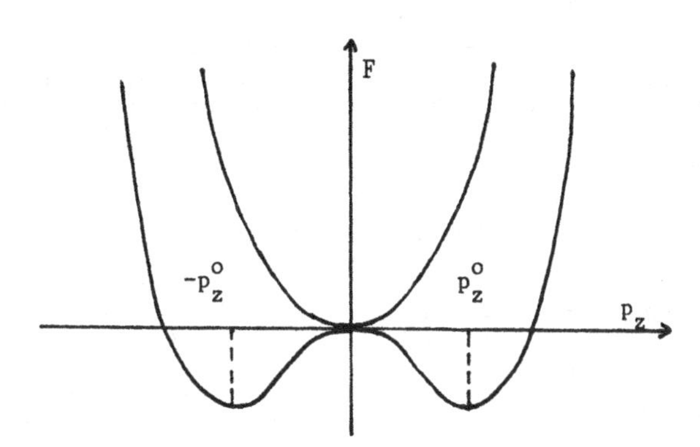

to obtain a minimum for finite values of p_z. However, as will be shown in chapter IV, in this case, one necessarily has a discontinuous transition, which does not enter strictly the framework discussed here.

For $\beta > 0$, the variations of $F(p_z)$ above and below T_c, are sketched on figure 3. Below T_c, the minima of F occur at $\pm p_z^o$, with :

$$p_z^o = \sqrt{\frac{a(T_c - T)}{\beta}} \qquad (2.5)$$

The OP has a "square-root" temperature dependence as function of (T_c-T). Note that one finds two minima corresponding to the same value of F (eq.2.4). The corresponding upward and downward displacements of the M^+ ion in fig. 1 are distinct states of the system possessing the same stability : there is <u>a twofold degeneracy</u> of states below the transition. Other physical consequences of the free-energy (2.4) can be drawn. They concern the absence of "latent heat" for the considered transition, and the existence of anomalies in the temperature dependence of the specific heat, and of the "susceptibility associated to the OP".

a) Absence of transition heat

The latent heat is $L = T_c \Delta S$, where ΔS is the difference of entropy between the two phases at T_c. We can derive S in each phase from the <u>equilibrium</u> free-energy F_{eq} $[T,P, p_z(T,P)]$: $S = -(\partial F_{eq}/\partial T)_P$. We also note that

$$\frac{\partial F_{eq}}{\partial T} = \frac{\partial F}{\partial T} + \frac{\partial F}{\partial p_z} \cdot \frac{dp_z}{dT} \qquad (2.6)$$

However, as $[\partial F/\partial p_z|_{p_z^o} = 0]$ we can write :

$$S = -\frac{\partial F}{\partial T}(T,P,p_z)|_{p_z^o} = -\frac{a}{2}(p_z^o)^2 - \frac{\partial F_o(T,P)}{\partial T} \qquad (2.7)$$

F_o (T,P) represents the "background" free-energy associated to degrees of freedom of the system not coupled to the OP. This function will be continuous at T_c. Moreover p_z^o is continuous at T_c. Thus $\Delta S = 0$, and, accordingly, there is no latent heat associated to the transition.

b) Anomaly of the specific heat at zero-field

As in the case of a fluid, two specific heats can be considered, depending of the quantity maintained constant, either p_z or the thermodynamically conjugated quantity which is proportional to the component ε_z of the electric field (since p_z is proportional to the macroscopic polarization component P_z). The specific heat can be derived from eq.(2.7) through the relation:

$$c_u = T\left(\frac{\partial S}{\partial T}\right)_u \tag{2.8}$$

If p_z^o is maintained constant, eq.(2.7) clearly shows that c_p only depends of the derivative of F_o(T,P) which has no anomaly at T_c. The specific heat at constant value of the OP varies smoothly through the transition. By contrast, in zero field, p_z^o is a function of temperature (eq.2.5), and we have,

$$T > T_c \qquad c_\varepsilon^o = -\frac{\partial^2 F_o(T,P)}{\partial T^2}$$

$$T < T_c \qquad c_\varepsilon = -T\frac{\partial^2 F_o}{\partial T^2} - \frac{a}{2}\frac{d[(p_z^o)^2]}{dT} = c_\varepsilon^o + \frac{a^2}{2\beta} \tag{2.9}$$

Above and below T_c, c_ε is a different, smoothly varying function of T, determined by the background free-energy F_o(T,P), and by the smooth variation of the coefficient β. The anomaly of c_ε is an upward step variation, on cooling through T_c (figure 4).

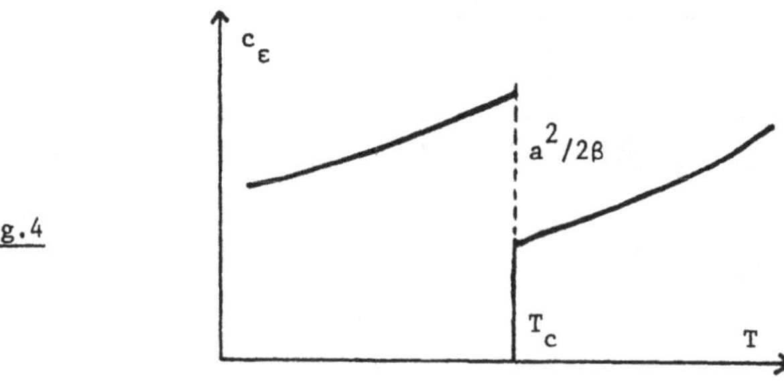

Fig.4

c) Anomaly of the susceptibility associated to the OP

The susceptibility χ is defined by :

$$\chi = \lim_{\varepsilon_z \to 0} \left[\frac{\partial p_z}{\partial \varepsilon_z} \right] \Big|_{p_z^o} \tag{2.10}$$

where ε_z is the field conjugate to p_z. χ is proportional to the dielectric susceptibility.

In order to calculate χ in (2.10), it is necessary to examine the behaviour of the system in the presence of the field ε_z. The appropriate variational thermodynamic potential to consider, in order to determine the equilibrium is not $F(T,P,\vec{p})$, but the potential $G = F - p_z \cdot \varepsilon_z$. Minimizing with respect to p_z, and using the expression (2.4) of F, we obtain :

$$p_z \left[a(T - T_c) + \beta p_z^2 \right] = \varepsilon_z \tag{2.11}$$

Deriving the two members of (2.11) with respect to ε_z and making ($\varepsilon_z \to 0$), we obtain :

$$\chi [a(T - T_c) + 3\beta (p_z^o)^2] = 1 \tag{2.12}$$

where p_z^o is the equilibrium value of p_z, at zero applied field. Replacing (p_z^o) by its values above ($p_z^o = 0$) and below (eq.2.5) T_c, we finally have :

$$\chi (T > T_c) = \frac{1}{a(T-T_c)} \quad ; \quad \chi(T < T_c) = \frac{1}{2a\,(T_c - T)} \tag{2.13}$$

The susceptibility goes to infinity when $T \to T_c$ from either sides of the transition. Figure 5 shows the form of the corresponding anomaly.

Fig.5

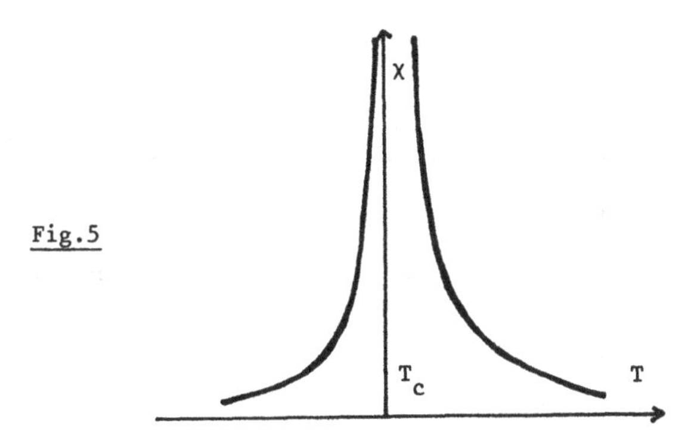

2.7.2. Order-parameter coinciding with (p_x, p_y)

As in the preceding case, one has first to find the form of the expansion of $F(T,P,p_x,p_y)$ up to the terms of lowest degree ensuring the existence of a minimum of F for finite (p_x,p_y) values.

The same type of argument as used in §.2.4.b shows the absence of 3rd degree terms, and provides the form of the 4th-degree ones.

Thus, the occurence of operations of G_o which reverse one coordinate p_i and preserve the two others (i.e. $\sigma_x,\sigma_y,\sigma_z$), forbid the invariance either of the third degree monomials $(p_x^3,p_x^2p_y,p_xp_y^2,p_y^3)$, or of their linear combinations.

This argument can also be used to predict the absence of fourth degree monomials containing odd powers of one coordinate p_i. Hence,the only possible invariants of 4th degree are linear combinations of the 3 monomials $(p_x^4, p_x^2p_y^2, p_y^4)$. Clearly, the two combinations $(p_x^4 + p_y^4)$ and $(p_x^2.p_y^2)$ are invariant by the generators of G_o indicated in §.2.4.b). Other 4th degree polynomials possess the same invariance (e.g. $(p_x^2+p_y^2)^2$) but they are linear combinations of the two former ones. Hence, the most general 4th degree polynomial of (p_x,p_y), invariant by G_o, is a linear combination, with arbitrary coefficients, of the two above polynomials : up to the 4th degree, the expansion of F can be written as :

$$F = F_o(T,P) + \frac{\alpha_2}{2}(p_x^2-p_y^2) + \frac{\beta_1}{4}(p_x^4+p_y^4) + \frac{\beta_2}{2}p_x^2.p_y^2 \tag{2.14}$$

where,again,we have $\alpha_2 = a(T - T_c)$ [the same notation for the coefficients a and β_1 is used, though the free-energies (2.14) and (2.4) are not related].

Unlike the situation in §.2.7.1, the signs or values of β_1 and β_2, ensuring the existence of a minimum of F for finite (p_x, p_y) values, are not obvious. They will be determined in the course of the minimization procedure of F.

For the two-variable function (2.14) a minimum is determined by the set of conditions :

$$\frac{\partial F}{\partial p_x} = 0 \quad ; \quad \frac{\partial F}{\partial p_y} = 0 \quad ; \quad \left[\frac{\partial^2 F}{\partial p_i \partial p_j}\right] > 0 \tag{2.15}$$

where the third condition represents the positiveness of the eigenvalues of the matrix built from the second derivatives of F.

a) Possible extrema of F.

Using eq.(2.14), the vanishing of the first derivatives is expressed by :

$$\begin{aligned} p_x(\alpha_2 + \beta_1.p_x^2 + \beta_2.p_y^2) &= 0 \\ p_y(\alpha_2 + \beta_1.p_y^2 + \beta_2.p_x^2) &= 0 \end{aligned} \tag{2.16}$$

Eqs; (2.16) have three sets of solutions:

i) $p_x = p_y = 0$

ii) $p_x = 0$ $p_y = \pm\sqrt{-\alpha/\beta_1}$; or $p_y = 0$, $p_x = \pm\sqrt{-\alpha/\beta_1}$. These 4 solutions correspond to the same value of the free-energy (2.14).

iii) $p_x \neq 0$ $p_y \neq 0$, p_x and p_y being determined by the vanishing of the two bracketted expressions in (2.16). Let us substract one of these expressions from the other. We obtain :

$$(\beta_1 - \beta_2)(p_x^2 - p_y^2) = 0 \qquad (2.17)$$

We can use an argument similar to the one in §.2.5.c) to show that the only acceptable solution of (2.17) is $p_x = \pm p_y$. Indeed, we can note that $(\beta_1 - \beta_2)$ is a function of temperature and pressure. For the considered pressure P_c, it cannot vanish in the entire temperature interval examined, above and below T_c. As a consequence (2.17) would yield $p_x = \pm p_y$ except at the discrete temperatures where $(\beta_1 - \beta_2)$ vanish (for which any direction of the $\vec{p}$ vector in the (p_x, p_y) plane is acceptable). Hence the symmetry of phase II would abruptly change in the vicinity of the vanishing points of $(\beta_1 - \beta_2)$. As in §.5.2.c) we exclude this "singular" type of behaviour and assume that $(\beta_1 - \beta_2) \neq 0$.

Reporting into (2.16) the condition $(p_x = \pm p_y)$ we obtain a third type of 4 solutions, all corresponding to the same value of F :

$$p_x = \pm p_y \qquad p_x = \pm\sqrt{-\alpha/(\beta_1 + \beta_2)}$$

We note that in phase II, the OP components vary as $(T_c - T)^{0.5}$, as in §.2.7.1.

b) conditions of stability of the solutions

The second derivatives of F are :

$$\frac{\partial^2 F}{\partial p_x^2} = \alpha_2 + 3\beta_1 . p_x^2 + \beta_2 . p_y^2$$

$$\frac{\partial^2 F}{\partial p_y^2} = \alpha_2 + 3\beta_1 . p_y^2 + \beta_2 . p_x^2 \qquad (2.18)$$

$$\frac{\partial^2 F}{\partial p_x \partial p_y} = 2\beta_2 . p_x p_y$$

One can check straightforwardly that the positiveness of the values of the symmetric matrix of second derivatives determines the following conditions of stability for the various extrema listed above :

Solution i) corresponds to the absolute minimum of F for $\alpha_2 > 0$, i.e. $T > T_c$.

Solutions ii) are stable for $\alpha_2 < 0$ and $\beta_2 > \beta_1 > 0$.

Solutions iii) are stable for $\alpha_2 < 0$ and $\beta_1 > |\beta_2|$.

When for $T < T_c$ β_1 and β_2 do not comply with any of the former conditions (e.g. $\beta_1 < 0$), the absolute minimum of F occurs for infinite values of (p_x, p_y). Similarly to the case involved in §.2.7.1, the addition to F of even degree terms of degree higher than 4 can determine a minimum for finite (p_x, p_y). However, the transition between phases I and II will then be a discontinuous one (cf.Chap.IV)

The above results can be summarized, for $T < T_c$, by a "phase diagram" in the (β_1, β_2) plane (fig.6).

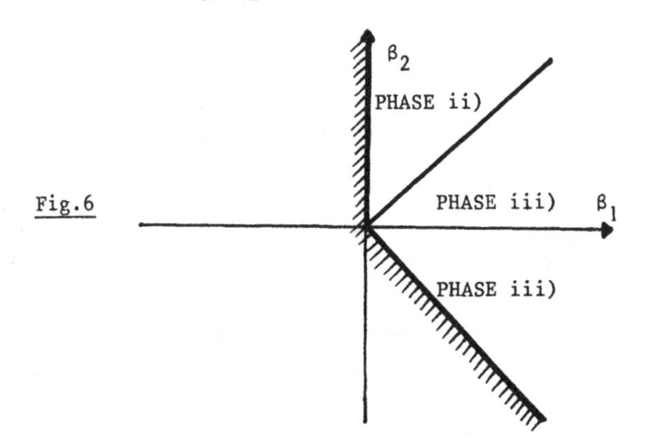

Fig.6

As in §.2.7.1, phase II displays a "degeneracy". Hence, for $\alpha_2 < 0$ and $\beta_2 > \beta_1 > 0$, the transition occuring at T_c will be towards one of four possible phases, corresponding to the 4 solutions ii) with the same free-energy : F has 4 minima possessing the same depth. The same situation occurs for solutions iii).

c) Physical anomalies at T_c

We will not repeat the derivation, performed in §.2.7.1, of the absence of latent heat at T_c, or of the shape of the specific heat anomaly induced by the transition.

By contrast, let us examine the behaviour of the susceptibility which is slightly more complex than §.2.7.1, since there are 3 components of the susceptibility tensor : $\chi_{xx} = (\partial p_x/\partial \varepsilon_x)$, $\chi_{yy} = (\partial p_y/\partial \varepsilon_y)$, and $\chi_{xy} = \chi_{yx} = (\partial p_x/\partial \varepsilon_y)$. Here again we form the potential $G = F - p_x \varepsilon_x - p_y \varepsilon_y$. Minimizing F with respect to p_x and p_y and applying the same method as in §.2.7.1.c), we obtain, in the case $\beta_2 > \beta_1 > 0$ (i.e. for a phase of type ii)) :

$$\chi_{xx} = \chi_{yy} = \frac{1}{a(T - T_c)} \; ; \quad \chi_{xy} = 0 \qquad \text{for } T > T_c \tag{2.19}$$

$$\chi_{xx} = \frac{1}{2a(T_c - T)} \; ; \quad \chi_{yy} = \frac{\beta_1}{\beta_2 - \beta_1} \cdot \frac{1}{a(T_c - T)} \; ; \quad \chi_{xy} = 0 \qquad \text{for } T < T_c$$

Both diagonal components diverge at T_c but their temperature dependence is different below T_c. This is due to the fact that the onset of a non-zero p_x^o value, below T_c, destroys the equivalence of the x and y directions : the "longitudinal" susceptibility χ_{xx} becomes different from the "transverse" one χ_{yy}.

2.8. Symmetry considerations

We have already stressed in §.2.2. that, for $\vec{p} \neq 0$, the group of invariance of the system is a subgroup of G_o, which is the invariance group of phase I. We can refer to table 1 to find the symmetry-groups corresponding to each of the possible phases II found in §.2.7.1. or 2.7.2.

2.8.1. Relationship between equally-stable phases

The fact that each minimum of F is associated to a phase of lower symmetry than G_o (i.e. invariant by a subgroup G of G_o) is directly related to the existence of several minima with the same value of the free-energy.

Consider, for instance, in §.2.7.2. the minimum of (2.14) consisting in a solution $\vec{p}_o$ of type ii) ($p_x \neq 0, p_y = 0$). According to table 1, the symmetry-group of the corresponding phase is the group $G_1 = (mm2_x)$. Consider, on the other hand, the fourfold rotation C_4. This geometrical transformation belongs to G_o and therefore leaves F (eq.2.14) invariant. As it does not belong to G_1, $\vec{p}_o$ is not preserved by it. It is transformed into a perpendicular vector $\vec{p}_o'$, of same modulus, and necessarily corresponding to a minimum of F with the same depth, <u>due to the invariance of F.</u> As sketched on figure 7, the various operations of G_o not belonging to G_1 will transform $\vec{p}_o$ into one of the 4 equivalent minima found directly in §.2.7.2. by minimizing F.

This argument shows that the "symmetry breaking" associated to a given minimum of F is necessarily compensated by the occurence of several "equivalent" minima corresponding to each other through operations of G_o not belonging to the "low-symmetry group".

Fig.7

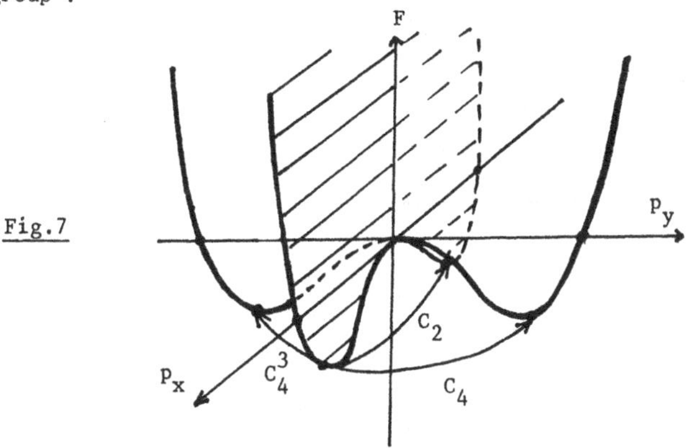

a) Conjugated character of the various "low-symmetry" groups

We note on table 1 that the symmetry groups relative to the 4 equivalent (symmetry related) minima of F ($\pm p_x^o$, 0) and (0, $\pm p_y^o$) are $mm2_x$ for the two first minima and $mm2_y$ for the two last ones. Thus the symmetry groups of symmetry related phases are not identical, in general. However, they are related to each other through a geometrical correspondance consisting in a <u>conjugation</u> with respect to the symmetry group G . G_1 and G_2 are said to be conjugated with respect to G_o if any element g_2^i of G_2 can be written as:

$$g_2^i = g.g_1^i.g^{-1} \qquad (2.20)$$

where g is a <u>given</u> element of G_o (necessarily external to G_1 and G_2) and g_1^i an element of G_1. Eq. (2.20) shows that the role of G_1 and G_2 can be interchanged in this definition. It is easy to check that in the example considered the groups $mm2_x$= and $mm2_y$ are conjugated with respect to G_o. Indeed we have :

$$(mm2_y) = C_4.(mm2_x).(C_4)^{-1} \qquad (2.21)$$

where the fourfold rotation C_4 belongs to G_o.

b) "Irreducibility" of the order-parameter

Table 2 hereunder indicates the manner $\vec{p} = (p_x, p_y, p_z)$ transforms under the action of the generators of G_o.

<u>Table 2.</u>

G_o	C_4	σ_x	I
p_x	p_y	$-p_x$	$-p_x$
p_y	$-p_x$	p_y	$-p_y$
p_z	p_z	p_z	$-p_z$

We note that p_z is transformed into itself or into its opposite. If we consider the direction p_z as a vector space, we can see that this vector space is globally invariant by the transformations belonging to G_o.

By contrast the direction p_x does not constitute an invariant vector space since p_x is transformed into ($\pm$ p_y) by certain elements of G_o. However the vector space constituted by the set of directions (p_x, p_y) is globally invariant by G_o.

Obviously the entire set (p_x, p_y, p_z) also carries a 3-dimensional vector space invariant by G_o. This invariant space differs on one point, crucial to our argumentation, from the two preceding ones. It contains smaller spaces possessing themselves the property of invariance by G_o i.e. the spaces carried by p_z and (p_x,p_y) . Clearly p_z being one-dimensional does not contain any smaller space invariant by G_o. It is easy to check that this is also the case of (p_x, p_y). Indeed, in this plane, the C_4 rotation does not leave any direction fixed (we restrict here to "real" directions, a more rigorous approach will be described in chapter II). Hence, (p_x, p_y) does not contain any one-dimensional space invariant by all the transformations of G_o (including C_4).

We characterize this situation by saying that (p_x, p_y, p_z) is a reducible invariant space with respect to G_o, while (p_z) and (p_x, p_y) are irreducible invariant spaces with respect to this group.

It is well known that if a vector space is invariant by a linear transformation g_o, this transformation can be represented by a matrix $M(g_o)$, expressing the action of g_o on the basic vectors of the space. Table 3 indicates the matrices representing the generators of G_o in each of the three vector spaces considered above.

Table 3.

G_o	C_4	σ_x	I
(p_x,p_y,p_z)	$\begin{vmatrix} 0 & -1 & 0 \\ 1 & 0 & 0 \\ 0 & 0 & 1 \end{vmatrix}$	$\begin{vmatrix} -1 & 0 & 0 \\ 0 & 1 & 0 \\ 0 & 0 & 1 \end{vmatrix}$	$\begin{vmatrix} -1 & 0 & 0 \\ 0 & -1 & 0 \\ 0 & 0 & -1 \end{vmatrix}$
(p_x,p_y)	$\begin{vmatrix} 0 & -1 \\ 1 & 0 \end{vmatrix}$	$\begin{vmatrix} -1 & 0 \\ 0 & 1 \end{vmatrix}$	$\begin{vmatrix} -1 & 0 \\ 0 & -1 \end{vmatrix}$
p_z	1	1	-1

The set of all the matrices of the group G_o (there are 16 such matrices) relative to one of the vector spaces constitute a representation of G_o in this vector space.

By definition, the set of 3-dimensional matrices relative to the space (p_x, p_y, p_z) constitutes a reducible representation of G_o, because the carrier space is itself a reducible invariant space. The two other sets constitute irreducible representations of G_o, since the corresponding spaces are irreducible invariant spaces by G_o.

On the basis of the preceding remarks, we are now able to formulate differently the result, obtained in §.2.5., showing that the OP of the considered transition can either coincide with p_z, or with (p_x,p_y) but not with the set (p_x, p_y, p_z) : the degrees of freedom constituting possible order-parameters are characterized, from the standpoint of symmetry, by the fact that the OP components span an irreducible representation of the group G_o of phase I (the more symmetric phase). We will see in chapter II that this property is the central symmetry property used in the Landau theory of phase transitions.

c) Groups G and invariant directions in the OP space

Consider the case (§.2.7.2) of the OP coinciding with the set (p_x, p_y), and one possible set of values of the OP component corresponding to phase II, e.g. $(p_x^o = \sqrt{-\alpha/\beta_1}, p_y^o = 0)$. The symmetry group G of this phase has been determined (Table 1) by enumerating the elements of G_o preserving the structure of the unit cell (including the displacement p_x^o of the central ion). Actually as the ions at the vertices of the cell are invariant by G_o, G is the subgroup of G_o which preserves the vector (p_x^o, o). The action of G_o on (p_x^o, p_y^o) is entirely determined by the set of matrices representing G_o in the irreducible space (p_x, p_y). Therefore the determination of G can be performed by enumerating the matrices which leave invariant the first basis vector of the OP space. In practice, G will correspond to the set of matrices which have (+ 1) as their first diagonal element. Table 4 hereunder indicates the complete list of 16 matrices associated to the 16 elements of G_o. One checks directly that G_1 is, as already indicated by table 1, the group $(mm2_x)$. The same method can be used for the other stables phases compatible with the OP(p_x, p_y) (e.g. $p_x^o = 0$, $p_y^o \neq 0$ or $p_x^o = \pm p_y^o$) : the symmetry-group of each phase can be obtained by selecting the set of matrices of the OP– representation leaving invariant the corresponding vector belonging to the OP space. Clearly, this group does not depend on the magnitude of the vector (i.e. $p_x = \sqrt{-\alpha/\beta_1}$) but only of its direction in the OP space.

We summarize this result by saying that the possible symmetries of phase II (stable below T_c) are constituted by the various invariance groups of certain directions in the OP space, these directions being associated to the minima of the free-energy.

Table 4.

G_o	E	C_4	C_2	C_4^3	σ_x	σ_y	σ_{xy}	$\sigma_{\bar{x}y}$
p_x p_y	$\begin{vmatrix}1&0\\0&1\end{vmatrix}$	$\begin{vmatrix}0&-1\\1&0\end{vmatrix}$	$\begin{vmatrix}-1&0\\0&-1\end{vmatrix}$	$\begin{vmatrix}0&1\\-1&0\end{vmatrix}$	$\begin{vmatrix}-1&0\\0&1\end{vmatrix}$	$\begin{vmatrix}1&0\\0&-1\end{vmatrix}$	$\begin{vmatrix}0&-1\\-1&0\end{vmatrix}$	$\begin{vmatrix}0&1\\1&0\end{vmatrix}$
G_o	I	S_4^3	σ_z	S_4	U_x	U_y	U_{xy}	$U_{\bar{x}y}$
p_x p_y	$\begin{vmatrix}-1&0\\0&-1\end{vmatrix}$	$\begin{vmatrix}0&1\\-1&0\end{vmatrix}$	$\begin{vmatrix}1&0\\0&1\end{vmatrix}$	$\begin{vmatrix}0&-1\\1&0\end{vmatrix}$	$\begin{vmatrix}1&0\\0&-1\end{vmatrix}$	$\begin{vmatrix}-1&0\\0&1\end{vmatrix}$	$\begin{vmatrix}0&1\\1&0\end{vmatrix}$	$\begin{vmatrix}0&-1\\-1&0\end{vmatrix}$

2.9. Secondary order-parameters

The only relevant degree of freedom considered up to now, in the above model, is the set of 3 components of $\vec{p}$. Let us allow the possibility that the c-parameter, equal to the distance along z between the planes of negative ions, can change with temperature and pressure. This change corresponds to a deformation of the unit-cell along z which can be represented by a component e' of a strain tensor.

In the same manner as for the components of $\vec{p}$, the equilibrium value of e' can be determined by minimizing a variational free-energy which will be a function of e', in addition to the variables already considered (cf. §.2.3) : $F = F(T,P,p_x,p_y,p_z,e')$. The form of F can be determined along the same symmetry principles used in §.2.4. and §.2.7 in order to find the 4th degree Taylor expansion of F as a function of (p_x, p_y, p_z). In this view, we note that e' (which is a component of a second rank symmetric tensor, left invariant by the inversion I) is invariant under the application of all the transformations of G_o. The part of the expansion of F depending exclusively on e' is therefore of the form :

$$\lambda e' + \frac{C'e'^2}{2} + \frac{De'^3}{3} + \ldots\ldots \qquad (2.22)$$

Terms of any degree depending on e', are allowed, by symmetry, to be present in F.

On the other hand, any power of e' multiplied by any invariant polynomial of (p_x, p_y, p_z) will be invariant by G_o, and will appear in the expression of F. The term of lowest degree of this type is :

$$e'[\delta_1\ (p_x^2 + p_y^2) + \delta_2\ p_z^2\ ] \qquad (2.23)$$

Let us examine the effect of the presence of the e'-dependent terms. The minimum of F with respect to e' yields :

$$\lambda + C'e' + De'^2 + \delta_1(p_x^2 + p_y^2) + \delta_2 p_z^2 = 0 \qquad (2.24)$$

a) Elimination of the thermal expansion

As $\lambda(T,P)$ is a parameter which, in general, will not be vanishingly small, e' cannot either be vanishingly small. The reason underlying this result is that λ is attached to the thermal expansion of the material along z, i.e. to a particular form of the strain e', which exists at any temperature and is not induced by the presence of a phase transition at T_c. We can eliminate the resulting "normal" component of e' by setting $e' = (e_o + e)$ where e_o is the non-zero value of e' at the transition temperature. e_o is defined by :

$$\lambda + C'e_o + De_o^2 + \ldots = 0 \qquad (2.25)$$

b) Secondary OP

Both λ and e_o will vary smoothly across the considered transition [actually (2.25) could be integrated into the background free-energy F_o (T,P)]. Eq. (2.24) can then be written as :

$$e = -\frac{1}{C} [\delta_1 (p_x^2 + p_y^2) + \delta_2 p_z^2] \qquad (2.26)$$

where the C coefficient differs from C' by corrections depending on e_o through the presence in (2.24) of terms such as De'^2. The C coefficient being the effective coefficient of the quadratic contribution in e to the free-energy, we know that, if α_1 or α_2 vanish at T_c, C necessarily remains strictly positive in the neighborhood of T_c (cf. §2.5). We can then deduce from (2.26) that e vanishes above T_c and becomes non-zero below T_c.

Hence, the OP (which is either p_z or (p_x, p_y) is not the only quantity possessing the property of being non-zero for $T < T_c$ and zero for $T > T_c$. Quantities such as e also possess this property. e will be called a <u>secondary OP</u>, while the OP defined formerly can be termed a <u>primary OP</u>. The latter OP can be distinguished from the former by the fact that the α_i coefficient, associated to it vanishes at T_c while the coefficient C remains strictly positive nearby T_c. The onset of a non-zero value for e is related to the occurence of a <u>coupling term</u> in the free-energy . If the coefficient δ of this term is zero, eq.(2.26) shows that e will remain equal to zero).

c) Influence of e on the symmetry properties of the system

Let us assume that $\alpha_2 (T_c, P_c) = 0$, i.e. that the primary OP of the transition is the set (p_x, p_y). Taking into account (2.26)(2.25) and (2.14) we can eliminate e from the free-energy and obtain an expression of F which only depends, as (2.14),of the components p_x and p_y. This expression is (omitting the background free-energy):

$$F = \frac{(\alpha_2 - 2e_o \cdot \delta_1)}{2}(p_x^2 + p_y^2) + \frac{[\beta_1 - \frac{2\delta_1^2}{C}]}{4}(p_x^4 + p_y^4) + \frac{[\beta_2 - \frac{2\delta_1^2}{C}]}{2} p_x^2 p_y^2 \quad (2.27)$$

We can see that F has the same functional form as (2.14) but different coefficients. The effect of the coupling between the primary and secondary OP is to modify the coefficients of the free-energy associated to the sole primary OP, by amounts increasing with the "strength" δ_1 of the coupling.

Eq.(2.27) establishes that none of the conclusions reached in §.2.7.2 is changed : the list of possible phases and the form of the anomalies associated to the transition, which only depend on the functional form of F, are the same as in the absence of coupling. Only the boundaries of these phases are displaced. For instance the temperature of the phase transition corresponds to the vanishing of $(\alpha_2 - 2.e_o.\delta_1)$. It will be shifted with respect to the temperature T_c appearing in α_2. However the latter point is not necessarily significant : one can consider that expression (2.14) is the free-energy obtained when the couplings to degrees of freedom such as e have been taken into account and already eliminated through procedures similar to the one leading to (2.27).

We also note that the symmetry of the various phases will be the ones worked out in table 1. Indeed, the onset of a non-zero deformation e preserves the symmetry G_o of the structure and thus the breaking of symmetry is imposed by the onset of (p_x, p_y).

In summary, the symmetry characteristics of the considered transition (possible phases and their symmetries) is exclusively determined by the primary OP.

Accordingly, the reduction of the number of relevant degrees of freedom performed in 2.5.d) though not strictly correct (since other quantities than the primary OP are affected by the transition) is entirely justified in the first place: the results obtained by limiting the description of the system to the primary OP are correct. In a further step of investigation of the transition, consideration of other degrees of freedom allows to study "secondary" anomalies. Still, the primary OP plays a central role in this second step, since the occurence of a coupling between the additional degrees of freedom and the primary OP is a necessary condition for the singular behaviour of the additional quantities.

d) Anomalies related to the secondary OP

Eq.(2.26) can be used to deduce the temperature dependence of e. As already stressed e = 0 above T_c (we use the same notation for the temperature of the transition renormalized by the coupling to e_o). Below T_c, for any of the minima of F considered in §.2.7.2, $(p_x^2 + p_y^2)$ is proportional to $(T_c - T)$. Inserting this temperature dependence in (2.26) yields :

$$e \propto (T_c - T) \tag{2.28}$$

Let us now derive the behaviour of the "susceptibility" associated to e.

The quantity which is thermodynamically conjugated to e is a component σ of the stress tensor, since e is a component of the strain tensor. The susceptibility $[s = \lim (\partial e/\partial \sigma)$ when $e \to 0]$ has the meaning of an "elastic compliance".

In the presence of a non-zero stress we minimize the thermodynamic potential $[F - \sigma.e]$ and obtain first as a substitute to eq. (2.26)

$$e = \frac{\sigma}{C} - \frac{1}{C}\,\delta_1 \cdot (p_x^2 + p_y^2) \tag{2.29}$$

Likewise, instead of (2.27), we obtain the following expression for $(F - \sigma.e)$

$$\frac{(\alpha_2 - 2\delta_1 \cdot e_o + \sigma\delta_1/C)}{2} \cdot \rho^2 + \frac{(\beta_1 - 2\,\delta_1^2/C)}{4} \cdot \rho^4 + \frac{(\beta_2 - \beta_1)}{2}\,\rho^4 \sin^2\theta \cos^2\theta \tag{2.30}$$

where we have set $p_x = \rho \cdot \cos\theta$ and $p_y = \rho . \sin\theta$

If we assume that the "mechanical susceptibility" s corresponds to conditions of zero applied field (i.e. no field conjugate to (p_x, p_y) is applied), we can write the minimum of (2.30) with respect to ρ and θ :

$$\rho\{[\alpha_2 - 2\delta_1 e_o + \sigma\delta_1/C] + [\beta_1 - 2\delta_1^2/C]\,\rho^2 + 2[\beta_2 - \beta_1]\rho^2 \sin^2\theta\cos^2\theta\} = 0$$

and (2.31)

$$\sin 4\,\theta = 0$$

From (2.29) we draw the expression of the susceptibility s :

$$s = \left(\frac{\partial e}{\partial \sigma}\right) = \frac{1}{C} - \frac{\delta_1}{C} \cdot \left(\frac{\partial \rho^2}{\partial \sigma}\right) \tag{2.32}$$

In (2.31), the solution $\rho = 0$ corresponds to $T > T_c$ while the vanishing of the expression between brackets corresponds to $T < T_c$. In the latter case, we derive the bracket with respect to σ [θ is independent from σ] and obtain :

$$\frac{\partial \rho^2}{\partial \sigma} = \frac{-\,\delta_1/C}{\beta_1 - 2\,\delta_1^2/C + b} \tag{2.33}$$

where b = 0 if $\theta = 0$ $(\pi/2)$, i.e. if solution ii) in §.2.7.2 is considered, or $b = (\beta_2 - \beta_1)/2$ if $\theta = \pi/4$ $(\pi/2)$, i.e. if solution iii) is retained below T_c.

Replacing (2.33) into (2.32) we finally obtain :

$$s = \frac{1}{C} \qquad \text{for} \qquad T > T_c$$

$$s = \frac{1}{C} + \frac{\delta_1^2/C^2}{\beta_1 - 2\delta_1^2/C} \quad \text{for } T < T_c \text{ and solution ii)}$$

$$s = \frac{1}{C} + \frac{\delta_1^2/C^2}{(\beta_1+\beta_2)/2 - 2\,\delta_1^2/C} \quad \text{for } T < T_c \text{ and solution iii)} \tag{2.34}$$

Results which are summarized by fig. 8 . We see that the susceptibility experiences an upward jump at T_c similar to the jump of the specific heat sketched on fig. 4.

We can also note on eq.(2.31) that α_2 is modified by the stress σ : the transition temperature is shifted linearly by the application of the stress. This shift, as well as the magnitude of the anomalies (2,34) is controlled by the "strength" δ_1 of the coupling between e and the primary OP

Fig.8

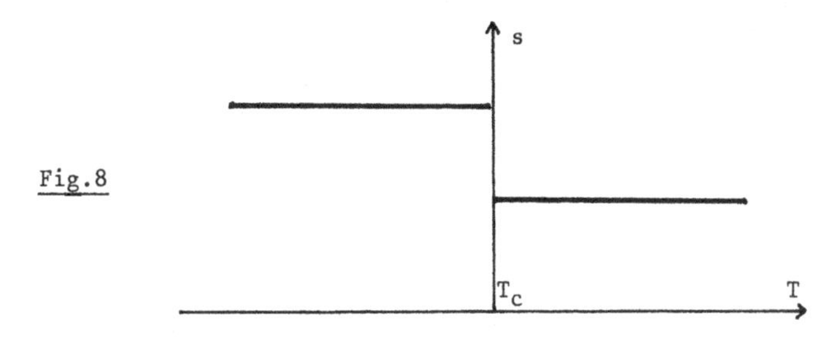

3. CONCLUSIONS

The analysis of the chosen type of "model" crystalline transition has illustrated the prominent features of the phenomenological theory of phase transitions.

1/ The description of the symmetry and physical characteristics of the transition is related to the identification of a specific set of degrees of freedom constituting the (primary) order-parameter of the transition. This set has well defined symmetry properties: its components span an irreducible representation of the group of transformations G_o leaving invariant the atomic structure of the "more symmetric" phase.

2/ The derivation of the physical anomalies accompanying the phase transition, and the enumeration of the possible stable phases of the system above and below T_c, rely on the form of the Landau free-energy , which is a variational thermodynamic potential. It is a polynomial expansion as a function of the components of the (primary) OP, and of the other degrees of freedom of the system which are coupled to the OP.

3/ The form of this polynomial expansion is entirely determined, on the one hand, by the fact that terms of a given degree are invariant by all the transformations belonging to G_o, and on the other hand, by the symmetry properties of the primary, and secondary order-parameters.

4/ The primary OP is particularized, with respect to other degrees of freedom, by the fact that the coefficient α of its quadratic contribution to the free-energy vanishes and changes sign at T_c.

As shown in chapter II, these features can be shown to possess a general validity. Their derivation constitutes the Landau theory of phase transitions.

CHAPTER II

FORMULATION OF LANDAU'S THEORY

1. INTRODUCTION

In this chapter, we generalize the considerations developped in chapter I and show that one can define, independently from any specific structural model, the concepts of order-parameter and transition free-energy.

As emphasized in chapter I, the method leading to the clarification of these concepts is based on three main ingredients : i) the use of a set of variational degrees of freedom among which the order-parameter (OP) will be selected. ii) the expression of the condition of continuity of the transition ; iii) the use of relevant symmetry groups to define the sudden qualitative change occuring at the phase transition.

The general formulation of this method, as well as the considerations pertaining to the form of the Landau free-energy rely partly on group-theoretical concepts and results. For this reason, we have placed in §.2 a brief summary of the group theoretical elements used in the theory and, in particular, an introduction to the notion of irreducible representation of a group. Fig. 1 shows the articulation of the paragraphs in this chapter.

Fig 1

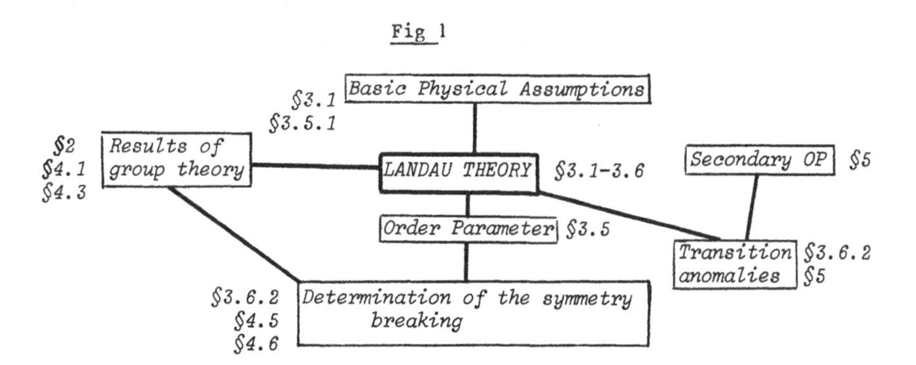

2. BASIC CONCEPTS OF GROUP REPRESENTATION THEORY.

In this paragraph, we describe some basic concepts of the theory of group representations. Technical aspects of the use of group-theoretical methods are exposed in § 4. and in chapter III.

More formal and complete accounts of the notions discussed here can be found in the large number of existing treatises devoted to the applications of group-theory to physics (e.g. references 1 and 2).

2.1. Irreducible representations of a group G_o

The concept of an irreducible representation (IR), which has already been encountered in chapter I (§.2.8.b), pertains to the classification of spaces which are left globally invariant by the action of the elements of a group G_o . This classification is based on the "simple" lack of symmetry of these spaces with respect to G_o. Here, the spaces of interest are sets of physical degrees of freedom.

2.1.1. The example of inversion

Consider, on the x-axis, the transformation $x \to (-x)$ which we denote I. Certain functions are symmetric (even) with respect to the action of I : f_1 satisfies the condition $f_1(x) = f_1(-x)$. Among the functions which are not symmetric, certain ones lack in a "simple" manner this symmetry. These are the antisymmetric (odd) functions, which satisfy the condition $f_2(-x) = -f_2(x)$. Considering each f_i as the basis of a vector space (i.e. all vectors of the form λf_i), we note that this vector space is invariant by the action of I, as well as by the action of the powers of I , since the iterated action of I is either the identical transformation E or is equal to that of I : the vector space f_i is invariant by the group G_o of two elements (E,I). In the space f_i, G_o can be represented by the set of two matrices expressing the action of the two elements of the group. These one-dimensional matrices are reproduced on table 1 for the two spaces f_i. Each set constitutes an irreducible representation of G_o, the irreducibility being defined by the fact that the carrier spaces f_i do not contain any smaller subspace itself invariant by the two elements of G_o.

TABLE 1

	E	I
f_1	+ 1	+ 1
f_2	+ 1	- 1

The prominent role of the two types of functions $f_1(x)$ and $f_2(x)$ arises from the fact that there are no irreducible invariant spaces other than the symmetric and antisymmetric ones : any set of functions $\phi_i(x)$ spanning a vector space invariant and irreducible with respect to G_o will give rise to matrices for the set (E,I) identical to one of the two sets listed on table 1. In addition, any function $\phi(x)$ can be expressed as the sum of two functions, a symmetric one and an antisymmetric one,

2.1.2. Generalization

a) Irreducible spaces and representations

The generalization of the situation illustrated in the preceding example consists in considering an m-dimensional vector space $\mathcal{E}$ (defined on complex numbers) which is left globally invariant by the action of every element g_i of a group G_o (i.e. g_i transforms any vector of $\mathcal{E}$ into a vector of $\mathcal{E}$).

Each g_i can be represented by a matrix $M(g_i)$ referred to a certain basis $\{\vec{n}_k\}$ of the space. The set $M(g_i)$ is an irreducible representation of G_o if $\mathcal{E}$ contains no subspace invariant by all the elements of G_o (otherwise, the set $M(g_i)$ is a reducible representation). In general, G_o will not act in the same manner on different irreducible spaces (e.g. I preserves f_1 and reverses f_2 in the above illustration). In order to classify the different action of G_o in different spaces, we compare the two sets of matrices $\{M_1(g_i)\}$ and $\{M_2(g_i)\}$ corresponding to the irreducible spaces $\mathcal{E}_1$ and $\mathcal{E}_2$. There is an inessential part in the difference between $\{M_1\}$ and $\{M_2\}$ which is due to the arbitrary choice of a basis in each space. We therefore consider as equivalent the sets $\{M_1\}$ and $\{M_2\}$ which can be brought to coincide by a suitable change of basis in $\mathcal{E}_1$ or $\mathcal{E}_2$. More precisely , $\{M_1\}$ and $\{M_2\}$ are equivalent if, for all the elements g_i of G_o, it is possible to write :

$$M_2(g_i) = S.M_1(g_i).S^{-1} \qquad (2.1)$$

where the same unitary matrix S is used for all the elements g_i of G_o. If (2.1) is obeyed, the irreducible representations carried by $\mathcal{E}_1$ and $\mathcal{E}_2$ are said to be equivalent. Otherwise they are unequivalent

b) Irreducible representation and classification of spaces

The classifying power of the concept of IR stems from the fact that the unequivalent IR's of a group G_o can be enumerated, once and for all, on the sole basis of the structure of the group (e.g. of the multiplication table of the group if we have to deal with a finite group), i.e. independently from the particular vector spaces which carry the representations. In this enumeration, each IR can be specified unambiguously, and independently from the choice of a basis, by the set of traces $\chi(g_i)$ of the matrices $M(g_i)$ of the IR. Clearly, these sets are distinct for unequivalent IR's In the context of the theory of group-representations, the traces $\chi(g_i)$ are termed "characters".

Hence, any vector space, invariant and irreducible under the action of G_o will carry a representation of G_o which is necessarily equivalent to one of the IR's enumerated beforehand for G_o . On the other hand, if a vector space $\mathcal{E}$ is invariant and reducible with respect to G_o one can decompose it into subspaces $\mathcal{E}_i$ which are invariant and irreducible. In this decomposition, distinct $\mathcal{E}_i$ can carry equivalent or unequivalent IR belonging to the established list of IR's of the group G_o. One says that the reducible representation Γ carried by $\mathcal{E}$ is the sum of IR's carried by the $\mathcal{E}_i$: $\Gamma = \Sigma\, \Gamma_i$.

One can show that if G_o is a finite group, it has a finite number ν of unequivalent IR. This number ν is generally smaller than the order $d(G_o)$ of the group (i.e. its number of elements). More precisely ν is equal to the number of C_i classes of equivalent elements g_j of the group, the equivalence between g_j and g_k being defined by $g_k = g g_j g^{-1}$ where g is also an element of G_o [1] . In this case, the enumeration of the IR's of G_o can be displayed in the form of a character table of the group (table 2). In such a table the various unequivalent IR are listed vertically, and specified by a symbol of the type

Table 2.

G_o	C_1	$C_i(g_j)$	C_ν
Γ_1	$\chi^{(1)}(C_1)$	$\chi^{(1)}(C_i)$	$\chi^{(1)}(C_\nu)$
Γ_2	....		
Γ_i	$\chi^{(i)}(C_1)$	$\chi^{(i)}(C_i)$	$\chi^{(i)}(C_\nu)$
Γ_ν	....		

Γ_i $(i = 1, 2, \ldots .\nu)$. Each line contains the characters relative to the considered IR, and associated to a given element g_i of G_o (or in a more compact form, associated to all the elements of G_o which belong to the same class of equivalence C_i, since all these elements correspond to the same character even though their matrices are different).

Table 1 is an illustration of such a table in the case of the group $G_o = (E, I)$ of order 2. In this case it can be noted, on the one hand, that the number ν of IR is equal to the order of the group. On the other hand, the two unequivalent IR are carried by one-dimensional spaces, a circumstance which is denoted, on the character table, by the fact that the character associated to E (i.e. the trace of the identical transformation in the carrier space) is $\chi(E) = 1$. The two preceding features are valid for all the groups G_o which are abelian (i.e. for which the product of any two elements commute : $g_i . g_j = g_j . g_i$).

In general, neither of these properties are true. For non-abelian groups there will always be certain IR's whose carrier spaces have dimensions $m > 1$. On the other hand for these groups a strict inequality $\nu < d(G_o)$ is obeyed.

By contrast, a property common to all groups is the existence of a one-dimensional IR for which all the characters satisfy the condition $\chi(g_i) = 1$. The unique vector spanning the corresponding one-dimensional carrier space is left unchanged (not only its direction as for any other one-dimensional IR) by the action of all the elements of G_o. This IR is termed the "identity" or "totally symmetric"

IR of G_o, and it is often denoted Γ_1. The totally symmetric irreducible spaces provide a generalization of the space spanned by a symmetric function f_1 (cf.§ a).

All the other IR's of a group, being unequivalent to Γ_1, are only globally invariant : none of their vectors remains fixed under the action of all the $g_i \in G_o$. The corresponding carrier spaces can be considered to display the various manners by which a space can have a "simple lack of symmetry with respect to G_o". For a group G_o with a less simple structure than the inversion group (E,I), there are, therefore, several "simple manners" by which an invariant space will lack symmetry while there will be a unique manner by which it can be symmetric.

c) Example of a non-abelian finite group

An illustration of this situation is found in the case of the group G_o of rotations and reflexions which preserve the equilateral triangle : rotations by $(2\pi/3)$ and $(4\pi/3)$ about the center of gravity of the triangle (denoted respectively C_3 and C_3^2), and reflexions σ_i about the three medians of the triangle (fig. 2). The order of G_o is $d(G_o) = 6$, while the number of its classes C_i is $\nu = 3$. The character table of G_o is reproduced on table 3. We

G_o	E	$C_2 \begin{bmatrix} C_3 \\ C_3^2 \end{bmatrix}$	$C_3\ (\sigma_1, \sigma_2, \sigma_3)$
Γ_1	1	1	1
Γ_2	1	1	-1
Γ_3	2 $\begin{vmatrix} 1 & 0 \\ 0 & 1 \end{vmatrix}$	-1 $\frac{-1}{2}\begin{vmatrix} 1 & +\sqrt{3} \\ -\sqrt{3} & 1 \end{vmatrix}$	0 $\begin{vmatrix} +1 & 0 \\ 0 & -1 \end{vmatrix}$

Fig.2

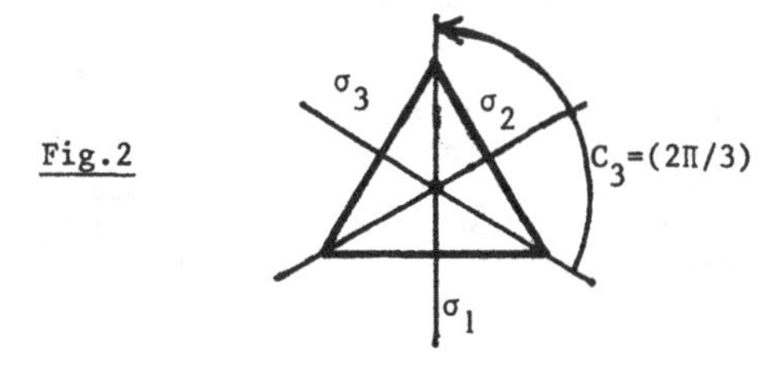

have completed it by showing the explicit form of the matrices E, C_3, and σ_1, relative to Γ_3, and referred to a certain basis. Examples of carrier spaces corresponding to the 3 unequivalent IR of G_o are provided by the z-axis perpendicular to the plane of the triangle, (Γ_1), by the difference of ordered pairs of coordinates [(x.y)-(y.x)] where x and y are coordinates in the plane of the triangle (Γ_2), and by the 2-dimensional space constituted by the plane (x,y) of the triangle (Γ_3). The first space is an example of totally symmetric space under the action of G_o, while the two others are examples of "simple non-symmetric spaces" related to IR's of G_o. One is one-dimensional, and the other two-dimensional. In agreement with the definition of irreducible spaces no real or complex direction can be found in the (x,y) plane which would be invariant by all the elements of G_o: (x,y) contains no invariant subspace.

2.1.3.) "Physically irreducible" representations

a) Example of a group with complex representations

Consider the set of 3 transformations in fig. 2 (E, C_3, C_3^2). This set forms a group G'_o which is a subgroup of the group G_o examined above.

It is an abelian group, associated to 3 one-dimensional IR's. Its character table is reproduced in table 4 [we have set $\omega = \exp(2 \cdot i\pi/3)$]

TABLE 4

G_o	E	C_3	C_3^2
Γ_1	1	1	1
Γ'_2	1	ω	ω^2
Γ'_3	1	$\bar{\omega} = \omega^2$	$\bar{\omega}^2 = \omega$

We note that Γ'_2 and Γ'_3 have characters (and matrices) relative to C_3 and C_3^2, which are complex numbers. These two representations are termed complex representations. Example of carrier spaces for Γ'_2 and Γ'_3 are respectively provided by the complex coordinates (x + iy) and (x - iy) where (x,y) is the plane perpendicular to the axis of the 3-fold rotations. Indeed, using the matrix of C_3 in the (x,y) plane, indicated on table 3, we find, for instance, that (x + iy) transforms under the action of C_3 into ω(x + iy).

Obviously, no unitary change of basis will transform these representations into real ones since the characters are not modified by such a change of the basis.

By contrast, if we consider the reducible vector space spanned by the set (x + iy ; x - iy) a unitary change to the basis (x,y) will transform the complex diagonal matrices of C_3^2 and C_3 into the real matrices indicated on table 3 for the same rotations. This reducible representation is real. If we do not allow complex combinations of x, and y to be taken as basis of the (x,y) space, this space will be irreducible : no real direction in it, is invariant by G_o'. The representation carried by the space (x,y) is said to be irreducible on the real numbers and reducible on the complex numbers. Note that the two complex IR in which it decomposes have complex conjugated matrix elements . They are termed conjugate representations, and one can denote $\Gamma_3' = \bar{\Gamma}_2'$.

In the context of the physical applications considered here, a different terminology is used. The two-dimensional real representation carried by the (x,y) space is termed physically-irreducible in contrast with Γ_2' and Γ_3' which are merely irreducible or mathematically irreducible.

b) Generalization

In general, a complex representation is defined by the fact that its set of matrices cannot be transformed to a set of real matrices by a unitary change of basis in the carrier space. If G_o is a group possessing such an IR, Γ_i, it will also possess the complex conjugate IR, $\bar{\Gamma}_i$. One can therefore always construct a physically IR from Γ_i and $\bar{\Gamma}_i$ which can be denoted $(\Gamma_i + \bar{\Gamma}_i)$. Several situations exist regarding the relationship between Γ_i and $\bar{\Gamma}_i$.(Cf. ref. 1).

2.2. Group theoretical properties used in the formulation of Landau's theory

2.2.1.) Decomposition of f(r) into a sum of irreducible functions

This property generalizes the decomposition of a function $\phi(x)$ into its symmetric and antisymmetric parts with respect to the inversion group.

Let G_o be a group of transformations acting in the 3-dimensional space $\{\vec{r}\}$. G_o will also act on any function $f(\vec{r})$: indeed we can define the transform of $f(\vec{r})$ by $g_i \in G_o$ as $f_i(\vec{r}) = f(g_i\vec{r})$.

The important property used in this chapter is that any function $f(\vec{r})$ can be written :

$$f(\vec{r}) = \sum_k \left[\sum_r \lambda_{kr} \cdot \psi_{kr}(\vec{r})\right] \tag{2.2}$$

where the set of functions $\psi_{kr}(\vec{r})$ (given k), carries the IR Γ_k of the group G_o.

The origin of this property can be easily understood if G_o is finite. In this case, one can construct all the transforms $f_i(\vec{r}) = g_i \cdot f(\vec{r})$ [$i = 1,2,\ldots, d(G_o)$] of $f(\vec{r})$ by the elements g_i of G_o. The set $f_i(\vec{r})$ is, clearly, globally invariant by G_o, and can be considered as generating an invariant vector space. If this space is not irreducible one can decompose it into irreducible spaces. Each irreducible space will be spanned by functions $\psi_{kr}(\vec{r})$ which are linear combinations of the $f_i(\vec{r})$. Conversely, each $f_i(\vec{r})$ can be expressed as a linear combination of the $\psi_{kr}(\vec{r})$. In particular the considered function $f(\vec{r})$ can be expressed as in eq.(2.2).

The decomposition in eq.(2.2) contains a part of arbitrariness. In the first place, each set $\psi_{kr}(\vec{r})$ (given k) is defined up to a multiplication by a coefficient c_k since this coefficient can be integrated in the λ_{kr}. Besides, if the decomposition of the vector space$[f_1(\vec{r}),\ldots, f_n(\vec{r})]$ contains several irreducible invariant spaces carrying equivalent representations Γ_k, an additional index should be used in eq.(2.2) to distinguish the various symmetry - equivalent sets of functions $\psi_{kr}(\vec{r})$. In this case, we can note that each set is not unambiguously defined : if two equivalent irreducible spaces are spanned by homologous bases $\psi^{(1)}_{kr}$ and $\psi^{(2)}_{kr}$(i.e. if these bases are such that the matrices of the two equivalent IR's coincide), then any set of functions $\psi^{(3)}_{kr} = [c_k\,\psi^{(1)}_{kr} + c'_k\,\psi^{(2)}_{kr}]$ carries an equivalent IR.

Fig.3 shows an example of the decomposition (2.2) by means of the IR's of the group G_o of invariance of the equilateral triangle (fig. 2 and table 3).

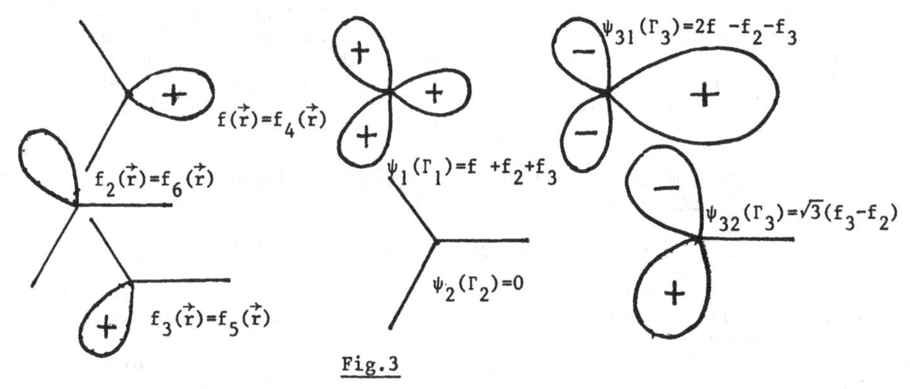

Fig.3

The decomposition of f(r) in this case is, for example :

$$f(\vec{r}) = (\frac{\psi_1}{3} + \frac{\psi_{31}}{3}) \qquad (2.2)'$$

Another example is provided by the infinite discrete group $\mathcal{E}_o$ of translations of a crystal (cf. Chapter III) which is generated by three non-coplanar elementary translations $(\vec{a}_1, \vec{a}_2, \vec{a}_3)$.

The irreducible representations of this group form an infinite set of one-dimensional IR's labelled by the vector $\vec{k}$. For an infinite crystal, the coordinates of $\vec{k}$ can have any value, and the irreducible functions $\psi_{\vec{k}}(\vec{r})$ carrying the IR $(\Gamma_{\vec{k}})$ can be taken as $\psi_{\vec{k}} = e^{-i\vec{k}.\vec{r}}$. The decomposition (2.2) can then be written as :

$$f(\vec{r}) \propto \int \lambda(\vec{k})\; e^{-i\vec{k}.\vec{r}}\; d\vec{k} \qquad (2.2)''$$

in which one recognizes in $\lambda(\vec{k})$ the Fourier transform of $f(\vec{r})$.

2.2.2.) Invariant polynomials

We have seen in §.2.2.21. how the action of G_o, defined in the space $\{\vec{r}\}$ can be extended to the functions $f(\vec{r})$.

More generally, consider a vector space $\mathcal{E}$ left globally invariant by G_o. The elements of G_o act linearly on the coordinates η_j in this space. We can consider functions $f(\eta_1,\ldots,\eta_m)$ and their transforms $f_i = g_i f = f(g_i \cdot \eta_j)$. An important case is constituted by functions which are homogeneous polynomials of a given degree p. These polynomials are linear combinations of C^p_{p+m-1} monomials of degree p. Thus, they can be considered to generate a vector space $\mathcal{E}^p$ of dimension C^p_{p+m-1}. Since G_o acts linearly on the η_j, a polynomial of degree p is transformed into a polynomial of the same degree. The group G_o therefore acts linearly on the space $\mathcal{E}^p$ and leaves this space invariant. $\mathcal{E}^p$ carries a representation of G_o which, in general, will be reducible. If we decompose $\mathcal{E}^p$ into irreducible vector spaces with respect to G_o, there <u>can be</u>, in this decomposition, one or several spaces which carry the totally symmetric representation Γ_1 of G_o. Each of the latter one-dimensional spaces is spanned by a polynomial of degree p, which has the property to be left invariant by all the elements of G_o. Such polynomials are termed <u>invariant polynomials</u> of the η_j coordinates.

Consider, for instance, the space $\mathcal{E} = (x,y)$ which, with respect to the group G_o in table 3, carries the IR Γ_3. Homogeneous polynomials of degree 2 form a 3-dimensional space $\mathcal{E}^2$ spanned by (x^2, y^2, xy). The decomposition of this space into irreducible spaces, by methods which will be detailed in §.4 , show that $\mathcal{E}^2$ contains one totally-symmetric subspace spanned by the polynomials (x^2, y^2). Likewise, $\mathcal{E}^3$ is generated by (x^3, y^3, x^2y, xy^2) which contains a single totally symmetric subspace spanned by the polynomial $x(x^2-3y^2)$. Finally, $\mathcal{E}^4$ is generated by the 5 monomials $(x^4, y^4, x^3y, xy^3, x^2y^2)$. It also contains a single Γ_1-space spanned by $(x^2+y^2)^2$.

2.2.3.) 1st and 2nd degree invariants for non-totally symmetric IR's

Important properties can be stated if the vector space $\mathcal{E}(\eta_1,\ldots,\eta_m)$ carries a physically—IR of G_o (i.e. a representation irreducible on the real numbers) which we denote Γ.

a) Absence of linear invariant

If Γ is not totally symmetric, <u>no linear combination of the η_i with real coefficients, can be invariant by G_o</u>. This is an obvious property since such a combination constitutes a (real) vector direction in $\mathcal{E}$. Its invariance would imply that ε contains an invariant subspace, in contrast with the irreducibility (on the real numbers) of $\mathcal{E}$.

b) Uniqueness of the second degree invariant

For any m-dimensional space $\mathcal{E}$ carrying a <u>physically-IR</u> of G_o, there is a single polynomial of degree 2, which is totally-symmetric. If the matrices of Γ are referred to a real basis of $\mathcal{E}$ (and are thus real), the form of this polynomial is, up to a multiplicative constant :

$$f_2 (\eta_1 \dots \eta_m) = \sum_{j=1}^{m} \eta_j^2 \qquad (2.3)$$

c) Absence of "cross" second-degree invariants for unequivalent IR's

If $\mathcal{E}_1 = (\eta_1 \dots \eta_m)$ and $\mathcal{E}_2 = (\eta'_1, \dots, \eta'_n)$ are two physically irreducible spaces which carry unequivalent IR's, <u>there is no</u> bilinear polynomial of the form $(\Sigma\, a_{ij} \eta_i \eta'_j)$ <u>invariant by G_o</u>.

d) Form of the "cross" second-degree invariant for two equivalent IR's

If $\mathcal{E}_1$ and $\mathcal{E}_2$ carry <u>equivalent</u> IR's, we can consider homologous coordinates $(\eta_1, \dots, \eta_m)$ and $(\eta'_1, \dots, \eta'_m)$ in the two spaces. It can then be shown that the only bilinear "cross-term" which can be invariant by G_o is : $(\Sigma \eta_i . \eta'_i)$

<u>**2.2.4) Standard form of the second degree invariant for a reducible set of degrees of freedom**</u>

Consider a space $\mathcal{E}$ which is reducible with respect to the action of G_o, and $\mathcal{E}_i$ its irreducible subspaces : $\mathcal{E}_i = (\eta_{i1}, \eta_{i2}, \dots, \eta_{im})$. From property c) in the preceding paragraph we can conclude that any <u>second degree</u> polynomial of the coordinates in $\mathcal{E}$, invariant by G_o, will contain no cross-terms between the coordinates of unequivalent $\mathcal{E}_i$. However it will contain cross terms relative to $\mathcal{E}_i$ carrying equivalent IR's. Consider, for instance, that $\mathcal{E}$ contains two subspaces $\mathcal{E}_1$ and $\mathcal{E}_2$ relative to the same IR, (Γ). Adopting, as in d) here-above, homologous coordinates in the two spaces, and using property d) and eq.(2.3), we obtain the following invariant polynomial of degree two as a function of the sets of coordinates η_{1j} and η_{2j} :

$$a_1 \left(\sum_{j=1}^{m} \eta_{1j}^2\right) + a_{12} \left(\sum_{j=1}^{m} \eta_{1j}\, \eta_{2j}\right) + a_2 \left(\sum_{j=1}^{m} \eta_{2j}^2\right) \qquad (2.4)$$

This quadratic form can be transformed into the sum of two squares by setting $\eta'_{1j} (\eta_{1j} + \lambda \eta_{2j})$ and $\eta'_{2j} (\eta_{2j} - \lambda \eta_{1j})$, with $\lambda(a_1, a_{12}, a_2)$ independent of j. As stressed in §.2.2.1, the spaces $\mathcal{E}'_1$ and $\mathcal{E}'_2$ spanned by the η'_{1j} and the η'_{2j} are also irreducible and relative to (Γ).

As a result of this type of transformation, any homogeneous polynomial of degree 2, associated to the reducible space $\mathcal{E}$, can be written in the form :

$$f_2 = \sum_i a_i \left(\sum_j \eta_{ij}^2\right) \qquad (2.5)$$

where we have relabelled η_{ij} the η'_{ij}, when relevant. In eq.(2.5) each sum ($\sum_j \eta_{ij}^2$) (and each coefficient a_i) is associated to an irreducible space with respect to G_o, these spaces being equivalent or unequivalent.

3. LANDAU'S THEORY

In this paragraph we define two functions, the particle density and the variational free-energy, which are the tools needed to develop the theory on more general terms than in chapter I. We show how these quantities allow to define the continuity of the considered phase transition and its symmetry characteristics. We then perform a transformation of the particle density, partly based on the group theoretical results just stated, in order to obtain a scalar set of small degrees of freedom which can serve as variational quantities, as well as a variational free-energy expandable as a function of the latter quantities. We then find ourselves in the same situation as examined in chapter I. The same arguments invoked in chapter I are applied and the concept of order-parameter is derived. Finally, the physical and symmetry implications of the theory are drawn.

3.1. Variational and equilibrium particle density and free-energy

With some generality we can consider that phase transitions correspond to a modification of the equilibrium spatial configuration of the particles (atoms, ions, electron clouds) composing the system, and that a state of the system (equilibrium or non-equilibrium) can be defined by specifying this spatial configuration. However, such a description will not account for magnetic transition which are mainly concerned with spin configurations. We will show in chapter VI, how the arguments exposed hereunder can be extended to spin-configurations. Certain other types of transitions can fall out of this description (e.g. certain models of metal-insulator transition, the superconducting transition). For each type (i) of particle (electrons, ions or atoms of a certain kind), we denote $\rho^{(i)}(\vec{r}).d^3\vec{r}$, the probability of finding a particle of type (i) in the elementary volume $d\vec{r}$ surrounding $\vec{r}$. For the sake of simplicity we can consider a single type of particles and drop the label (i) in the notation of the particle probability density $\rho(\vec{r})$.

$\rho(\vec{r})$ provide us with a convenient variational quantity to specify an arbitrary state of the system. We can consider a variational free-energy as a functional $F(T,P,\rho(\vec{r}))$ (since F is a scalar quantity, its dependence on $\rho(\vec{r})$ involves integrations over the volume of the system .Cf. Chapter III ,§3).

The equilibrium state of the system at a given temperature and pressure is described by the density $\rho_{eq}(T,P,\vec{r})$, corresponding to the minimum of $F(T,P,\rho(\vec{r}))$ with respect to $\rho(\vec{r})$. This minimum is $F(T,P,\rho_{eq})$.

3.1.1. Symmetry of the system

At each temperature and pressure, the symmetry of the

system considered is defined as the set of geometrical transformations which leave ρ_{eq} $(T, P, \vec{r})$ invariant : g_i is a symmetry-operation if :

$$g_i \cdot \rho_{eq}(T,P,\vec{r}) = \rho_{eq}(T,P,g_i\vec{r}) = \rho_{eq}(T,P,\vec{r}) \qquad (3.1)$$

Clearly, the set of g_i-symmetry operations forms a group, the symmetry-group of the system at given T and P. This group can be a continuous group (as in a fluid), or an infinite discrete group (as in a crystal), or a finite group (if crystal symmetry can be reduced to point-symmetry).

3.1.2. Single phase of a system. Continuous phase transition

When temperature and pressure change, ρ_{eq} will generally change. This change can be continuous, i.e. some mathematically defined distance between the functions $\rho_{eq}(T+dT, P+dP, \vec{r})$ and $\rho_{eq}(T,P,\vec{r})$ is infinitesimal. The system will be considered to remain in the same phase when T and P vary, if, besides this continuous variation of ρ_{eq}, the symmetry group of the system is the same in the entire interval of temperatures and pressures considered.

By contrast, the occurence of a continuous transition at T_c (or P_c) is defined by the conjunction of two conditions :

i) ρ_{eq} varies continuously across the transition point.

ii) The group of symmetry of the system changes at the transition point.

It is clear that if condition i) is not satisfied, the occurence of a phase transition can be defined without requiring condition ii) : the phase transition will correspond to the point of discontinuity of ρ_{eq} ; such a transition can occur with or without a symmetry change.

3.2. Decomposition of the density increment into irreducible parts

As compared to the set of quantities (p_x, p_y, p_z) used in the example treated in chapter I, the use of the variational function $\rho(\vec{r})$ has several drawbacks : i) It cannot be considered as a small quantity, since, near T_c, its equilibrium value is a finite density function $\rho_{eq}(T_c, P_c, \vec{r})$. ii) Its symmetry properties are not explicit iii) It is not a scalar quantity ; this will lead to a more complex mathematical formalism if we want to expand F as a function of $\rho(\vec{r})$. Those three difficulties can be removed by substituting to $\rho(\vec{r})$ appropriate quantities related to this function.

3.2.1. Definition of a small density increment

Let us denote $\rho_o(\vec{r})$ the equilibrium density at the transition point :

$$\rho_o(\vec{r}) = \rho_{eq}(T_c, P_c, \vec{r}) \qquad (3.2)$$

The <u>density increment</u> $\delta\rho'(\vec{r})$ is defined by :

$$\rho(\vec{r}) = \rho_o(\vec{r}) + \delta\rho'(\vec{r}) \qquad (3.3)$$

Due to the <u>continuous character of the transition,</u> $\rho_{eq}(T,P,\vec{r})$ is close to $\rho_o(\vec{r})$ when T and P lye in the vicinity of the transition point. Hence the equilibrium value $\delta\rho'_{eq}(T,P,\vec{r})$ of the density increment is a "small" function. We can restrict the study of the variational free-energy $F(T,P, \rho_o + \delta\rho')$ to small functions $\delta\rho'(\vec{r})$, since the equilibrium function $\delta\rho'_{eq}$, which we wish to determine by minimizing F, belongs to this class.

3.2.2. Symmetry properties of $\delta\rho'(\vec{r})$

Let G_o be the symmetry group of the system at the transition point. As defined in §.3.1.1., G_o is the group of invariance of the equilibrium density $\rho_o(\vec{r})$. The density increment $\delta\rho'(\vec{r})$, being an arbitrary function, is not invariant under the action of G_o. Its transformation properties can be made explicit by using eq.(2.2). We write :

$$\delta\rho'(\vec{r}) = \sum_k \left[\sum_r \lambda_{kr} \cdot \psi_{kr}(\vec{r}) \right] \qquad (3.4)$$

where the ψ_{kr} corresponding to a given value of the index k, carry the irreducible representation Γ_k. The transformation properties of the ψ_{kr}, under the action of G_o are determined by the matrices of Γ_k which are assumed to be known. Thus the manner in which $\delta\rho'(\vec{r})$ transforms under the action of G_o is entirely defined by (3.4).

It is worth pointing out that the functions ψ_{kr} are not specified, in general, independently from $\delta\rho'(\vec{r})$: i.e. they <u>do not constitute a</u> set of "basic irreducible functions" from which one can generate an arbitrary function $\delta\rho'(\vec{r})$. The construction of these functions performed in §.2.2.1, as well as the example underlying eq.(2.2)', show that each $\delta\rho'(\vec{r})$ will involve different ψ_{kr}. In certain cases, however, for instance if G_o is the infinite group of translations underlying eq.(2.2)", a large class of $\delta\rho'(\vec{r})$ can be expanded as the linear combination of a fixed set of ψ_{kr} functions (in the considered example we had $\psi_{\vec{k}} = \exp(-i\vec{k}\vec{r})$.

Eq.(3.4) has been written, as eq.(2.2), in the form of a <u>discrete sum</u> over the set of IR's of G_o. As illustrated by eq.(2.2)" it can be necessary to write an integral over a continuous (or quasi-continuous) set of IR's. In this chapter, we will use the discrete form. The case of the continuous sum will be considered in chapter III.§ 3 , and shown to be quite similar.

Let us now consider the equilibrium form of $\delta\rho'(\vec{r})$ at a given temperature and pressure :

$$\delta\rho'_{eq}(T, P, \vec{r}) = \sum_k \left[\sum_r \lambda^{eq}_{kr}(T,P) \cdot \psi_{kr}(T, P, \vec{r}) \right] \qquad (3.5)$$

If this sum contains functions ψ_{kr} transforming according to irreducible representations $\Gamma_k \neq \Gamma_1$, (i.e. non-totally-symmetric), then $\delta\rho'_{eq}$ is not invariant by <u>all</u> the elements of the group G_o, but only, by a subgroup of G_o. Indeed as stated in §.2.2.3.a) a linear combination of the form $(\sum_r \lambda^o_{kr} \psi_{kr})$ relative to $\Gamma_k \neq \Gamma_1$ cannot be invariant by G_o. However, we cannot exclude that such a combination (which is a vector in the carrier space of the IR will be invariant by a subgroup of G_o. Conversely if we require $\delta\rho'_{eq}$ to be invariant by G_o, the sum (3.5) is necessarily reduced to a totally-symmetric function $\delta\rho^o_{eq}$.

Hence, as compared to the set of variational quantities (p_x, p_y, p_z) used in chapter I, the variational increment $\delta\rho'(\vec{r})$ (eq.3.4) has the drawback to contain a totally-symmetric part $\delta\rho^o$, which will not vanish by symmetry at the transition point, and in the temperature interval where the symmetry is also described by the group G_o. $\delta\rho^o(\vec{r})$ is not characteristic of the symmetry change occuring at the transition point.

<u>We will therefore restrict the density increments considered, to functions whose sum (3.4) only contains non-totally symmetric irreducible functions</u> $\psi_{kr}(\vec{r})$. We denote $\delta\rho(\vec{r})$ (unprimed) the corresponding non-symmetric density increments.

This restriction is not entirely justified. As we have seen in chapter I, §.2.9, the phase transition can involve the onset of totally-symmetric quantities of physical interest due to the existence in the free energy, of coupling terms between these quantities and the non-totally symmetric order-parameter. Hence, strictly speaking, one should minimize the free-energy with respect to both the totally symmetric and non-totally symmetric increments. However, as also shown in §.2.9, the "secondary" quantities as $\delta\rho^o(\vec{r})$ can be taken into account in a second step of investigation of the phase transition (see § 5).

3.3. Scalar variational degrees of freedom

As pointed out in §.3.2.2., the form of the $\psi_{kr}(\vec{r})$ functions depends on $\delta\rho(\vec{r})$. In particular, the $\psi_{kr}(\vec{r})$ are <u>small</u> if $\delta\rho(\vec{r})$ is small. It would be desirable to transfer the expression of the magnitude, and of the symmetry properties of $\delta\rho(\vec{r})$, from the set of <u>functions</u> $\psi_{kr}(\vec{r})$ to a set of <u>scalar quantities.</u> This can be performed on the basis of two remarks.

In the first place, as noted in §.2.2.1, each irreducible set $\psi_{kr}(\vec{r})$ is defined up to a multiplicative factor which can be integrated to the coefficients λ_{kr}. We can therefore substitute to the <u>small functions</u> $\psi_{kr}(\vec{r})$ a set of <u>functions</u> $\phi_{kr}(\vec{r})$ <u>possessing "normalized" amplitudes</u>, and correlatively substitute to the λ_{kr}, <u>small parameters</u> η_{kr}.

On the other hand we can consider that the $\phi_{kr}(\vec{r})$ are individually invariant under the action of G_o, while the coefficients λ_{kr} carry the same irreducible representations Γ_k. Consider, for instance, the increment $\delta\rho = (\eta_1\phi_1 + \eta_2\phi_2)$ where ϕ_1 and ϕ_2 are transformed as $(\phi_1 \rightarrow \phi_2)$ and $(\phi_2 \rightarrow -\phi_1)$ by some transformation g. $\delta\rho$ transforms into $(\eta_1\phi_2 - \eta_2\phi_1)$. Clearly this transformation of $\delta\rho$ would be obtained if ϕ_1 and ϕ_2 were considered as fixed functions and if the set (η_2, η_1) transformed into $(\eta_1, -\eta_2)$. Applying, in the same manner, all the operations g_i of G_o, we see that the set (η_2, η_1) carries a representation of

G_o associated to the same matrices as those corresponding to the initial set (ϕ_1, ϕ_2).

In summary, by performing those two substitutions, we obtain an expression of the small, symmetry-breaking density increment $\delta\rho(\vec{r})$ as a linear combination of invariant functions $\phi_{kr}(\vec{r})$ with normalized amplitudes, the coefficients η_{kr} in this combination being sets of small scalar quantities, carrying the non-totally symmetric irreducible representations Γ_k of the group G_o.

3.4. Second degree Taylor expansion of $F(\eta_{kr})$.

The variational free-energy can now be written as :

$$F(T,P, \rho(\vec{r})) = F\,[T,P, \rho_o(\vec{r}) + \Sigma\, \eta_k \quad \phi_{kl}(\vec{r})] \qquad (3.6)$$

where the η_{kr} are small quantities. F can be expanded as a Taylor expansion of the η_{kr}, the coefficients of this expansion being functions of T, P, $\rho_o(\vec{r})$ and of the functions $\phi_{kr}(\vec{r})$. We will assume that the set $\phi_{kr}(\vec{r})$, besides having normalized amplitudes, have fixed forms (as in eq. (2.2.) in order to reduce the variational problem to the determination of the minimization of F with respect to the η_{kr}.

3.4.1. Invariance property of the Taylor expansion of F.

Let us apply to the system an arbitrary global displacement. This geometrical transformation can be looked upon as a change in the coordinate framework. Such a displacement will not change the value of the free-energy F. Let us now restrict the set of geometrical transformations to the elements g_i of the group G_o which leave invariant the equilibrium density $\rho_o(\vec{r})$. Among the arguments of $F(T,P,\rho_o(\vec{r}), \eta_{kr}$ the only ones which are not left invariant by g_i are the components η_{kr} These components are transformed linearly into each other. Hence, the invariance of F by each g_i, holds separately for each degree of the Taylor expansion of F as a function of the η_{kr}. Indeed, the linear character of the action of g_i on the η_{kr} warrants that terms of different degrees do no "mix" in the transformation. Accordingly, there is a constraint imposed to the form of the terms of successive degrees of the Taylor expansion of F : each term must be an invariant polynomial of the η_{kr}, invariant by the set of operations g_i of G_o.

3.4.2. Form of the second-degree expansion.

As the components η_{kr} carry a set of non-totally-symmetric IR's of G_o, the invariance property of each degree of the Taylor expansion, allows to use the two group-theoretical properties stated in §.2.2.3. and §.2.2.4. On the one hand, there is no linear combination of the η_{kr} which is invariant by G_o. On the other hand, the second-degree invariant polynomial can be written in the form (2.5). Up to the second degree ,the Taylor expansion of F is therefore :

$$F = F_o(T,P,\rho_o) + \sum_k \alpha_k \left[T,P,\rho_o(\vec{r}),\ \phi_{kr}(\vec{r})\right].(\sum_r \eta_{kr}^2) \qquad (3.7)$$

In the following discussion we will only retain the explicit dependence of α_k on T and P, and write $\alpha_k = \alpha_k(T,P)$. As assumed, the $\phi_{kr}(\vec{r})$ and $\rho_o(\vec{r})$ keep a fixed form in the vicinity of the considered transition, and will not be involved in the determination of the minima of F.

3.5. Definition of the order-parameter

3.5.1. Expression of the occurence of a symmetry change at T_c.

As stressed in §.3.2.2., the exclusion, from the density increment $\delta\rho(\vec{r})$, of its totally symmetric part, defines a correspondance between the occurence of the symmetry G_o and the vanishing of the equilibrium form of the density increment. Let us put $\delta\rho_{eq}(\vec{r}) = \Sigma\eta_{kr}^o.\phi_{kr}(\vec{r})$. The preceding correspondance implies that if the system is invariant by G_o, then all the components η_{kr}^o vanish. Conversely, if $\eta_{kr}^o = 0$, for all k and r, the equilibrium density is reduced to $\rho_{eq}(\vec{r}) = \rho_o(\vec{r})$, and the symmetry of the system is G_o. Furthermore, if we require that the symmetry of the system be different from G_o then, necessarily, one of the η_{kr}^o, at least, must be non-zero. We have assumed in §.3.1.2. that the transition at T_c involves a change of the symmetry of the system. The two phases adjacent to the transition are invariant by different symmetry groups G and G'. The relationship between these groups and the symmetry group G_o of interest at T_c, is not specified beforehand. In principle, G_o can be different from both G and G' or identical to one of these two groups. We will assume that it is the latter situation which is necessarily realized.

We choose $G \neq G_o$ to be the invariance group of phase II which is the equilibrium phase of the system below T_c. This will put the formulation of the theory in agreement with most experimental studies in real systems, which show that only a few exceptional transitions display the converse behaviour with a symmetry distinct from G_o, above T_c.

The correspondance stated above, leads to the following conditions imposed to the η_{kr}^o :

$$\begin{aligned} &\eta_{kr}^o\ (T>T_c) = 0 && \text{for all } (k,r). \\ &\eta_{kr}^o\ (T<T_c) \neq 0 && \text{for, at least, one } (k,r). \end{aligned} \qquad (3.8)$$

3.5.2. Order-parameter, and form of $\alpha(T,P)$.

Expressing the existence of a phase transition at T_c has been reduced to a set of conditions summarized by eqs.(3.7) and (3.8) :impose to the set of scalar quantities η_{kr}^o a zero-value above T_c ,and a non-zero value to,at least,one η_{kr}^o for $T < T_c$; Deduce the η_{kr}^o from the minimum of an expansion F in which each irreducible set of η_{kr} is associated to a single <u>coefficient</u> $\alpha_k(T,P)$. This formulation appears as an exact generalization of the situation encountered in chapter I, where the η_{kr} were represented by the two irreducible sets (p_z) and (p_x, p_y), and where the free-energy (3.7) was particularized in the form (chapter I, eq.(2.2)

We can therefore use, without modification, the sequence of arguments developped in chapter I, §.2.5, and leading to the definition of an order-parameter (OP) :

(1) positiveness of the $\alpha_k(T,P_c)$ for $T \geqslant T_c$ (chap. I §.2.5.a)

(2) vanishing of one $\alpha_k(T_c,P_c)$ at least (Chap.I, §2.5.b).

(3) vanishing of a single $\alpha_k(T_c,P_c)$ by excluding "singular" transitions which do not have preserved symmetry characteristics for a slight change of the pressure. (chap. I, §.2.5.c)

(4) strict positiveness of all the $\alpha_k(T,P_c)$ in the neighborhood of T_c, except for one α_k which vanishes with a change of sign at T_c (chap. I, §.2.5.d, and §.2.6). To lowest order the latter α_k can be written $\alpha_k \propto (T - T_c)$

Hence, among the irreducible sets η_{kr} $(k = 1,\dots)$, <u>one set</u> can be distinguished by the fact that its associated coefficient α_k satisfies $\alpha_k(T_c,P_c)=0$. We relabel $\alpha(T,P)$ the latter coefficient, and η_r the corresponding set of components η_{kr}, which carry the $IR(\Gamma_k)$ of G_o. The set η_r is the <u>order-parameter</u> (OP) of the transition. Each η_r is an OP-component. The <u>number</u> of η_r components, which is equal to the dimension of Γ_k, is the <u>OP-dimension.</u>

<u>In summary, we define the order-parameter of a continuous transition occuring at (T_c,P_c), as the set of scalar quantities η_r, carrying an IR of the group G_o (the symmetry group of the system for $T \geqslant T_c$), and associated, in the Taylor expansion of the free-energy F, to a coefficient $\alpha_k = \alpha(T,P)$ satisfying the condition $\alpha(T_c,P_c) = 0$.</u>

For η_{kr} distinct from the OP-components, the strict positiveness of the corresponding α_k on either sides of T_c, warrants, <u>on the basis of eq. (3.7)</u>, that their equilibrium values is $\eta^o_{kr} = 0$. As in chapter I (§.2.6), we can ignore, in the first place, these additional degrees of freedom and examine the symmetry and physical characteristics of the considered phase transition by restricting both the expansions of $\delta\rho(\vec{r})$ [eq.3.4] and of F [eq.3.7] to the sole components of the OP. However, the latter conclusion which is founded on the 2nd degree-expansion will be reexamined in §.5 , when higher-degree terms in F are taken into account.

3.6. Order-parameter expansion and nature of the symmetry change.

Reduced to a single second degree-term with negative coefficient for $T < T_c$, the free-energy expansion (3.7) is consistent with the instability of phase I ($\eta_r = 0$) below T_c. However this expansion is unsufficient to discuss the nature of the stable phase in this temperature-range, since it has no minimum for finite η_r. As in chap.I, §.2.6, it is necessary to expand F up to a dgree higher than 2, as a function of the OP components. The basis of the discussion of the possible minima of F and of the nature of the corresponding stable phases is then a set of two functions: the expression of $\delta\rho(\vec{r})$ as a linear combination of the η_r and the $\phi_r(\vec{r})$,

$$\delta\rho(\vec{r}) = \sum_r \eta_r \cdot \phi_r(\vec{r}) \tag{3.9}$$

and, the OP expansion

$$F = F_o(T,P) + \frac{a(T - T_c)}{2} (\sum_r \eta_r^2) + f_3(\eta_r) + f_4(\eta_r) + \ldots \tag{3.10}$$

where f_3, f_4,.... are the successive terms of the Taylor expansion of F. In particular, f_3 and f_4 are, respectively, homogeneous polynomials of degrees 3 and 4. As stated in §.3.4.1, each of these polynomials must be invariant by the transformation properties g_i belonging to G_o. The examples in chapter I (§.2.7.1, and 2.7.2), and by §.2.2.2, show that a third degree term $f_3(\eta_r)$ can be present or absent from the Taylor expansion, depending on the symmetry properties of the considered OP. In §.2.7.2. (chap. I), we found no third-degree term, function of the set (p_x,p_y) and invariant by G_o. In the present chapter §.2.2.1, we have indicated the form of the 3rd degree term, built from components carrying the IR (Γ_3) of table 3, and invariant by the group G_o of the equilateral triangle.

By contrast, whatever the symmetry of the OP, the Taylor expansion of F will always contain a 4th degree term $f_4(\eta_r)$. Indeed, for any symmetry, the term $(\Sigma\eta_r^2)^2$ is always invariant by G_o. Additional independent invariants of the same degree can also exist. Their number and form depend, as illustrated in chap. I, on the symmetry of the OP.

General symmetry considerations pertaining to the form of f_3 and f_4 will be developped in §. 4 . Let us examine, first, the physical rôle of these two terms.

3.6.1. Incompatibility of $f_3(\eta_r)$ with a continuous transition

Consider the case of a free-energy (3.10) which includes a third-degree term $f_3(\eta_r)$. For $T = T_c$ the minimum of (3.10) corresponds to $\eta_r = 0$. Hence, for any value of the η_r, close to $\eta_r = 0$ (i.e. infinitesimal), the free-energy $(F - F_o)$ must be positive. For such infinitesimal values of the η_r it is sufficient to consider the term of lowest degree in the expansion. At T_c, this term is $f_3(\eta_r)$. Clearly, this term cannot be positive for any set of infinitesimal η_r values: if f_3 is positive for a set $\{\eta_r\}$corresponding to a given "direction" in the OP-space, then, due to the fact f_3 is a homogeneous polynomial of degree-3, the set $\{-\eta_r\}$ corresponding to the "opposite direction" makes f_3 negative (fig. 4).

<u>In the presence of a third degree-term the value $\eta_r = 0$ does not correspond to a minimum of the free-energy at T_c, but to a saddle point of this function.</u> Hence, in order to ensure that the equilibrium state of the system at T_c corresponds to $\eta_r = 0$ (in agreement with the definition of a continuous transition), there must be no

third-degree term in the expansion of F.

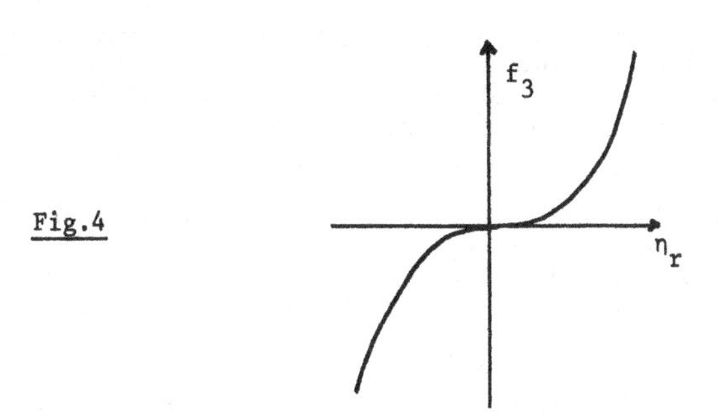

Fig.4

The absence of $f_3(\eta_r)$ can be due, as stated above, and discussed in §.4 , to the impossibility to find a 3rd degree polynomial invariant by the operations of G_o. This absence for "symmetry reasons" will hold for any temperature and pressure. A second possible reason for the absence of $f_3(\eta_r)$ at T_c can be the vanishing of temperature- and pressure-dependent coefficients in f_3. In general $f_3(\eta_r)$ will be of the form :

$$f_3(\eta_r) = b_1(T,P)\mathcal{R}_1(\eta_r) + b_2(T,P).\mathcal{R}_2(\eta_r) + \ldots \qquad (3.11)$$

where $\mathcal{R}_1, \mathcal{R}_2, \ldots$ are linearly independent polynomials of degree 3 separately invariant by G_o (f_3 is then the most general degree-3 polynomial possessing this property of invariance). The occurence of a continuous transition at (T_c, P_c) requires, in addition to $\alpha(T_c, P_c) = 0$, the fulfilment of the conditions $b_1(T_c,P_c) = b_2(T_c,P_c) = .. = 0$. The transition point (T_c, P_c) must therefore be at the intersection of the lines $\alpha(T,P) = 0$, and $b_i(T,P) = 0$. For the same reason as invoked in chap.I, §.2.5, all these functions of T, and P, are different, and (T_c, P_c) can, at most, be at the intersection of $\alpha(T,P)$ and one of the $b_i(T,P)$. If there are several independent $\mathcal{R}_i(\eta_r)$ terms in (3.11), the simultaneous vanishing of the $b_i(T,P)$ at (T_c,P_c) will not be possible, and no continuous transition exists.

If $f_3(\eta_r)$ includes a single $\mathcal{R}_i(\eta_r)$, associated to the coefficient $b(T,P)$, a continuous transition will be possible at (T_c,P_c) if the two lines $\alpha(T,P) = 0$ and $b(T,P) = 0$ cross at this point of the pressure-temperature plane. However, a slight shift of the pressure, dissociates the vanishing of α and b, and accordingly, no continuous transition can occur for $\alpha(T, P \neq P_c) = 0$. (T_c, P_c) corresponds to a continuous transition which is "singular" in the sense discussed in chap. I §.2.5.

Consistently with the choice made in §.3.5.2. herabove, we exclude such singular transitions from our investigation. In this case, the existence of a continuous transition requires that no 3rd degree polynomial of the OP-components is invariant by G_o.

3.6.2. Role of the fourth-degree term $f_4(\eta_r)$

The argument developped in the preceding paragraph and showing that F cannot have a minimum at T_c for $\eta_r = 0$, if a third degree term is present, is not applicable to the 4th degree term $f_4(\eta_r)$: such a term can be positive in every direction of the OP space. This has been illustrated by the examples of chap. I §.2.7.1 and §.2.7.2. (for $\beta_1 > 0$)

Let us put $\eta_r = \eta \cdot \gamma_r$, with $\eta^2 = (\Sigma \eta_r^2)$ [we therefore have $\Sigma\gamma_r^2 = 1$] . The γ_r specify the direction of a unit-vector in the OP space, while η is the "amplitude" of the OP. As a function of η and the γ_r, we can express $f_4(\eta_r)$, the density increment $\delta\rho(\vec{r})$, and the free-energy F in the forms :

$$f_4(\eta_r) = \sum_i \beta_i\ (T,P).\ Q_i(\eta_r) = \eta^4\ [\sum_i \beta_i(T,P).Q_i(\gamma_r)] \qquad (3.12)$$

$$\delta\rho(\vec{r}) = \eta\ .\ [\sum_r \gamma_r\ .\ \phi_r\ (\vec{r})] \qquad (3.13)$$

and

$$F - F_o = \frac{a(T - T_c)}{2}\ .\ \eta^2 + \eta^4\ [\sum_i \beta_i(T,P).\ Q_i(\gamma_r)] \qquad (3.14)$$

where the $Q_i(\eta_r)$ are the independent fourth-degree polynomials which are **separately** invariant by G_o , as defined in §.2.2.2

a) Positivity of the fourth-degree term

The occurence of a continuous transition at (T_c,P_c) [i.e. of a minimum of (3.14) for $\eta = 0$] requires that the sum $[\Sigma\beta_i\ .\ Q_i]$ in eq.(3.14) be positive at (T_c,P_c) in every direction (γ_r) of the OP space. This condition will impose certain inequalities between the $\beta_i(T_c,P_c)$. For instance, in the example of a two-component OP (p_x, p_y) examined in chap. I (§.2.7.2), the two relevant coefficients β_1 and β_2 had to comply with the set of conditions ($\beta_1 > 0$; $\beta_2 > -\beta_1$).

As no additional conditions are imposed to the $\beta_i(T,P)$ we will assume that these functions are constant nearby (T_c, P_c) and equal to their values at the transition point.

For a given set of β_i parameters compatible with the positivity of the fourth-degree term, the sum $(\Sigma\beta_i\ Q_i)$ has one absolute minimum (at least) corresponding to $\gamma_r = \gamma_r^o$. We can put :

$$\frac{\beta}{4} = \sum_i \beta_i\ .\ Q_i(\gamma_r^o) > 0 \qquad (3.15)$$

and write (3.14) as :

$$F - F_o = \frac{a(T - T_c)}{2}\ \eta^2 + \frac{\beta}{4}\ \eta^4 \qquad (3.16)$$

b) Physical anomalies

The form (3.16) is identical to the free-energy (2.4) in chap. I, with the additional feature that η is a positive quantity (unlike p_z). All the results derived in chapter I §.2.7.1 have a general validity. We summarize them :

i) square-root dependence of $\eta_o(T)$ as a function of $(T-T_c)$:

$$\eta_o(T) = \sqrt{\frac{a(T_c-T)}{\beta}} \quad , \quad \text{for} \quad T < T_c \tag{3.17}$$

ii) absence of transition heat.

iii) step-anomaly for the specific heat c_ζ corresponding to maintaining constant the quantity ζ thermodynamically conjugated to the amplitude η of the OP.

iv) divergence of the susceptibility $(\frac{\partial \eta}{\partial \zeta})$ on either sides of T_c, according to the laws :

$$\chi_\eta(T > T_c) = \frac{1}{a(T-T_c)} \quad ; \quad \chi_\eta\ (T < T_c) = \frac{1}{2a(T_c-T)} \tag{3.18}$$

c) Symmetry of the phase stable below T_c

The set of values $[\eta_o(T),\ \gamma_r^o]$, corresponding to the absolute minimum of F (eq. 3.14), defines the phase stable below T_c. As stated in §.3.1.1. the symmetry of this phase is the set of transformations leaving invariant the function $\rho_{eq}(\vec{r}) = (\rho_o(\vec{r}) + \delta\rho_{eq}(\vec{r}))$, with :

$$\delta\rho_{eq}(\vec{r}) = \eta_o(T)\ [\sum_r \gamma_r^o \cdot \phi_r\ (\vec{r})] \tag{3.19}$$

The set of invariance transformations of $\rho_{eq}(\vec{r})$, which forms a group G, will be the <u>intersection</u> of the groups of invariance of $\rho_o(\vec{r})$ and of $\delta\rho_{eq}(\vec{r})$. The density $\rho_o(\vec{r})$ is invariant by G_o. On the other hand, $\delta\rho_{eq}(\vec{r})$ is a linear combination of the γ_r^o and can be considered as a <u>vector</u> in the OP space, whose direction is defined by the γ_r^o (and is independent of $\eta_o(T)$). This vector belongs to the space (γ_r^o) which carries a <u>non-totally symmetric</u> IR (Γ_k) of G_o. Accordingly it cannot be invariant by all the elements of G_o, but only by a smaller group [this is also true if Γ_k is physically irreducible since the γ_r^o are real] . It can happen, as pointed out by Przystawa [4] that $\delta\rho_{eq}(\vec{r})$ will also be invariant by transformations <u>not belonging to G_o</u>. These transformations are irrelevant since we are only interested in the <u>intersection of G_o</u> with <u>the invariance group of $\delta\rho_{eq}(\vec{r})$</u>.

Hence, the symmetry of the phase which is stable below T_c, is defined by the <u>subgroup G of G_o</u> which leaves invariant the "vector" $\delta\rho_{eq}(\vec{r})$ of the physically-irreducible space spanned by the OP-components. This vector corresponds to the values γ_r^o which minimize the fourth degree term of the OP-expansion. Phase II, stable below T_c, can be termed the "low-symmetry" phase, while phase I, stable above T_c, is the "high-symmetry" phase.

d) Degeneracy of the absolute minimum of F

Let γ_r^o be a set of components corresponding to the absolute minimum of $(\Sigma\beta_i . Q_i)$, and, as a consequence, to the absolute minimum of F. This set defines a vector in the OP-space, which is invariant by G. As G is a subgroup of G_o, we can write :

$$G_o = G + h_1.G + h_2G + ... \qquad (3.19)$$

where the symbolic notation + means that G_o comprises the elements of the various sets (G, h_1G, h_2G,...) and where h_1, h_2,... are elements of G_o not belonging to G. (3.19) is called a decomposition of G_o into (left)-cosets with respect to G [5]. The h_p are not uniquely defined since each h_p can be replaced by any element of the set $(h_p.G)$. However, the number s of sets $h_p.G$ in the decomposition (3.19) is unambiguously defined (if this number has a meaning, i.e. if the number of sets is finite). Also note that the various sets $h_p.G$ have no elements in common.

Apply any operation h_p to the vector (γ_r^o) of the OP-space. One obtains a vector (γ_r^p) belonging to the same space (since h_p belongs to G_o, and since the OP-space is invariant by G_o). Necessarily, (γ_r^p) is a vector distinct from (γ_r^o) [if only its sense] otherwise h_p would be part of the invariance group G of (γ_r^o). On the other hand, the polynomial $Q_i(\gamma_r^o)$ will transform under the action of h_p, into $Q_i(\gamma_r^p)$. One has $Q_i(\gamma_r^p)=Q_i(\gamma_r^o)$, since each Q_i is invariant by all the operations of G_o (including the h_p). Accordingly, the free-energies associated to $[\eta_o(T), \gamma_r^o]$ and $[\eta_o(T), \gamma_r^p]$ are the same. The sets $[\eta_o(T), \gamma_r^p]$ (p = 0, 1,...), all correspond to the absolute minimum of F. They define different states which are equally likely to constitute the stable state of the system. The number of these states is equal to the number s of sets in the decomposition (3.19). This number is called the index of the subgroup G of G_o.

The degeneracy of the absolute minimum of F, previously noted (chap.I, §.2.8.1) in the case of particular examples of continuous transitions, is therefore a general feature of continuous transitions. Another feature noted in chap. I, §.2.8.1 has a general validity: the invariance group of (γ_r^p) is conjugated to the invariance group of (γ_r^o). Indeed the invariance group of the former direction is $h_p.G.h_p^{-1}$, as easily checked. In conclusion, we can consider that the states (γ_r^p) and (γ_r^o) represent the same physical situation in different frames of reference within the OP-space : their symmetries differ by the absolute orientation of the symmetry elements (the relationship between the orientations being defined by the h_p), and their free-energies are identical. On this basis they will be considered as equivalent states.

e) Enumeration of the possible, distinct, low-symmetry phases

The conditions imposed to the coefficients β_i by the positiveness of the fourth degree terms do not define uniquely a set of equivalent states, stable below T_c.

For instance, in the example studied in chap.I, §.2.7.2, for the OP (p_x, p_y), the condition of positiveness is expressed by $(\beta_1 > 0 ; \beta_2 > -\beta_1)$. Within the preceding range of β_i values we have distinguished two sub-ranges. For $\beta_2 > \beta_1 > 0$, a state equivalent to ($p_x \neq 0$, $p_y = 0$) is stable, while for ($\beta_1 > \beta_2 > -\beta_1$), the stable state is equivalent to ($p_x = p_y \neq 0$). In the range of stability of one type of state (eg. $p_x = p_y$), the other type of state is not stable (the latter state is not a minimum of F, and its free-energy is higher). Moreover, the symmetry groups G associated to each of these phases are not conjugated with respect to G_o.

In general, depending of the relative values of the β_i, within the range defined by the positivity of $f_4(\eta_r)$, several unequivalent states will constitute the stable phase of the system below T_c.

A symmetry classification of the sets of equivalent states, and of their associated symmetry groups, will be described in §.4.5 hereafter.

The various, physically distinct, phases which can onset in a given system, below a continuous transition T_c, correspond to different directions (γ_r^o) in the OP-space. Their symmetry groups G are subgroups of G_o. Each of these directions makes the free-energy F an absolute minimum, for a certain sub-range of values of the β_i coefficients, within the range of the β_i, defined by the positiveness of $f_4(\eta_r)$. Each kind of minimum is degenerate.

4. TECHNICAL GROUP-THEORETICAL ASPECTS OF LANDAU'S THEORY

In this paragraph, we discuss the methods, based on group-theoretical considerations which allow to assert the presence or absence of a third-degree term $f_3(\eta_r)$, to infer the number of linearly independent n^{th} degree terms separately invariant by G_o, and to determine their forms, once a basis is chosen in the space of the order-parameter. We also describe the methods which can be used to locate the degenerate minimum of F.

4.1. Reduction of an invariant space into irreducible spaces

Consider a space $\mathcal{E}$, spanned by the vectors $\vec{\phi}_i$, and carrying a reducible representation Γ of the group G_o. As pointed out in §.2.1.2, $\mathcal{E}$ can be decomposed into subspaces $\mathcal{E}_{k(m)}$ carrying irreducible representations Γ_k of G_o (the index (m) distinguishes spaces carrying equivalent IR). This decomposition can be specified at two distinct levels of detail.

For certain problems, it can be sufficient to know the nature of the IR's(Γ_k) involved into the decomposition of Γ. In this view the informations required are the characters $\chi(g_i)$ of Γ, i.e. the traces of the matrices representing (in the basis $\vec{\phi}_i$, for instance) the elements g_i of G_o. The practical rule is provided by the formula [1, 2]:

$$n_k = \frac{1}{d(G_o)} \sum_{g_i} \overline{\chi}(g_i) \cdot \chi^{(k)}(g_i) \qquad (4.1)$$

which specifies the number n_k of irreducible spaces of type Γ_k

contained in $\mathcal{E}$. $\chi^{(k)}(g_i)$ are the characters of Γ_k, to be found in the character table of G_o (cf. §.2.1.2).

In some cases, it is necessary to determine the linear combinations of the vectors $\vec{\phi}_i$ which span each irreducible space $\mathcal{E}_{k(m)}$. This problem can be solved only partly. Indeed, if we consider the set of spaces $\mathcal{E}_{k(m)}$ corresponding to a given IR (Γ_k), their union $[\cup_m \mathcal{E}_{k(m)}]$ is a space which, as stressed in §.2.2.1, can be decomposed in infinitely many ways into irreducible spaces $\mathcal{E}_{k(m)}$.

For instance an arbitrary linear combination of two distinct invariant vectors (i.e. carrying the IR Γ_1) will be an invariant vector. Hence, we can rely on group theoretical rules to determine the vectors spanning the spaces $\cup_m \mathcal{E}_{k(m)}$ but the further decomposition of ($\cup \mathcal{E}_{km}$) is arbitrary . The practical rule of decomposition relies on the existence of operators which "project" (in a broad sense) a vector $\vec{\phi}_i$ on the space $[\cup_m \mathcal{E}_{k(m)}]$. The simplest of the two types of such operators is :

$$P_k = \sum_{g_i} \bar{\chi}^{(k)}(g_i).g_i \tag{4.2}$$

P_k is a weighted sum of the elements of G_o, the weights being the complex conjugates of the characters of the considered irreducible representation Γ_k. The application of P_k to any of the $\vec{\phi}_i$ will either give zero, or a vector of the space $[\cup_m \mathcal{E}_{k(m)}]$. In order to generate distinct vectors of the latter space one has to apply P_k to different $\vec{\phi}_i$ (depending on the $\vec{\phi}_i$ considered, one obtains linearly dependent or independent vectors).

The second type of projection operator is :

$$\pi_k^{\alpha\beta} = \sum_{g_i} \bar{M}_{\alpha\beta}^{(k)}(g_i).g_i \tag{4.3}$$

In this case, the weights are the matrix elements (in a certain basis) of Γ_k. The set of operators $\pi_k^{\alpha\beta}$ associated to the different values of the index α , applied to a given vector $\vec{\phi}_l$, provide a set of distinct vectors $\psi_{k\alpha}$ generating a carrier-space of Γ_k.

4.2. Tensorial products of invariant spaces. n^{th} power of a representation

Let $\mathcal{E}^{(a)}$ and $\mathcal{E}^{(b)}$ be two spaces invariant under the action of G_o and respectively spanned by the vectors $\vec{\phi}_l^{(a)}$ and $\vec{\phi}_m^{(b)}$. $\mathcal{E}^{(a)}$ and $\mathcal{E}^{(b)}$ respectively carry representations Γ_a and Γ_b of G_o (which are reducible or irreducible).

The tensorial product $\mathcal{E}^{(a)} \times \mathcal{E}^{(b)}$ of the two spaces (taken in this order) is the vector space generated by the set of elements $(\vec{\phi}_l^{(a)}, \vec{\phi}_m^{(b)}) = \vec{\psi}_{lm}$, where ψ_{lm} is assumed to have a linear dependence with respect to each member of the pair : $(\Sigma\lambda_l \phi_l^{(a)}, \phi_m^{(b)}) = \Sigma\lambda_l\vec{\psi}_{lm}$.
We define, on the other hand, the action of $g_i \in G_o$ on the space $\mathcal{E}^{(a)} \times \mathcal{E}^{(b)}$ by :

$$g_i \vec{\psi}_{lm} = (g_i \phi_l^{(a)}, g_i \phi_m^{(b)}) \tag{4.4}$$

It results from this definition that $\mathcal{E}^{(a)} \times \mathcal{E}^{(b)}$ is invariant by G_o and that each ψ_{lm} is transformed linearly into the $\psi_{l'm'}$. This space is a carrier space for a representation Γ_{ab} of G_o which is called the product of the two representations Γ_a and Γ_b. This representation is reducible if either Γ_a or Γ_b are reducible. It can also be reducible if Γ_a and Γ_b are irreducible. It can be shown [1,2] that the characters $\chi^{ab}(g_i)$ of Γ_{ab} are :

$$\chi^{ab}(g_i) = \chi^{a}(g_i).\chi^{b}(g_i) \qquad (4.5)$$

The definition of a tensorial product applies to the case where $\mathcal{E}^{(a)}$ and $\mathcal{E}^{(b)}$ are identical. Γ_{aa} is then called the square of the representation Γ_a. It is straight-forward to extend this definition to the nth power of a representation Γ, which we will denote Γ^n. Eq.(4.5) shows that the characters of Γ^n will be $[\chi(g_i)]^n$ where $\chi(g_i)$ are the characters of Γ.

4.3. Symmetrized n^{th} power of a representation

4.3.1. Definition

Let Γ be a representation carried by the space $\mathcal{E}$ and $\vec{\phi}_l$ the basis of $\mathcal{E}$. The carrier space $\mathcal{E}^n$ of the nth power Γ^n of Γ is generated by the basis $(\vec{\phi}_{l_1}, \vec{\phi}_{l_2}, \ldots \vec{\phi}_{l_n})$. By performing a permutation of the indices l_i, we transform a vector of the basis of $\mathcal{E}^n$, into another vector of the basis. [e.g. $(\vec{\phi}_1, \vec{\phi}_2, \ldots) \rightarrow (\vec{\phi}_2, \vec{\phi}_1 \ldots)$] , generally distinct from the initial vector. Any vector of $\mathcal{E}^n$ can be written $\vec{V} = [\Sigma \lambda_{l_1, l_2, \ldots l_n} (\vec{\phi}_{l_1}, \ldots, \vec{\phi}_{l_n})]$. The symmetrized n^{th} power of $\mathcal{E}^n$ is the subspace, denoted $[\mathcal{E}^n]$, of $\mathcal{E}^n$, generated by all the vectors $\vec{V}$ which are invariant by any permutation of the n indices $(l_1, \ldots, l_n)$. It can be shown [1] that this space is invariant by G_o. Consequently $[\mathcal{E}^n]$ carries a representation of G_o which is called the symmetrized n^{th} power of Γ. We will denote it $[\Gamma^n]$, its character being denoted $[\chi^n](g_i)$. Table 5 indicates, following Lyubarskij [1], the characters of the representations $[\Gamma^2]$, $[\Gamma^3]$ and $[\Gamma^4]$ as functions of the characters $\chi(g_i)$ of the starting representation Γ.

Table 5 Characters of the symmetrized n^{th} power of a representation

Representation	character $[\chi^n](g)$
$[\Gamma^2]$	$\frac{1}{2}[\chi(g^2) + \chi^2(g)]$
$[\Gamma^3]$	$\frac{1}{3}\chi(g^3) + \frac{1}{2}\chi(g^2).\chi(g) + \frac{1}{6}\chi^3(g)$
$[\Gamma^4]$	$\frac{1}{4}\chi(g^4) + \frac{1}{3}\chi(g^3).\chi(g) + \frac{1}{8}\chi^2(g^2) + \frac{1}{4}\chi(g^2)\chi^2(g) + \frac{1}{24}\chi^4(g)$

4.3.2. Polynomials of nth degree ; reduction into irreducible sets

In §.2.2.2. we have considered the polynomials of nth degree defined on a set of coordinates $(\eta_1 \dots \eta_m)$ spanning an-m-dimensional space $\mathcal{E}$ invariant by G_o. We have noted that these polynomials form a space $\mathcal{E}^{(n)}$ generated by C^{n}_{n+m-1} monomials, $\mathcal{E}^{(n)}$ being invariant by G_o. Let us show that $\mathcal{E}^{(n)}$ actually coincides with the symmetrized n^{th} power of the space $\mathcal{E}$ spanned by the $(\eta_1 \dots \eta_m)$. Indeed, a basis of the latter space is constituted by vectors of the form [6,7]:

$$\vec{V} = \sum_j \mathcal{P}_j \, (\eta_{l_1}, \dots \eta_{l_n}) \tag{4.6}$$

where the sum is over all the p ! permutations $\mathcal{P}_j$ of the coordinates η, and where $(\eta_{l_1}, \dots, \eta_{l_n})$ comprises ν_1 times the coordinate η_1, ν_2-times the coordinate $\eta_2, \dots \nu_m$ times the coordinate η_m, and where $(\nu_1, \dots \nu_m)$ are a partition of n (i.e. $\Sigma \, \nu_i = n$). A given partition $(\nu_1, \dots, \nu_m)$ leads to an unambiguously defined vector $\vec{V}$ in (4.6), whatever the order of the coordinates $(\eta_{l_1}, \dots, \eta_{l_n})$. Conversely distinct partitions lead to distinct vectors $\vec{V}$. In the same way, each vector (4.6) can be associated to a monomial $\prod_{i=1}^{m} \eta_i^{\nu_i}$ (with $\Sigma \nu_i = n$).

The polynomial space $\mathcal{E}^{(n)} = [\mathcal{E}^n]$, being generally reducible, can be decomposed into irreducible spaces. In this view, we use the values of the characters $[\chi^n]$ (tabulated for the first values of n in table 5) and apply formulae (4.1) and (4.2) in order to enumerate the irreducible spaces of each type, and generate the polynomials belonging to each space.

4.3.3. Example of irreducible polynomial spaces

As an illustration of the foregoing considerations let us examine the case of the polynomials of degree 2 and 3 of the coordinates (x,y), which form an irreducible invariant space under the action of the group G_o of the equilateral triangle, already studied in §.2.1.2.c) and table 3. As stated in the latter paragraph, the set (x,y) carries the Γ_3 representation whose characters and matrices (in the (x,y) basis) are indicated in table 3.

a) 2nd degree polynomials

The space of degree-2 polynomials of x and y coincides, as pointed out in §.4.3.2 with the symmetrized-square of Γ_3. The characters of $[\Gamma_3^2]$ are straightforwardly computed from tables 5 and 3. They are indicated on the second line of table 6, hereunder [note that it is sufficient to compute the characters for a single element of each class, and that if $g = (E, C_3, \sigma_1)$, $g^2 = (E, C_3^2, E)$]

Table 6

	E	$\mathcal{C}_2 \begin{bmatrix} C_3 \\ C_3^2 \end{bmatrix}$	$\mathcal{C}_3 \begin{bmatrix} \sigma_1 \\ \sigma_2 \\ \sigma_3 \end{bmatrix}$
$[\chi^{(3)}(g)^2]$	3	0	1
x^2	x^2	$x'^2 = \frac{1}{4}(x^2+3y^2+2\sqrt{3}\ xy)$ $x''^2 = \frac{1}{4}(x^2+3y^2-2\sqrt{3}\ xy)$	x^2 x'^2 x''^2
xy	xy	$(xy)' = \frac{-1}{4}(\sqrt{3}(x^2-y^2)+ 2xy)$ $(xy)'' = \frac{1}{4}(\sqrt{3}(x^2-y^2)- 2xy)$	$-xy$ $-(xy)'$ $-(xy)''$

The space $[\Gamma_3^2]$ being 3-dimensional (as shown by the character of the identity E), is clearly reducible with respect to G_o, whose IR's have dimensions 1 and 2. Applying eq.(4.1) to $[\Gamma_3^2]$, and to each of the IR's in table 3, we obtain, for instance in the case of Γ_3:

$$n_3 = \frac{1}{6}\left[3.2 + 2.(-1)\ 0 + 3.0.1\right] = 1 \tag{4.7}$$

Likewise, formula (4.1) yields $n_1 = 1$ and $n_2 = 0$. The space $[\Gamma_3^2]$ decomposes into one totally-symmetric space Γ_1 and one space Γ_3. Each type of space is only present once in this decomposition, we know from §.4.1, that the projection operators (4.2) and (4.3) will provide us unambiguously with the polynomials carrying each of the preceding IR's.

The basic vectors of $[\Gamma_3^2]$ are $[x^2,\ y^2,\ xy]$. In order to apply the projector (4.2) to any of these monomials we need to determine their transformation properties under the action of the $g_i \in G_o$. In turn these properties are determined by the matrices of Γ_3 (table 3) which describe the transformations of (x,y).

Line 3 of table 6 shows the way x^2 transforms, as deduced from the matrices in table 3, under the action of the 6 elements g_i of G_o. The "projection" of x^2 on the totally symmetric space is (eq. 4.2) : $P_1.x^2$

$$P_1.x^2 = \sum_{g_i} g_i.x^2 = 3(x^2 + y^2) \tag{4.8}$$

Hence, up to an irrelevant multiplying factor, the totally symmetric polynomial of degree 2 is $(x^2 + y^2)$, in accordance with the general result stated in §.2.2.4. On the other hand, we obtain a polynomial belonging to the Γ_3-space, by means of the projection-operator P_3 (eq. 4.2):

$$P_3.x^2 = 2.E.x^2 - (C_3 + C_3^2).x^2 = \frac{3}{2}(x^2 - y^2) \tag{4.9}$$

In order to obtain another linearly independent polynomial of Γ_3 (which is 2-dimensional) we can either apply P_3 to xy or $\pi_3^{\alpha\beta}$ to x^2. (Application of P_3 to y^2 yields, up to a factor the same polynomial (4.9). The transformation properties of xy, indicated on table 6 lead to the expression:

$$P_3 xy = 2.E.xy - (C_3 + C_3^2)xy = 3xy \qquad (4.10)$$

The two linearly independant polynomials $[x^2-y^2, xy]$ span the subspace of $[\Gamma^2]$ of symmetry Γ_3. Actually, in this subspace the basis homologous to (x,y) (i.e) the one associated to the same matrices, is $(\frac{x^2-y^2}{2}, xy)$

Applying π_3^{11} and π_3^{21} to x^2 we obtain (eqs. 4.3)

$$\pi_3^{11} . x^2 = \frac{3}{2}(x^2 - y^2) \quad , \quad \pi_3^{21} . x^2 = - 3 xy \qquad (4.11)$$

Thus generating the same two independent functions (4.9) and (4.10) of Γ_3.

b) third degree polynomials

Table 7 shows the characters $[\chi^{(3)}(g)^3]$ of the symmetrized-third power of the representation (x,y) as deduced from table 5. As apparent from $[\chi(3)(E)^3] = 4$, the symmetrized third power Γ_3 is 4-dimensional, in agreement with its spanning by the 4-monomials (x^3, y^3, x^2y, xy^2). The same procedure as applied in eq.(4.7) yields :

$$n_1 = 1 \quad ; \quad n_2 = 1 \quad ; \quad n_3 = 1 \qquad (4.12)$$

There is one totally-symmetric combination of the 3rd degree monomials, the remaining space of degree-3 polynomials being decomposed

Table 7

	E	$C_2 \begin{bmatrix} C_3 \\ C_3^2 \end{bmatrix}$	$C_3 \begin{bmatrix} \sigma_1 \\ \sigma_2 \\ \sigma_3 \end{bmatrix}$
$[\chi^{(3)}(g)^3]$	4	1	0
x^3	x^3	$(x^3)' = - \frac{1}{8}[x^3+3\sqrt{3}\ y^3+3\sqrt{3}\ x^2y + 9\ xy^2]$ $(x^3)'' = - \frac{1}{8}[x^3-3\sqrt{3}\ y^3-3\sqrt{3}\ x^2y + 9\ xy^2]$	x^3 x'^3 x''^3

into one space of type Γ_2 and one space of type Γ_3. Let us determine the invariant,(Γ_1),combination, by applying P_1 to the x^3-monomial. Its transformation properties by the g_i are indicated on table 7. Application of the same formula as in eq.(4.8) yields :

$$P_1 x^3 = \sum_{g_i} g_i . x^3 = \frac{3x}{2}(x^2 - 3\ y^2) \qquad (4.13)$$

in accordance with the result stated in §.2.2.2.

The same method provides us with the basis of the spaces Γ_2 and Γ_3 : $y(y^2-3x^2)$, and $[x(x^2+y^2), y(x^2+y^2)]$. Note that the functions spanning Γ_3 do not appear "orthogonal" to those spanning Γ_1. This is due to the fact that the orthogonality has to be checked on the symmetrized products of coordinates (e.g. one should write $xy^2 = \frac{1}{3}[(x,y,y) + (y,x,y) + (y,y,x)]$). The latter form of orthogonality is, consistently, satisfied.

4.4. The Landau condition

The $\mathcal{R}_i$ polynomials of degree-3 in the OP components, considered in eq.(3.11), are the distinct invariant polynomials contained in the symmetrized third power of the OP irreducible representation Γ_k. Hence, the number of independent $\mathcal{R}_i$ is equal to the number of totally symmetric subspaces contained in $[\Gamma_k^3]$. This number is

$$n_1 = \frac{1}{d(G_o)} \sum_{g_i} [\chi^{(k)} (g_i)^3] \tag{4.14}$$

where $\chi^{(k)}(g_i)$ are the characters of the OP-representation. A necessary and sufficient condition for the absence of a third-degree term in Landau's free-energy is therefore $n_1 = 0$ in eq.(4.14). This condition imposed to the OP, is sometimes termed the Landau condition. It can also be expressed in the form :

$$[\Gamma^3] \not\supset \Gamma_1 \tag{4.14)'}$$

where $[\Gamma^3]$ is the symmetrized third power of the OP representation and Γ_1 the totally symmetric IR to G_o. The fulfilment of the Landau condition can be tested as soon as the symmetry of the OP is known.

4.4.1. Compliance of one-dimensional order-parameters with Landau's condition

If the OP-corresponds to a one-dimensional non-totally symmetric IR, the Landau condition is necessarily verified. Indeed, in this case the representation $[\Gamma^3]$ coincides with Γ^3 (the latter space is generated by (η,η,η) which is invariant by permutation). Applying (4.5), we find that $\chi^{(3)}(g) = \chi(g)$ (since $\chi(g) = \pm 1$) which implies that $[\Gamma^3]$ coincides with the IR representation Γ of the OP. This representation, being non-totally symmetric, (4.14)' is satisfied.

A different form given to the preceding statement is to point out that the Landau condition is necessarily verified if the group of invariance G of the low-symmetry phase is a subgroup of index 2 (cf.§.3.6.2.c) of G_o. In order to show that the latter situation is equivalent to the former one we can consider the density increment $\delta\rho_{eq}(\vec{r})$ (eq.3.19). This quantity is a vector in the OP-space, which is invariant by G. Let $G_o = G + hG$, and $\delta'\rho_{eq}(\vec{r}) = h\delta\rho_{eq}(\vec{r})$.

Any subgroup of index 2 is an invariant subgroup (i.e.Gh = hG)[5].

Accordingly, the elements of G will leave δ_{eq} and $\delta'\rho_{eq}$ invariant, and the elements of hG will exchange $\delta\rho_{eq}$ and $\delta'\rho_{eq}$. As h acts, in the OP space as an orthogonal operator, then, either $\delta\rho'_{eq}=\pm\delta\rho_{eq}$, or $\delta\rho'_{eq}$ is a vector having the same modulus as $\delta\rho_{eq}$ and noncolinear with it. Assume that the latter case is true. $\delta\rho_{eq}$ and $\delta\rho'_{eq}$ span a two dimensional space which carries a representation of G_o, with $\chi(g) = 2$ if $g \in G$, and $\chi(g) = 0$ if $g \in hG$. Applying eq.(4.2) we find easily that this representation contains once the totally symmetric representation of G_o, in contradiction with the irreducibility and non-totally symmetric character of the OP-space. We are left with the case $\delta\rho'_{eq}=\pm\delta\rho_{eq}$ which implies that $\delta\rho_{eq}$ carries a one-dimensional IR of G_o. Accordingly the OP-space, which is irreducible, is also one-dimensional. Necessarily, $\delta\rho'_{eq} = -\delta\rho_{eq}$, since the IR must be non-totally symmetric.

Hence, the occurence of a subgroup of index 2 is related to the one-dimensionality of the order-parameter, and the Landau condition is satisfied in this case, as shown above.

4.4.2. Non-compliance when G is a subgroup of index 3 of G_o

If G is a subgroup of index 3 of G_o, then at least one cubic term is allowed by symmetry in the free-energy. Several proofs have been given of this property [8 , 9]. We follow the lines of the proof given by Brown and Meisel [9]. It proceeds in two steps: an analysis of the structure of G_o ; an analysis of the action of G_o on the equilibrium density increment $\delta\rho_{eq}$.

a) Structure of G_o

The basic assumption made ($d(G) = \frac{d(G_o)}{3}$) allows to write :

$$G_o = G + hG + kG \tag{4.15}$$

where h and k do not belong to G, and where G, hG and kG are distinct left-cosets.

i) Let g be an element of G. The products gh and gk belong either to hG or to kG. It is easy to show that out of 4 possibilities, two only are consistent, either [$gh \in hG$, $gk \in kG$] (G is then an invariant subgroup of G_o) or [$gh \in kG$, $gk \in hG$] (if G is not an invariant subgroup).

ii) Consider the elements h^2 and hk. They belong to one of the two cosets G or kG. Here again, out of 4 possibilities two only are consistent, either [$h^2 \in G$, $hk \in kG$] or [$h^2 \in kG$, $hk \in G$]. Likewise if we consider k^2 and kh, we have two possibilities, either [$k^2 \in G$, $kh \in hG$], or [$k^2 \in hG$, $kh \in G$].

Table 8 summarizes the 8 situations which can arise from i) and ii).

Table 8

gh	hG	hG	hG	hG	kG	kG	kG	kG
gk	kG	kG	kG	kG	hG	hG	hG	hG
h^2	G	G	kG	kG	G	G	kG	kG
hk	kG	kG	G	G	kG	kG	G	G
k^2	G	hG	G	hG	G	hG	G	hG
kh	hG	G	hG	G	hG	G	hG	G

b) Action of G_o in $\delta\rho_{eq}$

As stressed in §.3.6.2.c) $\delta\rho_{eq}$ is a vector in the OP space. Let us take this vector as one of the coordinate axes in the OP space. We can write $\delta\rho_{eq} \propto \eta_1$, η_1 being one component of the OP. By definition, η_1 is invariant by the action of G ($g\eta_1 = \eta_1$). Let $\eta_2 = h.\eta_1$ and $\eta_3 = k\eta_1$. The components η_2 and η_3 belong to the OP-space (η_1, η_2, and η_3 are not necessarily linearly independent).

Table 8 entirely defines the action of G_o on each of the 3 components η_i. For instance, if the situation realized in G_o corresponds to the first column in this table, we have ($g\eta_1 = \eta_1$; $g\eta_2 = gh\eta_1 = \eta_2$; $g\eta_3 = gk\eta_1 = \eta_3$) and likewise ($h\eta_1 = \eta_2$; $h\eta_2 = \eta_1$; $h\eta_3 = \eta_3$) and ($k\eta_1 = \eta_3$; $k\eta_2 = \eta_2$; $k\eta_3 = \eta_1$). In all cases, every element of G_o, realizes a permutation of (η_1, η_2, η_3).

Accordingly, <u>the product $\eta_1 . \eta_2 . \eta_3$ is invariant by all the elements of G_o</u>, thus showing that a third degree term in the OP components is compatible with the G_o-symmetry.

c) Two-dimensionality of the order-parameter

Let Γ be the physically-IR of the order-parameter. Γ is necessarily carried by the space (η_1, η_2, η_3) since this space is invariant by G_o. (η_1, η_2, η_3) cannot be one-dimensional, otherwise there would be no invariant cubic term (§.4.4.1). On the other hand the η_i are <u>not</u> linearly independent. Indeed any element of G_o will transform the vector ($\eta_1 + \eta_2 + \eta_3$) into itself by permutation of the η_i. As no invariant vector exists in Γ, this vector satisfies the condition ($\eta_1 + \eta_2 + \eta_3$) = 0. <u>Hence, Γ is 2-dimensional.</u>

The representation Γ_3 in table 3 provides an illustration of such a case. We have seen in §.4.3.2.b, that this representation was compatible with the occurence of a third-degree term (eq.4.13), invariant by G_o. On the other hand, consider that $\delta\rho_{eq}$ is along the x-axis of the carrier space of Γ_3. The matrices of Γ_3 (table 3) allow to derive the invariance group G of the $\vec{x}$ vector. It is clearly composed of the 2 elements E and σ_1 : G is a subgroup of index 3 of G_o.

4.5. Invariant polynomials and image of G_o in the OP-representation

4.5.1. Definition of the image

Consider the set of matrices $M(g_i)$ defining a representation of G_o in a given carrier-space (Irreducible or not).These matrices are not necessarily distinct. For instance, in table 3, the matrices of the representation Γ_2 satisfy the equalities $M(E) = M(C_3) = M(C_3^2)$, and $M(\sigma_1) = M(\sigma_2) = M(\sigma_3)$. By contrast the matrices of Γ_3 are all distinct. The latter representation is a faithful representation of G_o, for which the correspondance between the g_i and $M(g_i)$ is an isomorphism. The former representation is not faithful, the correspondance between G_o and the set $M(g_i)$ being an homomorphism. Even in this case, the set of distinct $M(g_i)$ forms a group which is the image I_o of G_o in the homomorphism. The Kernel K of the homomorphism is the set $g_i \in G_o$, such as the matrices $M(g_i)$ coincide with the identity matrix in the space $\mathcal{E}$. K is an invariant subgroup of G_o (i.e. $gK = Kg$ for $g \in G_o$), and we can express G_o, in the form (eq. 3.19), as a sum of (left-or-right) cosets associated to K :

$$G_o = K + g_2K + \ldots\ldots \qquad (4.16)$$

where the g_i do not belong to K. The set of cosets form the quotient group of G_o and K, which is denoted G_o/K. This group is isomorphous to the image I_o, and we can write $G_o = I_o \times K$ [if G_o is finite, the orders of the 3 groups satisfy the equality $d(G_o) = d(I_o).d(K)$]. We will see in chapter III, §4 , that the order of I_o can be finite even if G_o in an infinite group.

4.5.2. Application to the OP-space

These considerations apply to the n-dimensional physically IR carried by the n-components of the OP. The set of distinct matrices of this IR constitutes a group I_o, which is the image of the group G_o of invariance of the high-symmetry phase.

a) Geometrical nature of the image of G_o

Each matrix in I_o, is orthogonal, since it is real and unitary as any of the representations considered. Its action on the components η_r of the OP can therefore be considered as a geometrical transformation in the n-dimensional space of the OP, which has the property to preserve the origin in this space as well as the length of vectors. In the 3-dimensional space such transformations comprise rotations about an axis passing through the origin, reflexions about a plane passing through the origin, as well as products of these transformations. The rotations are the orthogonal transformations whose matrices have a determinant equal to +1. The set of all rotations about any axis passing through the origin forms a group denoted SO(3), while the set of all orthogonal transformations is a group denoted O(3). More generally, in the n-dimensional space, the set of all orthogonal transformations with determinant +1 forms a group denoted SO(n) while the set of all orthogonal transformations is the group denoted O(n).

The image I_o of G_o in the OP-representation is isomorphous to a subgroup (in a broad-sense) of O(n) since it is composed of n x n orthogonal matrices. I_o also has an additional peculiarity, which is related to the irreducibility of the OP-space (on real numbers) under the action of I_o. Accordingly, I_o can only be a subgroup of O(n) which leaves no subspace of the n-dimensional space invariant. We will term the subgroups of O(n) possessing this property irreducible subgroups of O(n). Another possible terminology is based on the remark that the n-dimensional space $\mathcal{E}$ is spanned by the components of a vector $(\eta_1, \ldots, \eta_n)$. $\mathcal{E}$ thus spans the "vector representation" of the n-dimensional space. The preceding requirement can be expressed by stating that the vector-representation of the group I_o be irreducible. For instance if n = 3, the group G_o in table 3 leaves the z-direction, and the (x,y) plane separately invariant. This group is not an irreducible subgroup of O(3). By contrast the restriction of this group to the (x,y) plane is an irreducible subgroup of O(2).

b) Use of the image I_o in the application of Landau's theory

The action of G_o in the OP-space is effected by means of the elements of its image I_o. Hence, once the OP-symmetry and the related image I_o, are known, it is possible, on the basis of this sole knowledge, to determine the form of the free-energy and to work out the possible symmetries of the low-symmetry phases.

Each degree in the free-energy expansion will be constituted by the most general polynomial of this degree in the η_r-components, which is invariant by I_o. As I_o is, in general, a simpler group than G_o (e.g. a finite group instead of an infinite one), the consideration of I_o will simplify the procedure of determination of the successive terms of the Landau free-energy. Besides, consideration of the images shows that the determination of the forms of the free-energies which arise for any phase transition associated to a n-component OP, is a mathematically well defined, task.

This task consists in enumerating the irreducible subgroups of O(n) , and , for each subgroup, in working out the form of the invariant p-th degree polynomials constructed from the set η_r of coordinates in the n-dimensional space [6, 8] . Finally, those subgroups which leave invariant a 3rd degree polynomial must be discarded as incompatible with the Landau-condition (§.4.4)

A similar simplification will occur in the derivation of the symmetry group G of the low-symmetry phase. As pointed out in §.3.6.2.e), G is the subgroup of G_o which leaves invariant a given vector in the space $\mathcal{E}$ of the OP. In this space G is represented by its image I which is the subgroup of I_o leaving invariant the considered vector. If I is known, G can be determined as its reciprocal image : if $M(g_i)$ is an element of I, all the elements $g_i.K$ belong to G (K is the kernel of the homomorphism between G_o and I_o). Hence, in order to determine G, it is sufficient to determine the simpler group I, and use the relation G= I x K. Examples of this procedure will be described in chapter III , § 5.7.

4.5.3 Images of G_o for $n \leqslant 3$, and irreducible subgroups of O(n). [6]

The enumeration of irreducible subgroups of O(n) has been performed for $n \leqslant 4$. [6,10]

Thus, one has access to a list of the OP-symmetries (i.e. to the action of any group G_o in the OP-space) for OP-dimensions up to 4. Let us examine here the cases n = 1,2,3. The case n = 4 is considered in §.4.5.5.

The group O(1) is reduced to the identity and to the "inversion" in the OP-space. This group of order 2 has one strict-subgroup coinciding with the identity. There are two possible OP-symmetries for n = 1, which are associated to the group O(1) and to its subgroup. However, as will be pointed out in §.4.5.4 hereunder, only the former group is relevant to the study of continuous transitions.

a) Case n = 2

For n = 2, the group O(2) acts in a plane. This group comprises the set of all rotations about an axis perpendicular to the plane and passing through the origin, as well as all the reflexions about any plane containing the preceding axis. Besides SO(2), O(2) possesses an infinite set of irreducible subgroups which can be classified in two categories. One category is composed of groups denoted $\mathbb{C}_m$ $(m \geqslant 3)$. They are cyclic abelian groups generated by the rotation $(2\Pi/m)$. The other category comprises the groups denoted $\mathbb{C}_{mv} = (\mathbb{C}_m + \sigma . \mathbb{C}_m)$ where σ is the reflection in a given plane. The choice of σ is arbitrary: different choices generally lead (fig. 5) to groups which, though distinct, are conjugated to eachother in O(2), and can be considered as physically equivalent. Indeed, conjugate subgroups are related by $G' = SGS^{-1}$ where $S \in O(2)$. As S is an orthogonal transformation, it can be considered as associated to a change of reference axes in the 2-dimensional space. Two conjugate images (associated to G and G') thus represent the same group G_o, in a different setting of coordinates in the OP-space. Accordingly, <u>conjugated subgroups need not be distinguished in the enumeration of subgroups of O(n).</u>

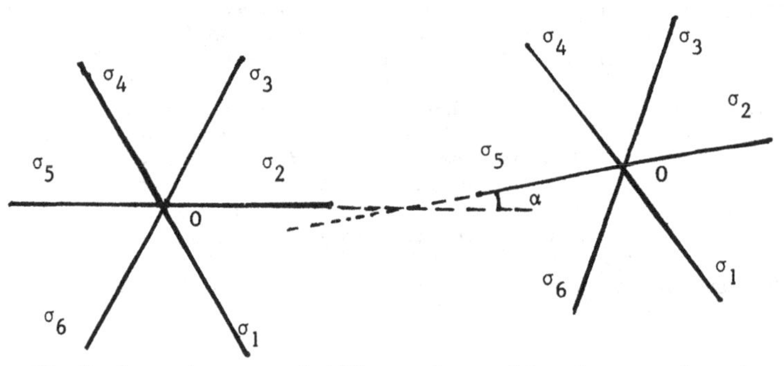

Fig.5. $\mathbb{C}_{6v}$ subgroups of O(2), conjugated by the rotation of angle α.

b) Case n = 3

The irreducible subgroups of O(3) form an infinite set. As in § a) hereabove, we can identify the subgroups which are conjugate to eachother as representing equivalent physical situations. One then finds [6] that, besides SO(3), there are 7 <u>irreducible groups among the subgroups</u> of O(3). These are, on the one hand, the 3 subgroups of SO(3) [i.e. of the set of all rotations] denoted T, O and Y, which leave invariant, respectively, the regular tetrahedron, the cube and the regular icosahedron. On the other hand, one finds the 3 subgroups of O(3) which are obtained by adjoining to T, O and Y the symmetry about the origin. These groups are denoted T_h, O_h and Y_h. Finally, the seventh subgroup T_d is obtained by adjoining to T the reflexion about a symmetry plane of the regular tetrahedron. The orders of these groups are d(T) = 12, $d(T_h)$ = $d(T_d)$ = d(O) = 24, $d(O_h)$ = 48, d(Y) = 60, $d(Y_h)$ = 120.

4.5.4. Invariant polynomials of p-th degree. Integrity bases

We have enumerated, in the preceding paragraph, the distinct irreducible subgroups of O(n), for n $\leqslant$ 3. For given n, the various possible forms of the Landau free-energy are constructed from the homogeneous polynomials of n variables $(\eta_1,..,\eta_n)$ invariant by each of the enumerated subgroup of O(n)(§.4.5.2.b).

a) n = 1

There are two groups to consider : O(1) = (E,I) and (E), where E is the identity and I the inversion ($I\eta = -\eta$). The polynomials invariant by O(1) comprise all <u>even</u> powers of η, while those invariant by its subgroup include all powers of η. In the latter case ,a third degree term will be allowed by symmetry in the free-energy. The corresponding image is not compatible with the Landau condition (§.4.4). Hence for n = 1, the only relevant OP symmetry is described by the image O(1), and the resulting free-energy is :

$$F = \frac{\alpha}{2}\eta^2 + \frac{\beta_1}{4}\eta^4 \qquad (4.17)$$

b) n = 2

The relevant images comprise two infinite groups O(2) and SO(2), and an infinite set of finite groups $\mathbb{C}_m$ and $\mathbb{C}_{mv}$. In order to describe the form of the invariant polynomials relative to each group we can choose polar coordinates (ρ,θ) in the plane of the two OP-components (η_1,η_2).

Since O(2) and SO(2) contain rotations by an arbitrary angle ϕ, the polynomials of (η_1,η_2) which are invariant by these two groups are only functions of $\rho^2 = (\eta_1^2 + \eta_2^2)$. The free-energy adapted to those two images is :

$$F = \frac{\alpha}{2}\cdot\rho^2 + \frac{\beta_1}{4}\cdot\rho^4 \qquad (4.18)$$

This free-energy is of particular interest in the study of incommensurate systems (cf. Chap. V, eq.2.19)

Let us examine the determination of the invariant-polynomials relative to the groups $\mathbb{C}_m$ and $\mathbb{C}_{mv}$.

i) In the first place, we can think of applying the method outlined in §.4.3.2., to derive the number and form of the invariant polynomials of p-th degree. The first step is to obtain the characters associated to a rotation (ϕ) or a reflexion in the two-dimensional space (ρ, θ). It is straightforward to show that $\chi(\phi) = 2\cos(\phi)$, and $\chi(\sigma) = 0$ (e.g. by taking one of the coordinate axes perpendicular to σ). We then derive the characters relative to the space $[\mathcal{E}^p]$ of polynomials of pth degree, i.e. the characters of the symmetrized p^{th}-power of the representation carried by the space (ρ, θ). For instance, using table 5, we obtain for p = 2, 3,4, the results indicated on table 9.

TABLE 9

	ϕ	σ
$[\chi^2]$	$1 + 2\cos(2\phi)$	1
$[\chi^3]$	$2(\cos\phi + \cos 3\phi)$	0
$[\chi^4]$	$1 + 2\cos 2\phi + 2\cos 4\phi$	1

Finally we use formula (4.1) to find the number of invariant polynomials in the former space. For instance let us obtain this number for p = 3, 4 in the case of the groups $\mathbb{C}_m$. These groups are constituted by m rotations $\phi = (2\pi k/m)$ (k = 1, 2,..., m). Noting that $[\Sigma \cos(2\pi lk/m) = 0]$ unless l = m, we find, for p = 3, and $m \geqslant 3$:

$$\frac{2}{m}\sum_{k=1}^{m}\left[\cos\frac{2\pi k}{m} + \cos\frac{6\pi k}{m}\right] = \begin{cases} 0 & \text{if } m \neq 3 \\ 2 & \text{if } m = 3 \end{cases} \qquad (4.18')$$

Likewise, for p = 4 and $m \geqslant 3$:

$$\frac{1}{m}\sum_{k=1}^{m}\left[1 + 2\cos\left(\frac{4\pi k}{m}\right) + 2\cos\left(\frac{8\pi k}{m}\right)\right] = \begin{cases} 1 & \text{if } m \neq 4 \\ 3 & \text{if } m = 4 \end{cases} \qquad (4.18'')$$

Thus, all the groups $\mathbb{C}_m$ except $\mathbb{C}_3$ are compatible with the Landau condition, since they do not leave invariant a 3rd degree polynomial. $\mathbb{C}_3$ must be discarded from the images associated to the OP of continuous transitions. Two independent 3rd degree polynomials are left invariant by it.

On the other hand, every group $\mathbb{C}_n$ except $\mathbb{C}_4$ gives rise to a single 4th degree polynomial. Clearly this polynomial is ρ^4. $\mathbb{C}_4$ corresponds to 3 independent invariant polynomials of 4th-degree (among which ρ^4).

The form of these polynomials will not be derived by the method outlined in §.4.3.2. Instead, this result will be obtained by an alternate method, which will also provide us with the form of the invariants of any degree p, for the groups $\mathbb{C}_m$ and $\mathbb{C}_{mv}$.

ii) An alternate method is based on the mathematical result stating that <u>for a finite group</u> (as $\mathbb{C}_m$ or $\mathbb{C}_{mv}$) <u>any invariant polynomial</u> can be obtained as a linear combination of the powers of a <u>limited number of basic polynomials</u>, forming <u>the integrity basis of the group</u> [11,13]. Methods of finding this integrity basis exist for certain types of groups [7]. They will not be discussed here. In the case of the groups $\mathbb{C}_m$ and $\mathbb{C}_{mv}$, the integrity basis is indicated on table 10.

TABLE 10 Integrity bases for invariant polynomials corresponding to the finite irreducible subgroups of O(2) and O(3) [from Refs.11 and 13].

$\mathbb{C}_m$	ρ^2 ; $\rho^m.\cos(m\phi)$; $\rho^m.\sin(m\phi)$	$\mathbb{C}_{mv}$	ρ^2 ; $\rho^m.\cos(m\phi)$
T	$\rho^2=(\eta_1^2+\eta_2^2+\eta_3^2)$; $B_2=\eta_1\eta_2\eta_3$	O	ρ^2 ; B_3 ; B_5 ; $B_2 \times B_4$
	$B_3=(\eta_1^4+\eta_2^4+\eta_3^4)$;	O_h	ρ^2 ; B_3 ; B_5
	$B_4=(\eta_1^2-\eta_2^2).(\eta_2^2-\eta_3^2).(\eta_3^2-\eta_1^2)$		
T_h	ρ^2 ; B_3 ; B_4 ; $B_5=\eta_1^2.\eta_2^2.\eta_3^2$	Y	ρ^2; $B_6=(a\eta_1^2-\eta_2^2)(a\eta_2^2-\eta_3^2)(a\eta_3^2-\eta_1^2)$ * $B_7=B_3 x (b\eta_1^2-\eta_2^2)(b\eta_2^2-\eta_3^2)(b\eta_3^2-\eta_1^2)$ * $B_8=B_2 x f(\eta_1,\eta_2,\eta_3) x f(\eta_3,\eta_1,\eta_2) x$ $x\ f(\eta_2,\eta_3,\eta_1)$ **
T_d	ρ^2 ; B_2 ; B_3	Y_h	^{2} ; B_6 ; B_7

* $a = (3+\sqrt{5})/2$; $b = (7+3\sqrt{5})/2$

** $f(\eta_1,\eta_2,\eta_3)=b\eta_1^4+\frac{1}{b}\eta_2^4+\eta_3^4-2\eta_1^2.\eta_2^2 -2\eta_3^2\left(\frac{(1+\sqrt{5})\eta_1}{2}+\frac{2\eta_2}{(1+\sqrt{5})}\right)$

This table shows that there are 3 polynomials in the integrity basis of $\mathbb{C}_m$ and 2 polynomials in that of $\mathbb{C}_{mv}$, for any value of m. For instance, for m = 3 the basis comprises ρ^2 and $\rho^3\cos(3\phi)$ for $\mathbb{C}_{3v}$, while for the group $\mathbb{C}_3$ it contains the additional polynomial $\rho^3\sin(3\phi)$. Hence, if we look for degree-three polynomials as linear combinations of powers of the former bases, we can see that $\mathbb{C}_{3v}$ will be associated with a single independent term ($\rho^3 . \cos 3\phi$) while $\mathbb{C}_3$ will correspond to <u>two independent terms</u> ($\rho^3 . \cos 3\phi$; $\rho^3 . \sin 3\phi$) in agreement with the result in eq.(4.18).

For m > 3, no 3rd degree invariant can be generated from the bases displayed on table 10, for $\mathbb{C}_m$ and $\mathbb{C}_{mv}$. This table also allows a derivation of the form of the <u>fourth degree</u> terms corresponding to $\mathbb{C}_m$ and $\mathbb{C}_{mv}$. For any value of m, except m = 4, the only 4th degree term is $(\rho^2)^2 = \rho^4$. For m = 4 there are two additional independent invariants relative to $\mathbb{C}_4$ [$\rho^4 . \cos 4\phi$ and $\rho^4 . \sin 4\phi$] in agreement with the result in eq.(4.18)". Likewise, there is one additional term ($\rho^4 . \cos 4\phi$) in the case of $\mathbb{C}_{4v}$. In the two latter cases, the free-energies associated to the OP are, up to the 4th degree :

$$F = \frac{\alpha}{2}\rho^2 + \frac{\beta_1}{4}\rho^4 + \frac{\beta_2}{4}\rho^4 \cos 4\phi + \frac{\beta_3}{4}\rho^4 \sin 4\phi \qquad (4.19)$$

$$F = \frac{\alpha}{2}\rho^2 + \frac{\beta_1}{4}\rho^4 + \frac{\beta_2}{4}\rho^4 \cos 4\phi \qquad (4.19')$$

Returning to "cartesian coordinates" in the OP space, it is straight-forward to show that (4.19') takes the form :

$$F = \frac{\alpha}{2}(\eta_1^2 + \eta_2^2) + \frac{(\beta_1 + \beta_2)}{4}(\eta_1^4 + \eta_2^4) + \frac{(\beta_1 - 3\beta_2)}{2}\eta_1^2\eta_2^2 \qquad (4.19'')$$

which is formally identical, within a change of the notations of the coefficients and of the OP-components, to the free-energy thoroughly discussed in §.2.7.2. of chapter I (eq.2.14).

c) **n** = 3

The two continuous groups O(3) and SO(3) give rise, for the same reason as invoked for n = 2, to the free-energy :

$$F = \frac{\alpha}{2}\rho^2 + \frac{\beta}{4}\rho^4 , \qquad (4.20)$$

where

$$\rho^2 = (\eta_1^2 + \eta_2^2 + \eta_3^2)$$

On the other hand, for the 7 finite subgroups of O(3) which can represent an OP-symmetry, the integrity bases of their invariant polynomials are known. They are reproduced on table 10 [11 , 13]

It appears that the groups T, and T_d are compatible with the presence of a third degree term in the free-energy ($\eta_1 . \eta_2 . \eta_3$ is a basic invariant). These groups cannot describe the symmetry of the OP of a continuous transition. The other groups are consistent with continuous transitions. For three of them, T_h, O and O_h, the <u>fourth-degree</u> free-energy expansion is the same. We can give to this expansion two equivalent forms :

$$F = \frac{\alpha}{2}\rho^2 + \frac{\beta_1}{4}\rho^4 + \frac{\beta_2}{4}(\eta_1^4 + \eta_2^4 + \eta_3^4) \qquad (4.21)$$

$$F = \frac{\alpha}{2}\rho^2 + \frac{\beta'_1}{4}(\eta_1^4 + \eta_2^4 + \eta_3^4) + \frac{\beta'_2}{2}(\eta_1^2 . \eta_2^2 + \eta_2^2\eta_3^2 + \eta_3^2\eta_1^2) \qquad (4.21')$$

These free-energies express the occurence of a "cubic-symmetry" in the OP-space.

For the two remaining groups Y and Y_h, the fourth degree expansion is identical to (4.20). However, as will be argued in §.4.6, in this case, the fourth degree expansion is insufficient to draw conclusions concerning the stable states of the system. The term of lowest degree which has to be taken into account is the sixth-degree term labelled B_6 in table 10.

4.5.5. Images and invariant polynomials for n = 4

For n $\geqslant$ 4, the enumeration of irreducible subgroups of O(n) and of their invariant polynomials is only partial at present. Systematic results are only available for n = 4. For higher OP-dimensions certain results exist in the restricted context of structural transitions. They will be mentioned in chapter III, § 4 . We examine here the situation for 4-components OP. For this value of n a list of irreducible subgroups of O(4) has been derived [10,14,15] . The integrity bases of invariant polynomials have been discussed in ref.15.

a) Irreducible subgroups of O(4)

As in the case n = 2, the number of non-conjugated irreducible subgroups of O(4) is infinite. Aside from SO(4), there are continuous subgroups of O(4) as well as discrete (finite or infinite) groups.

The specification of these groups is technically more complex than for n $\leqslant$ 3, since their elements pertain to the specialized field of 4-dimensional geometry. In order to perform this specification let us give some indications on the nature of the orthogonal transformations acting in the 4-dimensional space, and on the notations of the subgroups of O(4). A more detailed description of the structure

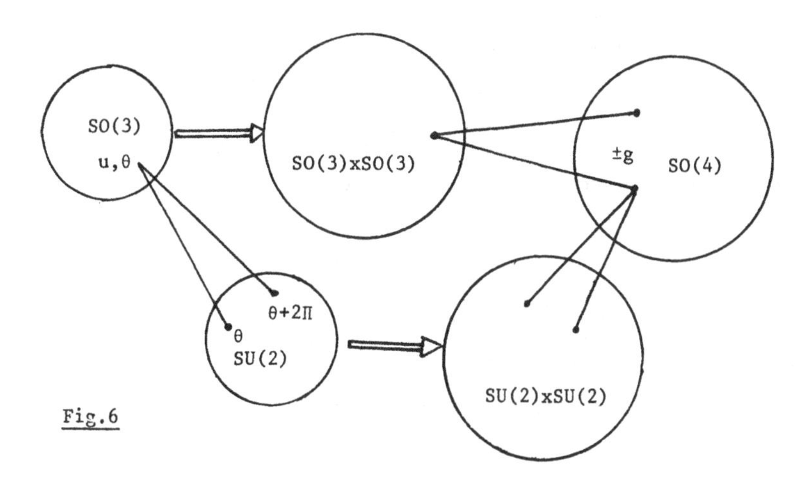

Fig.6

of O(4) as well as the proofs of the properties stated can be found in references [16,17,10].

i) It can be shown [16 , 10] that a transformation g belonging to SO(4) is in correspondance with a set of two 3-dimensional rotations :

$$g = [\vec{u}_1,\ \theta_1 | \vec{u}_2,\ \theta_2] \tag{4.22}$$

where $\vec{u}_i$ and θ_i are respectively the axes and angles of the rotations. This correspondance is an homomorphism (fig. 6). Each set of rotations in (4.22) corresponds to two distinct elements of SO(4) which are denoted (± g).

g is defined by its action on the current vector (η_1, η_2, η_3, η_4) in the 4-dimensional space (whose components η_r are the OP-components in the physical problem considered here). This action is specified by a complicated rule. Thus, let us put $\vec{\eta}_a = (\eta_1,\ \eta_2,\ \eta_3)$ and $\eta_b = \eta_4$, and on the other hand, $S_1 = (\sin\frac{\theta_1}{2}\vec{u}_1, \cos\frac{\theta_1}{2})$, $S_2 = (\sin\frac{\theta_2}{2}.\vec{u}_2\ ,\ \cos\frac{\theta_2}{2})$. The action of ±g on $(\vec{\eta}_a, \eta_b)$ is:

$$\pm g\ (\vec{\eta}_a,\ \eta_b) = S_1\ .\ (\vec{\eta}_a, \eta_b)\ .\ S_2^{-1} \tag{4.23}$$

where the multiplying rule between the elements $(\vec{V},t)$ is defined by :

$$(\vec{V},t)(\vec{V}',t') = (t\vec{V}' + t'\vec{V} + \vec{V}\wedge\vec{V}',\ tt' - \vec{V}.\vec{V}') \tag{4.23'}$$

In particular, the inverse of S_2 is :

$$S_2^{-1} = (-\sin\frac{\theta_2}{2}.\vec{u}_2\ ,\ \cos\frac{\theta_2}{2}) \tag{4.23''}$$

For instance, let us consider $g = [\vec{u}_1, \theta_1 | \vec{u}_2, \theta_2] = [\vec{k}, \pi\ |\ \vec{i}, \pi]$, where $(\vec{i},\ \vec{j},\ \vec{k})$ are the unit vectors in the 3-dimensional space. We have, first :

$$S_1 = (\vec{k},\ 0) \quad ; \quad S_2^{-1} = (-\vec{i},\ 0) \tag{4.24}$$

Applying (4.23)(4.23'), with $\vec{\eta}_a = (\eta_1\vec{i} + \eta_2\vec{j} + \eta_3\vec{k})$, we find :

$$\pm g(\eta_1\vec{i} + \eta_2\vec{j} + \eta_3\vec{k}, \eta_4) = \pm(-\eta_3\vec{i} - \eta_4\vec{j} - \eta_1\vec{k},\ \eta_2) \tag{4.24'}$$

Thus, the 4 x 4 orthogonal matrix associated to g in the basis $(\eta_1, \eta_2, \eta_3, \eta_4)$ is :

$$\left|\begin{array}{cc|cc} 0 & 0 & -1 & 0 \\ 0 & 0 & 0 & 1 \\ \hline -1 & 0 & 0 & 0 \\ 0 & -1 & 0 & 0 \end{array}\right| \tag{4.24''}$$

We can check on eq.(4.23) that rotations by θ_1 or by $(\theta_1 + 2\ \pi)$, which are identical in SO(3), give rise to two opposite elements $\pm$ g of SO(4) in agreement with the homomorphic character of the correspondance, pointed out above.

We can also note that the elements S_i involved in the procedure, form a group with respect to the multiplying rule (4.23'). This group can be shown [16] to be isomorphous to the group of unitary matrices 2 x 2, with determinant +1, which is denoted SU(2). The homomorphisms involving SU(2), SO(4) and SO(3) are schematized on figure 6.

ii) The group O(4) can be expressed as :

$$O(4) = SO(4) + \mathcal{S}.SO(4) \qquad (4.25)$$

where the element $\mathcal{S}$ is defined by $\mathcal{S}(\vec{n}_a, n_b) = (-\vec{n}_a, n_b)$. The subgroups of O(4) are either subgroups of SO(4) or groups G* of the form [16]

$$G^* = G + g.\mathcal{S}.G \qquad (4.25')$$

where G is a subgroup of SO(4) and g an element of SO(4).

iii) To specify the subgroups of O(4), it is sufficient, considering (4.25), (4.25'), to specify the subgroups of SO(4). The notation and construction of these subgroups derive from the correspondance (4.22). To define a subgroup G of SO(4), the following procedure must be carried out :

1) - Consider two subgroups L and R of SO(3), satisfying the conditions that they respectively possess an invariant subgroup (L_k, R_k), and that the quotient groups (cf. eq. 4.16) L/L_k and R/R_k are isomorphous. This implies :

$$L = L_k + h_2 L_k + \ldots, \quad R = R_k + h_2' R_k + \ldots. \qquad (4.26)$$

with an isomorphic correspondance between the cosets in the two decompositions.

2) - Consider homologous cosets $h_i.L_k$ and $h_i'.R_k$, and form all (ordered) pairs of elements $(h_i l_k, h_i' r_k)$, where l_k and r_k run over the groups L_k and R_k.

3) - Put $S_1^{(ik)} = h_i l_k$ and $S_2^{(ik)} = h_i' r_k$, and construct, according to eqs.(4.23) - (4.23"), the two elements $\pm g_{ik}$ of SO(4).

The set of all the elements $\pm g_{ik}$ forms a subgroup of SO(4), and all subgroups are obtained by this procedure. A subgroup is denoted

$$G = \left(\frac{L}{L_k} ; \frac{R}{R_k}\right) \qquad (4.26')$$

This notation is not unambiguous because, in certain cases, which will not be analyzed here [16], several distinct isomorphisms can be established between the cosets in (4.26) which lead to the construction of groups which are not equivalent up to a conjugation in SO(4). By contrast, a different setting of reference axes in either of the two 3-dimensional spaces relative to L and R, leads to conjugated subgroups of SO(4).

As an illustration of the construction of G, consider $L = R = \mathbb{D}_2$, where the group $\mathbb{D}_2$ of SO(3), of order 4, contains besides the identity, the 3 rotations U_x, U_y, U_z by the angle Π, about the three coordinate axes. Consider, on the other hand, the subgroup of $\mathbb{D}_2$, denoted $\mathbb{C}_2^x$, with $\mathbb{C}_2^x = (E, U_x)$. It is an invariant subgroup of $\mathbb{D}_2$, and, we can put $L_k = R_k = \mathbb{C}_2^x$. The quotient group $(\mathbb{D}_2/\mathbb{C}_2^x)$ has two elements :

$$\mathbb{D}_2 = \mathbb{C}_2^x + U_y.\mathbb{C}_2^x \qquad (4.27)$$

We have, therefore, to consider the following 8 ordered pairs :

$$(E,E)(E,U_x)(U_x,E)(U_x,U_x) \; ; \; (U_y,U_y)(U_y,U_z)(U_z,U_y)(U_z,U_z) \qquad (4.27')$$

To each member of a pair we associate one element $S_i = (\sin\frac{\theta}{2}\vec{u}, \cos\frac{\theta}{2})$ by taking the determination of the rotation angle $0 < \theta < 2\pi$. We obtain :

$$E \to (0,1) \; ; \; U_x \to (\vec{i}, 0) \; ; \; U_y \to (\vec{j},0) \; ; \; U_z \to (\vec{k}, 0) \qquad (4.27'')$$

Finally, we form the 16 elements $\pm S_1(\vec{\eta}_a,\eta_b)S_2^{-1}$, which constitute the group $G = (\mathbb{D}_2/\mathbb{C}_2 \; ; \; \mathbb{D}_2/\mathbb{C}_2)$. For instance the elements of G corresponding to the pair (U_y, U_z) have the orthogonal matrices :

$$\left|\begin{array}{cc|cc} 0 & 0 & 0 & -1 \\ 0 & 0 & 1 & 0 \\ \hline 0 & 1 & 0 & 0 \\ -1 & 0 & 0 & 0 \end{array}\right| \qquad (4.27''')$$

Ref. 10 contains the list of the irreducible subgroups of O(4). According to the considerations in §4.5.2, this list represents the "intrinsic symmetries" of all possible order parameters with 4 components.

b) Invariant polynomials for n = 4

As mentioned above, the integrity bases relative to the finite irreducible groups of O(4) are partly specified in ref.15.

Besides, third-and-fourth-degree invariants have been worked out for all the irreducible subgroups of O(4) in refs.10 and 16. Unlike the situation for $n \leqslant 3$, which could be summarized in table 10, the results for n=4 cannot be displayed in a very concise form. For instance, a subgroup of O(4) can be associated to as many as 11 independent fourth degree terms. We will not reproduce the results of the preceding references. The introduction to the structure of O(4) provided by this paragraph will allow the reader to use the content of these references.

4.6. Minimization of the free-energy, and symmetry breaking.

In this paragraph, we describe the methods which can be used, in order to locate the absolute (degenerate) minimum of the Landau-free-energy. We also examine the determination of the symmetry group G associated to the absolute minimum. G is the symmetry of the phase stable below T_c.

4.6.1. Standard minimization of F.

The standard method consists, once the form of the fourth degree expansion has been determined, in searching the values of the OP components η_r which satisfy the set of conditions:

$$\frac{\partial F}{\partial \eta_r} = 0 \qquad ; \qquad \left[\frac{\partial^2 F}{\partial \eta_r . \partial \eta_s}\right] > 0 \qquad (4.28)$$

where the second condition stands for the positiveness of the eigenvalues of the symmetric matrix of second derivatives. In order to express the latter condition, one can proceed for instance, in imposing the positiveness of the set of determinants obtained by eliminating the last p lines and columns of the preceding matrix (p=0,1,...,n-1) (fig.7).

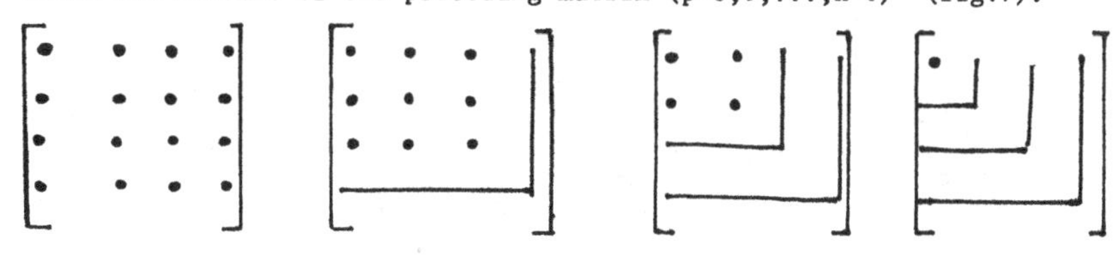

Fig.7

Such a method has been applied without difficulty to the two examples examined in chapter I (§.2.7.1, and §.2.7.2). However, the latter examples correspond to situations in which the number of OP-components is small (n = 1, and n = 2) and in which the fourth-degree term in the free-energy has a simple form. In the general case, technical difficulties are to be expected, related either to the complexity of the fourth degree terms (number and form of the independent invariants), or to the necessity, in certain cases, to take into account terms of degree higher than 4. As will be shown in §.4.6.5 hereunder, this necessity arises in three situations: i) whenever there is a single independent 4th degree-invariant (which will necessary be $(\Sigma\eta_r^2)^2$; ii) if one needs to specify more accurately the temperature dependence of physical quantities away from the immediate vicinity of T_c (i.e. for (T_c-T) not infinitesimally small) ; iii) if one wants to extend the applicability of Landau's theory to first order (discontinuous) transitions (cf. Chapter IV).

In order to cope with those complex cases, several methods, mathematically more elaborate than the standard one, have been described. They sometimes have the advantage to yield results valid for an entire class of transitions, i.e. independent from the specific form of the free-energy corresponding to a given OP-symmetry.

These methods have a common feature : they take advantage of the symmetry properties of the minima η_r^o satisfying eqs (4.28). Prior to the description of these methods, let us summarize some useful symmetry concepts.

4.6.2. Isotropy groups in the OP-space [6,7,13,19]

We have seen in §.4.5., that the OP-space $\mathcal{E} = (\eta_1, \eta_2, \eta_3, \ldots, \eta_n)$ is an irreducible space (on the real) under the action of the image $I \subset O(n)$. No subspace of $\mathcal{E}$ is invariant by I_o. However every vector direction $\vec{\eta}$ in $\mathcal{E}$ is left invariant by some subgroup $I_{\vec{\eta}}$ of I_o. I_η, which is the complete set of elements of I_o leaving $\vec{\eta}$ unchanged is called the isotropy-group of $\vec{\eta}$ (or the little-group of $\vec{\eta}$). Refering to §.4.5.2.b and §.3.6.2.c) we note that the symmetry-group G of the phase stable below T_c is related to the isotropy group I of the vector $\vec{\eta}_o = (\eta_1^o \ldots \eta_n^o)$ corresponding to the absolute minimum of the free-energy, by $G = I \times K$, where K is the kernel of the homomorphism $G_o \rightarrow I_o$.

a) orbit

Applying the elements of the group I_o to a given vector $\vec{\eta}$, we generate a set of distinct vectors $\vec{\eta}^{(p)}$ (since $\vec{\eta}$ is not invariant by all the elements of I_o). This set forms the orbit (or star) of $\vec{\eta}$. The isotropy-groups $I_\eta^{(p)}$ and $I_\eta^{(q)}$ of any two vectors $\vec{\eta}(p)$ and $\vec{\eta}(q)$ of the orbit are conjugated in I_o, i.e. :

$$I_\eta^{(p)} = S_{(pq)} \cdot I_\eta^{(q)} \cdot S_{(pq)}^{-1} \quad ; \quad S_{(pq)} \in I_o \tag{4.29}$$

This result applies to the orbit of the absolute minimum $\vec{\eta}_o$ of F. It has the consequence, already pointed out in §.3.6.2.d), that the groups $G^{(p)}$, such as $(G^{(p)}/K) = I^{(p)}$, are conjugated in G_o.

In paragraph 3.6.2.d), it was also shown that the vectors of the orbit of $\vec{\eta}_o$ all correspond to the absolute minimum of F, and that they are associated to a degenerate set of low-symmetry phases of the system. More generally, all the vectors of the orbit of a vector $\vec{\eta}$, will satisfy $F(\vec{\eta}(p)) = F(\vec{\eta})$, due to the invariance of F by I_o.

b) stratum

Two vectors whose isotropy groups are conjugated in I_o do not necessarily belong to the same orbit. If they belong to different orbits, these distinct orbits are said to be of the same type. The set of all orbits of a given type (i.e. characterized by isotropy groups which are conjugated in I_o) is called a stratum. Distinct orbits of a stratum contain the same number of vectors, and are associated to the same degeneracy of the free-energy. However those distinct orbits do not correspond to the same value of the free-energy.

c) **Isotropy-manifold of I**

The isotropy group I_η of the vector $\vec{\eta}$ can leave invariant other vectors of $\mathcal{E}$. These vectors (and $\vec{\eta}$) generate a subspace $\mathcal{E}_\eta$ of $\mathcal{E}$, each vector of which is invariant by I_η. This subspace can be termed the isotropy manifold of I_η. If I_o is finite, $\mathcal{E}_\eta$ belongs to the stratum of $\vec{\eta}$.

d) **Minimal isotropy-group, and symmetry-manifolds**

The isotropy groups I_η associated to all the vector directions in $\mathcal{E}$ possess a common subgroup I_k, possibly reduced to the identical transformation. I_k is an invariant subgroup of I_o and of all the I_η. Consequently (as it is self-conjugate) it is the isotropy group of all the vectors of a stratum. This stratum contains a continuous set of distinct vector directions. If a vector $\vec{\eta} \in \mathcal{E}$ does not belong to this stratum, it constitutes a <u>symmetry direction</u> in $\vec{\eta}$: its isotropy group will contain I_k as a <u>strict subgroup</u>. The isotropy manifold of I_η (defined in ii)) will be a symmetry-manifold.

e) **Maximal isotropy-groups**

A partial order exists between the isotropy-groups I_η, up to a conjugation in I_o. Thus, for two directions $\vec{\eta}$ and $\vec{\eta}'$, the isotropy-groups I_η and $I_{\eta'}$ can satisfy an inclusion relation $I_\eta \subset I_{\eta'}$, up to a conjugation [i.e. I_η belongs to a group conjugated to $I_{\eta'}$]. If the direction $\vec{\eta}$ is such that I_η is the subgroup of no other isotropy-group $I_{\eta'}$, I_η is called a <u>maximal-isotropy-group</u> and $\vec{\eta}$ can be termed a maximal-symmetry direction in $\mathcal{E}$.

f) **illustration**

We can examine the application of these concepts to the example treated in chapter I, §.2.7.2. In this example the OP-space $(p_x, p_y) = (\eta_1, \eta_2)$ is two-dimensional, and the action of the group G_o in this space, is represented, as pointed out in §.4.5.4. by the image $I_o = C_{4v}$. The kernel of the homomorphism $G_o \to I_o$ is a group K of order 2 comprising the identity and the symmetry σ_z about the plane (x,y) [G_o is a group of order 16 and C_{4v} is of order 8] . In table 1 of chapter I, we had derived the invariance groups G of every direction (p_x, p_y) in the OP-space. The isotropy-groups of these directions, subgroups of I_o, could be derived by taking the quotient groups (G/K). It is straightforward to work them out directly, by simple inspection of figure 8.

It appears that there are 3 types of orbits (i.e. 3 strata):

i) The orbit of the vectors $(\eta, 0)$ which comprises 3 additional vectors : $(-\eta, 0)$, $(0, \eta)(0, -\eta)$. Their isotropy group is, up to a conjugation, the group of order 2, $I_1 = (E, \sigma_2) \subset C_{4v}$, where σ_2 is the symmetry-plane perpendicular to the η_2-axis. The isotropy-manifold of I_1 is reduced to the $\pm(\eta, 0)$ direction : it is a line.

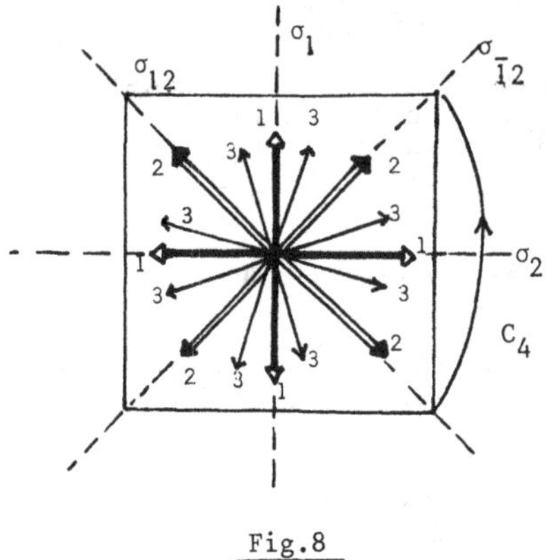

Fig.8

ii) The orbit of the vectors (η, η) which comprises the vectors $(\eta,-\eta)$, $(-\eta,-\eta)$ and $(-\eta,\eta)$ [along the 2 bissectors] The isotropy-group is also of order 2, $I_2 = (E,\sigma_{\bar{1}2})$ where $\sigma_{\bar{1}2}$ is a symmetry-plane perpendicular to one of the bissectors. Here again the isotropy manifold of I_2 is a line coinciding with one of the bissectors.

iii) The orbit of the vector (η_1, η_2) with $\eta_1 \neq 0$ $\eta_2 \neq 0$ and $\eta_1 \neq \pm\eta_2$. There are 8 distinct vectors in the orbit $(\pm \eta_1, \pm \eta_2)$ and $(\pm\eta_2, \pm\eta_1)$. The isotropy group of any vector of the orbit is the identical transformation E. This isotropy group is obtained for all the orbits whose representative vector (η_1, η_2) is in a "general direction of the plane". The stratum corresponding to this isotropy-group is composed of all the directions in the plane except the directions i) and ii). This is the situation pointed out in §d), and corresponding to the minimal isotropy-group.

The isotropy-lines in i) and ii) are symmetry lines. The vectors $(\eta, 0)$ and $(0,\eta)$ as well as the vectors in their orbits are maximal symmetry directions in $\mathcal{E}$. Their isotropy-groups are maximal, in agreement with the definition in e).

g) Determination of the isotropy-groups - subduction criterion [6 ,10,20-22]

In the preceding example we have determined the isotropy groups of every direction in the 2-dimensional space by checking, in each case, which ones of the 8-elements of $I_o = \mathbb{C}_{4v}$, leave the considered vector-direction invariant. An alternate method, which does not require the detailed knowledge of the action of I_o on the OP-space, is applicable. This method, which is termed the "subduction-criterion" (or "rule") or the "chain-criterion" relies on the knowledge of the characters of the representation of I_o, carried by the space $\mathcal{E}$, and on the application of the decomposition formula eq.(4.1) to the subgroups of I_o.

Assume that a list of the subgroups of I_o has been established up to a conjugation in I_o. To select, among these subgroups, the isotropy-groups of the various vector directions in $\mathcal{E}$, one can note that $\mathcal{E}$ carries a representation of any subgroup of I_o ($\mathcal{E}$ being invariant). While $\mathcal{E}$ is irreducible under the action of I_o it can be reducible under the action of its subgroup I. In particular, in the decomposition of $\mathcal{E}$ into irreducible spaces with respect to I, there can be one or several totally symmetric spaces. Their number $\nu(I)$ is (eq.4.1) :

$$\nu(I) = \frac{1}{d(I)} \cdot \sum_g \chi(g) \qquad (4.30)$$

where the sum is over the elements of I and, where the $\chi(g)$ are the characters of the representation carried by $\mathcal{E}$. $\nu(I)$ is the dimension of the subspace $\mathcal{E}_I$ of $\mathcal{E}$, each vector of which is left invariant by I.

A subgroup $I \subset I_o$ can be discarded from the list of possible isotropy groups if another subgroup I' of I_o, exists, satisfying the set of conditions :

$$I \subset I' \qquad \nu(I) = \nu(I') \qquad (4.31)$$

Indeed, the equality in (4.31) implies that the spaces $\mathcal{E}_I$ and $\mathcal{E}_{I'}$ coincide (since the inclusion relation $I \subset I'$ has the consequence that $\mathcal{E}_{I'}$ is invariant by I). Thus, the isotropy group I_v of any vector of $\mathcal{E}_I$ satisfies the relation :

$$I_v \supseteq I' \supset I \qquad (4.32)$$

If I_o is <u>a finite group</u>, a positive criterion, converse of the preceding one can be formulated. A subgroup I of I_o will be an isotropy group if, <u>for every I'</u> such as $I' \supset I$, one has, the strict inequality :

$$\nu(I) > \nu(I') \qquad (4.33)$$

If I_o is a continuous group, the condition (4.33) does not necessarily define isotropy-groups [6]. This situation can occur if every vector $\vec{\eta}$ in $\mathcal{E}_I$ has an isotropy-group $I_\eta \supset I$, with the set of I_η constituting an infinite set of groups, conjugated to each other in I_o[6]and whose isotropy spaces $\mathcal{E}_\eta$ satisfy :

$$\mathcal{E}_I = \cup \mathcal{E}_\eta$$

4.6.3. Absolute minimum of F and maximal isotropy-groups

Refering to chapter I (§.2.7.2), and to §.4.6.2.f) hereabove, we can notice that the two phases which are likely to be stable below a continuous transition [(±p, 0)(0, ±p)] or [(±p, ±p)] both correspond to vector directions in the OP-space whose isotropy-groups are <u>maximal</u>. By contrast, the "general" direction in the OP-space ($px \neq py$; $px \neq 0$ $py \neq 0$), which has the minimal isotropy-group, does not describe one of the relevant stables phases.

As shown in §.4.6.4, a similar result exists for any OP with 2 or 3 components ($n = 2$ or $n = 3$) : the vector directions which correspond to the absolute minimum of F (just below a continuous transition), are along maximal symmetry directions in the OP-space.

a). The maximal isotropy-group-conditions

A generalization of the preceding situation can be expressed in two different manners :

i) Below a continuous transition the symmetry of the stable phase <u>necessarily</u> corresponds to a vector $\vec{\eta}$ in the OP-space, whose isotropy group $I_\eta \subset I_o$ is maximal in the sense defined in §.4.6.2.e).

In this formulation, I_o is the image of G_o in the OP-representation.

ii) Let I'_o be the largest subgroup of O(n) which leaves invariant the <u>fourth-degree free-energy</u> expansion F associated to the n-component OP. With reference to I'_o, we can define isotropy-groups for every $\vec{\eta}$-direction in the OP-space : I'_η is the largest subgroup of I'_o which leaves $\vec{\eta}$ invariant. The claimed property is that the absolute minimum of F occurs in a direction $\vec{\eta}$ whose isotropy-group I'_η is maximal.

This second formulation refers to the symmetry-group I'_o of the 4 th degree free-energy expansion which is, in general, a larger group than I_o : $I_o \subseteq I'_o$ (the image of G_o leaves the full-expansion invariant, while I'_o only leaves invariant the 4th degree-expansion).

We will show in § c and d) that counter-examples (corresponding to $n = 4$) exist to both the statements i) and ii). Nevertheless these statements are interesting, despite their lack of absolute validity. Indeed, even for OP-dimensions larger than 3, both conditions are satisfied by <u>most</u> of the relevant cases. On the other hand, general theoretical considerations allow to show that directions $\vec{\eta}$ having maximal-isotropy groups I'_η (formulation ii) are necessarily extrema (or saddle points) of the free-energy expansion (whatever its degree).

The existence of counterexamples to statement ii) means that the maximal-symmetry directions are not the <u>only extrema</u> of the fourth-degree-expansion, and that, in certain cases these additional extrema, can constitute the absolute minimum of F.

b) Maximal isotropy-groups and extrema of F

The connection between the location of the extrema of F and the isotropy-groups of the various directions in the OP-space relies on the fact that, since F is invariant under the action of $g \in I'_o$, the set of derivatives $(\partial F/\partial\eta_1, \ldots, \partial F/\partial\eta_n) = \text{grad } F(\vec{\eta})$, transforms under the action of g in the same way as the components of the vector $\vec{\eta}$ of the OP-space.

Let $\vec{\eta}_o$ be a direction in the OP-space $\mathcal{E}$, and I'_η its isotropy-group refered to the invariance-group I'_o of the free-energy $F(\vec{\eta})$. Let us place ourselves in the <u>particular case</u> where the subspace of $\mathcal{E}$ invariant by I'_η is reduced to the single vector direction $\vec{\eta}_o$.

In this case, grad $F(\vec{\eta}_o)$ is parallel to $\vec{\eta}_o$ since the two vectors have the same symmetry properties, and that, accordingly, grad $F(\vec{\eta}_o)$ belongs to the isotropy-space of I'_η, which is reduced to the direction $\vec{\eta}_o$. A consequence of this property is that the occurence of an extremum (or a saddle-point) along $\vec{\eta}_o$ can be deduced from the study of the one-variable function $F(\eta_o)$, equal to the restriction of the free-energy $F(\vec{\eta})$ to the $\vec{\eta}_o$ axis ($F(\eta_o)$ can be obtained by adopting $\vec{\eta}_o$ as one of the coordinate axes in $\mathcal{E}$ and putting all the other variable equal to zero). $F(\eta_o)$ is, like $F(\vec{\eta})$, a fourth-degree polynomial. This polynomial has a local maximum for $\eta_o = 0$ (we are considering the temperature range below T_c, for which the second degree-term in F has a negative coefficient), and is positive for large values of η_o (cf. §.3.6.2). Clearly (see fig. 3, chapter I) $F(\eta_o)$ has a minimum for some, finite, non-zero value of η_o. This minimum is either a local minimum or a saddle-point of $F(\vec{\eta})$.

As, on the other hand, it can be shown that the one-dimensional isotropy-spaces of the type considered are necessarily maximal-symmetry directions, the preceding argument establishes, in a particular situation, that maximal-symmetry directions correspond to extrema of the free-energy. This proof can be extended [7] to the general situation where the isotropy-manifold associated to I'_η is not restricted to a single $\vec{\eta}_o$ direction. We will not develop this proof here.

c) Counter example to the rule i) of maximal symmetry [24]

Consider the set I_o^o of 32 orthogonal matrices generated by the 3 matrices hereunder :

$$a = \left[\begin{array}{cc|cc} -1 & 0 & 0 & 0 \\ 0 & 1 & 0 & 0 \\ \hline 0 & 0 & -1 & 0 \\ 0 & 0 & 0 & 1 \end{array}\right] \qquad b = \left[\begin{array}{cc|cc} 0 & 0 & 0 & 1 \\ 0 & 0 & -1 & 0 \\ \hline 1 & 0 & 0 & 0 \\ 0 & 1 & 0 & 0 \end{array}\right] \qquad c = \left[\begin{array}{cc|cc} 1 & 0 & 0 & 0 \\ 0 & 1 & 0 & 0 \\ \hline 0 & 0 & -1 & 0 \\ 0 & 0 & 0 & -1 \end{array}\right] \qquad (4.33)$$

This set is an irreducible subgroup of O(4) which can be considered to act in the space of a 4-component OP $(\eta_1,\eta_2,\eta_3,\eta_4)$. The isotropy-groups of the various directions $\vec{\eta}$ are indicated on table 11, up to a conjugation in I_o, each group being specified by its elements as a function of a, b, c in (4.33).

There are six non-conjugated isotropy-groups labelled (I - VI). Clearly I and II are maximal isotropy-groups, while the four other groups are not maximal. VI is the minimal group (§ 4.6.2.d).

The fourth degree expansion invariant by I_o has the form :

$$F = \frac{\alpha}{2}(\Sigma\eta_r^2) + \frac{\beta_1}{4}(\Sigma\eta_r^4) + \frac{\beta_2}{2}(\eta_1^2\eta_2^2 + \eta_3^2\eta_4^2) + \frac{\beta_3}{2}(\eta_1^2\eta_3^2 + \eta_1^2\eta_4^2 + \eta_2^2\eta_3^2 + \eta_2^2\eta_4^2) \qquad (4.34)$$

Table 11

$\vec{\eta}$	number of vectors in the orbit	isotropy-group I_η	dimensions of ε_η	label
$(\eta_1\eta_1\ 0\ 0)$	8	$I\eta_I = \{a^2,\ ab^6,\ c,\ acb^6\}$	1	I
$(\eta_1\ 0\ 0\ 0)$	8	$I\eta_{II} = \{a^2,\ ab^4,\ c,\ acb^4\}$	1	II
$(\eta_1\eta_1\eta_3\eta_3)$	16	$I_{III} = \{a^2,\ ab^6\}$	2	III
$(\eta_1\eta_2\ 0\ 0)$	16	$I_{IV} = \{a^2,\ c\}$	2	IV
$(\eta_1\ 0\ \eta_3\ 0)$	16	$I_V = \{a^2,\ ab^4\}$	2	V
$(\eta_1\eta_2\eta_3\eta_4)$	32	$I_{VI} = \{a^2\}$	4	VI

The absolute minima of F are straightforwardly determined for the various ranges of the β_i coefficients by the standard method (§.4.6.1). The results are summarized on table 12.

Table 12

Stable orbit	orbit label	Stability range	Comment
$(\eta_1\eta_1\ 0\ 0)$	I	$\beta_1 > \|\beta_2\|\ ;\ \ 2\beta_3 > (\beta_1 + \beta_2)$	maximal
$(\eta_1\ 0\ 0\ 0)$	II	$\beta_2 > \beta_1 > 0\ ;\ \beta_3 > \|\beta_1\|$	maximal
$(\eta_1\eta_1\eta_1\eta_1)$	III	$\begin{cases}\beta_1 > \|\beta_2\| \\ \beta_1 > \|\beta_3\|\end{cases}\ ;\ \beta_1(\beta_1+\beta_2) > 2\beta_3^2$	non-maximal
$(\eta_1\ 0\eta_1\ 0)$	V	$\beta_2 > \ \beta_1 > \|\beta_3\|$	non-maximal

The two orbits III and V correspond to the absolute minimum of F for some range of the β_i coefficients and are acceptable to describe the symmetry of a system below a continuous transition. Their occurence contradicts the formulation i) of the maximal-isotropy-group-rule since their isotropy groups are non-maximal (they are subgroups of the isotropy-groups I and II, respectively). However, this example does not contradict formulation ii). The invariance group I' of the 4th degree expansion (4.34) is generated, in addition to the matrices (4.33) by the two matrices :

$$d = \begin{vmatrix} -1 & 0 & 0 & 0 \\ 0 & -1 & 0 & 0 \\ 0 & 0 & -1 & 0 \\ 0 & 0 & 0 & 1 \end{vmatrix} \qquad e = \begin{vmatrix} 0 & 0 & -1 & 0 \\ 0 & 0 & 0 & 1 \\ 1 & 0 & 0 & 0 \\ 0 & -1 & 0 & 0 \end{vmatrix} \qquad (4.35)$$

I'_o is a group of order 128 containing I_o. With respect to I'_o the decomposition of the space $\mathcal{E}$ into orbits is indicated on table 13. It appears that the two minima of F corresponding to the directions $(\eta_1\eta_1\eta_1\eta_1)$ and

Table 13

$\vec{\eta}$	number of vectors in the orbit	Isotropy-group I'_η	label	order or I'_η
$(\eta_1\eta_1\ 0\ \ 0)$	8	$(1 + db^4 + deb + eb^5)I\eta_I$	I	16
$(\eta_1\ 0\ 0\ 0)$	8	$(1 + db^4 + deb + eb^5)I\eta_{II}$	II	16
$(\eta_1\eta_1\eta_1\eta_1)$	16	$(1 + cb^2e + adb + adce^3b)I\eta_{III}$	III'	8
$(\eta_1\eta_1\eta_3\eta_3)$	32	$(1 + adce^3b)I\eta_{III}$	III	4
$(\eta_1\eta_2\ 0\ 0)$	16	$(1 + db^4 + ab^3e + adb^3e^3)I\eta_{IV}$	IV	8
$(\eta_1\ 0\ \eta_1\ 0)$	16	$(1 + ad + ace + cde)I\eta_V$	V'	8
$(\eta_1\ 0\ \eta_3\ 0)$	32	$(1 + ad)I\eta_V$	V	4
$(\eta_1\eta_2\eta_3\eta_4)$	128	$I\eta_{VI}$	VI	1

$(\eta_1\ 0\ \eta_1\ 0)$, which were non-maximal symmetry directions with respect to I_o, become maximal-symmetry directions with respect to I'_o.

d) Counter example to the rule ii) of maximal-symmetry [25,26]

Two different examples in which the absolute minimum of F occurs for a non-maximal symmetry direction with respect to I'_o, have been pointed out.

One, studied by Mukamel and Jaric [25] corresponds to n = 6, while the other, investigated by Jaric [26] corresponds to n = 4. We describe the latter one, as it refers to subgroups of O(4) already discussed in §.4.5.5.

Following Jaric, we consider $I'_o = \pm\ (D_3/C_1\ ;\ O/D_2)$ irreducible subgroup of O(4) defined by (4.26'). Its order is 48. The decomposition of the 4-dimensional space $(\eta_1, \ldots, \eta_4)$ into orbits with respect to I'_o is represented on table 14 (the primed and double primed I'_η groups correspond to non-conjugated subgroups of O(4) associated to different isomorphisms between the two quotient

Table 14

$\vec{\eta}$	number of vectors in the orbit	I'_η	label	order of I'_η
$(0\ 0\ 0\ \eta_4)$	8	$(D_3/C_1\ ;\ D_3/C_1)'$	I	6
$(\eta_1\eta_1\eta_1\ 0)$	8	$(D_3/C_1\ ;\ D_3/C_1)''$	II	6
$(\eta_1\eta_1\eta_1\ 4)$	16	$(C_3/C_1\ ;\ C_3/C_1)$	III	3
$(\eta_1\eta_2\eta_2\ 0)$	24	$(C_2/C_1\ ;\ C_2/C_1)$	IV	2
$(\ _1\ _2\ _3\ _4)$	48	$(C_1/C_1\ ;\ C_1/C_1)$	V	1

groups). The isotropy-groups I and II are maximal while III-IV and V are not maximal.
It can be shown [10,18,26] that I'_o is the invariance group of the fourth degree free-energy expansion :

$$F = \frac{\alpha}{2}(\Sigma\eta_r^2) + \frac{\beta_1}{4}(\Sigma\eta_r^2)^2 + \frac{\beta_2}{4}(\Sigma\eta_r^4) + \beta_3\Sigma[\ \eta_r\eta_s(\eta_t^2-\eta_4^2) - \eta_4\eta_r(\eta_s^2-\eta_t^2)] \quad (4.36)$$

where the last term of the second member is a sum over the permutations (r.t) of (123).

Let us consider the directions labelled III in table 14. For these directions,the cancellation of $(\partial F/\partial\eta)$ yields:

$$\begin{aligned} &\alpha + (3\beta_1+\beta_2+4\beta_3)\eta_1^2 + (\beta_1 - 2\beta_3)\ \eta_4^2 = 0 \\ &\alpha + (3\beta_1 - 6\beta_3)\eta_1^2 + (\beta_1+\beta_2)\ \eta_4^2 = 0 \end{aligned} \quad (4.37)$$

which determine

$$\eta_1^2 = \frac{-\alpha(\beta_2+2\beta_3)}{\Delta} \qquad \eta_4^2 = -\frac{\alpha(\beta_2+10\ \beta_3)}{\Delta} \quad (4.37')$$

with :

$$\Delta = 4\ \beta_1\ (\beta_2 + 4\ \beta_3) + (\beta_2+6\beta_3)(\beta_2-2\beta_3) \quad (4.37'')$$

The positiveness of η_1^2 and η_4^2 requires :

$$\Delta\ .\ (\beta_2+2\beta_3) > 0 \quad \text{and} \quad \Delta\ .\ (\beta_2 + 10\ \beta_3) > 0 \quad (4.38)$$

while the positiveness of the eigenvalues of the matrix $(\partial^2F/\partial\eta_r\partial\eta_s)$ requires :

$$\Delta > 0 \quad ; \quad (\beta_1\beta_2 + 2\beta_1\beta_3 + 2\beta_2\beta_3) > 0 \quad ; \quad \beta_1+\beta_2 > 0 \quad (4.38')$$

It is straightforward to check that this set of conditions is fulfilled, for instance, by coefficients β_i satisfying :

$$\beta_1 > 0 \quad ; \quad \beta_2 > 2\beta_3 > 0 \quad (4.38'')$$

For this range of coefficients, the free-energy (4.36) has a local minimum in the non-maximal directions labelled III. To complete the proof we need to show that no other minimum of F exists for the same range of coefficients (4.38"). We will point out in §.4.6.4, hereunder, that if F has minima in the directions III, it cannot have additional minima in the directions IV and V. Let us show here that, likewise, no additional minima exist in the directions I and II, for the range of coefficients (4.38"). One can easily establish by expressing the positiveness of the eigenvalues of $[\partial^2 F/\partial\eta_r \partial\eta_r]$, that the occurence of minima in directions I and II are respectively submitted to the necessary conditions (4.39) and (4.39') :

$$\beta_2 < 0 \qquad (4.39)$$

and

$$3\beta_1 + \beta_2 + 7\beta_3 < 0 \qquad (4.39')$$

Clearly, these conditions are incompatible with (4.38").

4.6.4. Minima of F and Morse theory [7, 27]

Michel and Mozrzymas[27] have described a method which allows in certain cases, to determine the possible minima of the Landau free-energy, without solving the set of equations (4.28). In the cases where the method is not able to perform a complete specification of the minima, it provides, nevertheless, useful indications for the search of the minima.

The method relies on two ingredients. On the one hand, it uses inequalities (termed the Morse relations) which are applicable to the numbers of extrema of various types, of the Landau free-energy. These inequalities express the facts that if a polynomial of a given degree has a number μ_o of minima, there are constraints on its number of saddle points, and maxima, and that the total number of its extrema is bounded. Strictly speaking the inequalities are valid when the variables belong to a "closed" space such as a sphere. However it is possible to use them in the usual situation where the η_r can take infinite values by including $|\eta_r| = \infty$ among the extrema .

On the other hand, the method uses the property that each type of extremum is degenerate in a manner which is specified by the symmetry of the orbits in the OP-space.

a) Morse relations

These relations apply to the pth-degree Landau free-energy, for $T < T_c$. In this temperature range, F has a local maximum for $\eta_r = 0$. Also, due to its positiveness, F is maximum (and infinite) for $|\eta_r| \to \infty$.

In addition to these extrema, F possesses generally several other extrema. Some are minima, characterized by n positive eigenvalues for the matrix $(\partial^2 F/\partial\eta_r \partial\eta_s)$. The others are saddle points or maxima for which the former matrix has k negative eigenvalues and (n - k) positive ones (k is the Morse-index of the extremum). Let μ_k be the number of extrema with given Morse index k, for a specified choice of the coefficients in F. μ_o is the number of minima of F, and μ_n the number of maxima. The numbers μ_k satisfy a set of inequalities, and one equality:

$$\left.\begin{array}{l} \mu_o \geqslant 2 \qquad \mu_n \geqslant 2 \\ \sum_{k=0}^{m} (-1)^{m-k}.\mu_k \geqslant (-1)^m \qquad (m=0,1,2,\dots,n-1) \end{array}\right\} \qquad (4.40)$$

$$\sum_{k=0}^{n} (-1)^{n-k}.\mu_k = 1 + (-1)^n \qquad (4.40')$$

If F is expanded to the p-th degree,its first derivatives are of degree (p-1). The extrema are determined by a system of n equations of degree (p-1). Counting the solutions of the system corresponding to $|\eta_r| = \infty$,we have an additional inequality:

$$\sum_{k=0}^{n} \mu_k \leqslant (p-1)^n + 1 \qquad (4.40'')$$

b) The method of Michel and Mozrzymas.

As pointed out in § 4.6.2.a) ,the absolute minimum $\vec{\eta}$ of F is q_η-fold degenerate, q_η being the number of vectors of the orbit of $\vec{\eta}$.

More generally if $\vec{\eta}$ corresponds to an extremum of F of a given type (i.e. to a given index k), the other vectors of its orbit will be extrema of F of the same type [27]

The consequence of this remark is that the numbers μ_k are necessarily sums of integers q_η , where the q_η are the numbers of vectors in the different orbits. The q_η can be enumerated on the basis of the symmetry arguments described in §.4.6.2. The relationship between the μ_k and the q_η does not hold for μ_n, as the origin and the point $|\eta| = \infty$, do not give rise to q_η distinct directions under the action of I_o.

The search of the minima of F consists in determining the various possible decompositions of the μ_k as a function of the q_η in order to satisfy the set of relations (4.40-4.40"). In certain cases this will determine unambiguously the nature of the orbits associated to the minima of F, i.e. the directions in the OP-space which correspond to these minima. One is then left with the task of determining the modulus of the OP along these directions, by a trivial minimization of F with respect to the sole modulus.

c) Illustration of the method

Let us consider the example of the 4th degree expansion (p = 4) relative to the 2-dimensional OP (n = 2) associated to the image $I_o = I_o' = C_{4v}$. (§.4.5.4.b, and §.4.6.2.f)

The set of relations (4.40-4.40") takes the form :

$$\left.\begin{array}{l} \mu_o \geqslant 1 \quad ; \quad \mu_1 - \mu_o \geqslant -1 \quad ; \quad \mu_o - \mu_1 + \mu_2 = 2 \\ \mu_o \geqslant 2 \qquad \mu_2 \geqslant 2 \end{array}\right. \qquad (4.41)$$

$$\mu_o + \mu_1 + \mu_2 \leqslant 10 \qquad (4.41')$$

The decomposition of the two dimensional space (η_1,η_2) into orbits, with respect to C_{4v}, has been performed in §.4.6.2.f). We have seen that there are three types of orbits. Two have $q_\eta = 4$ (4 vectors in the orbit) and one type of orbit has $q_\eta = 8$ (this is the orbit with minimal symmetry). The possible values of μ_1 and μ_2 are therefore (0, 4, 8, 12,...) while the possible values of μ_o are (4, 8, 12,...). Replacing $\mu_1 = (\mu_o + \mu_2 - 2)$ into the inequality (4.41'), we obtain :

$$\mu_o + \mu_2 \leqslant 6 \tag{4.41''}$$

which yields as only possible solution

$$\mu_o = 4 \qquad \mu_2 = 2 \qquad \mu_1 = 4 \tag{4.41'''}$$

The possible minima of F are along one of the two orbits of maximal-directions as obtained by direct minimization of F (§.4.6.1)

As additional illustration of the method, let us derive the result stated in §.4.6.3. If the directions III in table 14, are minima of F, then directions IV and V cannot be additional minima. The set of Morse relations in this case (n = 4) takes the forms (for the 4th-degree expansion) :

$$\mu_o \geqslant 2 \qquad \mu_4 \geqslant 2 \tag{4.42}$$

$$\mu_1 \geqslant \mu_o - 1 \;;\; \mu_2 \geqslant \mu_1 - \mu_o + 1 \tag{4.42'}$$

$$\mu_3 \geqslant \mu_2 - \mu_1 + \mu_o - 1 \;;\; \mu_o + \mu_2 + \mu_4 = \mu_1 + \mu_3 + 2$$

$$\mu_o + \mu_1 + \mu_2 + \mu_3 + \mu_4 \leqslant 82$$

With the exception of μ_4, the μ_i are either zero, or equal to sums of q numbers indicated on table 14. Thus the possible values of the μ_i are (0, 8, 16, 24, 32,...)

If both an orbit of type III and an orbit of type IV or V corresponded to minima of F, we would have $\mu_o > 40$. The Morse relations then imply :

$$\begin{aligned} &\mu_1 \geqslant 39 \quad , \text{ hence } \quad \mu_1 = (40, 48, \ldots) \\ &\mu_2 \geqslant \mu_1 - \mu_o + 1, \text{ hence } \mu_2 = (8, 16, \ldots\ldots) \end{aligned} \tag{4.42''}$$

Besides, taking into account the equality $(\mu_1 + \mu_3) = (\mu_o + \mu_2 + \mu_4 - 2)$, we obtain :

$$\mu_o + \mu_2 + \mu_4 \leqslant 42 \tag{4.42'''}$$

Clearly, the latter inequality is incompatible with the minimum values of μ_o, μ_2, and μ_4, respectively 40, 8 and 2.

4.6.5. Exclusion of orbits with minimal symmetries [7]

The two examples considered in c) suggest that the success of the method essentially relies on the constraint (4.40"). Hence, as measure as the number n of OP-components increases, the set of Morse relations will become less conclusive (for p = 4). It has been shown [24] that these relations permit to determine completely the stable orbits with respect to a fourth-degree Landau expansion for n = 2 and for

$n = 3$. By contrast, the inequality (4.40") becomes unsufficiently restrictive for $n \geqslant 4$ [18]. However, one can hope to use the method in order to examine the possible stability of orbits containing a large number of vectors (such as the 16-fold and 24-fold orbits considered in c)), since the μ_i numbers will also be large in these cases.

The orbits containing the largest number of vectors are associated to the vector directions with minimal symmetry. In §.c) we have seen that in the case of the image $\mathbb{C}_{4v}$ ($n = 2$) this orbit could not be a minimum of the fourth-degree free-energy. In the example of the image considered in table 14 ($n = 4$), the orbit with minimal symmetry contains 48 vectors. We can repeat for this orbit the argument expressed by eqs. (4.42"-4.42"'), with $\mu_1 = \mu_o = 48$, and show that it cannot correspond to a minimum of the free-energy (4.36).

This result can be generalized. It has been shown by Michel [7] that if we consider the fourth-degree Landau free-energy associated to an irreducible OP, the directions with minimal-symmetry cannot correspond to extrema of F.

4.6.6. Determination of the minima of F by means of the decomposition into orbits and of the integrity basis of I_o

It has been first noted by Gufan ,and illustrated by Gufan and Sakhnenko[11] that the expression of F in terms of the integrity basis of invariant polynomials (cf. §.4.5.4) leads naturally to a classification of the extrema of F based on their symmetry, i.e. on the type of orbit associated to them. Developments along this line and various refinements of the method have been pointed out by Michel [7] by Jaric[28] and by Kim[29].

a) Principle of the method

The method of Gufan aims at obtaining results concerning the extrema of a free-energy which is not necessarily restricted to the fourth-degree, and also, results which are independent from the degree of the expansion. It relies on two elements. On the one hand it uses the possibility to expand F as a function of a limited number of basic polynomials B_m : the members of the integrity basis relative to the image I_o of the invariance group of the system, in the OP-representation. On the other hand ,it implies a prior enumeration of the maximal isotropy-groups and of their isotropy-spaces. The use of these two elements allow a separation of the OP-components corresponding to an extremum into two sets : i) a set of components which is determined by symmetry considerations and which is, accordingly, independent both of the temperature and of the degree of the free-energy expansion studied. ii) The remaining set for which an actual calculation of the extremum must be performed. This set concerns temperature dependent OP-components, whose behaviour depends on the degree and on the coefficients of the free-energy expansion.

Let $B_1, B_2, \ldots, B_k$ be the k elements of the integrity basis of polynomials relative to the image I_o (§.4.5.4). We denote $B_1 = (\Sigma\eta_r^2) = \rho^2$, the second-degree invariant polynomial which is part of the basis for any group I_o (§.2.2.3, and 4.5.4). As any term in the free-energy expansion can be expressed as a function of the B_i, we can write :

$$F = F(B_1, \ldots, B_k) \tag{4.43}$$

where each B_m is a homogeneous polynomial of degree p_m in the OP-components η_r. Let us put, as in §.3.6.2., $\eta_r = \rho\gamma_r$, where the γ_r specify a direction in the OP-space. On the other hand, as B_m is homogeneous, we have :

$$B_m(\eta_1, \ldots, \eta_n) = \rho^{p_m} \cdot D_m(\gamma_1, \ldots, \gamma_n) \tag{4.44}$$

We can substitute ρ^2 and the D_m ($m = 2, \ldots, k$) in (4.43). We obtain:

$$F = F_1(\rho^2, D_2, \ldots, D_k) \tag{4.43'}$$

The extrema of F are determined by the set of equations :

$$\frac{\partial F}{\partial \rho} = 2\rho \cdot \frac{\partial F_1}{\partial \rho^2} = 0 \tag{4.45}$$

$$\frac{\partial F}{\partial \gamma_r} = \sum_{m=2}^{k} \frac{\partial F_1}{\partial D_m} \cdot \frac{\partial D_m}{\partial \gamma_r} = 0 \tag{4.45'}$$

(Note that there are only (n-1) independent equations of the type (4.45'))

Let us consider a symmetry direction (γ_r^o). As stated in §.4.6.3.b), this is a direction whose isotropy-group I_η has an invariant-manifold reduced to the sole direction (γ_r^o). Repeating the argument in §.4.6.3.b, we can observe that the gradient (with respect to the η_r, or the γ_r) of any function invariant by I_ρ, at a point of the axis (γ_r^o), is parallel to this axis. Hence, all the basic invariants D_m necessarily satisfy the set of conditions :

$$\left.\frac{\partial D_m}{\partial \gamma_r}\right|_{\gamma_r^o} = 0 \tag{4.46}$$

The equations (4.45') are identically satisfied along the direction (γ_r^o). In order to determine the location of the extrema of F_1 along this direction we have to solve eq. (4.45), and specify the temperature dependent value of the modulus $\rho_o(T)$.

In this case, the actual computation is reduced to the determination of $\rho_o(T)$. It is only the result of the latter computation which depends on the coefficients of F or on its degree. The values of the γ_r^o derive from the decomposition of the OP-space into orbits with respect to I_o, or, alternately, from a set of equations (4.46) bearing on a small number of polynomials in the γ_r (i.e. the polynomials of the integrity basis). These polynomials have well defined numerical coefficients and their extremalization does not determine a temperature dependence for the γ_r^o.

We can generalize the preceding situation by considering the case where the isotropy-group I_η has an isotropy-space of dimension $1 < n' < n$. Let us choose, in the OP-space, the n' first coordinate axes ($\gamma_1, \ldots, \gamma_{n'}$) within the isotropy-manifold of I_η. Using, once more the argument (§.4.6.3.b) that the gradient of any invariant function D_m, at a point lying within the manifold is a vector belonging to the isotropy-manifold of I_η , we conclude that the components of the gradient perpendicular to the considered manifold, vanish identically. This implies :

$$\left.\frac{\partial D_m}{\partial \gamma_{r'}}\right|_{\gamma_r^o} = 0 \qquad (4.47)$$

with $(r' = n' + 1,\ldots, n)$ and $(\gamma_r^o = 0$ for $r = n' + 1,\ldots,n)$

In (4.45'), (n-n') equations vanish identically. The remaining "angular" equations (4.45') and the "radial" equation (4.45) determine a temperature dependent direction (γ_r^o) $(r = 1,\ldots,n')$ within the isotropy-manifold of I_η .

It is worth pointing out, following Jaric[28], that the n'-dimensional manifold considered, can contain symmetry directions, whose isotropy-groups are larger than I_η . The extrema of F_1 corresponding to these directions (of the type (4.46)) must be part of the extrema existing in the n'-dimensional manifold. Thus, if one has solved beforehand eqs. (4.46) for these directions, it is possible to separate the corresponding solutions out of eqs.(4.45) (4.45') and lower the degree of the latter equations.

In order to locate the minima among the extrema thus determined, one can make use, again, of the form (4.43') of the free-energy, and express the second derivatives of F with respect to ρ and the γ_j, in the form :

$$\left.\frac{\partial^2 F}{\partial \rho^2}\right|_{\rho_o,\gamma_r^o} = 4\,\rho_o^2 \cdot \left.\frac{\partial^2 F_1}{\partial(\rho^2)^2}\right|_{\rho=\rho_o} \qquad (4.48)$$

$$\left.\frac{\partial^2 F}{\partial\rho\,\partial\gamma_r}\right|_{\rho_o,\gamma_r^o} = 2\rho_o \cdot \sum_m \frac{\partial^2 F_1}{\partial(\rho^2)\,\partial D_m} \cdot \left.\frac{\partial D_m}{\partial \gamma_r}\right|_{\rho_o,\gamma_r^o} \qquad (4.48')$$

$$\left.\frac{\partial^2 F}{\partial\gamma_r\,\partial\gamma_{r'}}\right|_{\rho_o\gamma_r^o} = \left[\sum_{m,m'} \frac{\partial^2 F_1}{\partial D_m\,\partial D_{m'}} \cdot \frac{\partial D_m}{\partial\gamma_r} \cdot \frac{\partial D_{m'}}{\partial\gamma_{r'}}\right] + \left.\left[\sum_m \frac{\partial F_1}{\partial D_m} \cdot \frac{\partial^2 D_m}{\partial\gamma_r\,\partial\gamma_{r'}}\right]\right|_{\rho_o,\gamma_r^o} \qquad (4.48'')$$

In particular, if we consider a symmetry direction for which condition (4.46) is satisfied, (4.48') is identically zero, and the second member of (4.48") reduces to its second term.

Finally, if one wishes to associate the worked-out minima of F with possible low-symmetry phases below a continuous transition, one must check that the conditions defining these minima remain valid for infinitesimal values of ρ .

b) Application to 2-and 3-dimensional order-parameter [7,11]

Following Gufan and Sakhnenko [11], let us consider first the case of the 2-dimensional OP whose symmetries are described by the image $I_o = \mathbb{C}_{mv}$ with $m \geqslant 4$ (cf. §.4.5.3.a). The integrity basis is composed of 2 polynomials (§.4.5.4.b) : ρ^2 and $B = \rho^m \cdot \cos(m\varphi)$. On the other hand, one can easily check (figure a and b) that, depending if m is odd or even there are 2 or 3 types of isotropy groups, unequivalent up to a conjugation in $\mathbb{C}_{mv}$.

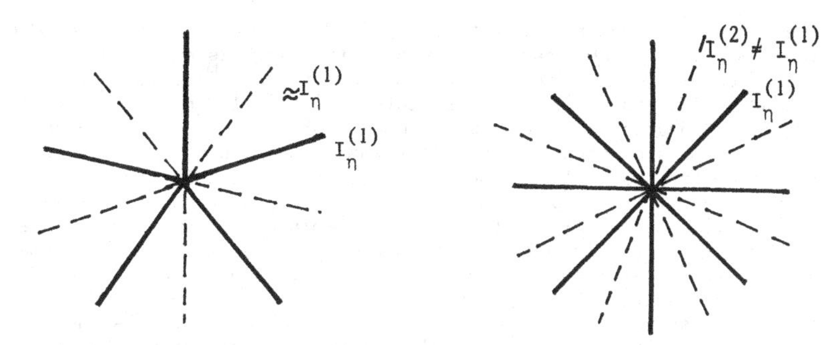

Fig.8. Symmetry directions and isotropy groups corresponding to $I_o = \mathbb{C}_{mv}$ with m odd (left) and m even (right).

In the case where m is odd, the first type $I_\eta^{(1)}$ corresponds to directions $\vec{\eta}$ lying within the symmetry planes. These isotropy groups, of order 2, are composed of the elements (E,σ_v), and their manifold is reduced to a single symmetry direction. The second type corresponds to the general direction (i.e. away from symmetry planes) in the OP-space, and is reduced to the identity (E). Its manifold is 2-dimensional.

If m is even, two types of isotropy groups $I_\eta^{(1)}$ and $I_\eta^{(2)}$ correspond to directions $\vec{\eta}$ lying in either of two families of symmetry planes rotated by the angle (π/m) with respect to eachother. Their respective manifolds are reduced to a single symmetry direction. The third isotropy group is, again, the invariance group of the general direction in the OP-space. It consists of the sole identity (E) and its manifold is 2-dimensional.

In agreement with (4.43') let us write $F = F_1\ [\rho^2,\ D = \cos m\,\phi]$, and use the variable ϕ, instead of the set (γ_1,γ_2), to specify a direction in the OP-space

Eq. (4.46) takes the particular form :

$$\sin m\,\phi = 0 \tag{4.49}$$

Its solutions, $\phi = (k\pi/m)$, correspond to the directions lying within the symmetry planes of the image $\mathbb{C}_{mv}$, as expected from the considerations following eq.(4.46). The value of the modulus ρ_o along these directions is determined by eq.(4.45). Below T_c, we have :

$$\frac{\partial F_1}{\partial(\rho^2)} = \frac{1}{2\rho} \cdot \frac{\partial F}{\partial \rho}\Bigg|_{D = \pm 1} = 0 \tag{4.49'}$$

These extrema are minima if (4.48)-(4.48") determine a positive matrix of second derivatives. The conditions are expressed as :

$$\frac{\partial^2 F}{\partial \rho^2} > 0 \qquad \text{and} \qquad \mp \frac{\partial F_1}{\partial D} > 0 \tag{4.49''}$$

where the ± sign corresponds to D = ± 1. We can see that the second of the conditions (4.49') can only be satisfied if F is expanded up to terms of mth degree (otherwise F is independent of B = $\rho^m \cos m\phi$, and $\partial F_1/\partial D_2 = 0$). Thus, for m > 4, as stated in §.4.6.1, in the case where the fourth-degree term in F is reduced to $\rho^4 = (\Sigma\eta_r^2)^2$, it is necessary to take into account higher-degree terms in the expansion of F, in order to ensure the stability of the low-symmetry phase. This circumstance does not preclude the possibility of a continuous transition. Indeed the incriminated condition $(\partial F/\partial D) > 0$, pertains to the "angular stability" of the considered symmetry directions. As for the radial stability (the first condition in (4.49')), it is controlled by a condition independent of the angular stability for m > 4 and $\rho \to 0$. Indeed we can restrict to the 4th-degree expansion which has the standard form $[(\alpha/2)\rho^2 + (\beta/4)\rho^4]$, which is clearly compatible with a continuous transition.

In the case m > 4, and of the symmetry directions in the OP-space, the continuity of the transition is ensured by the radial terms, which differentiate the high-symmetry phase from the low symmetry one. The angular stability condition, which can be expressed independently from the preceding one, ensures the preference of the system for one set of symmetry directions (see fig. 9).

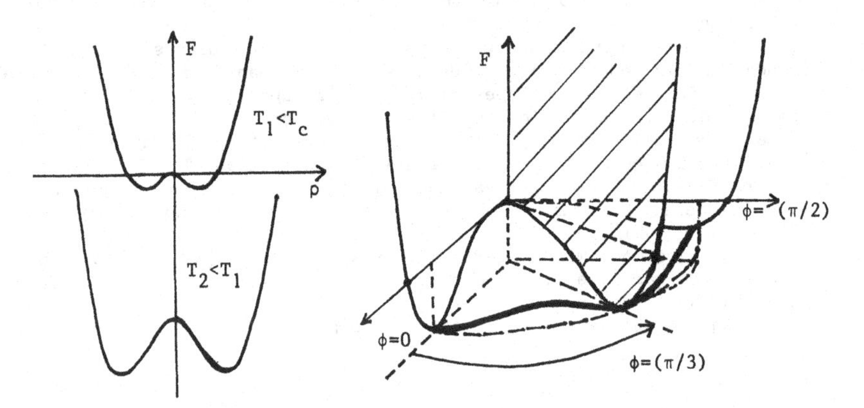

Fig.9. Radial and angular dependence of F below T_c for a symmetry C_{6v}. The different depth of the minimum for φ=(0,π/3) and for φ=(π/2) is controlled by the 6th degree term in F.

Thus, a continuous transition is possible between the high-symmetry phase and either of the phases corresponding to the isotropy-group (E,σ_v).

Let us now consider the second type of solutions in (4.45)(4.45'). These are determined by the set of equations :

$$\frac{\partial F}{\partial \rho} = 0 \quad ; \quad \frac{\partial F_1}{\partial D} = 0 \qquad (4.50)$$

In this case, an expansion up to the mth-degree is insufficient to determine the direction ϕ in the OP-space, since, up to this degree, F_1 depends linearly on D. The minimum required degree is (2m) (A 2m-th degree expansion contains terms D^2). Eqs.(4.50) then provide a value of D, which is a function of ρ, unlike eq.(4.49). Hence the corresponding ϕ value will specify a general direction in the OP-space, whose isotropy-group coincides with the identity (E). Besides, this direction will be temperature dependent.

A necessary condition for the stability of this phase is provided by the positiveness of (4.48"). The corresponding inequality reduces to

$$\frac{\partial^2 F_1}{\partial D^2} > 0 \qquad (4.50')$$

which also requires expanding F up to the (2m)-th degree, at least. Let us show that a continuous transition between the high-symmetry phase and this second type of low-symmetry phase is impossible. In this view we consider the expansion F up to the 2m-th degree-term and examine the behaviour of the solutions of (4.50) when $\rho \to 0$. We have :

$$F_1 = \frac{\alpha}{2}\rho^2+\ldots+\beta_{2m}\rho^{2m}+\rho^m.D(\delta_o+\delta_2\rho^2+\delta_4\rho^4+\ldots+\delta_m\rho^m) + \frac{\gamma_{2m}}{2}\rho^{2m}.D^2 \qquad (4.51)$$

The second equation (4.50) takes the form (for $\rho \neq 0$) :

$$\frac{\partial F_1}{\partial D} = \delta_o + \delta_2\rho^2 + \delta_4\rho^4 + \ldots+\delta_m\rho^m + \gamma_{2m}.\rho^m.D = 0 \qquad (4.52)$$

For this equation to be satisfied when $T \to T_c$, and $\rho \to 0$, we must have $\delta_o(T_c) = 0$. In addition, since $|D| = |\cos m\phi| < 1$, for this inequality to be satisfied when $T \to T_c$ and $\rho \to 0$, we must either have $\delta_2(T_c)$, $\delta_4(T_c) = 0\ldots$ [the value of D then tends towards $(-\delta_m/\gamma_{2m})$], or have $\gamma_{2m}(T_c) = 0$. Since the transition at T_c also implies $\alpha(T_c) = 0$, three coefficients, at least, of the expansion (4.51) must vanish at T_c. In general this condition cannot be met, and a continuous transition leading to the low-symmetry phase determined by eqs.(4.50) is therefore impossible.

The analysis along the same lines, of the phase transitions corresponding either to a 2-dimensional OP with the $\mathbb{C}_m$-symmetry (§.4.5.3.a), or to a 3-dimensional OP with a "cubic" symmetry (§.4.5.3.b) has been performed in references [7, 11, 28]. We have summarized, in table 15, the results of these studies concerning the low-symmetry phases which can be reached from the high-symmetry phase through a continuous transition.

Table 15

Image of G_o	$\mathbb{C}_m$	O	T_h	O_h	O	T_h	O_h
Direction in the OP-space	General	$(\gamma_1 = 1\ ;\gamma_2=\gamma_3= 0)$			$(\gamma_1=\gamma_2=\gamma_3= \frac{1}{\sqrt{3}})$		
Isotropy-group	E	$\mathbb{C}_4$	$\mathbb{C}_{2v}$	$\mathbb{C}_{4v}$	$\mathbb{C}_3$	$\mathbb{C}_3$	$\mathbb{C}_{3v}$

c) Kim's refinement of the method [29]

The general principles exposed in a) emphasize the role of the integrity basis in sorting the "symmetry-extrema" from the other extrema. They also provide some simplification in discussing the nature of the extrema (minima, maxima, or saddle points) independently from the specific degree of the expansion. However, the determination of the absolute minimum still requires carrying out lengthy calculations.

Kim has described [29] a refinement of Gufan's method which provides a simpler working-out of the absolute minimum. The method relies on a further use of the integrity basis, in order to locate geometrically the absolute minimum.

i) In the first place the various directions in the OP-space are specified by the values of the $D_m(\gamma_o)$ invariants. The correspondance $\gamma_i \rightarrow D_m(\gamma_i)$ is a mapping of the unit-circle in the OP-space, into a contour traced in the space spanned by the D_m. Actually, each point of the contour represents all the directions of an orbit (since D_m is invariant by I_o, while each vector of an orbit is transformed to the other vectors of the orbit). Besides, the contour in the D_m-space possesses cusps corresponding to the symmetry-directions in the OP-space (since in the corresponding directions one has $\partial D_m/\partial\gamma_i = 0$). Fig. 10 a) shows such a contour in the case of the image $I_o^m = \mathbb{C}_{4v}$. In this case there is a single D_m : $D = \cos(4\phi)$ (cf. table 10). The contour is a segment ($\pm$ 1) on the coordinate axis D. The extremities ($\phi = 0$ (mod. $\pi/2$), and $\phi = \pi/4$ (mod. $\pi/2$) correspond to the symmetry directions (with isotropy group (E,σ_v)), while the internal points correspond to the general directions in the OP-space (with isotropy-group E) (cf. Fig. 8)

Fig.10

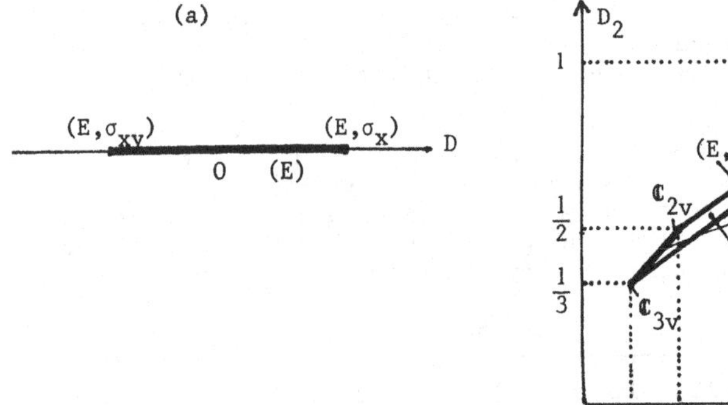

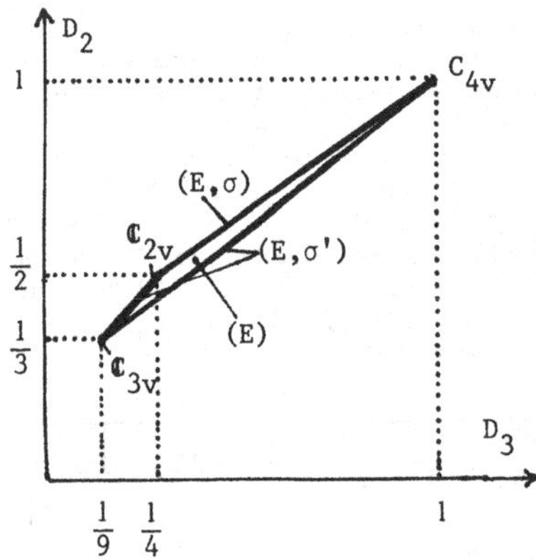

Fig.10 b) shows the contour relative to the image O_h, pertaining to a 3-component OP [29]. In this case, table 10 shows that there are, in addition to ρ^2, two members B_3 and B_5 in the integrity basis. B_5 can be replaced by $(\eta_1^6 + \eta_2^6 + \eta_3^6)$, which is a linear combination of ρ^6, B_5, and $\rho^2.B_3^2$.. The contour is traced in the space spanned by $D_2 = (\gamma_1^4+\gamma_2^4+\gamma_3^4) = B_3/\rho^4$, and $D_3 = (\gamma_1^6+\gamma_2^6+\gamma_3^6)$. The isotropy-groups of the various orbits are indicated for the cusps, and the lines joining them.

ii) The second ingredient of Kim's method is to search for the extrema of F in a given direction of the D_m space. These extrema yield values $\rho_o(D_m)$ of the modulus ρ of the OP. Using these values, we can express the extremal value of the free-energy as a function $F_o(D_m)$, and study the variations of $F_o(D_m)$ in the D_m-space. Examination of the location of the contour defined in i) with respect to these variations allows to specify the nature of the absolute minimum of F. It is worth pointing out that the variations of $F_o(D_m)$ depend on the degree of the OP-expansion and, these variations will become more difficult to study as measure as this degree is increased.

iii) As an illustration of the method, let us again consider the two-component OP corresponding to the image $I_o = \mathbb{C}_{4v}$, and study a free-energy expanded to the 4th-degree.

$$F = \frac{\alpha}{2}\rho^2 + (\beta_1 + \beta_2 D)\frac{\rho^4}{4} \qquad (4.53)$$

with $D = \cos 4\phi$. For fixed D the extremum of F occurs for $\rho^2 = -\alpha/(\beta_1+\beta_2 D)$, and the function $F_o(D)$ takes the form :

$$F_o(D) = \frac{-\alpha^2}{(\beta_1 + \beta_2 D)} \qquad (4.54)$$

The positiveness of ρ^2 requires $(\beta_1 + \beta_2 D) > 0$. When D varies between (-1) and (+1), $F_o(D)$ varies <u>monotonously</u> from $-\alpha^2/(\beta_1-\beta_2)$ to $-\alpha^2(\beta_1+\beta_2)$.

This monotonous variation implies (whatever the sign of β_2, that the minimum of F_o is at one of the extremities of the segment (± 1) in the D-space. Considering the contour in fig.10a), we see that this minimum will correspond to either of the orbits with maximal-isotropy groups (as already known, Cf. §.4.6.3).

We can see that, in order to obtain the absolute minimum of F in a general direction in the OP-space (i.e. <u>inside</u> the interval ±1, in the D-space). $F_o(D)$ must have a non-monotonous variation on the interval (±1). Such a variation can arise, in particular, if we expand F up to a degree such as certain coefficients of ρ^{2p} depend non-linearly on D. For instance, in the present example, it can be checked, along the lines just developped, that a sixth-degree expansion of F has coefficients depending linearly on D and that the absolute minimum of F remains at D = ±1. By contrast an eighth-degree expansion involves a non-linear dependence of the coefficients on D and is compatible with an absolute minimum <u>within</u> the interval (D = ±1) for certain ranges of the expansions coefficients [11]. However, such a minimum is not stable just below a continuous transition, since it is not obtained for the 4th degree-expansion.

5. SECONDARY ORDER PARAMETERS

Starting from §.3.6, we have considered that the free-energy F of the system and the particle density $\rho(\vec{r})$ describing its structure, were functions of a single irreducible set of degrees of freedom of the system, the components η_r of the OP. The rejection of other degrees of freedom, corresponding to other irreducible representations of G_o (equivalent or inequivalent) was based on the form of the second-degree expansion of F (eq. 3.7), which imposed that only the η_r can take non-zero equilibrium values below T_c.

As already pointed in chapter I, §.2.9, this conclusion is no longer valid if the expansion of F to terms of degree higher than 2 is considered.

In this paragraph we analyze the general conditions which preside over the onset, below T_c, of non-zero equilibrium values for degrees of freedom distinct from the OP-components η_r. These degrees of freedom will be termed secondary OP, while the set η_r can be termed the primary OP.

We will examine the form of the contributions of the secondary OP to the free-energy and show that the symmetry characteristics of the transition (e.g. the possible symmetries of the phases stable below T_c) are entirely determined by the primary OP.

Relying on the decomposition (eqs. 3.5, 3.7) we can concentrate on the case of a secondary OP carrying an irreducible representation of the group G_o.

5.1. Coupling between secondary and primary OP

Let ζ_s (s = 1,...,m) be a set of degrees of freedom carrying an m-dimensional IR of G_o, and η_r(1,..,n) the components of the primary OP

We have stated in §.2.2.4 and 3.6, that the second-degree invariant polynomials constructed from distinct irreducible degrees of freedom were "decoupled" (one sum of squared η_r^2 or ζ_s^2 for each IR). This decoupling does not persist for higher-degree invariant polynomials. For instance for any two irreducible sets, the 4th degree homogeneous polynomial :

$$(\Sigma\eta_r^2)(\Sigma\zeta_s^2) \qquad (5.1)$$

is obviously invariant since it is the product of 2 invariant terms.

More generally, one can find "mixed invariants" $\phi(\zeta_s,\eta_r)$ which are homogeneous polynomials of the set of variables ζ_s and η_r. These invariants are linear combinations of monomials of the form :

$$\zeta_1^{p_1} \ldots \zeta_m^{p_m} \cdot \eta_1^{f_1} \ldots \eta_2^{f_2} \qquad (5.1')$$

with $p = (\Sigma p_s) \neq 0$ and $f = (\Sigma f_r) \neq 0$. Unlike (5.1), the mixed invariants ϕ are not necessarily the products of polynomials of ζ_s and of η_r, separately invariant by G_o. Let us consider the free-energy expansion F of the system as a function of the 2 sets of degrees of freedom ζ_s and η_r :

$$F = \frac{\alpha}{2}\eta^2 + f_4(\eta_r) + \ldots + \frac{b}{2}\zeta^2 + h_3(\zeta_s) + h_4(\zeta_s) + \ldots + \Sigma_{p,f}\phi_{p,f}(\zeta_s,\eta_r) \qquad (5.2)$$

where $\eta^2 = (\Sigma\eta_r^2)$, $\zeta^2 = (\Sigma\zeta_s^2)$, the f_q and h_q are the homogeneous invariant polynomials of degree-q, in the η_r or in the ζ_s, and the $\phi_{p,f}$ are the various "mixed invariants" of degree p in the ζ_s and of degree f in the η_r. We know that $\alpha \propto (T-T_c)$, while $b > 0$ (cf. §.3.5.2, condition (4)).

Taking into account the degrees of the various polynomials in (5.2), we can write the ζ_s contribution to F in the form :

$$\zeta^2 \left[\frac{b}{2} + \zeta h_3\left(\frac{\zeta_s}{\zeta}\right) + \zeta^2 h_4\left(\frac{\zeta_s}{\zeta}\right) + \ldots + \Sigma\zeta^{p-2}.\eta^f\phi_{p,f}\left(\frac{\zeta_s}{\zeta}, \frac{\eta_r}{\eta}\right)\right] \qquad (5.3)$$

where, due to their homogeneous character, the polynomials, $h_3(\zeta_s/\zeta)$, $h_4(\zeta_s/\zeta)$ and $\phi_{p,f}(\zeta_s/\zeta, \eta_r/\eta)$, remain finite when $\zeta \to 0$ and $\eta \to 0$.

<u>If $p > 2$</u>, the coefficient of ζ^2 in (5.3) tends to $(b/2) > 0$ when η and $\zeta \to 0$. As a consequence, F has a minimum for $\zeta = 0$, on either sides of a continuous transition at T_c. In this case, the conclusion reached in §.3.5.2, is not modified by the consideration of terms of degree higher than 2, in the expansion : the ζ_s components <u>keep a zero equilibrium value below T_c</u>.

By contrast, if the "mixed invariant" of smallest degree in the ζ_s, is a <u>linear function of the</u> ζ_s, the preceding argument does not hold ; the effective coefficient of ζ^2 is determined by the limit of the "mixed invariant" :

$$\frac{\eta^f}{\zeta} \cdot \phi_{1,f}\left(\frac{\zeta_s}{\zeta}, \frac{\eta_r}{\eta}\right) \qquad (5.4)$$

It is worth pointing out that, since the total degree (f+1) of the mixed-invariant ϕ, is strictly larger than 2 (the bilinear terms have been eliminated, as shown in §.2.2.4), the exponent f satisfies the condition :

$$f \geqslant 2 \qquad (5.5)$$

In conclusion, we can see that a <u>necessary condition for</u> certain ζ_s to acquire a non-zero equilibrium value below a <u>continuous</u> transition at T_c, is the possibility to construct from the set ζ_s and the components η_r of the primary OP, a "mixed" homogeneous polynomial $\phi(\zeta_s, \eta_r)$, <u>invariant by G_o, linear in the ζ_s, and of degree $f \geqslant 2$ in the η_r</u>.

Eq.(2.23) in chapter I, provides an illustration of such a mixed invariant. Other examples will be examined in chapter III, since the concept of secondary OP plays an important role in the chapter devoted structural transitions.

5.2. Possible symmetries of secondary OP, and form of $\phi(\zeta_s,\eta_r)$

We have seen that, if ζ_s are the components of a secondary OP, then, the free-energy $F(\eta_r,\zeta_s)$ will contain, at least, one mixed-invariant of the form :

$$\phi(\zeta_s,\eta_r) = \Sigma\ \zeta_s.\phi_s(\eta_r) \qquad (5.6)$$

where the $\phi_s(\eta_r)$ are homogeneous polynomials of degree $f \geqslant 2$. Let us denote Γ_ζ the IR of G_o carried by the ζ_s-components, and Γ_η the primary-OP-representation. The set of fth-degree polynomials can be considered as a vector of the space carrying the symmetrized f-th power $[\Gamma_\eta^f]$ of the OP-representation. In order that $\phi(\zeta_s, \eta_r)$ be invariant by G_o, it is necessary and sufficient that the ϕ_s carry Γ_ζ, and, that the two sets $\{\zeta_s\}$ and $\{\phi_s\}$ constitute homologous bases of this IR [§.2.2.3.d].

Therefore, the decomposition of the symmetrized f^{th}-power of Γ_η, with $f > 2$, into irreducible representations will provide all the possible symmetries of secondary-order-parameters.

A possible method for performing this decomposition has been described in §.4.3.2. It consists in using eqs.(4.1) and (4.2) and "project" the symmetrized space $|\Gamma_\eta^f|$ on the various irreducible representations of G_o.

As an illustration of the method, let us consider the group G_o of the equilateral triangle (table 3) and take the primary OP-symmetry as $\Gamma_\eta = \Gamma_3$. For $f = 2$ (the lowest relevant f-value), we must decompose the symmetrized-square of Γ_η into irreducible representations of G_o. Such a decomposition has already been performed in §.4.3.3.a). It yields two irreducible spaces. One, $(\eta_1^2+\eta_2^2)$, is totally symmetric. The other, $(\eta_1^2 - \eta_2^2 ; 2\eta_1\eta_2)$, has the symmetry Γ_3.

Thus, we can conclude that, for f=2, and in a phase transition whose primary OP has the symmetry Γ_3, the secondary OP have necessarily, either the symmetry Γ_1 or the symmetry Γ_3. The two "mixed invariants" (5.6) respectively associated to the two types of secondary OP can be written :

$$\phi_1(\zeta_s, \eta_r) = \zeta . (\eta_1^2 + \eta_2^2) \tag{5.7}$$

$$\phi_2(\zeta_s, \eta_r) = \zeta_1 . (\eta_1^2 - \eta_2^2) + 2.\zeta_2\eta_1\eta_2$$

where ζ carries Γ_1, and (ζ_1, ζ_2) transform according to the matrices of Γ_3 in table 3.

For f = 3 the decomposition of $[\Gamma_3^3]$ has been performed in §.4.3.3.b). It yields :

$$[\Gamma_3^3] = \Gamma_1 + \Gamma_2 + \Gamma_3 \tag{5.8}$$

Let ζ, ζ', and (ζ_1, ζ_2) be the secondary OP-components associated to the three preceding representations. The corresponding mixed-invariants are, using the expressions determined in §.4.3.3.b) :

$$\phi_1 = \zeta . \eta_1 (\eta_1^2 - 3 \eta_2^2)$$

$$\phi_2 = \zeta' . \eta_2 (\eta_2^2 - 3 \eta_1^2) \tag{5.9}$$

$$\phi_3 = \zeta_1 . \eta_1 (\eta_1^2 + \eta_2^2) + \zeta_2\eta_2 (\eta_1^2 + \eta_2^2)$$

We can note that for a secondary-OP of Γ_3 symmetry, a bilinear term of the form $(\zeta_1\eta_1 + \zeta_2\eta_2)$ is also invariant by G_o. However, such a bilinear term has been assumed to be suppressed from the free-energy expansion through a "renormalization" of the second-degree terms (cf. §.2.2.4).

Also note that a dimensionality condition applies to the primary-OP, if one looks for a secondary-OP which is both non-totally-symmetric, and of a symmetry unequivalent to the primary-OP (e.g. the Γ_2-case in eq.5.9) : the dimension n of the primary-OP must satisfy :

$$n \geqslant 2 \tag{5.10}$$

Indeed the decomposition of $[\Gamma_\eta^{\,f}]$ for n = 1, will yield, either the totally symmetric representation (for f even), or the representation Γ_η (for f odd).

5.3. Equilibrium value of the secondary OP, nearby T_c

Let us consider expression (5.2) of the free-energy, in which we substitute to the set of functions $\phi_{p,f}$, a single function in the form (5.6), the degree f of the functions $\phi_s(\eta_r)$ being the smallest degree involved in the mixed invariants of the type (5.6). We obtain, by expressing $\partial F/\partial \zeta_s$ and $\partial F/\partial \eta_r$:

$$\frac{\partial F}{\partial \zeta_s} = b\zeta_s + \frac{\partial h_3}{\partial \zeta_s} + \frac{\partial h_4}{\partial \zeta_s} + \ldots + \phi_s(\eta_r) = 0 \tag{5.11}$$

$$\frac{\partial F}{\partial \eta_r} = \eta_r\,[\alpha + \frac{1}{\eta_r}\frac{\partial f_4}{\partial \eta_r} + \ldots + \sum_s \frac{\zeta_s}{\eta_r} \cdot \frac{\partial \phi_s}{\partial \eta_r}\,] = 0 \tag{5.11'}$$

These two equations determine the extrema of F. Since $\partial h_s/\partial \zeta_s$ is of degree (q-1) $\geqslant$ 2 in the ζ_s, eq.(5.11) yields for small ζ_s and η_r :

$$\zeta_s = -\frac{\phi_s(\eta_r)}{b} \tag{5.12}$$

Using this expression, we can eliminate ζ_s from (5.11') :

$$\eta_r\,[\alpha + \frac{1}{\eta_r}\frac{\partial f_4}{\partial \eta_r} - \frac{1}{2b\eta_r}\sum_s (\frac{\partial \phi_s^2}{\partial \eta_r})\,] = 0 \tag{5.12'}$$

In eq.(5.12'), $\frac{1}{\eta_r} \cdot \partial f_4/\partial \eta_r$ is of degree 2, while $\frac{1}{\eta_r} \cdot (\partial \phi_s^2/\partial \eta_r)$ is of degree (2f-2). If f is strictly larger than 2, the latter term is negligible as compared to the former one, for small η_r. Equation (5.12') is identical to the equation providing the equilibrium value of η_r in the absence of a secondary-OP. The equilibrium is not modified by ζ_s.

If f = 2, the part of eq.(5.12') between brackets is modified, for small η_r. The modification is equivalent to the replacement of $f_4(\eta_r)$ by $[f_4 - (\sum_s \phi_s^2)/2b]$ in the primary-OP-expansion. Note that, since the ϕ_s carry the irreducible representation Γ_ζ, $(\sum_s \phi_s^2)$ is the quadratic invariant associated to the ϕ_s. Accordingly, both f_4 and its modified form are fourth-degree invariants. Referring to eq. (3.12), we can write:

$$\begin{aligned} f_4 &= \Sigma \beta_i \cdot Q_i\,(\eta_r) \\ f_4 - \Sigma \phi_s^2/2b &= \Sigma \beta_i' \cdot Q_i\,(\eta_r) \end{aligned} \tag{5.13}$$

The two terms are formally identical, but they differ in their coefficients. The effect of the secondary OP is therefore to modify the coefficients of the fourth-degree terms in the primary-OP-expansion. If this modification preserves the positive character of $f_4(\eta_r)$, a continuous transition will still occur for $\alpha = 0$. i.e. at the same temperature T_c as in the absence of the "coupling term" $\zeta_s.\phi_s(\eta_r)$. If, however, the coefficients β_i' belong to a range for which $\Sigma\beta_i'.Q_i$ is not positive, a continuous transition will no longer be possible.

Let us place ourselves in the first case. Eq.(5.12) shows us that, below T_c, the equilibrium values ζ_s^o are related to the equilibrium values η_r^o. It can happen that certain, or all functions $\phi_s(\eta_r^o)$ vanish identically for $\eta_r = \eta_r^o$. In this case ζ_s does not acquire a non-zero value below T_c. If, on the other hand, certain $\phi_s(\eta_r^o)$ are non-zero, the equilibrium values ζ_s^o will satisfy the relation :

$$\zeta_s^o \propto (\eta_r^o)^f \tag{5.14}$$

with $f \geqslant 2$. The ζ_s^o are therefore small quantities as compared to the η_r^o. The terms $h_3(\zeta_s^o)$, $h_4(\zeta_s^o)$ are respectively small quantities of order $3f \geqslant 6$ and $4f \geqslant 8$, and can be neglected. Just below a continuous transition at T_c, the contribution of the secondary-OP to the free-energy can be restricted to two terms : the quadratic invariant $(b/2)(\Sigma\zeta_s^2)$, and the mixed-invariant $(\Sigma\zeta_s.\phi_s(\eta_r))$.

As stressed in §.3, the primary OP-expansion must be developped up to, at least, the fourth-degree terms. We will see in chapter III how the resulting free-energy $F(\eta_r, \zeta_s)$ can be used in order to study the behaviour of physical quantities related to the secondary OP.

5.4. Irrelevance of the secondary-OP, to the symmetry below T_c

Taking into account both the η_r^o and the ζ_s^o, and using equation (3.5), we can express the density increment describing the structure of the system, below T_c, as :

$$\delta\rho_{eq}(T,P,\vec{r}) = \sum_r \eta_r^o.\phi_r(\vec{r}) + \sum_s \zeta_s^o.\psi_r(\vec{r}) \tag{5.15}$$

where the $\phi_r(\vec{r})$ and $\psi_r(\vec{r})$ are fixed normalized functions, and where the η_r^o and ζ_s^o are the equilibrium primary-and secondary-OP components.

Let G be the invariance group of the "vector" η_r^o in the space of the primary OP. The action of G leaves each η_r^o unchanged.

Eq.(5.14) shows that likewise G will leave the "vector" ζ_s^o invariant. As a consequence, the full invariance-group G of the "primary-increment"

$$\delta\rho_{eq}^{(1)} = \sum_r \eta_r^o\,\phi_r(\vec{r}) \tag{5.16}$$

is also the full invariance-group of the "total-increment" (5.15).

Hence the symmetry of the phase stable below T_c is entirely determined by the symmetry-properties of the primary OP.

If secondary order-parameters exist, it is not necessary to consider them for the study of the nature of the symmetry-change.

It can happen that the "secondary-density increment" **expressed** by,

$$\delta\rho_{eq}^{(2)}=\sum_s \zeta_s^o.\psi_r(\vec{r}) \qquad (5.17)$$

will have additional elements leaving it invariant. Hence, in general, the modification of the structure of the system represented by $\rho_{eq}^{(2)}$ will have a higher symmetry than G.

6. CONCLUSIONS

In this chapter, while establishing the foundation's of Landau's theory, and discussing the group theoretical methods pertaining to it, we have implicitly defined the procedure to be used in applying the theory to a given system, and in exploiting its consequences.

The starting information needed consists of two elements : the group of invariance G_o of the "most-symmetric" of the two phases surrounding the transition point, and the symmetry-properties of the (primary) order-parameter $\{\eta_r\}$ of the transition. The latter symmetry is specified by an irreducible representation Γ of G_o (§.3.5).

Using this information one aims at determining two types of results. One type consists in the temperature (and pressure) dependence of certain physical quantities, either directly related to the primary OP (§.3.6.2.b, and chapter I, §.2.7), or related to secondary-order-parameters (§.5, and chapter I §.2.9). These results derive from the form of the free-energy F expanded, as a function of the primary, and secondary OP.

The second type of results consists in the determination of the symmetry-properties (i.e. the invariance group G) of the low-symmetry phase, stable below T_c. This result is rooted, on the one hand, in the OP-symmetry (G is the isotropy-group of a certain direction $\vec{\eta}$ in the OP-space)(§.3.6.2.c), and, on the other hand in the nature of the absolute minimum of the free-energy expansion F (§.3.6.2.c)

In consequence, the two important tasks, preliminary to the working out of the above results, are the construction of the free-energy expansion F, and the enumeration of the isotropy-groups (subgroups of G_o) of the various directions in the OP-space.

The construction of the free-energy proceeds through the determination of the homogeneous polynomials of various degrees in the η_r, which are invariant under the action of G_o. In this view, one can either use standard group theoretical methods directly applied to the group G_o (§.4.1-4.3), or take advantage of the simplifications provided by the uses of the image I_o of G_o in the OP-representation and of the availability of integrity bases of invariant polynomials (§.4.5).

The determination of the polynomials invariant by G_o, or alternately, the application of the Landau criterion (§.4.4) will show if an invariant third-degree polynomial is allowed by G_o and by the OP-symmetry. On this basis, the impossibility of a continuous transition can be asserted (§.3.6.1).

The enumeration of the isotropy-groups of the various directions in the OP-space, most conveniently proceeds through the use of the image I_o (§.4.6.2). In order to clarify the symmetry-equivalence (and the degeneracy of the free-energy) between different possible low-symmetry groups G_i, it is of interest to rely on the concepts of orbit, and stratum (§.4.6.2)

One has then to select among the isotropy-groups by finding out which orbit corresponds to the absolute minimum of the free-energy expansion.

In this respect, we have pointed out that the various methods recently suggested are based on the existence of "obligatory extrema", in symmetry-directions of the OP-space, and also on the use of the integrity basis of polynomials, adapted to the OP-symmetry (§.4.6.3-4.6.5)

These methods have clarified the nature of the mathematical problem set by the Landau theory : find the absolute minimum of a fourth-(and occasionally higher)-degree polynomial in several variables η_r, which has the following characteristics :

i) it has a local maximum at the origin ($\eta_r = 0$) and it is positive and infinite at infinity ($\eta_r \to \pm \infty$) in any direction.

ii) its extrema are degenerate, the degeneracy being equal to the number of points in either of the orbits defined in the OP-space by the action of G_o.

iii) the directions of its extrema, in the OP-space, are constrainted by the OP-symmetry (e.g. some extrema necessarily lye along symmetry directions).

REFERENCES

|1| G. Ya. Lyubarskii. *The application of group theory in physics.* (Pergamon press; London 1960).

|2| M. Tinkham. *Group theory and quantum mechanics.* (Mc Graw Hill; New-York. 1964).

|3| L.D. Landau. *Collected Papers.* (Pergamon; Oxford 1965). L.D. Landau and E.M. Lifschitz. *Statistical Physics.* (Pergamon, London 1958).

|4| J. Przystawa. Preprint; (unpublished).

|5| R.D. Charmichael. *Groups of finite order.* (Dover publications, New-York 1956).

|6| L. Michel. Reviews of Modern Physics. 52,617 (1980)

|7| L. Michel. in *Regards sur la Physique Contemporaine.* (CNRS ,Paris 1980) p. 157

|8| N. Boccara. *Symétries Brisées.* (Herrmann, Paris 1973)

|9| L.V. Meisel, D.M. Gray, and E. Brown. Journ. of Mathematical Physics. 16,2520 (1975). E. Brown, and L.V. Meisel. Phys. Rev. B13,213 (1976).

|10| L. Michel, J.C. Tolédano, and P. Tolédano, in *Symmetries and Broken Symmetries in Condensed Matter Physics.* Ed. N. Boccara. (IDSET, Paris 1981) p.263.

|11| Yu. M. Gufan, Soviet Physics, Solid State 13, 175 (1971). Yu. M. Gufan and V.P. Sakhnenko. Soviet Physics JETP 36, 1009 (1973).

|12| V. Kopsky. Ferroelectrics. 24,3 (1980).

|13| L. Michel. in *Group Theoretical Methods in Physics*.Ed. R. Sharp, and B. Kolman (Academic Press, New-York, 1977). p.75.

|14| Yu.M. Gufan, and V.V. Popov. Soviet Physics. Crystallography. 25, 527 (1980).

|15| J.S. Kim, H.T. Stokes, and D.M. Hatch. Phys. Rev. B 33 ,6210 (1986)

|16| P. Du Val. *Homographies, Quaternions, Rotations* .(University Press, Oxford 1964).

|17| J. Mozrzymas,and A. Solecki. Reports on Mathematical Physics. 7,363 (1975).

|18| J.C. Tolédano, L. Michel, P. Tolédano, and E. Brézin. Phys. Rev. B31, 7171 (1985),

|19| E. Ascher. J.Physics C10, 1365 (1977)

|20| J.L. Birman. Phys. Rev. Letters 17, 1216 (1966).

|21| M.V. Jaric. Phys. Rev. B23, 3460 (1981).

|22| J. Lorenc, J. Przystawa,and A.P. Cracknell. J. Physics C13,1955 (1980).

|23| E. Ascher. Physics Letters, 20,352 (1966).

|24| J.C. Tolédano,and P. Tolédano. Journ. de Physique. 41,189 (1980).

|25| D. Mukamel, and M.V. Jaric. Phys. Rev. B29, 1465 (1984).

|26| M.V. Jaric. Phys. Rev. Letters. 51, 2073 (1983).

|27| L. Michel, and J. Mozrzymas. Lecture Notes in Physics.79, 447 (1978).

|28| M.V. Jaric. Phys. Rev. Letters. 48, 164 (1982).

|29| J.S. Kim. Phys. Rev. B31 , 1433 (1985).

CHAPTER III

CONTINUOUS STRUCTURAL TRANSITIONS BETWEEN PERIODIC PHASES

1. INTRODUCTION

Structural transitions (ST) are a fraction of the phase transitions which occur in crystals and involve a modification of the crystal's atomic configuration (i.e. the structure). ST are sometimes formally defined by the fact that the structural modification taking place across the transition does not break chemical bonds, but only slightly changes their length and orientation (or alternately, that the ST does not modify the coordination numbers of the various atoms in the structure). This definition is not very useful, in practice, because the detailed structural investigation which would allow to check the preservation of the characteristics of the bonds requires accurate measurements which are often difficult to undertake.

At the experimental level, ST are rather defined by a set of properties which are easier to check : e.g. a transition latent heat smaller than a few calories/gramme, small discontinuities of the physical quantities at the transition point (scaled with respect to the over-all temperature dependence of these quantities which is induced by the existence of the transition), and most importantly, a symmetry relationship between the two phases (in the simplest case, a group-subgroup relationship between the invariance groups of the two phases) which is infered from the evolution of the X-ray or Neutron diffraction spectrum across the transition.

From a theoretical point of view, structural transitions can be singled out among the phase transitions taking place in crystals by the fact that an order-parameter (OP) can be defined for them on the basis of the symmetry principles of the Landau theory. More precisely, the set of their properties (including the symmetry-change across the transition) should display an overall consistency with the Landau theory described in chapter II, or with some extension of this theory retaining similar symmetry and thermodynamic principles.

From this theoretical point of view, ST can be classified into 3 categories which are examined separately in 3 chapters of the book. In this chapter we analyse the ST in which the two phases present a strict crystalline periodicity, and which comply closely with the framework exposed in chapter II : irreducibility of the OP, and continuity of the transition. Chapter IV is devoted to ST displaying discontinuous transitions between strictly periodic phases. Chapter V focuses on ST where the high-symmetry phase is strictly periodic, and the low-symmetry one is "incommensurate".

In accordance with the general principles described in chapter II, the investigation of a ST requires, as a starting information, the specification of the invariance group G_0 of the more-symmetric of the two phases surrounding the transition, and also the specification of the symmetry of the OP.

For an ST, G_0 is the invariance group of a crystalline phase : i.e. G_0 is one of the so-called "space-groups". In paragraph 2, we will summarize the characteristics of the space-groups and describe the handling of their irreducible representations. In paragraph 3, we will show that only certain of these representations can provide the required OP-symmetry, for a ST between strictly crystalline phases. We will then be in a position to select, among the OP-symmetries and free-energies listed in chapter II, §4 for various OP-dimensions, the ones which are relevant to structural transitions. This aspect will be developped in §4. In §5 the technical aspects of the working out of the "low-symmetries" compatible with a given OP, will be described. § 6 will be devoted to the concept of "ferroicity" which is, in great part, related to the existence of secondary-OP introduced in chapter II, § 5. In § 7 we will rely on an example to illustrate in details the application of Landau's theory to a ST. Finally, we will review in § 8 the results of the use of Landau's theory in the study of the structural transitions investigated up to now, and also discuss the physical nature of the order-parameter of these transitions.

2. CRYSTALLOGRAPHIC SPACE-GROUPS AND THEIR REPRESENTATIONS

In this paragraph we outline the main features of crystallographic space-groups, specify their notations, and describe their irreducible representations. Brief justifications only are given for the stated characteristics. For more detailed developpments on these subjects the reader can refer to one of the many available text-books (e.g. refs [1] [2] and [3]).

2.1. Crystallographic space-groups

2.1.1. Geometrical elements composing a space-group

In a crystalline solid the equilibrium configuration of the average positions of the atoms (which is called the structure) has a three dimensional spatial periodicity. This definition means that the preceding configuration is transformed into a configuration undistinguishable from the starting one by translations of vectors :

$$\vec{T}\,(m_1,m_2,m_3) = m_1\vec{a}_1 + m_2\vec{a}_2 + m_3\vec{a}_3 \qquad (2.1)$$

where the m_i are arbitrary integers (positive, negative, or equal to zero), and $(\vec{a}_1, \vec{a}_2, \vec{a}_3)$ are 3 non-coplanar vectors which are called the basic-translations of the crystal's structure.

The set of the translations $\vec{T}\,(m_1, m_2, m_3)$, which constitutes an infinite (discrete) abelian group $\mathcal{T}$ of geometrical transformations is called the group of primitive translations of the crystal's structure, or more abreviatedly, the group of the crystal's translations.

In certain cases, the set of geometrical transformations leaving the structure invariant is reduced to $\mathcal{T}$. In general, however, there will be, besides $\mathcal{T}$, additional displacements which leave the structure invariant. The set of all such displacements (including $\mathcal{T}$) form a group G which is called the space-group of the crystal.

In addition to $\mathcal{T}$, the group G generally comprises three types of geometrical displacements (fig. 1) [as well as their products].

i) Rotations by an angle Θ about an axis of direction $\vec{u}$ passing through a point 0 of the crystal (fig. 1a).

ii) Symmetries about a plane perpendicular to the direction $\vec{u}$ and passing through a point 0 of the crystal (fig. 1b).

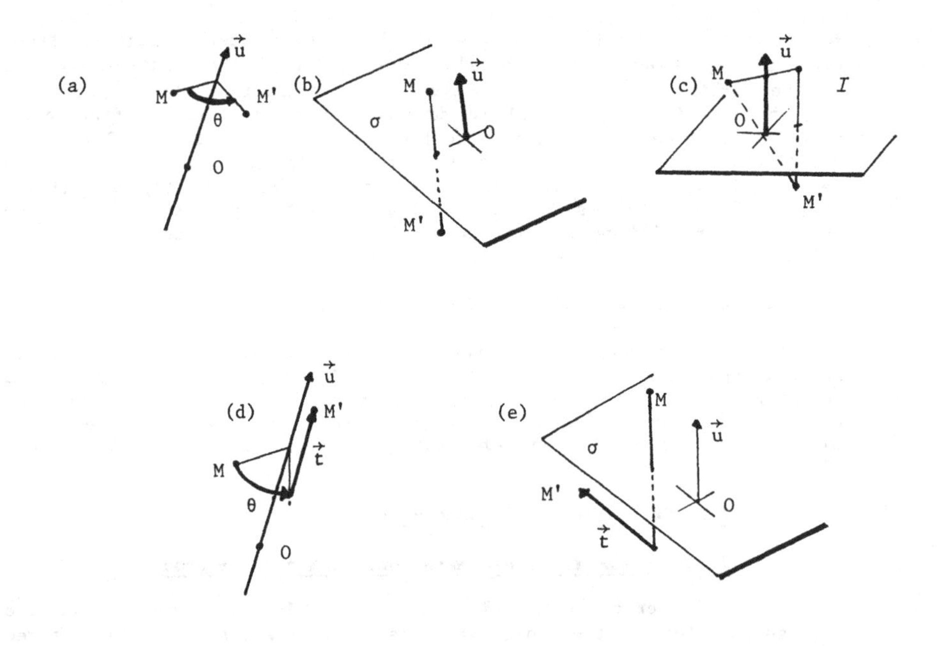

Fig. 1

The symmetry about a point O of the crystal (inversion) is a combination of two of the above defined displacements : e.g. a rotation by $\Theta = \Pi$ around an axis passing through O and parallel to the direction $\vec{u}$, followed by a symmetry in the plane perpendicular to $\vec{u}$ and passing through O (fig. 1c).

The displacements i) and ii) as well as the inversion are called point-symmetry displacements as they have the common property to leave, at least, one point O unmoved in the displacement (unlike translations which displace all the points of the system). We will denote by the symbol R a point-symmetry operation.

iii) The displacements obtained by applying subsequently a point -symmetry operation R and a translation $\vec{t}$. If R is a rotation (Θ, $\vec{u}$, O), the interesting case corresponds to $\vec{t}$ parallel to $\vec{u}$: one obtains a type of displacement called an helicoidal-rotation, or screw-axis (fig. 1d) ($\vec{t}$ perpendicular to $\vec{u}$, generates an ordinary rotation about an axis translated by $\vec{t}/2$). If R is a plane-symmetry, the interesting case corresponds to $\vec{t}$ parallel to the plane : one obtains a displacement called a glide-plane (t perpendicular to $\vec{t}$he plane generates an ordinary plane-symmetry about a plane translated by $\vec{t}/2$ (fig. 1e).

It is worth pointing out that any space-group containing pure rotations or plane-symmetries R will contain, trivially, elements of the type iii) generated by the products of R by the primitive translations $\vec{T}$ of the crystal. However, certain space-groups can contain in addition non-trivial products of the type iii), with $\vec{t}$ a non-primitive translation (i.e. not belonging to $\mathfrak{T}$).

The notation of the set of elements constituting a space-group is customarily denoted by :

$$\{R|\vec{t}\} \tag{2.2}$$

where $\vec{t}$ is, either, the zero-translation (the symbol (2.2) then represents a pure point-symmetry displacement), or a primitive translation $\vec{T}$ (if R is reduced to the identity, (2.2) represents the elements of $\mathfrak{T}$), or a non-primitive translation (R is then necessarily a non-identical point-symmetry element, and (2.2) represents a displacement of the type iii)).

Generally, the symbols R provide explicitely the angles θ of the various rotations contained in G, and sometimes the directions $\vec{u}$ of the rotation-axes, or of the normal to the symmetry planes. However they do not specify the points 0 in the system, which lye on the axes or on the planes. This is due to the fact that the notation (2.2) refers to a common origin 0 (a definite point of the crystal) which does not necessarily lye at the intersection of all the rotation axes, and of the symmetry-planes (clearly such an intersection does not necessarily exist). As a consequence, a "pure-rotation" or a "pure-plane-symmetry" of the types i) or ii) can be, nevertheless, represented by a symbol (2.2) involving a non-zero translation $\vec{t}$: indeed if the origin 0 is not on the axis of the rotation, or on the symmetry-plane R considered, and if 0' is a point of this axis, or of this plane, the symbol (2.2) referred to 0, for the point-symmetry element R will be :

$$\{R|R.\vec{\alpha} - \vec{\alpha}\} \tag{2.3}$$

where $\vec{\alpha} = \vec{0'0}$, and where $R.\vec{\alpha}$ is the transformed of the vector $\vec{\alpha}$ by the R transformation.

More generally, if the origin chosen to represent by (2.2) all the elements of the space-groups is shifted from 0' to 0, each element of G will have its notation changed as following :

$$\{R|\vec{t}\} \rightarrow \{R|\vec{t} + R\,\vec{\alpha} - \vec{\alpha}\} \tag{2.4}$$

Note that, in spite of this incidence of the choice of an origin on the form of the notation (2.2), a pure point-symmetry element can be easily recognized because the associated non-primitive translation $\vec{t}$, induced by the change of origin, is necessarily perpendicular to the axis of the rotation, or to the symmetry-plane. The notation (2.2) has the conveniency to provide a relatively simple mathematical rule for finding, with reference to the same origin, the product of any two elements of the space-group. One has :

$$\{R_1|\vec{t}_1\}\cdot\{R_2|\vec{t}_2\} = \{R_1R_2|\vec{t}_1 + R_1\vec{t}_2\} \tag{2.5}$$

This rule shows, in particular, that even if R_1 and R_2 commute, the product of the two space-group elements associated to R_1 and R_2 will not commute.

2.1.2. Bravais lattice and point-group $\hat{G}$ of the crystal

a) Bravais lattice

If one chooses an arbitrary point 0 in space, and if, with this point as origin, one constructs the infinite-discrete set of vectors $\vec{T}(m_1,m_2,m_3)$ specified by eq. (2.1), one obtains as extremities of these vectors a 3-dimensional lattice of points which is called the Bravais lattice associated to the crystal. Hence, the Bravais lattice provides a geometrical picture of the translation group $\mathfrak{T}$ of the crystal.

b) point-group

Equation (2.5) shows that the set of all the R_i point-symmetry elements referred to the chosen origin 0 [i.e. rotations and symmetries whose axes and planes intersect at 0] form a group which is called the point-group of the crystal and is denoted $\hat{G}$. In general, the space-group G will contain screw-axes or glide-planes $\{R_i|\vec{t}_i\}$ among its elements (G is said to be non-symmorphic). In this case, the R_i associated to the screw-axes and glide-planes do not leave the crystal invariant. Accordingly the point-group $\hat{G}$ (referred to any origin in the crystal) will not be a group of invariance of the crystal. Conversely, if G contains neither screw-axes nor glide-planes, (G is said to be symmorphic) it is possible to refer $\hat{G}$ to an origin 0, within the crystal, such as $\hat{G}$ will be a group of invariance of the crystal. With respect to this origin, the symmorphic-group G can be considered as the product of the translation-group $\mathfrak{T}$ by the group of elements $\{R_i|\vec{o}\}$, isomorphous to $\hat{G}$.

c) Invariance of the Bravais lattice by $\hat{G}$. 32 point-groups

An important property of all space-groups is the invariance of their Bravais lattice by the point-group $\hat{G}$, referred to one of the nodes of the lattice. This property results from the fact that the translation group $\mathfrak{T}$ of the crystal is an invariant subgroup of G [i.e. $\{R|\vec{t}\}.\mathfrak{T} = \mathfrak{T}.\{R|\vec{t}\}$]. Accordingly, if we choose a set of coordinate axes along the three basic vectors ($\vec{a}_1$, $\vec{a}_2$, $\vec{a}_3$) of the lattice, the 3-dimensional orthogonal matrices representing the R_i, in this frame , will necessarily have all their elements equal to integers (since the R_i exchange the nodes of the Bravais lattice). In particular, the traces ($1 \pm 2.\cos\theta_i$) of the 3-dimensional rotations R_i have to be integers, and this requirement imposes to the θ_i to be one of the angle-values shown on the first line of table 1. The second and third lines of this table show two conventional symbols for the corresponding rotations (whenever necessary, the direction of the rotation axis has to be specified by an additional indication). We have also included in the table the conventional symbols for the plane-symmetry and the inversion.

TABLE 1. Schoenflies and international symbols for point-symmetry elements.

Θ	0	π	$2\pi/3$	$4\pi/3$	$\pi/2$	$3\pi/2$	$\pi/3$	$5\pi/3$	plane-symmetry	Inversion
R	E	$\{C_2$, U_2	C_3	C_3^2	C_4	C_4^3	C_6	C_6^5	σ	I
R	1	2	3		4		6		m	I

$\hat{G}$ is a subgroup of O(3) [cf. § 4.5.2 chapter II] composed of the elements present in table 1. Up to a conjugation in O(3) [i.e. up to the setting of the global orientation of the directions of the axes and of the planes] the point-groups which can be constructed from elements of the above table (the relative orientation of the axes and planes is not specified before-hand) form a limited set of <u>32 distinct groups</u> which are the <u>crystallographic point-groups</u>. Their list can be found in any of the quoted text-books [1-3].

d) full invariance group of the Bravais lattice. Crystal systems

Due to the fact that the Bravais lattice is a configuration of points (the nodes of the lattice), it generally possesses, besides $\hat{G}$, additional point-symmetry elements leaving it invariant. Its invariance point-group $\hat{H}$ is generally larger than $\hat{G}$. Hence, if $\hat{G}$ does not contain the inversion I, $\hat{H}$ necessarily contains the inversion about the origin (a node of the lattice). Also, if there exist planes joining lattice nodes and intersecting at right-angle, the lattice will be invariant by symmetries about these planes. The groups $\hat{H}$ are part of the 32 crystallographic point-groups (since they leave a Bravais lattice invariant). They coincide with the groups in this list which contain the inversion, and, if relevant, symmetry-planes intersecting at right-angle. There are 7 crystallographic groups (called the holohedries) which satisfy these conditions. These groups are the full invariance groups of possible Bravais lattices.

As usual, a partial order can be defined among the 7 holohedries by means of the group-subgroup relationship. The symmetry of a crystal will be said to pertain to a given <u>crystal system</u> (one also uses the term syngony) defined by $\hat{H}$, if $\hat{H}$ is the smallest holohedry containing the point-group $\hat{G}$ of the crystal. The list of the 7 crystal systems, of the symbols for their defining holohedries, and of the orders of these groups, is indicated on table 2.

TABLE 2

SYSTEM	{Cubic	tetragonal	hexagonal	orthorhombic	trigonal	monoclinic	triclinic
$\hat{H}$	O_h	D_{4h}	D_{6h}	D_{2h}	D_{3d}	C_{2h}	C_i
order	48	16	24	8	12	4	2
Bravais lattices	P, I, F	P, I	P	P, C, I, F	R	P, C	P

e) Enumeration of unequivalent Bravais lattices

It is desirable to define an equivalence between Bravais lattices in order to express the identity of their shapes. Thus two "square" lattices only differing in the size of the squares should be considered as equivalent. This equivalence will permit to identify the Bravais lattices associated to the same crystal at different temperatures (thus discarding the variations induced by the thermal expansion).

More generally, two Bravais lattices are equivalent if : i) they have the same invariance group $\hat{H}$; ii) if one can be transformed into the other by continuous deformation of the lattice, without reaching, at any stage of the deformation, a lattice of lower symmetry than $\hat{H}$.

Hence, different holohedries $\hat{H}$ will define unequivalent Bravais lattices. However, the specification of $\hat{H}$ does not always define completely, up to an equivalence, a Bravais lattice. An enumeration of unequivalent lattices compatible with the seven holohedries $\hat{H}$, shows that there exist 14 unequivalent types of lattices. Their symbols are listed on the last line of table 2.

This restriction on the number of unequivalent of Bravais lattices can be expressed by saying that there are 14 distinct types of translation groups $\mathfrak{T}$ which leave invariant a crystalline structure.

2.1.3. Restrictions on non-primitive translations. Enumeration of space-groups. Notations

If G contains screw-axes or glide planes $\{R|\vec{t}\}$, the repeated applications of these geometrical transformations must belong to G. This requirement can be shown to restrict the possible vectors $\vec{t}$ defining the non-primitive translations involved in a screw-axis or glide-plane. If R is a rotation C_m^p (cf table 1), $\vec{t}$ can only be of the form $(\ell\vec{T}/m)$ with $\vec{T}$ a primitive translation (2.1), and ℓ an integer $(0 < \ell < m)$. If R is a plane-symmetry, one has $\vec{t} = \vec{T}/2$.

In summary the elements $\{R|\vec{t}\}$ composing a space-group-G comply with a 3-fold set of restrictions : i) the point-symmetry elements form one of the 32 crystallographic point-groups. ii) the primitive translations $\vec{T}$ constitute, up to an equivalence, one of the 14 translation groups represented by the 14 Bravais lattice. iii) the non-primitive translations associated to a point-symmetry element can be equal to one of a few specified fractions of the primitive translations.

As a consequence, the number of distinct space-groups, i.e. differing by either of the characteristics i) to iii) herebefore is limited. 230 space-groups are listed in refs (1-4). Some of the groups listed have in common the three features i) - iii), but the relative orientations of the Bravais lattice basic vectors, and of the elements of the point-group are different in the distinct groups. 72 among the 230 groups are symmorphic, while the other ones contain screw-axes and/or glide-planes.

Two different types of symbols are usually used, to identify space-groups. The first one uses the Schoenflies symbol for the point-group $\hat{G}$ and an additional number referring to a standard list of the space-groups having in common this point-group [1-6] : O_h^6 is the 6th space-group with point-group O_h. The international symbol shows explicitely the type of Bravais lattice, the point-group, and the possible occurence of glide-planes or screw-axes : P4mm has a P Bravais lattice, the point-group denoted 4mm [1-6], and no glide-planes or screw-axes (the symmetry planes are denoted m). P4bm has the same characteristics but one : the symbol b refers to a glide-plane with non-primitive translation parallel to the b-axis ($\vec{a}_2$) : this group is non-symmorphic while the former one is symmorphic. The drawback of the international symbol is that, depending of their absolute orientation, equivalent space-groups can be represented by different symbols [4].

2.1.4. Unit-cells

A primitive unit-cell of a crystal is the smallest volume which, displaced by the translations $\vec{T} \in \mathfrak{T}$, allows an exact filling of space. Clearly, unit cells can be found by considering the Bravais lattice of the crystal. An obvious example of a primitive unit-cell is provided by the parallelepiped constructed on the basic vectors ($\vec{a}_1$, $\vec{a}_2$, $\vec{a}_3$). In general, several primitive unit-cells exist, all having the same volume. One of them, the Wigner-Seitz cell, displays the same symmetry as the entire Bravais lattice. This cell coincides with the one constructed on ($\vec{a}_1$, $\vec{a}_2$, $\vec{a}_3$) for P-Bravais lattices (see table 1) while these primitive unit-cells are different for other types of Bravais lattices. For the latter lattices, it is customary to define a "multiple-unit-cell", whose volume is a multiple of the volume of the primitive-one, and which, like the Wigner-Seitz cell, displays the full symmetry of the lattice. An illustration of these various unit-cells is shown on figure 2.

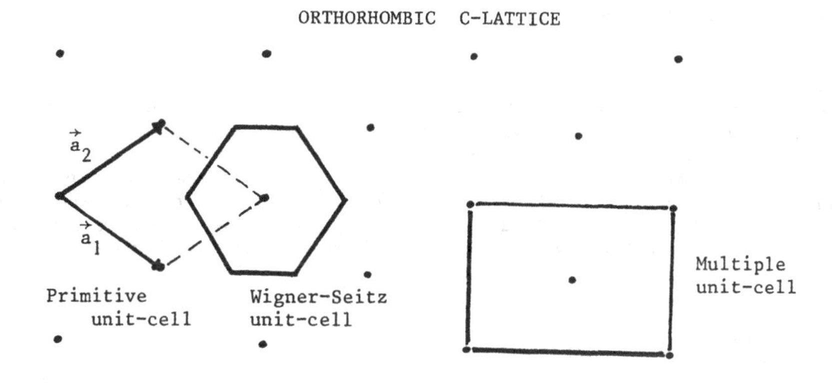

Fig. 2

2.2. Irreducible representations of the space-groups

Space-groups are discrete infinite groups (since they contain the discrete infinite group $\mathfrak{T}$ of translations) possessing non-commuting elements (Cf. eq. 2.5). They possess an infinite set of irreducible representations which cannot be tabulated in a concise manner as described for finite groups (chapter II 2.1.2). Actually, the existing tabulations do not provide explicitely the representations, but only elements which are needed to construct these representations according to a well-defined procedure [3, 5, 6]. Let us describe the elements involved in the procedure.

2.2.1. Irreducible representations of the translation-group. Brillouin-zone

The translation group $\mathcal{T}$ is abelian, and, therefore, its irreducible representations (IR) are one-dimensional. Given an arbitrary vector $\vec{k}$, the set of one-dimensional matrices,

$$M(\vec{T}) = e^{-i \vec{k}.\vec{T}} \tag{2.6}$$

clearly constitutes a one-dimensional IR, which can be denoted $\gamma(\vec{k})$, of $\mathcal{T}$. However 2 distinct vectors $\vec{k}_1$ and $\vec{k}_2$ do not necessarily define unequivalent IR's of $\mathcal{T}$. It can easily be checked that $\gamma(\vec{k}_1)$ is equivalent to $\gamma(\vec{k}_2)$ if, and only if, one has :

$$\vec{k}_1 - \vec{k}_2 = \vec{K} = m_1 \vec{a}_1^* + m_2 \vec{a}_2^* + m_3 \vec{a}_3^* \tag{2.7}$$

(m_i : positive, negative, or zero integers), with :

$$\vec{a}_1^* = \frac{2\pi\, \vec{a}_2 \wedge \vec{a}_3}{(\vec{a}_1, \vec{a}_2, \vec{a}_3)} \qquad \vec{a}_2^* = \frac{2\pi\, \vec{a}_3 \wedge \vec{a}_1}{(\vec{a}_1, \vec{a}_2, \vec{a}_3)} \qquad \vec{a}_3^* = \frac{2\pi\, \vec{a}_1 \wedge \vec{a}_2}{(\vec{a}_1, \vec{a}_2, \vec{a}_3)} \tag{2.8}$$

The $\vec{a}_i^*$ are the reciprocal-basic-vectors. The lattice constructed on the $\vec{K}$-vectors (2.7), starting from an arbitrary origin, is the reciprocal Bravais lattice (or more usually <u>the reciprocal lattice</u>) associated to $\mathcal{T}$. It has the same invariance group $\bar{H}$ as the Bravais lattice of $\mathcal{T}$, but it is generally unequivalent to it.

Eq. (2.7) shows that $\vec{k}$ vectors contained in <u>one unit-cell</u> of the recriprocal lattice allow to index all the unequivalent IR's of $\mathcal{T}$. Besides, two $\vec{k}$ vectors <u>inside</u> a unit cell correspond necessarity to unequivalent IR's. By contrast, several vectors of the surface limiting the unit-cell can correspond to equivalent IR's (fig. 3).

Any of the various possible types of unit-cells has this property, and, in particular, the Wigner-Seitz cell of the reciprocal lattice. The latter cell is called the <u>Brillouin-zone</u> (fig. 2). The preceding considerations show that the $\vec{k}$ vectors of the Brillouin-zone constitute a

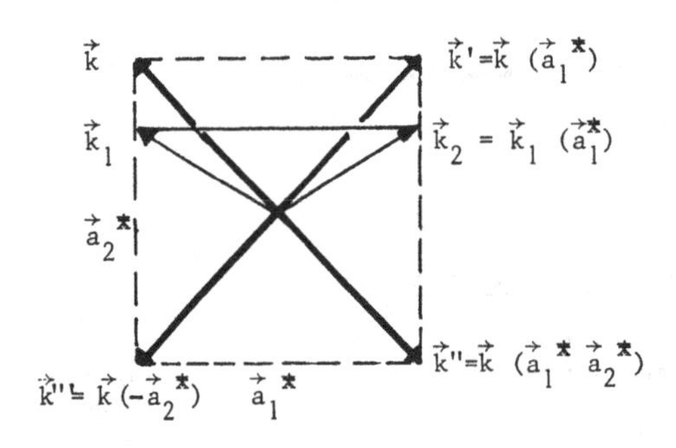

Fig. 3

geometrical picture of the set of irreducible representations of the translation group $\mathcal{T}$ of the crystal (with some redundancy when one takes into account all the limiting surfaces of this unit-cell).

One can set a character table for the infinite number of IR's of $\mathcal{T}$, in a similar manner as shown in table 1, chapter II (table 3 hereunder). Here, the group $\mathcal{T}$ is infinite (discrete) and the IR's form a continuous set, indexed by a $\vec{k}$ vector of the Brillouin-zone. The totally-symmetric IR of $\mathcal{T}$, which has all its characters equal to 1, corresponds to $\vec{k} = 0$.

	0	$\vec{a}_1$	$\vec{a}_2$	$\vec{a}_3$		$\vec{T}(m_1,m_2,m_3)$
$\gamma(0):\Gamma_1$	1	1	1	1		1
$\gamma(\vec{k})$	1	$e^{-i\vec{k}.\vec{a}_1}$	$e^{-i\vec{k}.\vec{a}_2}$	$e^{-i\vec{k}.\vec{a}_3}$	...	$e^{-i\vec{k}.\vec{T}}$

Table 3 .

2.2.2. Group of $\vec{k}$; Star of $\vec{k}$; high-symmetry points.

Let G be a space-group and $\vec{k}_1$ a vector of the Brillouin-zone (BZ). If we apply to $\vec{k}_1$ the transformations belonging to the point-group $\hat{G}$ (referred to the center of the BZ), this vector is transformed into vectors $\vec{k}_2 \ldots \vec{k}_d$ (d is the order of $\hat{G}$) of same modulus, and all belonging to the BZ.

a) point-group $\hat{G}(\vec{k})$ and space-group G $(\vec{k})$.

Some of these vectors represent the same IR of $\mathcal{T}$ as $\vec{k}_1$ (i.e. they satisfy eq. 2.7) : these vectors can be considered as equivalent to $\vec{k}_1$. Their existence implies that certain elements of $\hat{G}$, forming a subgroup of $\hat{G}$, transform $\vec{k}_1$ into an equivalent vector. This subgroup, which we denote $\hat{G}(\vec{k}_1)$, is called the point-group of the vector $\vec{k}_1$, or the little-point-group of $\vec{k}_1$ (the meaning of $\hat{G}(\vec{k}_1)$ is broader than that of the isotropy-groups defined in chap.II, § 4, since, here, the invariance of $\vec{k}_1$ is up to an equivalence).

Starting from each element R of $\hat{G}(\vec{k}_1)$, we can consider all the elements $\{R|\vec{t}\}$ of G which involve R as a point-symmetry element. The set of elements $\{R|\vec{t}\}$ associated to all the elements R of $\hat{G}(\vec{k}_1)$ constitute a subgroup $G(\vec{k}_1)$ of G. This subgroup contains the group of translations $\mathcal{T}$ (which is associated to the element R = E of $\hat{G}(\vec{k}_1)$). Thus $G(\vec{k}_1)$ is one of the 230 space-groups mentioned in § 2.1.3. It is called the space-group of the vector $\vec{k}_1$, or, more concisely, the little group of $\vec{k}_1$.

b) Star of $\vec{k}$.

The set of unequivalent vectors in $(\vec{k}_1,\ldots\vec{k}_d)$, which are generated by the application of $\hat{G}$, constitute the star of $\vec{k}_1$ which we denote k_1^*. Each $\vec{k}_i$ is an arm of k_1^* (the concept of star is closely related to the concept of orbit defined in chap.II § 4. However, here again, the two concepts differ by the consideration of the equivalence between $\vec{k}$ vectors generated by the action of $\hat{G}$ on $\vec{k}_1$). The number of arms in the star of $\vec{k}_1$ is equal to the index of the subgroup $\hat{G}(\vec{k}_1)$ with respect to the group $\hat{G}$. Note that the same star is generated by the application of $\hat{G}$ on any arm of k_1^*. Similarly to the property stated in chapter II § 4 for the points of an orbit, the groups $\hat{G}(\vec{k}_i)$ of the different arms $\vec{k}_i$ are conjugated in $\hat{G}$, to the group $\hat{G}(\vec{k}_1)$. Likewise $G(\vec{k}_i)$ and $G(\vec{k}_1)$ are conjugated in G.

Fig.4, hereunder, illustrates the definitions of $\hat{G}(\vec{k}_1)$ and of k_1^*, on, the example of vectors belonging to the BZ of a simple (P)-tetragonal reciprocal lattice.

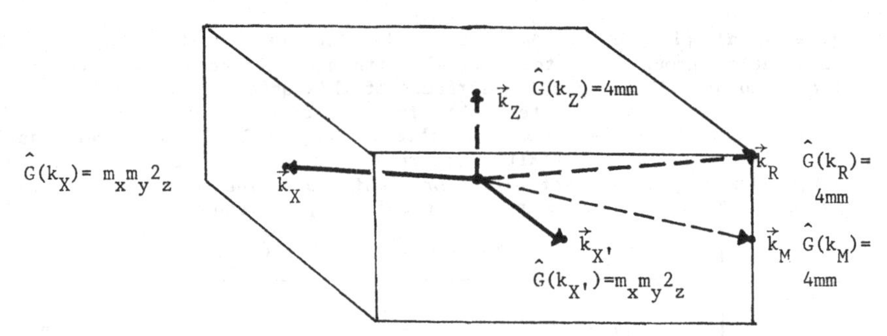

Fig. 4. Stars and little groups for some vectors in the BZ of the space-group P4mm. The stars relative to $\vec{k}_Z$, $\vec{k}_M$, and $\vec{k}_R$ are reduced to a single vector. $\vec{k}_X$ and $\vec{k}_{X'}$ form a two arm star.

c) high-symmetry points

Consider a point A of the BZ, and $\vec{k}_o$ the vector of the reciprocal space, joining the BZ center to A. Similarly, let $\vec{k}$ be the vector associated to a point located in the neighborhood of A. If $\hat{G}(\vec{k}) \subset \hat{G}(\vec{k}_o)$ (strict relationship), for any point in the neighborhood of A, then A is a high-symmetry point of the BZ.

Likewise, a line or a plane of high-symmetry are composed of points associated to $\hat{G}(\vec{k})$, satisfying $\hat{G}(\vec{k})=\hat{G}(\vec{k}_o)$, while any point located in the neighborhood of the line, or plane (but outside it) corresponds to $\hat{G}(\vec{k}) \subset \hat{G}(\vec{k}_o)$.

Clearly, a point which does not belong to a symmetry manifold (point, line, or plane) corresponds to a general direction in reciprocal space and has $\hat{G}(\vec{k})$ reduced to the identity E. In agreement with § b, the corresponding vector $\vec{k}$ defines a star k^* whose number of arms equals the order of the point-group $\hat{G}$ of the crystal.

Customarily, high-symmetry points, and points on high-symmetry manifold are labelled by letters (e.g. A, Z or Λ). Some variations occur in this labelling, according to the reference used [1-6]. There is, however, agreement on the labelling of the center of the BZ ($\vec{k}$ = o) by the letter Γ. The Γ-point is a high-symmetry point : its group $\hat{G}(o)$ coincides with the point-group $\hat{G}$ of the crystal while any neighbouring non-zero vector has a smaller $\hat{G}(\vec{k})$ group. Clearly, the star k^* is reduced to the single arm $\vec{k}$ = o.

2.2.3. Small representation of $G(\vec{k}_1)$; weighted representation of $G(\vec{k}_1)$

The little-group $G(\vec{k}_1)$ has irreducible representations which play an essential role in the construction of the IR's of the entire space-group G.

Though $G(\vec{k}_1)$ is like G, a space-group, its IR's are simpler than those of G, because it preserves, up to an equivalence, the vector $\vec{k}_1$ of the Brillouin-zone. Its IR's associated to $\vec{k}_1$ are in finite number.

These IR's of $G(\vec{k}_1)$ are called the small-representations, and they can be denoted $\tau_i(\vec{k}_1)$ [i^{th} IR of G $(\vec{k}_1)$]. We will also use the abreviated notation τ_i. In that case, the matrices of τ_i for the elements of $G(\vec{k}_1)$ will be denoted :

$$\tau_i(\{R|\vec{t}\}) \tag{2.9}$$

Starting from the preceding matrices, we can define another set of matrices [1] :

$$\tau_i'(R) = \tau_i(\{R|\vec{t}\}).e^{+i.\vec{k}_1.\vec{t}} \tag{2.10}$$

With respect to the nature of τ_i',3 different situations can be distinguished:

(1) $\vec{k}_1$ is a vector whose extremity lies inside the BZ (excluding its limiting surfaces). $\tau_i'(R)$ coincides with the set of matrices of an IR of the finite point-group $\hat{G}(\vec{k}_1)$. Equation (2.10) then provides us straight-forwardly with the matrices of the small representation $\tau_i(\{R|\vec{t}\})$, deduced from the IR's of $\hat{G}(\vec{k}_1)$.

(2) $\vec{k}_1$ is any vector of the BZ (including the limiting surfaces), but the group G is symmorphic (cf. § 2.1.2.b). The same rule as in case (1) applies, with the peculiarity that only primitive translations $\vec{T}$ are involved in the elements of G.

(3) $\vec{k}_1$ is a vector whose extremity lies on the limiting surfaces of the BZ, and the space-group G is non-symmorphic. In this case τ'_i is not an IR of $\hat{G}(\vec{k}_1)$ in the usual sense. It is a so-called "weighted"-or "loaded"-irreducible representation of the finite point-group $\hat{G}(k_1)$. A weighted IR, unlike an ordinary IR, does not only depend on the structure of the point-group $\hat{G}(k_1)$ but also on the $\vec{k}_1$ vector considered, and on the nature of the translations $\vec{t}$ existing in the space-group G. The number of weighted IR's associated to a given $\hat{G}$ and $\vec{k}_1$, is finite. These weighted IR's have been tabulated for all the space-groups and relevant $\vec{k}_1$-vectors [6]. From these tables, using eq. (2.10), one can easily construct the small-representation τ_i associated to each weighted IR τ'_i.

2.2.4. Construction of the irreducible representations of G

A space-group G has an infinite number of irreducible representations. A given representation can be specified by two elements : i) a star k^*, generated by applying $\hat{G}$ to a given vector, $\vec{k}$, of the BZ. Unequivalent stars will define unequivalent IR's of G. ii) one of the small representations $\tau_i(\vec{k})$ of the group $G(\vec{k})$. Unequivalent τ_i define unequivalent IR's of G. Hence an IR of G can be denoted $\Gamma_i(\vec{k})$. There is an infinite set of $\vec{k}$-vectors in the BZ , and, for each $\vec{k}$-vector, a finite set of possible values for the index i .

$\Gamma_i(\vec{k})$ is an IR of dimension r.s ,where r is the dimension of the small - representation τ_i ,and s is the number of arms in the star k^* associated to $\vec{k}$.

a) standard basis of the representation $\Gamma_i(\vec{k})$

Assume that the carrier space of $\tau_i(\vec{k})$ is $\mathcal{E}(\vec{k})$, spanned by vectors $\phi_1(\vec{k})\ldots\phi_r(\vec{k})$. Let, on the other hand, $g_p = \{R_p | t_p\}$ be s elements of the space-group G, chosen in such a way that the p-th vector in the star k* is :

$$\vec{k}_p = \hat{g}_p.\vec{k} = R_p.\vec{k} \qquad (2.11)$$

A standard basis of r x s vectors for the representation $\Gamma_i(\vec{k})$ is provided by the set of vectors :

$$\phi_1(\vec{k}_p) = g_p.\phi_1(\vec{k}) \qquad (2.12)$$

In (2.11) and (2.12) it is implicity assumed that $\vec{k}_1 = \vec{k}$. Note that several standard bases exist, because the choice of the set of elements g_p is not unique.

b) Construction of the matrices of $\Gamma_i(\vec{k})$ in the standard basis

The vectors of the above basis can be ordered as follows :

$$\phi_1(\vec{k})\ldots\ \phi_r(\vec{k})\ ;\ \phi_1(\vec{k}_2)\ldots\ \phi_r(\vec{k}_2),\ldots\ \phi_1(\vec{k}_s)\ldots\ \phi_r(\vec{k}_s) \qquad (2.13)$$

In that case, the matrices $M(\{R|\vec{t}\})$ of the representation $\Gamma_i(\vec{k})$ can be considered as formed of s x s blocks, each block being related to a pair $(\vec{k}_m, \vec{k}_n)$ of vectors of k*. Let $\mu(\vec{k}_m, \vec{k}_n|\{R|\vec{t}\})$ be the r x r matrix constituting such a block. It can be shown [1,2,5,6] that :

$$\begin{aligned} &\mu(\vec{k}_m, \vec{k}_n|\{R|\vec{t}\}) = 0 \text{ if } R.\vec{k}_m \neq \vec{k}_n \\ &\mu(\vec{k}_m, \vec{k}_n|\{R|\vec{t}\}) = \tau_i\ [g_n^{-1}.\{R|\vec{t}\}.g_m] \text{ if } R\vec{k}_m = \vec{k}_n \end{aligned} \qquad (2.14)$$

where g_m and g_n are defined by (2.11) and τ_i is the matrix of the <u>small-representation</u> corresponding to the element between brackets (which is an element of $G(\vec{k})$). Figure 5 hereunder summarizes the characteristics of a representation $\Gamma_i(\vec{k})$.

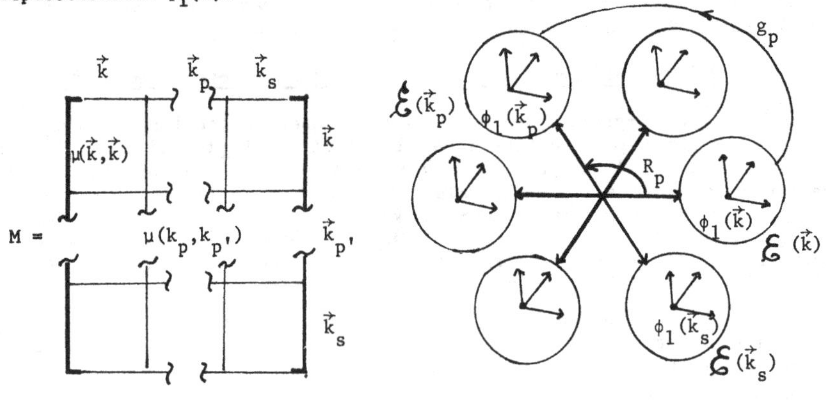

Fig. 5

c) Summary

To summarize, $\Gamma_i(\vec{k})$ is defined by the specification of $\vec{k} = \vec{k}_1$, and of the small-representation τ_i of $G(\vec{k}_1)$. In order to determine the matrix $M(\{R|\vec{t}\})$ associated to an arbitrary element $\{R|\vec{t}\}$ of G, the following sequence of operations must be performed :

1) Consider the vector $\vec{k}_1$ and find the group $\hat{G}(\vec{k}_1) \subset \hat{G}$ which leaves $\vec{k}_1$ invariant, up to an equivalence.

2) Determine the star k* associated to $\vec{k}_1$. In this view, apply to $\vec{k}_1$ the elements of $\hat{G}$ not belonging to $\hat{G}(\vec{k}_1)$ and determine the s unequivalent vectors $\vec{k}_p$ generated in this manner. Check that s is equal to the index of $\hat{G}(\vec{k}_1)$ in $\hat{G}$.

3) Choose a set $(E, R_2, \ldots R_s)$ of elements of $\hat{G}$ satisfying $\vec{k}_p = R_p \vec{k}_1$. For each R_p, choose an element $g_p = \{R_p|\vec{t}_p\}$ of the space-group G.

4) Find the form of the matrices of the small representation τ_i of $G(\vec{k}_1)$. The realization of this stage of the procedure depends if the situation considered pertains to cases (1) (2) or (3) in § 2.2.3. In cases (1) or (2), (i.e. for symmorphic groups G, or for $\vec{k}_1$ corresponding to a point inside the BZ), the matrices τ_i are obtained through eq. (2.10), from the matrices τ'_i of an IR of the point-group $\hat{G}(\vec{k}_1)$. The latter group being one of the 32-crystallographic groups, its IR's are tabulated in a number of text-books [1-3, 5, 6]. In case (3), τ_i is still obtained through equ. (2.10), but, this time, τ'_i is a weighted-representation of $\hat{G}$ The matrices of these representations have been tabulated by Kovalev [5] .

If, for certain problems, the knowledge of the characters of the small-representation τ_i is sufficient, one can refer to the tables by Zak and al [6] or by Miller and Love [7] where these characters are directly provided (consideration of the τ'_i in then superfluous).

5) Construct the matrix of the IR of $G, (\Gamma_i(\vec{k}))$, for any element $\{R|\vec{t}\}$ of G, by determining the blocks $\mu(\vec{k}_m, \vec{k}_n|\{R|\vec{t}\})$ constituting this matrix, through eq. (2.14).

d) Remarks

If one considers the subgroup $\mathcal{T}$ of translations, eq. (2.14) shows that the matrices of the elements $\{E|\vec{T}\}$ will only contain diagonal blocks $\mu(\vec{k}_m, \vec{k}_m)$. Besides, these blocks will have a specified form. Indeed, we can write :

$$\mu(\vec{k}_m, \vec{k}_m|\{E|\vec{T}\}) = \tau_i[\{E|R_m^{-1}\vec{T}\}] \tag{2.15}$$

Using eq. (2.10), we have :

$$\tau_i[\{E|R_m^{-1}\vec{T}\}] = e^{-i\vec{k}_1\cdot(R_m^{-1}\vec{T})}.\tau'(E) = e^{-i\vec{k}_m\cdot\vec{T}}.\tau'(E) \quad , \tag{2.16}$$

since the scalar product $\vec{k}_1.(R_m^{-1}\vec{T})$ is invariant by application of R_m, and that $\vec{k}_m = R_m.\vec{k}_1$. As $\tau'(E)$ is the r-dimensional identity matrix (where r is the dimension of the small representation), we see that the matrix of $\{E|\vec{T}\}$ is a diagonal matrix constituted by s sets or r elements $e^{-i\vec{k}_m\cdot\vec{T}}$. This form reveals that, with respect to the subgroup $\mathcal{T}$ of G, $\Gamma_i(\vec{k})$ is a reducible representation involving r-times each of the s unequivalent representations of $\mathcal{T}$ defined by the s arms of the star k*.

On the other hand, if we consider the IR's of G associated to $\vec{k}$ = o, we can note that as $\hat{G}(\vec{k}) = \hat{G}$, the star k* is reduced to the single vector $\vec{k}$ = o.

The matrices of $\Gamma_i(\vec{k})$ coincide with those of the small-representation. Eq. (2.10), with $\vec{k}_1$ = o, shows that τ_i coincides with τ'_i which is merely an IR of the point-group $\hat{G}$ ($\vec{k}$ = o is inside the BZ).

Hence, the matrices of the IR's of a space-group with $\vec{k}$ = o, coincide with the matrices of the IR's of its point-group $\hat{G}$ (all the translations $\{E|\vec{T}\}$ are represented by the identity matrix).

An example of the use of the procedure described in § c hereabove will be described in § 7.

3. THE LIFSCHITZ CRITERION

In chapter II § 3.5 we have seen that the order-parameter (OP) of the considered transition can be distinguished from the other irreducible degrees of freedom by the fact that the coefficient α multiplying the quadratic polynomial invariant ($\Sigma\, \eta_{kr}^2$) of the OP components, vanishes at T_c, while the similar coefficients associated to the other degrees of freedom are strictly positive at T_c. Hence, in the neighborhood of T_c, α is the smallest of the coefficients α_k (eq. 3.7, chapter II).

In the case of a structural transition, Eq. (3.7) of chapter II must be expressed differently. Indeed, since the high-symmetry group G_o is a crystallographic space-group, the set of its irreducible representations Γ_i ($\vec{k}$) is a continuous set, where $\vec{k}$ is any vector of the Brillouin-zone. The discrete sum (chap. II eq. 3.7) over the IR's of G_o must therefore be replaced by a continuous sum over the volume of the BZ, as well as by a discrete summation over the small representations τ_i of G ($\vec{k}$) [actually the continuous summation is over distinct stars k*]. Eq. (3.7) chapter II is replaced by :

$$F = F_o + \int_{BZ} \{ \sum_i \alpha_i [\vec{k}, T, P, \rho_o(\vec{r}), \phi(\vec{k}', 1'|\vec{r})] \, [\sum_1^2 \eta_1^2(\vec{k}, \tau_i)] \} \, d\vec{k} \qquad (3.1)$$

The normalized functions $\phi(\vec{k}, 1|\vec{r})$ carry the n-dimensional IR $[\Gamma_i(\vec{k})]$ of G_o. The coefficients $\alpha_i(\vec{k})$ constitute "branches", each branch being associated to a given index i, and a continuouly varying vector $\vec{k}$.

Adapting the definition of the OP to this "continuous" situation, we can say that the OP is the irreducible set η_1 ($\vec{k}$) corresponding to the minimum, with respect to $\vec{k}$, of the lowest branch α_i ($\vec{k}$) in the vicinity of T_c, this minimum being equal to zero at T_c (see figure 6).

Symmetry considerations impose to the various branches $\alpha_i(\vec{k})$, the occurence of extrema at certain points of the BZ. If the minimum of α_i ($\vec{k}$) coincides with one of these extrema, the OP will possess a symmetry determined $\vec{k}$-vector, whose location in the BZ will not be submitted to variations when temperature or pressure are varied : the entire line of transitions T_c (P), considered (Cf. § 2.1.c. chapter I) will be described by the same irreducible OP.

By contrast, if the minimum of α_i ($\vec{k}$) does not coincide with a symmetry-extremum, one can have, in principle, a continuous variation of the OP-symmetry with T and P.

In this paragraph, we examine the symmetry requirements imposed to the OP, which imply the occurence of a symmetry determined extremum for the associated coefficient α. These requirements are expressed by the Lifschitz-criterion.

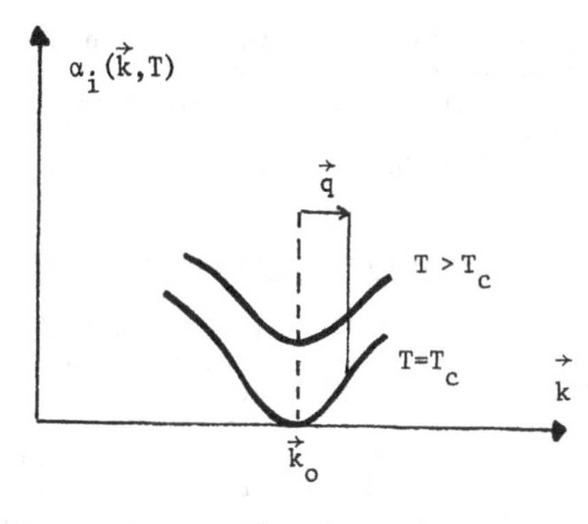

Fig. 6

We will see that the satisfaction of this criterion warrants that the invariance groups G of the possible low-symmetry phases (stable below T_c) are, like G_o, crystallographic space-groups. We will restrict, in the rest of the chapter, to the structural transitions complying with this criterion. The case of an OP not complying with the Lifschitz criterion will be examined in chapter V.

3.1. Derivation of the Lifschitz-criterion

The initial derivation of the relevant symmetry criterion has been performed by Lifschitz [8]. This derivation, which considers spatially modulated components $\eta(\vec{r})$ of the OP, is not rigorous from the standpoint of group-theory [the OP components being, by definition scalar quantities]. However, the concept of a modulated OP plays an essential role in the Landau theory of incommensurate phases. Accordingly, we indicate the principle of Lifchitz's derivation in chapter V, § 2.3.2.
The principle of a different derivation has been described by Dzialoshinskii [9] and by Haas [10], and worked out in detail by Goshen, Mukamel, and Shtrikman [11], and, by Onodera and Tanabe [12]. We follow here the presentation of the latter authors.

3.1.1. Existence of an extremum of α ($\vec{k}$)

Let Γ_i ($\vec{k}_o$) be the irreducible representation of the OP. It is associated, in eq. (3.1) to the coefficient α_i ($\vec{k}_o$). This equation contains degrees of freedom corresponding to $\vec{k}$ = ($\vec{k}_o$ + $\vec{q}$) whose coefficients α_j ($\vec{k}_o$ + $\vec{q}$) tend towards α_i ($\vec{k}_o$) when $|\vec{q}| \to o$. Depending if $\vec{k}_o$ is a general vector of the BZ, or a vector ending at a point of high-symmetry, or on a line or plane of high-symmetry (Cf § 2.2.2), there can be one or several branches meeting at $\vec{k}_o$ for a given direction $\vec{q}$. For each branch j and direction $\vec{q}$ we can write :

$$\alpha_j\ (\vec{k}_o + \vec{q}) = \alpha_i\ (\vec{k}_o) + \sum_m \left(\frac{\partial \alpha_j}{\partial q_m}\right|_{\vec{q}} = o).q_m + \dots \qquad (3.2)$$

Clearly, if α_i $(\vec{k}_o)$ is an extremum, the term in the second member of (3.2), which is linear in the q_m, must vanish. If α_i $(\vec{k}_o)$ is a symmetry-determined extremum of α_j $(\vec{k})$, then the presence of this linear term must be incompatible with the symmetry of the OP.

Let us obtain the group theoretical condition expressing this incompatibility.

3.1.2. Functional form of the Landau free-energy

As apparent from eq. (3.1), the coefficients α_j $(\vec{k})$ are functionals of the density increment $\delta\rho$ $(\vec{r})$, or of the expression of this increment in terms of the irreducible functions $\phi(\vec{k}, 1|\vec{r})$. It is through their dependence on the $\phi(\vec{k}, 1|\vec{r})$ that the coefficients α_j $(\vec{k})$ depend on $\vec{k}$. Hence in order to determine this $\vec{k}$-dependence in the vicinity of $\vec{k}$, and specify the terms in eq. (3.2), we must work out the function form of the $\alpha_j(\vec{k})$. In this view, we write the second-degree expansion of F as a function of $\delta\rho$ $(\vec{r})$. Omitting the explicit indication of the temperature and pressure, we have.

$$F(\rho_o + \delta\rho) = F(\rho_o) + \int h(\vec{r}, \vec{r}').\delta\rho(\vec{r})\ \delta\rho(\vec{r}')\ d\vec{r}\ d\vec{r}' + \ldots \qquad (3.3)$$

The term, linear in $\delta\rho(\vec{r})$, in the expansion of F with respect to $\delta\rho(\vec{r})$ is absent for the reasons stated in § 3.4.2 of chapter II (symmetry breaking character of $\delta\rho(\vec{r})$). In (3.3) the function $h(\vec{r}, \vec{r}')$ is a coordinate-dependent "coefficient" of the "quadratic term" in this functional expansion of F. As it plays the same role as the second-derivative of F in a Taylor expansion $[$i.e. $h = (1/2)[\partial^2 F/\partial\delta\rho(\vec{r}).\partial\delta\rho(\vec{r}')]$ we have :

$$h(\vec{r}, \vec{r}') = h(\vec{r}', \vec{r}) \qquad (3.3')$$

3.1.3 Form of ϕ $(\vec{k} + \vec{q}, 1)$

Let us assume that the functions $\phi(\vec{k}, 1|r)$ constitute a standard basis (§ 2.2.4.a) of the irreducible representation $\Gamma_i(\vec{k}_o)$ of the OP, with $\vec{k}$ a vector of the star k_o^*, and 1 running over the basis of the small representation τ_i. If we consider now the set of IR's whose coefficients $\alpha_j(\vec{k}_o + \vec{q})$ tend towards $\alpha_i(\vec{k}_o)$, when $|\vec{q}| \rightarrow o$, we can find a set of functions $\phi(\vec{k}+\vec{q}, 1|\vec{r})$ spanning these representations and tending towards $\phi(\vec{k}, 1|\vec{r})$ when $|\vec{q}| \rightarrow o$. Under the action of a primitive translation $\vec{T}$ of G_o, we have :

$$\begin{aligned} \phi(\vec{k}, 1|r) &\rightarrow e^{-i\vec{k}.\vec{T}}\ \phi(\vec{k}, 1|r) \\ \phi(\vec{k} + \vec{q}, 1|\vec{r}) &\rightarrow e^{-i(\vec{k}+\vec{q})\vec{T}}.\ \phi(\vec{k} + \vec{q}, 1|\vec{r}) \end{aligned} \qquad (3.4)$$

These transformation properties suggest to take the normalized functions $\phi(k + q, 1|r)$ in the form :

$$\phi(\vec{k} + \vec{q}, 1|\vec{r}) = e^{-i\vec{q}.\vec{r}}.\ \phi(\vec{k}, 1|\vec{r}) \qquad (3.5)$$

A translation $\vec{T}$ of the system changes $\vec{r}$ into $(\vec{r} + \vec{T})]$. Equation (3.4) shows that, in the basis $\phi(\vec{k}, 1|\vec{r})$, the matrices of the primitive translations $\vec{T}$ are diagonal matrices with elements exp $(-i\vec{k}.\vec{T})$, $\vec{k}$ being a vector of k_o^*. Though, in general, these elements are complex numbers, the matrices can be equivalent to a real matrix : this requires, in particular, that k_o^* contains both the vector $(+\vec{k})$ and $(-\vec{k})$. Let us place ourselves in the case where $\Gamma_i(\vec{k}_o)$ is a real IR, a circumstance which implies the conditions $(\pm \vec{k}) \in k_0^*$.

Taking the complex conjugate of the two members in eq. (3.4) we see that $\overline{\phi}(\vec{k}, 1|\vec{r})$ is a function associated to the vector $(-\vec{k})$. This function can therefore be expressed as a linear combination of the $\phi(-\vec{k}, 1'|\vec{r})$.

3.1.4. Form of the density increment

In order to determine $\alpha_j(\vec{k}_0 + \vec{q})$ by means of eq. (3.3), we express the density increment $\delta\rho(\vec{r})$ as a linear combination of the functions $\phi(\vec{k} + \vec{q}, 1|\vec{r})$ with $|\vec{q}|$ small ($|\vec{q}| \ll |\vec{a}_1^*|$). Particularizing eq. (3.4) of chapter II, we have :

$$\delta\rho\ (\vec{r}) = \sum_{\vec{k} \in k_0^*} \sum_{1} \int_{|\vec{q}| \ll |\vec{a}_1^*|} \eta_1(\vec{k}+\vec{q}) . \phi(\vec{k}+\vec{q},\ 1|r) . d\vec{q} \qquad (3.6)$$

where a linear relation exists between $\eta_1(\vec{k}+\vec{q})$ and the $\eta_{1'}(-\vec{k}-\vec{q})$, in order to ensure the reality of $\delta\ (\vec{r})$ [since $\phi(-\vec{k}-\vec{q},\ 1|\vec{r})$ is related to the complex conjugate of $\phi(\vec{k}+\vec{q},\ 1|\vec{r})$].

3.1.5. Functional form of $\alpha_1(\vec{k}+\vec{q})$

The free-energy (3.3) can be specified by reporting the density increment (3.6) into (3.3) and keeping the terms invariant by the space-group G_0. In particular, these terms must be translationally invariant. This requirement imposes the form :

$$F_0 = \sum_{\vec{k},1,1'} \int \eta_1(\vec{k}+\vec{q})\eta_{1'}(-\vec{k}-\vec{q}) \{ \int h(\vec{r},\vec{r}')\phi(\vec{k}+\vec{q},1|\vec{r})\phi(-\vec{k}-\vec{q},1'|\vec{r})d\vec{r}.d\vec{r}' \} . d\vec{q} \qquad (3.7)$$

Comparison of (3.7) with (3.1) shows that these two expression are quadratic forms of $\eta_1(\vec{k})$, eq. (3.1) being in the diagonal form and (3.7) in the general form. The coefficients $\alpha_1(\vec{k}+\vec{q})$ are the eigenvalues of the matrix $A(\vec{k}+\vec{q})$ whose elements are :

$$A_{11'}(\vec{k}+\vec{q}) = \int h(\vec{r},\vec{r}') . \phi(\vec{k}+\vec{q},\ 1|\vec{r}) . \phi(-\vec{k}-\vec{q},\ 1'|\vec{r})\ d\vec{r} . d\vec{r}' \qquad (3.8)$$

Using the expression (3.5) of $\phi(\vec{k}+\vec{q},\ 1|\vec{r})$, and expanding (3.5) to first order in $\vec{q}$, we obtain :

$$A_{11'}(\vec{k}+\vec{q}) = A_{11'}(\vec{k}) + \vec{q} . \vec{B}_{11'}(\vec{k}) \qquad (3.8')$$

with,

$$A_{11'}(\vec{k}) = \int h(\vec{r},\ \vec{r}') . \phi(\vec{k},\ 1|\vec{r}) . \phi(-\vec{k},\ 1'|\vec{r})\ d\vec{r} . d\vec{r}' \qquad (3.9)$$

and

$$\vec{B}_{11'}(\vec{k}) = i \int (\vec{r}-\vec{r}')\ h(\vec{r},\ \vec{r}') . \phi(\vec{k},\ 1|\vec{r}) . \phi(-\vec{k},\ 1'|\vec{r})\ d\vec{r} . d\vec{r}' \qquad (3.9')$$

Using the symmetry of $h(\vec{r},\ \vec{r}')$ (eq. 3.3'), we can write :

$$A_{11'}(\vec{k}) = \frac{1}{2} \int [\phi(\vec{k},1|\vec{r}) . \phi(-\vec{k},1'|\vec{r}') + \phi(-\vec{k},1'|\vec{r}) . \phi(\vec{k},1|\vec{r}')] h(\vec{r},\vec{r}') . d\vec{r} . d\vec{r}' \qquad (3.10)$$

$$0 = \int \{\phi(\vec{k},1|\vec{r})\cdot\phi(-\vec{k},1'|\vec{r}') - \phi(-\vec{k},1'|\vec{r})\cdot\phi(\vec{k},1|\vec{r}')\}\, h(\vec{r},\vec{r}')\cdot d\vec{r}\cdot d\vec{r}' \quad (3.10')$$

and,

$$\vec{B}_{11'}(k) = -i\int \vec{r}\{\phi(\vec{k},1|\vec{r})\phi(-\vec{k},1'|\vec{r}') - \phi(-\vec{k},1'|\vec{r})\phi(\vec{k},1|\vec{r}')\}\, h(\vec{r},\vec{r}')\cdot d\vec{r}\cdot d\vec{r}' \quad (3.10'')$$

In order that the q-expansion of $\alpha_1(\vec{k}+\vec{q})$ does not contain terms linear in $\vec{q}$, the vectors $\vec{B}_{11'}(\vec{k})$ must vanish. $\alpha_1(\vec{k})$ is the lowest eigenvalue of the matrix $A_{11'}(\vec{k})$ [eqs. (3.9) and (3.10)]. As $\alpha_1(\vec{k})$ does not vanish by symmetry (it is zero, only at T_c), comparison of (3.10) and (3.10''), shows that the required vanishing by symmetry of $\vec{B}_{11'}(\vec{k})$ implies the vanishing of each component of :

$$\vec{r}\{\phi(\vec{k},1|\vec{r})\ \phi(-\vec{k},1'|\vec{r}') - \phi(-\vec{k},1'|\vec{r})\ \phi(\vec{k},1|\vec{r}')\} \quad (3.11)$$

3.1.6. Antisymmetrized square of a representation

In chapter II, § 4.3, we have defined the symmetrized m^{th}-power of a representation Γ of the group G. We can define, similarly, the antisymmetrized m^{th} power of Γ. Using the notations of chapter II, § 4.3.1, we select in $\mathcal{E}^m$ the vectors which are invariant by even permutations of the m indices $(1_1, \ldots 1_m)$, and which are transformed into their opposite by odd permutations. The antisymmetrized subspace $\{\mathcal{E}^m\}$ of $\mathcal{E}^m$ is generated by these vectors. It can be shown that $\{\mathcal{E}^m\}$ carries a representation, denoted $\{\Gamma^m\}$ of G. In particular, the antisymmetrized square $\{\Gamma^2\}$, is spanned by the antisymmetric sets of ordered pairs $\{(\phi_1, \phi_m) - (\phi_m, \phi_1)\}$, where the ϕ_1 span Γ. The characters $\{\chi^2\}(g)$ of $\{\Gamma^2\}$ are equal to [1] :

$$\{\chi^2\}(g) = \frac{1}{2}\,[\chi^2(g) - \chi(g^2)] \quad (3.12)$$

where g are the elements of G, and $\chi(g)$ are the characters of the representation Γ. The knowledge of these characters permits, in the same manner as shown in chapter II, § 4.3.2, the decomposition of $\{\Gamma^2\}$ into irreducible representations of G.

3.1.7. Group-theoretical expression of the Lifschitz-criterion

Let us consider the action of the space-group G_o on the expression (3.11). In studying the transformations properties of the vector $\vec{r}$, it is sufficient to consider the action of the point-group $\hat{G}_o$. Indeed, replacing $\vec{r}$ by $(\vec{r}+\vec{t})$ in eq. (3.10'') will not change the value of this integral, as shown by eq. (3.10'). Under the action of G_o, $\vec{r}$ generates the carrier space of a representation V of $\hat{G}_o$ (which is 3-dimensional).

We have pointed-out in § 2.2.4.d, hereabove, that representations of $\hat{G}_o$ can be considered as representations of the space-group G_o, corresponding to $\vec{k} = o$.

Accordingly, we term V <u>the vector-representation of G_o</u>.

On the other hand, the set of functions between brackets { } in (3.11), clearly belongs to the carrier space of the antisymmetrized-square $\{\Gamma_i(\vec{k}_o)^2\}$ of the OP-representation. The expressions (3.11) transform under the action of G_o as vectors of the space which carries the <u>product</u>

of the two representations V ,and $\{\Gamma_i(\vec{k}_o)^2\}$ (Cf. § 4.2 in chap. II) . The group-theoretical condition for the vanishing of (3.11) is, therefore, that this product does not contain the totally symmetric representation Γ_1 of G_o :

$$V \times \{\Gamma_i(\vec{k}_o)^2\} \not\supset \Gamma_1 \qquad (3.13)$$

The relation (3.13) has also been shown [12] to be valid if $\Gamma_i(\vec{k}_o)$ is not a real IR but a physically irreducible representation, i.e. the sum of 2 complex conjugate IR's.

This relation constitutes the Lifschitz-criterion, applicable to the OP-symmetry. As pointed out in the introduction of §3, this criterion expresses the fact that the coefficient $\alpha_i(\vec{k}_o)$ associated to the quadratic invariant of the OP corresponds to symmetry-determined extremum of the function $\alpha_i(\vec{k})$. Its mathematical meaning is similar to the "maximal isotropy-group condition" discussed in chapter II § 4.6.3.

3.2. Procedure of application of the Lifschitz-criterion

The method of application of condition (3.13) to the IR's $\Gamma_i(k_o)$ has been elaborated by Lyubarskii [1]. This author has shown that one could determine the compliance of an IR with (3.13) on the basis of the knowledge of the characters of the small representation τ_i. Let $\chi_o(g)$ be the character of the small representation τ_i relative to the element $g = \{R|\vec{t}\}$ of $G(\vec{k}_o)$, and V(R) the character of the vector-representation of the group $\hat{G}(\vec{k}_o)$. It is relevant to distinguish 3 cases.

3.2.1. Stars containing $(\vec{k} = -\vec{k})$

If each vector $\vec{k}$ in the star k_o^* satisfies the relation $(-\vec{k}) = \vec{k}$, condition (3.13) is expressed by :

$$\sum_R [\chi_o^2(g) - \chi_o(g^2)] . V(R) = o \qquad (3.13')$$

where the sum is over the elements R of $\hat{G}(k_o)$ [g being any element of $G(\vec{k}_o)$ whose point symmetry element is R]. (3.13') also expresses the condition that the antisymmetrized-square $\{\tau_i^2\}$ of τ_i, and the vector-representation of $G(\vec{k}_o)$ have no IR of $G(\vec{k}_o)$ in common. We have previously seen that if a representation is real and one-dimensional (cf. chap. II, § 4.4) its square τ_i^2 coincides with its symmetrized-square $[\tau_i^2]$. In this case the antisymmetrized-square is reduced to the zero-vector.

Hence, if the small representation τ_i is real, and one-dimensional, and if $(\vec{k} = -\vec{k})$, the Lifschitz criterion is always satisfied by $\Gamma_i(\vec{k}_o)$.

Let us illustrate (3.13') by applying it to 2 examples. In this view,we can note that the characters V(R) of the vector-representation can be specified, once and for all, for each point-symmetry element R involved in point-group [cf. refs. 1, 13].

i) Consider the example of a representation $\Gamma_i(o)$ whose star is reduced to $\vec{k}_o = o$. In that case (§ 2.2.4.d), τ_i coincides with a representation of the point-group $\hat{G}_o$. Let us take, for instance, the 2-dimensional representation of the point-group $\hat{G} = \mathbf{D}_4$ (422).

In eq. (3.13'), we have g = R. Table 4 shows the values of $\chi_o(R)$, $\chi_o(R^2)$, V(R) (which is relative to the transformation properties of V = (x, y, z) by the elements of $\hat{G}_o$), as well as the value of $A(R) = [\chi_o^2(R)-\chi_o(R^2)]V(R)$ appearing in eq. (3.13'). It is easy to check that the sum condition (3.13') is not satisfied by this representation ($\sum A(R) = 16$).

TABLE 4

R	E	C_4	C_4^3	U_z	U_x	U_y	U_{xy}	$U_{\bar{x}y}$
R^2	E	U_z	U_z	E	E	E	E	E
$\chi_o(R)$	2	o	o	-2	o	o	o	o
$\chi_o(R^2)$	2	-2	-2	2	2	2	2	2
V(R)	3	1	1	-1	-1	-1	-1	-1
A(R)	6	2	2	-2	2	2	2	2

The same procedure can be straighforwardly applied to the other IR's of the 32 crystallographic point groups [13, 14]. These representations coincide with the IR's of the space-groups whose star is reduced to $\vec{k}_o$=o. One finds that, in most cases, eq. (3.13') is satisfied, in contrast to the preceding example. The only cases of IR's not complying with the Lifschitz criterion are found in the groups denoted 4, 422, 3, 32, 6, 622, 23, and 432.

ii) Consider the example of the space-group Pccn(D_{2h}^{10}) and of the representation $\Gamma_1(\vec{k}_o)$ with $\vec{k}_o = (\vec{a}_1^* + \vec{a}_2^*)/2$. The star k_o^* has one arm ($\vec{k}_o$) and the little-point-group is $\hat{G}_o$ = mmm (D_{2h}). The small-representation τ_1 has the characters indicated on the third line of table 5. The vector $\vec{k}_o$ being equivalent to ($-\vec{k}_o$), we can apply eq. (3.13'). Table 5 shows the values of the elements needed to apply this formula (here $A(g) = [\chi_o^2(g) - \chi_o(g^2)].V(R)$). One can check that the Lifschitz-condition is obeyed by $\Gamma_1(k_o)$: $\sum A(g)$ = o.

TABLE 5. $\vec{t}_1 = (\vec{a}_1 + \vec{a}_2)/2$; $\vec{t}_2 = (\vec{a}_2 + \vec{a}_3)/2$; $\vec{t}_3 = (\vec{a}_1 + \vec{a}_3)/2$

g	$\{E\|o\}$	$\{U_z\|\vec{t}_1\}$	$\{U_y\|\vec{t}_2\}$	$\{U_x\|\vec{t}_3\}$	$\{\sigma_z\|\vec{t}_1\}$	$\{\sigma_y\|\vec{t}_2\}$	$\{\sigma_x\|\vec{t}_3\}$	$\{I\|o\}$
g^2	$\{E\|o\}$	$\{E\|o\}$	$\{E\|\vec{a}_2\}$	$\{E\|\vec{a}_1\}$	$\{E\|\vec{t}_1\}$	$\{E\|\vec{a}_3\}$	$\{E\|\vec{a}_3\}$	$\{E\|o\}$
$\chi_o(g)$	2	2	o	o	o	o	o	o
$\chi_o(g^2)$	2	2	-2	-2	2	2	2	2
V(R)	3	-1	-1	-1	1	1	1	-3
A(g)	6	-2	-2	-2	-2	-2	-2	6

3.2.2. Stars containing $\vec{k}$ and $(-\vec{k}) \neq \vec{k}$

If the star $\vec{k}_o$ contains 2s arms : s vectors $\vec{k}$ and the s vectors $(-\vec{k}) \neq k$, the Lifschitz criterion (3.13) is expressed by :

$$\sum_{R} \{\chi_o(g)\cdot\chi_o(g_1^{-1}\cdot g\cdot g_1)\cdot V(R) - \chi_o(g_1\cdot g\cdot g_1\cdot g)\cdot V(R_1 R)\} = o \qquad (3.13'')$$

where, again, the sum is over the elements of $\hat{G}(k_o)$, and where $g_1 = \{R_1 | \vec{t}_1\}$ is an element of G_o transforming $\vec{k}_o$ into $(-\vec{k}_o)$ [such an element necessarily exists since $(-\vec{k}_o)$ belongs to the star k_o^*].

As an illustration to eq. (3.13'') consider an IR whose vector $\vec{k}_o$ has its extremity inside the BZ (necessarily, $\vec{k}_o \neq (-k_o)$). Assume that the space-group G_o contains the inversion I among its point-symmetry elements, and that $\vec{k}_o$ is in a general position (i.e. outside symmetry-planes or lines). As $\vec{k}_o \neq (-\vec{k}_o)$, we can apply eq. (3.13''). For $\vec{k}_o$ in a general position, $G(\vec{k}_o)$ reduces to the identity E. The sum in (3.13'') contains a single term :

$$\chi_o(E)\cdot\chi_o(I.E.I).V(E) - \chi_o(I.E.I.E.).\ V(I) \qquad (3.14)$$

Clearly $V(I) = -V(E) = -3$, and (3.14) is equal to $6.\chi_o(E) = 6$ (since $\hat{G}(\vec{k}_o)$ has a single, one-dimensional, small-representation).
Hence (3.13'') is not satisfied : For a group containing the inversion, the IR's corresponding to $\vec{k}_o$ in a general position in the BZ, do not satisfy the Lifschitz criterion. This property can be easily shown to hold for vectors $\vec{k}_o$ lying on symmetry lines or planes, but whose extremity is inside the BZ [1].

3.2.3. Stars not containing $(-\vec{k})$

The third situation of interest arises when vectors $\vec{k}$ in the star k_o^* satisfy the relation $\vec{k} \neq -\vec{k}$, and when $(-\vec{k})$ does not belong to k_o^*.

The form of the translations matrices of the IR $\Gamma_i(k_o)$ (cf. 2.2.4.d) reveals that these matrices are not real : they contain diagonal elements exp $(-i\vec{k}.\vec{T})$ and not their complex conjugates. The relevant representation for the OP is the physically irreducible IR : $(\Gamma_i(\vec{k}_o) + \overline{\Gamma_i}(\vec{k}_o))$. For such a representation, the Lifschitz criterion takes the form :

$$\sum_{R} |\chi(g)|^2.V(R) = o \qquad (3.13''')$$

We can apply this formula in the situation where $\vec{k}_o$ is a general vector of the BZ, and where $\hat{G}_o$ does not contain elements transforming $\vec{k}_o$ into $(-\vec{k}_o)$. The first member of (3.13''') reduces to the term :

$$|\chi(E)|^2.V(E) = 3 \neq o \qquad (3.14')$$

The Lifschitz criterion is not satisfied.

3.3. Selection of $\vec{k}_o$ vectors resulting from the Lifschitz-criterion

As illustrations of eqs. (3.13'') and (3.13'''), we have seen that if $\vec{k}_o$ is a general vector of the BZ, the representations $\Gamma_i(\vec{k}_o)$ do not comply with the Lifschitz condition. Actually a more general result can be shown to hold [1, 8, 10] : unless k_o^* belongs to a specified set of "high-symmetry" stars, the Lifschitz-condition is not satisfied.

3.3.1. Proper symmetry $\hat{P}(\vec{k}_o)$ of $\vec{k}_o$. Polar groups

Let $\hat{H}$ be the invariance-group (holohedry) of the reciprocal lattice of the considered crystal, and $\vec{k}_o$ a vector of the BZ. The proper symmetry-group $\hat{P}(\vec{k}_o)$ of $\vec{k}_o$ is the largest subgroup of $\hat{H}$ leaving $\vec{k}_o$ invariant, up to an equivalence. As the point-group of the crystal satisfies the relation $\hat{G} \subset \hat{H}$, we have $\hat{G}(\vec{k}_o) \subset \hat{P}(\vec{k}_o)$ [we can write $\hat{G}(k_o) = \hat{P}(k_o) \cap \hat{G}$] For instance, if we take $\hat{G}$ = P4mm, and the vector $\vec{k}_o = (\vec{a}^*/2)$, we have $\hat{G}(\vec{k}_o) = \hat{G}$ = 4mm. On the other hand, the holohedry is $\hat{H}$ = 4/mmm, and the proper-symmetry-group of $\vec{k}_o$ is $\hat{P}(\vec{k}_o) = \hat{H}$.

It can happen that $\hat{P}(\vec{k}_o)$ is one of the "polar" crystallographic point-groups. A polar point-group is defined by the property that the vector-representation V(R) of this group is a reducible representation containing the totally-symmetric IR of the group : the component of a vector which carry Γ_1 is then invariant by all the elements of the polar-group.

Among the 32-crystallographic groups, 10 are polar. These are the groups whose international symbols are 1, 2, m, mm2, 4, 4mm, 3, 3m, 6, 6mm. The remaining 22 groups either contain the inversion I, or intersecting rotation axes, and thus possess an invariant "central point".

3.3.2. Allowed $\vec{k}_o$ vectors

It can be shown [1, 8, 10] that if $\vec{k}_o$ has a polar-proper-symmetry-group $\hat{P}(\vec{k}_o)$, the irreducible representation $\Gamma_i(\vec{k}_o)$ does not satisfy the Lifschitz-condition.

Conversely, a necessary condition imposed to IR's compatible with this condition is that the vectors $\vec{k}$ of their star k_o^* have $\hat{P}(\vec{k})$ non-polar (Another formulation is to say that $\hat{P}(\vec{k})$ possesses a "central point" [1]). This condition on the groups $\hat{P}(\vec{k})$ is only obeyed by a few $\vec{k}$-vectors ending <u>on high-symmetry points of the surface</u> of the BZ, or by the Γ-points (i.e. the centers of the various BZ). These vectors are of the form :

$$\vec{k} = q_1.\vec{a}_1^* + q_2.\vec{a}_2^* + q_3.\vec{a}_3^* \qquad (3.15)$$

where the q_i are rational numbers which, in their reduced form, have denominators equal to 2, 3, or 4. Fig. 7 shows one of these $\vec{k}$ vectors, for each distinct star, in the cases of cubic space-groups with a P-Bravais lattice.

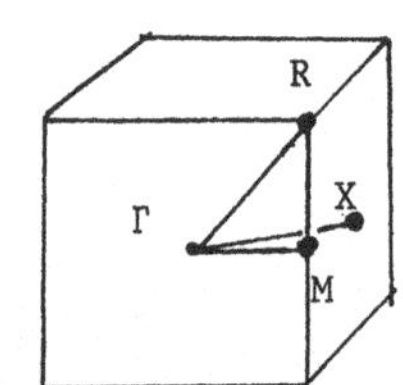

Fig.7

On the other hand, table 6 indicates the coordinates q_i of a representative vector of each relevant star for the entire set of Bravais lattices.

The fact that $\vec{k}_o$ is one of the acceptable vectors does not warrant the fulfilment of the Lifschitz condition by the representation $\Gamma_i(\vec{k}_o)$. This fulfilment depends on the characteristics of the small representation τ_i, and also of the space-group G_o considered.

The working out of the IR's satisfying the Lifschitz condition, by means of the procedure described in § 3.2, has been performed in Refs. 14 and 15. It has been found that this condition eliminates a significant fraction of the IR's whose $\vec{k}_o$-vector appears on table 6.

TABLE 6 Column (a) : label of the high-symmetry BZ-point. (b) coordinates of $\vec{k}$. The symbols in column (b) have the following meaning : s = (1/2) ; $\bar{s}$ = (- 1/2) ; t = (1/4) ; $\bar{t}$ = (- 1/4). u = (1/3) ; $\bar{u}$ = (- 1/3)

BRAVAIS LATTICE (a)	(b)
MONOCLINIC P	
Γ	o o o
A	s o o
B	s s o
Y	o s o
Z	o o s
E	s o s
D	s s s
C	o s s
MONOCLINIC B	
Γ	o o o
Z	s s o
B	s $\bar{s}$ s
A	s $\bar{s}$ o
Y	o o s
C	s s s
F	s o o
F"	s o s
ORTHORHOMBIC P	
Γ	o o o
Z	o o s
X	s o o
Y	o s o
R	s s s
U	s o s
T	o s s
S	s s o
ORTHORHOMBIC C	
Γ	o o o
Y	s $\bar{s}$ o
Z	o o s
T	s $\bar{s}$ s
S	s o o
R	s o o
ORTHORHOMBIC F	
Γ	o o o
Z	o s s
X	s s o
Y	s o s
T	s s 2s
R	s s s
ORTHORHOMBIC I	
Γ	o o o
X	s s s
U	s o o
T	s o s
S	s s o
R	3t t $\bar{t}$
TETRAGONAL P	
Γ	o o o
A	s s s
M	s s o
Z	o o s
R	o s s
X	o s o
TETRAGONAL I	
Γ	o o o
Z	s $\bar{s}$ $\bar{s}$
A	3t t $\bar{t}$
X	s s o
X	s s o
N	s o o
HEXAGONAL P (RHOMBOEDRAL R)	
Γ	o o o
A(Z)	o o s
K	2u $\bar{u}$ o
H	2u $\bar{u}$ s
M(X)	s o o
L(A)	s o s
CUBIC P	
Γ	o o o
R	s s s
X	o s o
M	s s o
CUBIC I	
Γ	o o o
H	s s s
N	s o o
P	t t t
CUBIC F	
Γ	o o o
X	s s o
W	3t s t
L	s s s

3.4. Lifschitz-criterion and periodicity of the low-symmetry phases

The simple forms of the $\vec{k}_o$-vectors compatible with the Lifschitz-criterion (eq. 3.15 and table 6) have an important physical consequence : the group of invariance G of the low-symmetry phase, stable below T_c, is, like G_o, a space-group. Its subgroup of primitive translations is generated by simple multiples of the primitive translations $\vec{a}_i$ of G_o.

In other terms, if G_o is the space-group of a crystal and η_r an OP carrying an IR of G_o complying with the Lifschitz criterion, this OP necessarily describes a transition towards a strictly crystalline phase.

In order to establish the preceding property we can use the fact that G is the invariance group of the equilibrium density increment $\delta\rho_{eq}(r)$ [eq. 3.19 in chapter II). Adapting the form of $\delta\rho_{eq}(r)$ to the characteristics of the space-group representations, we can write :

$$\delta\rho_{eq}(r) = \sum_{\vec{k},l} \eta_l^o(\vec{k}).\phi_l(\vec{k}) \qquad (3.16)$$

where $\vec{k}$ runs over the arms of k_o^*, and l over the basis of the small representation. $\eta_l^o(\vec{k})$ is the equilibrium value of the OP-component. Let $\vec{T}$ be any translation of the subgroup $\mathcal{T}_o$ of translations of G_o(eq. 2.1). Under the action of $\vec{T}$, $\delta\rho_{eq}$ transforms into :

$$\delta\rho'_{eq} = \sum_{\vec{k},l} e^{-i\vec{k}.\vec{T}}.\eta_l^o(\vec{k}).\phi_l(\vec{k}) \qquad (3.17)$$

Assume that the non-zero equilibrium values of the $\eta_l^o(\vec{k})$ are associated to p of the s vectors of the star k_o^*. We denote these vectors $\vec{k}^{(1)}$, $\vec{k}^{(2)}$, $\vec{k}^{(p)}$. Clearly, $\delta\rho_{eq}$ is invariant by any translation $\vec{T}$ verifying :

$$e^{-i\vec{k}^{(t)}.\vec{T}} = 1 \quad (t = 1, 2, \ldots p) \qquad (3.18)$$

Since $\vec{k}^{(t)}$ and $\vec{T}$ are respectively of the forms (3.5) and (2.1), we can write (3.18) as :

$$\sum_i^3 q_i^{(t)}.m_i = 1 \text{ (modulo 1) } (t = 1, 2, .. p) \qquad (3.19)$$

The $q_i^{(t)}$ are rational numbers with denominators smaller than 4 (table 6). Let $q_i = (m_i'/n_i')$ be the coordinate $q_i^{(t)}$ with the largest denominator.

Clearly the set of three independent translations $\vec{T}_i = n_i'.\vec{a}_i$ ($n_i' \leqslant 4$) satisfies eq. (3.9) and leaves $\delta\rho_{eq}$ invariant. This set generates a 3-dimensional translation group which is a subgroup of G.

As the group G possesses a subgroup of translations it is one of the crystallographic space-groups (cf. § 2.1).

In general the translations $\vec{T}_i$ defined hereabove are not the basic translations $\vec{a}_i'$ of the group G. The $\vec{a}_i'$ can be defined as the smallest translations $\vec{T}$, which are non-coplanar, and satisfy the set of equations (3.19). They define a unit-cell of the Bravais lattice of G whose volume is a simple multiple of the volume of the unit-cell $(\vec{a}_1, \vec{a}_2, \vec{a}_3)$ relative to G_o.

Eq. (3.19) shows that the relevant multiplying factor of the unit-cell will have its smallest value if p = 1, and its largest value if p equals the number of arms in the star k_o^*.

By considering the stars k_o^* represented by any vector in table 6, and by solving for each star and each possible value of p (p = 1, 2 ... s) the set of equations (3.19), one can determine the possible changes of translational symmetry accross a structural transition compatible with the Lifschitz-criterion.

Such a work has been partly performed by Lifschitz [8] and later completed by Naish and Syromiatnikov [16]. We illustrate the results made available in these works by the example of G_o possessing a cubic P-Bravais lattice, for an OP transforming according to an IR $\Gamma_i(k_o)$ of G_o, with $\vec{k}_o = (\vec{a}_1^*/2)$ and an unspecified small representation τ_i. The star of $\vec{k}_o$ contains 2 additional vectors : $(\vec{k}_2 = (\vec{a}_2^*/2)$ and $\vec{k}_3 = (\vec{a}_3^*/2)$. The possible translational groups of the low-symmetry phases are generated by $\vec{a}_i'$, determined by eqs. (3.19) with one, two, or three vectors of k_o^*. Up to an overall rotation of the system, we can choose these three situations to correspond respectively to $(\vec{k}_o)$, $(\vec{k}_o, \vec{k}_2)$ and $(\vec{k}_o, \vec{k}_2, \vec{k}_3)$. In the first case, we have a single equation :

$$\frac{1}{2} \cdot m_1 + o.m_2 + o.m_3 = 1 \ (1) \qquad (3.19')$$

The set of non-coplanar combinations of the basic translations $\vec{a}_i$ of G_o, which corresponds to the smallest unit-cell is clearly determined by (m_1 = 2, m_2 = 1, m_3 = 1). In the second case we have two equations associated to $\vec{k}_o$ and $\vec{k}_2$:

$$\frac{m_1}{2} = 1 \ (1) \text{ and } \frac{m_2}{2} = 1 \ (1) \qquad (3.19")$$

which is satisfied by (m_1 = 2, m_2 = 2, m_3 = 1). Finally, in the third case, (3.9) gives rise to 3 equations :

$$\frac{m_1}{2} = \frac{m_2}{2} = \frac{m_3}{2} = 1 \ (1) \qquad (3.19''')$$

which is satisfied by $m_1 = m_2 = m_3 = 2$. Hence, depending if the non-zero values of the equilibrium OP-components correspond to one arm, two arms, or three arms of the star k_o^*, the translation group $\mathcal{T}$ is respectively generated by $(2\vec{a}_1, \vec{a}_2, \vec{a}_3)$ $(2\vec{a}_1, 2\vec{a}_2, \vec{a}_3)$ and $(2\vec{a}_1, 2\vec{a}_2, 2\vec{a}_3)$. The unit-cell of the crystal, below T_c, is, respectively, 2-times, 4-times, and 8-times larger than the unit-cell relative to G_o.

4. IMAGES AND FREE-ENERGIES FOR ACTIVE REPRESENTATIONS

4.1. Active representations of a space-group

We have seen that if the OP of a transition has a symmetry complying with the Lifschitz criterion two characteristics of the transition are warranted : i) this OP is apt to describe an entire line of phase transitions possessing the same symmetry properties and not only a "singular" transition in the pressure-temperature plane (cf. introduction of § 3). ii) the low-symmetry-group G is a crystallographic space-group whose basic translations $\vec{a}_i'$ are simply related to those, $\vec{a}_i$ of the group G_o (§ 3.14).

We know, on the other hand, that a necessary condition imposed to the OP-symmetry of a continuous transition is the Landau condition discussed in chapter II, § 4.4.

Hence, continuous structural transitions between periodic phases are associated to irreducible representations $\Gamma_i(\vec{k}_o)$ complying with both, the Lifschitz condition :

$$V \times \{\Gamma_i(\vec{k}_o)^2\} \not\supset \Gamma_1 \quad , \tag{4.1}$$

and the Landau condition :

$$[\Gamma_i(\vec{k}_o)^3] \not\supset \Gamma_1 \tag{4.2}$$

Such IR's are termed active representations of G_o [1]. The procedure relevant to the checking of (4.1) has been described in § 3.2. The general method to be used in order to check the fulfiment of (4.2) by an IR has been provided in § 4.4 of chapter II.

In the case of IR's of a space-group a simple rule involving consideration of the star k_o^* can often be used. This rule is based on the fact that the totally symmetric IR of $G_o(\Gamma_1)$ corresponds to $\vec{k}$ = o. Hence, if we decompose the representation $[\Gamma_i(k_o)^3]$ into IR's of G_o, (4.2) will be satisfied if this decomposition contains no IR with $\vec{k}$ = o. Note that, if $\Gamma_i(k_o)$ is spanned by the OP-components $\eta_1(\vec{k})$ with $\vec{k} \in k_o^*$, the third power $\Gamma_i(k_o)^3$ is carried by combinations of products $\eta_1(\vec{k}_1).\eta_{1'}(\vec{k}').\eta_{1''}(\vec{k}'')$. Under the action of a translation $\vec{T}$ of G_o, such products transform as :

$$e^{-i(\vec{k}+\vec{k}'+\vec{k}'').\vec{T}}.\eta_1(\vec{k}).\eta_{1'}(\vec{k}').\eta_{1''}(\vec{k}'') \tag{4.3}$$

Accordingly, (cf. § 2.2.4.d) the decomposition of $[\Gamma_i(\vec{k}_o)^3]$ will only contain IR's of G_o with $\vec{k}$-vectors satisfying the relation :

$$\vec{k} = \vec{k}_1+\vec{k}_2+\vec{k}_3, \text{ with } \vec{k}_i \in k_o^* \tag{4.4}$$

In conclusion, the Landau condition will be fulfilled by $\Gamma_i(\vec{k}_o)$ if the sum of any 3 vectors $\vec{k}_i$ (distinct or equal) in the star k_o^* is different from zero (up to a vector of the reciprocal lattice).

For most of the stars whose representative $\vec{k}$-vector appears on table 6 (i.e. which are compatible with the Lifschitz criterion (4.1)) this rule is conclusive, and the Landau condition is satisfied. Aside from the Γ-points, there are 4 points only for which it is necessary to determine the possible fulfilment of condition (4.2) : the M and K points of the BZ relative to hexagonal crystals, the M-point relative to crystals with a P-cubic lattice, and the X-point relative to crystals with a F-cubic lattice.

In these cases, the effective decomposition of $[\Gamma_i(\vec{k}_o)^3]$ is, most of the time, unecessary : the identification of the image of G_o in the representation space (cf. § 4.2 hereafter) will provide the required information.

4.2. Images of G_0 for active representations

In chapter II, §4.5.1, the image of G_0 in the OP-space representation has been defined as the group I_0 whose elements are the distinct matrices of the OP-representation.

The use of I_0 has been described in chapter II, § 4.5 and 4.6 : the image I_0 can be substituted to the group G_0 for the determination of the form of the Landau free-energy as well as for the enumeration of the possible invariance groups G of the low-symmetry phase. The free-energy F is a polynomial expansion invariant by I_0. The groups G satisfy $G = K \times I$, where K is the kernel of the homomorphism $G_0 \to I_0$, and where $I \subset I_0$, is one of the isotropy-groups of directions in the OP-space. Furthermore, the isotropy-groups, can be used to simplify the procedure of minimization of F (§4.6 chap. II).

It is worth pointing out, however, that, due to its group-theoretical expression involving two representations [V and $\Gamma_i(\vec{k}_0)$], the Lifschitz criterion cannot be worked out on the basis of the sole image I_0. Its satisfaction depends on the form of the image of G_0 in the reducible-space $V \times (\Gamma_i(\vec{k}_0)^2)$.

Conversely, the compliance with the Landau condition, which relies on the OP-representation alone is entirely determined by I_0.

4.2.1. Finite character of the images, for active representations

A space-group G_0 can be expanded into cosets with respect to its translation subgroup $\mathcal{T}_0$:

$$G_0 = \{E \,|\, \vec{0}\}\, \mathcal{T}_0 + \{R_2 \,|\, \vec{t}_2\}\, \mathcal{T}_0 + \dots + \{R_d \,|\, \vec{t}_d\}\, \mathcal{T}_0 \qquad (4.5)$$

where ($E \;\; R_2 \dots R_d$) are the elements of the point-group $\hat{G}_0$, and $\vec{t}_1, \dots \vec{t}_d$, a set of non-primitive translations associated to the R_i. Equation (4.5) shows that if the image of $\mathcal{T}_0$ in the OP-representation is finite, the image of G_0 will also be finite.

We know (§ 2.2.4.d) that the matrix of a translation $\vec{T}$, relative to the OP-representation $\Gamma_i(k_0)$, is a diagonal matrix whose elements are of the form :

$$e^{-i\vec{k}p.\vec{T}} \qquad (4.6)$$

with $\vec{k}_p$ the various arms of the star k_0^*.

For an active representation, we have seen in § 3.4, hereabove, that the Lifschitz criterion imposes $\exp(-i\vec{k}p.\vec{T}) = 1$, for an infinite set of translations $\vec{T}$ forming a subgroup $\mathcal{T}$ of $\mathcal{T}_0$. The index of $\mathcal{T}$ in $\mathcal{T}_0$ (which is equal to the multiplying factor of the unit-cell discussed in 3.4) is generally equal to a few units (2, 4, 8, ...). As explained in § 3.4, this index is at most equal to the product $n'_1 . n'_2 . n'_3$, where the n'_i are the largest denominators found in the i-th components of the $\vec{k}_p$ vectors ($n'_i \leq 4$). It is therefore finite. Let $\mathcal{T}$ be the group of translations $\vec{T}$ satisfying ex(-$i\vec{k}p.\vec{T}$) = 1 for <u>all</u> the arms $\vec{k}_p$ of k_o^*. We can write $\mathcal{T}_o$ as :

$$\mathcal{T}_o = \mathcal{T} + (\vec{T}_1).\mathcal{T} + \ldots + (\vec{T}_m).\mathcal{T} \tag{4.7}$$

where m is the finite index of $\mathcal{T}$ in $\mathcal{T}_o$. Eq.(4.7) reveals that, since the matrices of $\mathcal{T}$, in the OP-representation, are all equal to the identity matrix, there are only m distinct matrices associated to the infinite group $\mathcal{T}_o$.

Accordingly, the image I_o of G_o is finite. Its order is, at most, equal to dxm, where d is the order of $\hat{G}_o$, and m the index of $\mathcal{T}$ in $\mathcal{T}_o$ appearing in eq.(4.7).

From table 6, it can be deduced that in most cases $m \leq 8$. A larger index m is only found in space-groups of the cubic system. The largest value is m = 32, occuring for the W-point of the BZ (F-cubic lattice). As, in this system, the largest order d of the point-group $\hat{G}_o$ is 48, we conclude that the order of the images I_o of space-groups, for order-parameters complying with the Lifschitz criterion, is less than 48x32 = 1536.

4.2.2. Enumeration of the images for active representations [17]

In chapter II, § 4.5.3, we have listed the images I_o representing all possible OP-symmetries, for OP-dimension n = 2, 3, 4. We have also singled out the images compatible with the Landau-condition. Some of the listed images have an infinite number of elements (e.g. $\mathbb{C}_{\infty v}$ = O(2) for n = 2, or $\mathbb{D}^*_\infty \times \mathbb{D}^*_\infty$ for n = 4). These images are not relevant to active representations of space-groups since the latter representations give rise to finite images only (cf. § 4.2.1). Actually the $\vec{k}$-vector selection determined by the Lifschitz criterion imposes a selection among the images listed in chapter II.

Let us consider, for instance, an IR whose star is composed of vectors $\vec{k}$ of the form (3.15), with each component $q_i = (m_i'/2)[\,m_i' = \pm 1$ or $m_i' = o]$. If $\vec{T}$ is a primitive translation of the crystal (eq. 2.1), the matrix of $\vec{T}$ in the space of the IR has all its (diagonal) elements of the form :

$$e^{-i\vec{k}.\vec{T}} = (e^{-i\pi})^{(m_1m_1' + m_2m_2' + m_3m_3')} = \pm 1 \tag{4.8}$$

Hence, the square of any translation matrix is equal to the identity matrix. Assume now that the considered IR is 2-dimensional. The image of G_o in the representation is necessary one of the cyclic groups $\mathbb{C}_m$, or one of the groups $\mathbb{C}_{mv}$ which contain $\mathbb{C}_m$ as a subgroup of index 2 (cf. chapter II, § 4.5.3). Let us show that eq. (4.8) imposes the condition $m \leq 12$. Indeed, let $\{R|\vec{t}\}$ be one of the elements of G_o whose image in I_o is the rotation by $(2\Pi/m)$ belonging to $\mathbb{C}_m$. Clearly, the images of the successive powers $[\{R|\vec{t}\}]^p$, are the rotations $(2\Pi p/m)$ of $\mathbb{C}_m$. If R is a reflection σ, the square of $\{R|\vec{t}\}$ is a primitive translation, and, in agreement with (4.8), the matrix of $[\{R|\vec{t}\}]^4$ is the identity matrix. In this case, we have $m \leq 4$. If R is a rotation of angle $(2\Pi l/p)$, $[\{R|t\}]^p$ is a pure translation and $[\{R|\vec{t}\}]^{2p}$ corresponds to the identity matrix. In this case we have $m \leq 2p$. As the maximum value for crystallographic space-groups is p = 6, we have, in all cases, $m \leq 12$.

Using the preceding argument in a more accurate way for all the vectors $\vec{k}$ in table 6 which are relevant to consider for 2-dimensional IR's, it is easy to show that the images of G_o for active representations are $\mathbb{C}_m$ and $\mathbb{C}_{mv}$, with m = 4, 6, 8, and 12 (the restriction $m > 3$, is the result of the Landau condition, as shown in chapter II, § 4.5.4).

Similarly, for 3-dimensional order-parameters, certain of the images considered in chap. II , are incompatible with active representations. Of the 7 finite images listed in chapter II, 3 only are compatible with the Landau condition and the $\vec{k}$-vectors in table 6. They are the groups denoted T_h, O, and O_h.

For higher-OP-dimensions a larger number of "active images" exist. They have been enumerated for n = 4, 6, and 8 [14, 18-19]. The latter value of OP-dimension (n = 8) is the largest dimension compatible with the Lifschitz-criterion [14].

4.3. Free-energies for active representations

The Landau-free-energy F is a sum of homogeneous polynomials of the OP-components invariant by the image I_o[chap. II. § 4.5.4]. The procedure of construction of the invariant polynomials has been described in chap. II, and a list of the "basic invariant polynomials" has been indicated for 2- and 3-dimensional images. Thus, for 2- and 3-dimensional active representations, the information required to determine the form of Landau's free-energy is entirely available in table 10 of chapter II. For 4-dimensional active images the form of the 4th-degree Landau expansion can be found in refs. 14,18,and 19.

Finally, for 6 and 8-dimensional active IR's, the corresponding fourth-degree expansions have been listed in refs. 14,and 20. As an illustration, we indicate the results for the 6-dimensional image I_o, which possesses the largest number of elements of all active images (1536).

This image arises for the space-group G_o = Fm3m (cubic system). The relevant irreducible representation $\Gamma_1(\vec{k}_o)$ has a 6-arm star k^*_o, whose representative vector $[\vec{k}_o = (3\vec{a}^*_1/4) + (\vec{a}^*_2/2) + (\vec{a}^*_3/4)]$ ends at the W-point of the BZ (Cf. table 6). G_o is a symmorphic-group and, accordingly, the small-representations τ_i coincide with the ordinary IR's of $\hat{G}(\vec{k}_o)$. Inspection of the tables in refs [5] or [6] show that $\hat{G}(\vec{k}_o) = \bar{4}2m$. This point-group has a two-dimensional IR which corresponds to an <u>inactive</u> IR of G_o [14], and 4 one-dimensional τ_i, giving rise to active 6-dimensional IR's of G_o. The image I_o of G_o in the 4 representations is the same. Since G_o is symmorphic, it is isomorphous to the product $\hat{G}_o \times \mathcal{T}_o$ where $\mathcal{T}_o$ is the group of primitive translations. Accordingly I_o can be written $I_o = I_1 \times I_2$, where I_1 is the image of $\mathcal{T}_o$ and I_2 is the image of the group of elements $\{R|\vec{\sigma}\}$ isomorphous to $\hat{G}_o$. I_1 is a group of order 32 generated by the three 6x6 matrices :

$$A_1 = \begin{vmatrix} a & o & o \\ o & -a & o \\ o & o & -1 \end{vmatrix} \quad A_2 = \begin{vmatrix} -a & o & o \\ o & -1 & o \\ o & o & a \end{vmatrix} \quad A_3 = \begin{vmatrix} -1 & o & o \\ o & -1 & o \\ o & o & -1 \end{vmatrix} \tag{4.9}$$

where each element is a 2x2 matrix, with :

$$a = \begin{vmatrix} o & 1 \\ -1 & o \end{vmatrix} \quad 1 = \begin{vmatrix} 1 & o \\ o & 1 \end{vmatrix} \text{ and } o = \begin{vmatrix} o & o \\ o & o \end{vmatrix} \tag{4.9'}$$

On the other hand, I_2 is a group of order 48, isomorphous to $\hat{G}_o$ = m3m, and generated by the 6x6 matrices :

$$B_1 = \begin{vmatrix} o & o & 1 \\ 1 & o & o \\ o & 1 & o \end{vmatrix} \quad B_2 = \begin{vmatrix} b & o & o \\ o & o & 1 \\ o & b & o \end{vmatrix} \quad B_3 = \begin{vmatrix} 1 & o & o \\ o & o & 1 \\ o & 1 & o \end{vmatrix} \quad B_4 = \begin{vmatrix} b & o & o \\ o & b & o \\ o & o & b \end{vmatrix} \qquad (4.10)$$

where o and 1 stand for the same matrices as above, and where,

$$b = \begin{vmatrix} 1 & o \\ o & -1 \end{vmatrix} \qquad (4.10')$$

As the basic polynomial invariants are not available for this group, we have to work out the form of the invariant polynomials of every degree through the decomposition of $[\Gamma^m]$, as indicated in chapter II, § 4.3.2. However, as the group I_o considered here has a large order (1536), the general method can be conveniently replaced by a procedure adapted to the structure of I_o. Note that the various terms of the Landau free-energy must be invariant by any subgroup of I_o. This is true, in particular, for the set of matrices generated from the three matrices of I_1 (eq. 4.9) and the last matrix in (4.10). All these matrices consist of 3 diagonal blocks acting separately on the three sets of OP-components (η_1, η_2), (η_3, η_4) and (η_5, η_6) spanning the 6-dimensional carrier space of the representation. Consequently, we can form the required invariant polynomials by constructing, in the first place, the polynomials of two variables (η_i, η_{i+1}) (i = 1, 3, 5) which are left invariant by the set of 2x2 matrices a, b, and ± 1. Inspection of (4.9') and (4.10') reveals that these matrices generate a group isomorphous to $\mathbb{C}_{4v}$. For the latter group, the basic invariants are known (cf. chapter II, table 10). If we replace (η_i, η_{i+1}) by the variables (ρ_i, ϕ_i), these invariants are expressed as :

$$\rho_i^2 \;;\; \rho_i^4 . \cos 4\phi_i \qquad (4.11)$$

Any function of the invariants in (4.11), with (i = 1, 3, 5), will be invariant by the subgroup of I_o considered. We are interested in the polynomials of mth degree which are also invariant by the three remaining matrices of (4.10). Clearly the set generated by these matrices operates a permutation of the invariants of index i. Hence the solutions to the investigated problem are polynomials of ρ_i^2 and $\rho_i^4 . \cos 4\phi_i$, which are invariant by permutation of the i-indices.

In agreement with the active-nature of I_o, no such polynomial of 3rd degree can be formed. Polynomials of 4th degree are easily determined, thus allowing to write the expression of the 4th degree-expansion corresponding to I_o :

$$F = \frac{\alpha}{2} (\sum \rho_i^2) + \frac{\beta_1}{4} (\sum \rho_i^4) + \frac{\beta_2}{2} (\sum \rho_i^2 . \rho_j^2) + \frac{\beta_3}{4} (\sum \rho_i^4 \cos 4\phi_i) \qquad (4.12)$$

It is worth pointing out, that the set of OP-components $(\eta_1, \eta_2, \eta_3, \eta_4, \eta_5, \eta_6)$ does not coincide with the standard basis of the IR, defined in § 2.2.4.a).

Indeed, the matrix associated to the primitive translation $\vec{a}_1$, in the standard basis $\eta(\vec{k}_p)$, where $(\vec{k}_1, \vec{k}_2, \ldots, \vec{k}_6)$ are the 6-arms of k_o^*, has the diagonal form :

$$\begin{vmatrix} i & & & & & \\ & -i & & & & \\ & & -i & & & \\ & & & +i & & \\ & & & & -1 & \\ & & & & & -1 \end{vmatrix} \qquad (4.13)$$

[we have taken $\vec{k}_3 = C_3\,\vec{k}_o$, $\vec{k}_5 = C_3^2\,\vec{k}_o$, $\vec{k}_2 = -\vec{k}_o$, $\vec{k}_4 = -\vec{k}_3$, and $\vec{k}_6 = -\vec{k}_5$]. The set of matrices representing $\mathfrak{T}$ is equivalent to a real representation. However the basis in which the matrices are real is not a standard basis. It is defined by :

$$\eta_p = \frac{1}{\sqrt{2}}\,\{\eta(k_p) + i.\eta(k_{p+1})\}\ ;\ \eta_{p+1} = \frac{-i}{\sqrt{2}}\,\{\eta(k_p) - i\ \eta(k_{p+1})\}\,[p = 1,\ 3,\ 5] \qquad (4.14)$$

and each η_p is associated to the pair of vectors $(\pm\,\vec{k}_p)$ of the star k_o^*. It is the basis η_p which corresponds to the form (4.9) (4.10) of the OP-representation, the matrix (4.13) being transformed, in the change of basis, into the first matrix of the set (4.9).

5. DETERMINATION OF THE LOW-SYMMETRY SPACE-GROUP G

If the absolute minimum of the Landau-free-energy, below T_c, corresponds to the set of OP-components $(\eta_1^o, \ldots, \eta_n^o)$, the invariance group G of the low-symmetry phase is the largest subgroup of G_o which leaves invariant the vector $\vec{\eta}_o = (\eta_1^o, \ldots, \eta_n^o)$ of the OP-space (cf. chapter II § 3.6.2). We have pointed out (chap. II, § 4.5.2) that G = IxK, where I is the isotropy-group of the $\vec{\eta}_o$-direction, and where K is the kernel of the isomorphism $G_o \rightarrow I_o$. In the case of structural transitions, the advantage of making use of the image I_o to determine G is clear : I_o is a finite group which will be simpler to handle than the infinite group G_o. As shown in § 3.4 hereabove, G is one of the 230 space-groups. Accordingly, once the vector $\vec{\eta}_o$ is specified, the determination of G can be performed in three steps : i) Determination of K and I_o. ii) Determination of the isotropy-group I. iii) Identification of the symbol of the space-group $\underset{\sim}{G}$ = K x I. The last step requires the identification of the point-group $\hat{G}$, of the Bravais lattice of G, and of the screw-axes or glide-planes contained in G. Moreover, it can happen that the set of elements composing G does not coincide with one of the space-groups listed in standard tables [4, 5, 6] but that this coincidence can be obtained by a change of coordinate framework [G is not in a "standard setting" but is equivalent to a listed space-group].

To illustrate this procedure, consider the IR of G_o = Fm3m defined by $\vec{k}_o = [(3\vec{a}^*/4) + (\vec{a}^*/2) + (\vec{a}^*/4)]$ and by the totally symmetric representation τ_1 of the group $G(\vec{k}_o) = \bar{4}2$ m. The image I_o of this representation, already considered in § 4.3, is generated by the matrices (4.9) and (4.10).

a) Kernel of the homomorphism $G_o \to I_o$.

As stressed in § 4.3, the image I_o contains a subgroup isomorphous to $\hat{G}_o$. Hence, the kernel of the homomorphism $G_o \to I_o$ necessarily coincides with a subgroup of the translation group $\mathcal{T}_o$ of G_o [G_o is a symmorphic group]. In a cartesian coordinate-frame , parallel to the cube-axes, the basic translations of G_o are :

$$\vec{a}_1 = \frac{a}{2}(110) \quad \vec{a}_2 = \frac{a}{2}(011) \quad \vec{a}_3 = \frac{a}{2}(101) \tag{5.1}$$

The 3 matrices A_1, A_2, A_3 in (4.9) are respectively the images of the primitive translations $\vec{a}_1$, $\vec{a}_3$, and $(\vec{a}_1 + \vec{a}_2 + \vec{a}_3)$. The image of $\mathcal{T}_o$ being of order 32, the kernel K is a subgroup of $\mathcal{T}_o$ of index 32 : K is a group of translations whose associated unit-cell is 32 times larger than the unit-cell $(\vec{a}_1, \vec{a}_2, \vec{a}_3)$. The translations generating the infinite group K are :

$$\vec{b}_1 = 2(\vec{a}_1 + \vec{a}_3 - \vec{a}_2) = 2a(100); \vec{b}_2 = 2(\vec{a}_1 + \vec{a}_2 - \vec{a}_3) = 2a(010)$$
$$\vec{b}_3 = 2(\vec{a}_2 + \vec{a}_3 - \vec{a}_1) = 2a(001) \tag{5.2}$$

Indeed, one can easily check that the volume $(\vec{b}_1, \vec{b}_2, \vec{b}_3) = 32\ (\vec{a}_1, \vec{a}_2, \vec{a}_3)$, and that $\vec{b}_1$, $\vec{b}_2$, $\vec{b}_3$ are represented by the identity matrix in the OP-space.

b) Isotropy group I

In the 6-dimensional space spanning the matrices (4.9)-(4.10) let us assume that, for a certain range of the coefficients the absolute minimum of the free-energy (4.12) corresponds to the direction [11 00 00] of the OP-space. (it is straightforward to show that this range is defined by $\beta_3 > o$, $\beta_2 > \beta_1 - \beta_3 > o$).

I is composed of the elements in I_o which preserve the manifold (η_1, η_2) of the OP-space, and which, in this manifold, leave the first bissector invariant. As I_o acts, within this manifold, as the group C_{4v}. Among the 8-elements of the latter group only two, the identity, and the symmetry about the first bissector, leave the [11] direction unmoved. These two elements are respectively represented by the 2x2 matrices (1) and (b x a = - a x b) in (4.9')-(4.10'). Inspection of the matrices (4.9) (4.10), which generate I_o, show that I is composed of the elements generated by the set (A_1, A_2, A_3, B_2, B_3, B_4) which possess as first diagonal-block one of the two former 2x2 matrices.

Consider first the set (B_2, B_3, B_4) which generates a group I' of order 16. Any matrix of I' has its first diagonal block either equal to 1, or equal to b (since $b^2 = 1$). A matrix of the first type belongs to the isotropy-group I. If B is a matrix of the second type, the product $A_2 \times B$ clearly belongs to I. Hence, I contains 16 elements in correspondance with 16 elements of I'. On the other hand, in the group of order 32 generated by (A_1, A_2, A_3) one can find 8 matrices whose first diagonal block is 1. I is therefore constituted by the product of these two sets : its order is 8 x 16 = 128. It is a subgroup of index 12 of I_o. Accordingly G is a subgroup of index 12 of G_o.

Point-group $\hat{G}$

We have indicated in § 4.3, that the group generated by (B_1, B_2, B_3, B_4) in (4.10), is isomorphous to the cubic point-group m3m. In this isomorphism, B_1, B_2, B_3, and B_4 correspond respectively to the 3-fold rotation C_3, to the 4-fold rotation C_4^x about the x-axis, to the 2-fold rotation U_{yz} about the diagonal of a face, and to the inversion I. The isotropy-group I of the considered direction in the OP-space contains B_3, $A_2 \times B_2$, and $A_2 \times B_4$. Accordingly, the low-symmetry group G contains their reciprocal images $(U_{y\bar{z}} \mid \vec{o}) \times K$, $(C_4^x \mid \vec{a}_3) \times K$, and $(I \mid \vec{a}_3) \times K$. Since K is a group of primitive translations (eq. 5.2), and since the reciprocal images of the other generators of I are also primitive translations of G_o, we can infer that the point-group $\hat{G}$ is generated by the 3 elements I, C_4^x and $U_{y\bar{z}}$. $\hat{G}$ is a crystallographic point-group of order 16, which is denoted 4/mmm, the 4-fold axis being directed along the x-axis.

d) Bravais lattice of G

The point group $\hat{G}$ = 4/mmm belongs to the tetragonal system. The translation group $\mathcal{T}$ of the crystal must therefore define one of the Bravais lattices (P or I) of this system (Cf. table 2). The specification of this Bravais lattice involves two steps. The first one consists in the working out of the basic translations of $\mathcal{T}$. The second one is the identification of the set of basic translations with the standard set indicated in the tables [5, 6] for the relevant Bravais lattice.

i) determination of $\mathcal{T}$.

The group $\mathcal{T}$ can be determined by 2 equivalent methods. $\mathcal{T}$ is the reciprocal image of the 8 matrices generated from (A_1, A_2, A_3) (eq. 4.9) and belonging to I. Equivalently, $\mathcal{T}$ is the set of translations belonging to G_o and complying with condition (3.19), with t = 1 (since the direction [11 oo oo] in the OP-space involves a single pair ($\pm \vec{k}_o$) of vectors of k^*_o. The basic translations of $\mathcal{T}$ are any set of non coplanar vectors ($\vec{a}'_1$, $\vec{a}'_2$, $\vec{a}'_3$) complying with (3.9), and defining a unit-cell 4-times larger than the one defined by ($\vec{a}_1$, $\vec{a}_2$, $\vec{a}_3$). Indeed, since the index of G in G_o is 12 (Cf. §b) hereabove) and the index of $\hat{G}$ in $\hat{G}_o$ is 3, the index of $\mathcal{T}$ in $\mathcal{T}_o$ must be 4. One set of basic translations is :

$$\vec{c}_1 = \vec{a}_2 + \vec{a}_3 - \vec{a}_1 = a(o\ o\ 1) \qquad \vec{c}_2 = \vec{a}_2 - (\vec{a}_3 - \vec{a}_1) = a(o\ 1\ o)$$

$$\vec{c}_3 = \vec{a}_1 + \vec{a}_3 = a\ (1\ \tfrac{1}{2}\ \tfrac{1}{2}) \tag{5.3}$$

ii) identification of the Bravais lattice

When the 4-fold axis is along x, the two Bravais lattices of the tetragonal system are respectively defined by the following basic translations (referred to a cartesian coordinate framework) [5, 6] :

$$P \rightarrow C(1\ o\ o) \qquad A(o\ 1\ o) \qquad A(o\ o\ 1)$$

$$I \rightarrow (\tfrac{C}{2}, \tfrac{A}{2}, \tfrac{A}{2}) \qquad (-\tfrac{C}{2}, \tfrac{A}{2}, \tfrac{A}{2}) \qquad (-\tfrac{C}{2}, \tfrac{A}{2}, -\tfrac{A}{2}) \tag{5.4}$$

In order to identify the Bravais lattice generated by the $\vec{c}_i$ in eq. (5.3) we must find a combination of these vectors, defining a unit-cell of same volume as ($\vec{c}_1$, $\vec{c}_2$, $\vec{c}_3$) and displaying the same geometrical configuration as one of the two sets of vectors in (5.4). We note that the set :

$$\vec{a}'_1 = \vec{c}_3 = (a, \frac{a}{2}, \frac{a}{2}),\ \vec{a}'_2 = \vec{c}_1 + \vec{c}_2 - \vec{c}_3 = (-a, \frac{a}{2}, \frac{a}{2}) \qquad (5.5)$$

$$\vec{a}'_3 = \vec{c}_2 - \vec{c}_3 = (-a, \frac{a}{2}, -\frac{a}{2})$$

is identical to the standard set for the centered (I) tetragonal Bravais lattice, provided C = 2a and A = a.
The translations $\vec{a}'_i$ are the 3 basic translations of the low-symmetry phase.

d) Non-primitive translations and identification of the symbol for G.

We have seen, in §c that G contains the elements :

$$(U_{\bar{y}z} \mid \vec{0})\ ;\ (C_4^x \mid \vec{a}_3) = (C_4^x \mid \frac{C}{4}\ o\ \frac{A}{2})\ ;\ (I \mid \vec{a}_3) = (I \mid \frac{C}{4}\ o\ \frac{A}{2}), \qquad (5.6)$$

as well as the product of these elements. Finding the symbol for a group G, with $\hat{G}$ = 4/mmm, requires specifying the nature of the 4-fold axis (pure- or helicoidal-rotation) and the nature of the 3-types of reflection-planes σ_x, σ_y, and $\sigma_{\bar{y}z}$ (pure- or glide-reflections). The first line of table 7 shows the elements of G associated to these 4 points-symmetry elements (up to a primitive translation (5.5)).

TABLE 7

$(C_4^x \mid \frac{C}{4}\ 0\ \frac{A}{2})$	$(\sigma_x \mid \frac{C}{4}\ 0\ \frac{A}{2})$	$(\sigma_y \mid 0)$	$(\sigma_{\bar{y}z} \mid \frac{C}{4}\ 0\ \frac{A}{2})$
$(C_4^z \mid \frac{A}{2}\ 0\ \frac{C}{4})$	$(\sigma_z \mid \frac{A}{2}\ 0\ \frac{C}{4})$	$(\sigma_y \mid 0)$	$(\sigma_{\bar{x}y} \mid \frac{A}{2}\ 0\ \frac{C}{4})$

We can see that the fourfold -axis is a screw-axis whose non-primitive translation is one fourth of the period C along the axis-direction ; σ_x is a glide plane, the associated translation parallel to the plane, (A/2), being along the z-axis ; σ_y is a pure reflection, while $\sigma_{\bar{y}z}$ is a glide-plane with a translation (C/4) parallel to the plane. Following standard rules [4] the symbol of G deduced from table 7 is $I4_1$/cmd. This symbol does not coincide with any of the standard symbols listed for space-groups [6]. This is due to the fact that the standard setting of the fourfold axis is along-z. Making the appropriate reorientation of the axes we find the conjugated space-group whose elements are listed on the second line of table 7. The corresponding standard symbol identifying G is $I4_1$/amd.

6. FERROIC CLASSIFICATION OF STRUCTURAL PHASE TRANSITIONS

Though the OP describes entirely the symmetry-change at T_c (cf. chap. II, §5.4), it concerns degrees of freedom of the system which do not always have a clear relationship with quantities most directly determined in experiments. This is particularly true in the case of structural phase transitions, where it is interesting to know how the transition influences the behaviour of macroscopic quantities, represented by tensors, such as the dielectric polarization $\vec{P}$, the thermal expansion λ_{ij}, the piezoelectric response $[d_{ijk}]$, the elastic compliance $[C_{ijkl}]$, etc...[21]. These tensors, though they undergoe anomalies at T_c, do not possess, in general, the same symmetry properties as the OP. In fact, some of their components constitute "secondary order-parameters" in the sense given in chapt. II §5. As will be recalled below, the form of these tensors is entirely determined by the point-symmetry ($\hat{G}_0$ or $\hat{G}$) of the crystalline phases considered. Consequently, the study of the change of tensorial properties provoked by a structural transition must be based, in the first place, on the point-symmetry-change $\hat{G}_o \rightarrow \hat{G}$ across the transition. It is this point of view which underlies the ferroic classification of structural transitions exposed in §6.1 hereunder. The latter classification is, in general, unsufficient to account for the qualitative differences between the anomalies induced by the transition in the tensorial quantities, since the primary OP, alone, provides a complete description of a phase transition (cf. chap. II, §5). In this respect, the crucial question which must be examined, is the symmetry-relationship between the tensorial components and the primary OP (cf. § 6.2 hereunder).

6.1. Ferroicity and point-symmetry-change at T_c

6.1.1. Point-symmetry and macroscopic tensors

A number of text-books discuss the symmetry-properties of the tensors which provide the mathematical description of the macroscopic properties of crystals [1, 22, 23, 24]. With respect to these properties the crystal can be considered as a continuum : the atomic structure is smoothed out and, accordingly, these properties are invariant by any microscopic translation. The latter invariance holds, in particular, for the non-primitive translation $\vec{t}$ involved in the elements $\{R|\vec{t}\}$ of the crystal's space-group. Since $\{R|\vec{t}\}$ preserves the crystal's structure, and it also preserves its macroscopic properties. As a consequence R will leave unchanged the macroscopic properties of the crystal : these properties are invariant by the point-group $\hat{G}$ of the crystal [this is usually expressed by saying that macroscopic tensors are determined by the orientational properties of the crystal].

Macroscopic properties of crystals are classified according to the rank of the tensor which is associated to them. In a Cartesian coordinate framework a tensor S of rank p has 3^p components indexed by p indices : $S = [S_{ij..l}]$. These components are modified by a point-symmetry operation R. Their transformation properties is identical to that of the product $(v_i \cdot v_j \cdot \ldots v_l)$, of p components of a vector $V = [v_i]$.

The invariance of S by the point-group $\hat{G}$ implies certain relations between the components of $S_{ij\ldots l}$. In addition, thermodynamic requirements impose relations such as, for instance, the equality of components differing by a permutation of certain indices. As a result of these two sets of conditions, certain components of S are zero, while others are equal or opposite. For a given point-symmetry $\hat{G}$, the tensor S will have a specified number of independent elements [22-24].

From, a group-theoretical point of view, the components v_i of a vector V carry a 3-dimensional representation of $\hat{G}$. According to the transformation rule of the $S_{ij\ldots l}$, stated above, the components of S will carry a 3^p-dimensional representation of $\hat{G}$. Its decomposition into IR's of $\hat{G}$ yields a certain number of totally symmetric combinations of the $S_{ij\ldots l}$. Taking into account the relations imposed by thermodynamics, certain of these combinations vanish identically while ν combinations remain. These ν totally symmetric representations are relative to ν independent combinations of the $S_{ij\ldots l}$, invariant by $\hat{G}$: hence, ν is equal to the number of independent elements of S, mentioned above.

For instance, consider the dielectric polarization $\vec{P}$ which is described by a tensor of rank one (i.e. a vector), and take $\hat{G}$ = 4 mm. $\hat{G}$ is composed of the 8 elements indicated on the first line of table 8. The second line shows the transformation properties of the components of $\vec{P}$, referred to axes perpendicular to σ_x and σ_y, and parallel to the fourfold rotation axis C_4. Clearly, the macroscopic polarization compatible with the group $\hat{G}$ is reduced to the non-zero-component P_z.

TABLE 8

$\hat{G}$	E	C_4	U_z	C_4^3	σ_x	σ_y	σ_{xy}	$\sigma_{\bar{x}y}$
P_x	1 o o	o -1 o	-1 o o	o 1 o	-1 o o	1 o o	o -1 o	o 1 o
P_y	o 1 o	1 o o	o -1 o	-1 o o	o 1 o	o -1 o	-1 o o	1 o o
P_z	o o 1	o o 1	o o 1	o o 1	o o 1	o o 1	o o 1	o o 1

6.1.2. Ferroic and non-ferroic transitions

The strict group-subgroup relationship which exists between the space-groups G_o and G of the two phases adjacent to a structural transition implies a relationship $\hat{G}_o \supset \hat{G}$ between the point-groups of these phases.

We have seen in chap. II, § 3.6.2.c that below T_c, the system is degenerate : several phases with the same free-energy exist. Their symmetries G_i are conjugated with respect to G_o : $G_2 = hG_1h^{-1}$, with $h \in G_o$, $h \notin G$.

In the case of structural transitions, the two crystalline structures associated to two degenerate states G_1 and G_2 will correspond to each other through the action of $h = \{R|\vec{t}\}$. This implies that the macroscopic tensors of the two low-symmetry states correspond to eachother through the action of the point-symmetry element R.

a) Non-ferroic transitions

If the point-groups $\hat{G}_o$ and $\hat{G}_1$ are identical, all the degenerate states existing below T_c possess the same point-symmetry ($\hat{G}_i = \hat{G}_o$), since the point-symmetry operation establishing the correspondance between $\hat{G}_1$ and the $\hat{G}_i$ is reduced to the identity (the space-group element h involved in the conjugation relation between $\hat{G}_1$ and $\hat{G}_i$ is necessarily a primitive translation of G_o). In any of the degenerate states, the various macroscopic tensors have the same form as in the high-symmetry phase.

The structural transitions complying with the condition $\hat{G}_o = \hat{G}$ are termed <u>non-ferroic</u> [14]. For these transitions, the symmetry-breaking occuring at T_c is a decrease of the translational symmetry (i.e. $\mathfrak{T} \subset \mathfrak{T}_o$). At macroscopic level, the components of macroscopic tensors whose zero-value is imposed by the crystal's symmetry are the same above and below T_c. Hence no such component can be a symmetry-breaking (primary or secondary)-order-parameter, i.e. no tensorial component can be zero, by symmetry above T_c and non-zero below T_c. Likewise, no component of a macroscopic tensor can serve to distinguish one degenerate state from another.

b) Ferroic transitions

We assume now $\hat{G} \subset \hat{G}_o$. In that case, some of the degenerate states existing below T_c are transformed from the state s_1, of symmetry $\hat{G}_1 \subset \hat{G}_o$, by operations $\{R_i|\vec{t}_i\}$ of G_o, with $R_i \neq E$. We term these states s_i, <u>orientation states</u> (they differ from s_1 by a global reorientation specified by the non-trivial point-symmetry element R_i). The orientation state s_i has a point group $\hat{G}_i = R_i.\hat{G}_1.R_i^{-1}$ which differs from $\hat{G}_1$. If we adopt a common frame of reference for all the states s_i, the tensors representing certain macroscopic quantities will have different values in the different states (e.g. distinct non-zero components). If a certain macroscopic tensor has components differing in two state s_i and s_j, these components are transformed by the action of the operation $R \in \hat{G}_o$ which transforms one state into the other. They are not invariant by $\hat{G}_o$ and their value is necessarily zero above T_c. Hence, the macroscopic tensors which allow to distinguish the various degenerate states, also have the property to possess components which are zero above T_c, and non-zero below T_c. Such components coincide either with components of the primary-OP, or with components of the secondary order-parameters.

Transitions complying with this scheme are termed ferroic transition. They possess a threefold characteristics : i) they are associated to a lowering of crystallographic point-symmetry. ii) components of certain macroscopic tensors have non-zero values onsetting below T_c. iii) the same tensors allow to distinguish, at macroscopic level, the various orientation states arising below T_c (these states constitute, in general, part of the degenerate states).

For instance, in chapter I (table 1, and § 2.2), we have seen that the point-symmetry change 4/mmm → 4mm gives rise to two degenerate states respectively corresponding to polarizations + P_z and - P_z. In this example the P_z component of the tensor $\vec{P}$ of rank one is zero above T_c and non-zero below T_c, and its value distinguishes the two degenerate states below T_c.

More generally, one can classify ferroic transitions on the basis of the rank of the tensors characterizing the different orientation states of the crystal below T_c.

The concept of a ferroic, and the classification of ferroics is due to Aizu [25]. The definition given by this author differs is certain aspects of the one given above. In particular a ferroic crystal can be defined independently from the occurence of a structural phase transition. However for our aim, which is to discuss the anomalies of macroscopic quantities induced by a phase transition, the less general definition adopted here is sufficient.

6.1.3. Classification of ferroics

The classification of ferroics is a crystallographic and physical classification of structural transitions. The crystallographic aspect is based on the type of point-symmetry change occuring at T ,while the physsical aspect focuses on the nature of the physical quantity (electrical, mechanical, ...) related to the relevant tensor affected by the transition. This dual aspect is summarized on figure 8 .

a) Ferroelectrics

A ferroelectric transition is defined by the fact that part or all of the different orientation states can be distinguished by means of a macroscopic quantity represented by a vector. Crystallographically, the point-group $\hat{G}$ must be compatible with non-zero components of a vector. Referring to § 6.1.1 and to § 3.3.1, we see that the vector representation [V] of $\hat{G}$ must contain, at least once, the totally symmetric IR of $\hat{G}$: $\hat{G}$ is one of the polar groups. As for $\hat{G}_o$ it is either a non-polar group (all the components of [V] will be zero by symmetry), or a polar-group compatible with a smaller number of independent components for [V]. In agreement with the example considered in chapter I, the prominent macroscopic quantity which transforms as a vector, is the dielectric polarization $\vec{P}$. A ferroelectric transition is therefore characterized by spontaneous polarization components, i.e. components of $\vec{P}$ which are zero above T_c, and whose non-zero value onset at T_c.

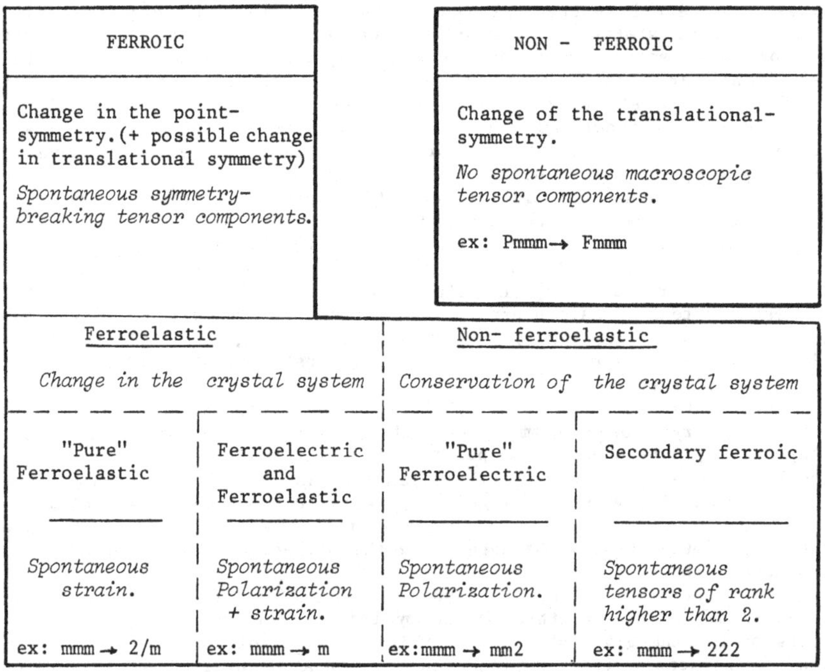

FERROIC	NON - FERROIC
Change in the point-symmetry.(+ possible change in translational symmetry) *Spontaneous symmetry-breaking tensor components.*	Change of the translational-symmetry. *No spontaneous macroscopic tensor components.* ex: Pmmm→ Fmmm

Ferroelastic		Non- ferroelastic	
Change in the crystal system		*Conservation of the crystal system*	
"Pure" Ferroelastic	Ferroelectric and Ferroelastic	"Pure" Ferroelectric	Secondary ferroic
Spontaneous strain.	*Spontaneous Polarization + strain.*	*Spontaneous Polarization.*	*Spontaneous tensors of rank higher than 2.*
ex: mmm → 2/m	ex: mmm → m	ex:mmm → mm2	ex: mmm → 222

Fig.8. Crystallographical and physical classification of structural transitions.

b) Ferroelastics

A ferroelastic transition is defined by the fact that part or all of the different orientation states can be distinguished by means of a macroscopic property represented by a symmetric tensor of rank 2. It has been pointed out [14, 26] that a necessary crystallographic requirement is that the point-groups $\hat{G}_o$ and $\hat{G}$ belong to different crystal systems (cf. § 2.1.2.d). This condition is also sufficient if the hexagonal and trigonal systems are considered as a single system (i.e. a transition from a hexagonal to a trigonal point-group is not ferroelastic). A prominent quantity which is represented by a symmetric second-rank tensor is the thermal expansion [22]. In accordance with the general considerations of § 6.1.2.b), the thermal expansion tensor of the low-symetry phase has more independent components than that of the high-symmetry phase. These components are strain components. Those which vanish, by symmetry, above T_c constitute a symmetric second-rank tensor which is termed, for obvious reasons the spontaneous-strain tensor. This tensor is necessarily traceless since its trace, which is an invariant by any point-symmetry operation, would not vanish by symmetry above T_c. Figure 9 illustrates the definition of the spontaneous strain tensor for a transition between a tetragonal phase and an orthorhombic one.

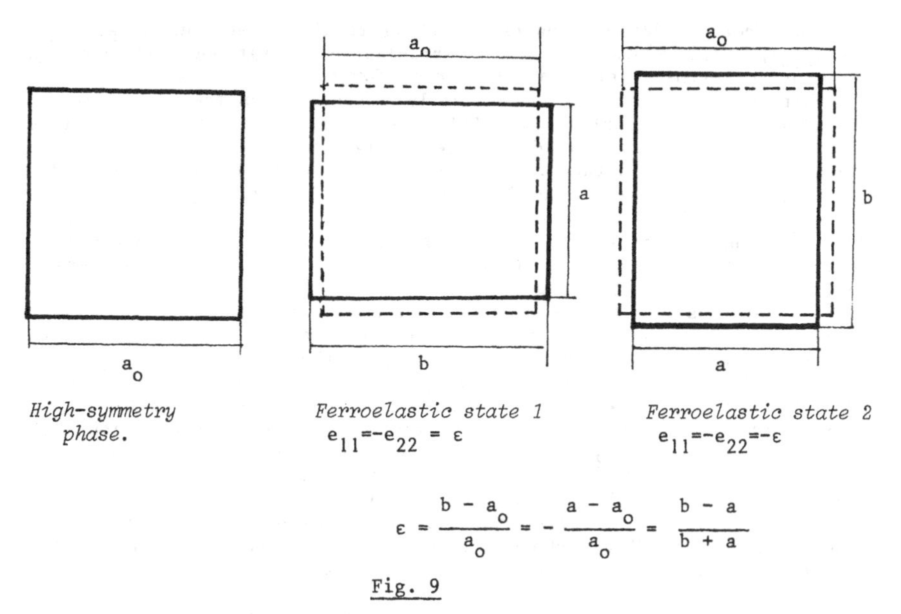

Fig. 9

Ferroelectricity and ferroelasticity can occur in conjunction or alone in a structural transition. Thus, a transition occuring between a non-polar point-group and a polar point-group both belonging to the same crystal system (e.g. $\hat{G}_o$ = mmm, $\hat{G}$ = mm2) will be ferroelectric non-ferroelastic. A transition between non-polar phases of different crystal systems will be ferroelastic, non-ferroelectric ($\hat{G}_o$ = 4/mmm, $\hat{G}$ = mmm). A transition from a non-polar phase of one crystal system to a polar phase of another system will be both ferroelectric and ferroelastic. (e.g. $\hat{G}_o$ = $\bar{4}$2m $\hat{G}$ = mm2).

The enumeration of the sets of point-groups ($\hat{G}_o$, $\hat{G}$) giving rise to each of the former combinations of properties has been performed by Aizu [25] and other authors [14], [14, 26, 27]. The relevant spontaneous components have also been specified [14, 25, 26].

c) Secondary and higher order ferroics

Certain sets of point-groups ($\hat{G}_o$, $\hat{G}$) correspond to identical forms for the physical quantities represented by vectors and by symmetric second rank tensors. Likewise the degenerate orientational states s_i cannot be distinguished by the values taken in each state by components of the two former types of tensors.
These states can nevertheless be distinguished by components of tensorial quantities of rank higher than 2.

The corresponding transitions belong to the categories of secondary ferroics, while ferroelectric and ferroelastic transitions pertain to the category of primary ferroics. The reason for this distinction is that the quantities which are thermodynamically conjugate to the characteristic tensors of ferroelectrics (i.e. Polarization) and of ferroelastics (i.e. strain) are "primary" fields, respectively the electric field and the mechanical stress. In the other cases, the conjugate fields, described by tensors of ranks larger than 2, correspond to products of primary fields [28].

Various classes of secondary ferroics have been distinguished on the basis of the rank of the spontaneous tensorial quantity. The sets of point-groups ($\hat{G}_0$, $\hat{G}$) pertaining to each class have been listed [30, 28, 14] as well as the rature of the spontaneous components of the tensors [28, 13]. Due to the non-ferroelastic character of the related transitions, the two groups $\hat{G}_0$ and $\hat{G}$ necessarily belong to the same crystal system. An example of secondary ferroic symmetry change corresponds to the set of point-groups ($\hat{G}_0$ = 4/mmm, $\hat{G}$ = 4/m) . he symmetric tensors of lowest rank which have a different number of zero-elements for the two groups are fourth-degree tensors, e.g. the elasticity tensor : the component C_{1112} of this tensor is compatible with the 4/m group and it vanishes by symmetry in a crystal of 4/mmm point symmetry (the two first reference axes are taken along reflection planes of $\hat{G}_0$ and the third one is along the 4-fold axis). Fig.10 shows schematically the type of structural modification which can give rise to such a secondary ferroic transition.

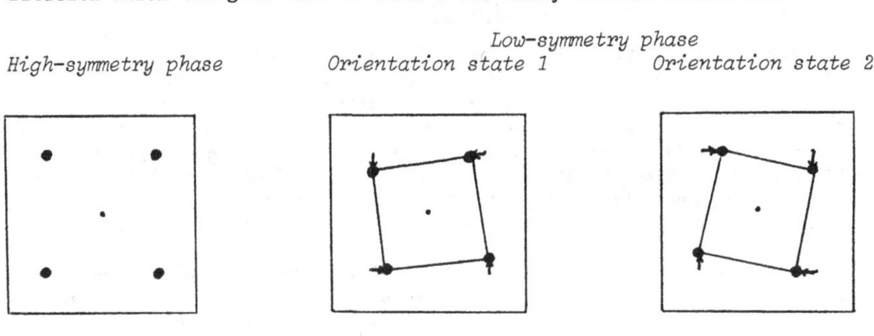

Fig. 10

It is worth pointing out that an infinite number of tensors of different ranks will acquire new "spontaneous" components for any symmetry-lowering $\hat{G}_0 \rightarrow \hat{G}$. For instance in a ferroelectric transition, tensors of odd rank will all acquire spontaneous components. This is, in particular, the case of the piezoelectric tensor [22] of rank three.
However, in primary ferroics (i.e. ferroelectric and/or ferroelastics), tensors of rank higher than two can generally be ignored, in the Landau theory of the considered transition. Likewise, in secondary ferroics, the account of the physical anomalies at the transition can be based on the consideration of the tensors of lowest ranks. Generally speaking, measured quantities do not often involve tensors of rank higher than four.

6.1.4. Use of the ferroic classification

The recognition that a transition is ferroelectric (resp. ferroelastic) directs the investigation of the transition towards the examination of the dielectric (resp. mechanical) properties of the system in the expectation that these will be the quantities mainly affected by the transition. This expectation is grounded on the fact that the dielectric polarization (resp. the thermal strain tensor) acquire spontaneous components across the transition.

Conversely, if none of these two classes of ferroics are involved in the considered transition, one knows that one must focus the study on components of higher rank macroscopic tensors in order to reveal the characteristic anomalies associated to the transition.

On the other hand, the characteristic tensors are different in the different orientation states s_i. Whenever these states coexist in the sample investigated, they constitute a "domain pattern" (in the same sense as the Weiss domains in ferro-magnetic substances). The knowledge of the ferroic class of a transition specifies the nature of the macroscopic tensorial quantity which must be measured in order to reveal the domain pattern. For instance, ferroelastic domains correspond to different values of symmetric second rank tensors. Aside from the spontaneous strain tensor, we can consider the permitivity tensor at optical frequencies. The latter tensor determines the optical indicatrix , which will be differently oriented in space for the distinct domains. Consequently, with suitably polarized light one should always be able to "visualize" ferroelastic domains. Conversely, such a visualization will never be possible by the same method, for a non-ferroelastic system.

6.2. Proper and improper ferroics [30, 31, 13, 14]

For each class of ferroics, there are spontaneous tensorial components sharing with the equilibrium-order-parameter the property of being zero above T_c and non-zero below T_c. However, as pointed out in chap. II §5, two different situations can arise depending if the tensorial components constitute the primary OP, or a set of secondary OP-components. If the tensorial components relative to the considered ferroic class coincide with the primary-components, the transition is a proper ferroic transition. If these tensorial components transform as IR's of G_o which are unequivalent to the OP-representation the considered transition is an improper ferroic transition.

The phase transition examined in chapter I, § 2.7.1, is a proper ferroelectric transition since the spontaneous polarization components P_z coincide with the OP-components § 7 hereunder will provide an example of a transition which is an improper ferroelectric and ferroelastic transition (a transition can be proper with respect to a certain ferroic class and improper with respect to another).

In the case of a proper ferroic transition, the temperature dependence of the spontaneous-tensorial components, and of their related susceptibilities corresponds to the behaviour described in the general context of chapter II for the primary OP and for its related quantities [Chap. II, § 3.6.2.b] : square root dependence of the spontaneous components as a function of $(T_c - T)$, divergence of the associated susceptibilities at T_c.

Since the tensorial components are, by definition (cf. § 6.1.1), invariant by the microscopic translations, these components carry IR's of G_o corresponding to $\vec{k}_o = o$ (Cf. eq. 2.6) : the OP of a proper ferroic transition transforms according to an IR of G_o with $\vec{k}_o = o$, i.e. of an IR of the point-group $\hat{G}_o$ (cf. § 2.2.4.d).

Conversely any non-totally-symmetric IR of G_o, with $\vec{k}_o = o$, determines a low-symmetry phase satisfying the condition $\hat{G} \subset \hat{G}_o$. Accordingly, the corresponding transition is ferroic, and it can be shown [13] that it is a proper ferroic transition with respect to one ferroic class at least. The nature of the ferroic transitions arising from each of the IR's of the 32 crystallographic point-groups $\hat{G}_o$ has been investigated in detail in refs. 29, 32 and 33.

Two types of questions are of interest when investigating the general characteristics of improper ferroic transitions. On the one hand, one should clarify the symmetry properties of the primary OP of each of the classes of ferroic transitions listed on figure 8. On the other hand, one should derive the qualitative and quantitative differences between the physical behaviour of the tensorial quantities across an improper ferroic transition and a proper one.

6.2.1. Characteristics of the OP of improper ferroics

For an improper ferroic transition, the spontaneous tensorial components constitute secondary order-parameters. Their peculiarity as compared to the secondary OP discussed in chapter II, § 5, is that they necessarily carry non-totally symmetric IR's of $\hat{G}_o$ (since they vanish by symmetry above T_c).

In accordance with the considerations developped in chap. II, § 5.2, the primary OP of an improper ferroic transition has a symmetry $\Gamma_i(\vec{k}_o)$ such as the symmetrized f-th power $[\Gamma_i(\vec{k}_o)^f]$ contains the representations of $\hat{G}_o$ pertaining to the relevant tensorial-components. This condition imposes certain constraints to $\Gamma_i(\vec{k}_o)$ and allows to specify the smallest value of the power f for each ferroic class. These constraints have been investigated in refs. 14. Let us briefly recall here the main results relative to the primary-OP of the non-ferroic transitions and of the main classes of improper ferroic transitions.

a) Non-ferroic transitions

Three necessary conditions are imposed to the OP-symmetry $\Gamma_i(k_o$

i) one must have $\vec{k}_o \neq o$

ii) The small representation τ_i must be one-dimensional (real or complex).

iii) If the star k_o^* has s-arms, and if $\eta(k_p)$ is a standard basis of the OP-space (no index i is necessary since τ_i is one-dimensional the equlibrium values of the OP-components , below T_c, must satisfy the condition :

$$\eta_{eq}(\vec{k}_o) = \eta_{eq}(\vec{k}_2) = \ldots = \eta_{eq}(\vec{k}_s) \qquad (6.1)$$

The symmetry-change $G_o \rightarrow G$, is reduced to a change $\mathfrak{T}_o \rightarrow \mathfrak{T}$ in the group of translations of the two phases, since, by definition $\hat{G}_o = \hat{G}$. This change is determined by the set of equations (3.19). Taking into account (6.1), we can see that the index of $\mathfrak{T}$ in $\mathfrak{T}_o$ is the largest compatible with the star k_o^*, since all the arms of k_o^* are involved in eqs. (3.19) : A non-ferroic transition is a associated with the "minimal" point-symmetry-change ($\hat{G}_o = \hat{G}$) and the maximal decrease of the translational symmetry.

Conditions i) to iii) are only sufficient if the small representation τ_i is real, and if k_o^* has a single arm, $\Gamma_i(k_o)$ is then one-dimensional [14d]. A systematic investigation of the symmetries of active representations of the 230 crystallographic space-groups shows [14d] that non-ferroic transitions are essentially related, to one-dimensional OP complying with the latter symmetry-condition. One finds, however, a certain number of non-ferroic symmetry changes related to multidimensional order-parameters [14d].

b) Improper non-ferroelastic transitions

Referring to figure 7, we see that these classes of ferroics comprise ferroelectric-non-ferroelastic-transitions, as well as secondary ferroic transitions.

For these transitions, the point-groups $\hat{G}$ and $\hat{G}_o$ belong to the same crystal system. This crystallographic requirement has been shown to imply the following properties of the OP representation $\Gamma_i\ (k_o)$ [14 a,b] :

i) One must have $\vec{k}_o \neq o$. Accordingly, these improper ferroics are necessarily associated (cf. eq. 3.19) to a change of the translational group of the system $\mathcal{T} \subset \mathcal{T}_o$, across the transition.

ii) As for non-ferroic transitions (cf. a), every arm $\vec{k}_p$ of the star k_o^* is associated to, at least, one non-zero equilibrium value of OP-component $\eta_1^o(k_p)$. Consequently, the loss of translational symmetry, across T_c, is the maximum loss compatible with the considered star k_o^*. However, by contrast to the case of non-ferroics, the small representations need not be one-dimensional.

A systematic investigation of the symmetry-changes determined by active IR's [14 a, b] has shown that improper ferroelectric non-ferroelastic transitions are associated exclusively with 2- and 3-dimensional OP, corresponding, essentially, to IR's of the trigonal and hexagonal space-groups [14 a]. The index f, for these transitions, is equal either to 2 or to 3.

In the case of improper secondary-ferroics the possible dimensions of the primary-OP are 2, 3, 4, 6, while the possible values of the f-indices are 2 or 3 [14 b].

c) Improper ferroelastics

The class of ferroelastics (with or without the simultaneous occurence of ferroelectricity) involves less stringent crystallographical constraints than the other ferroic classes. In particular, there is no imposed preservation of the crystal system across T_c. As a consequence, the OP of the feroelastic transitions pertain to a wider variety of symmetry types. The corresponding active IR's have dimensions ranging from 2 to 8 [14 c] The equilibrium values of OP-components $\eta_1^o(k_p)$ need not be non-zero for all the arms of the star k_o^*. Accordingly the loss of translational symmetry across T_c is not necessarily the maximum loss determined by eqs (3.19) for the considered star k_o^*. In an improper ferroelastic transition the index f can take the values (2, 3, or 4) [14 c].

It can also be noted that, by contrast to the situation in a) and b), improper ferroelastic transitions can possess a primary OP corresponding to $\vec{k}_o = o$. For instance, the point-symmetry-change m3m $\rightarrow$ 4/m is related to a 3-dimensional representation of the point-group m3m, whose symmetry properties are the same as the components of a 4th-rank tensor. This transition, which has the latter tensor components as primary OP, is nevertheless ferroelastic (since it involves a change of crystal system). However, the strain components constitute secondary OP-components and the transition is improper ferroelastic [Cf. ref. 14 c].

6.2.2. Physical behaviour of improper ferroics. Faintness index

We examine in this paragraph the temperature dependence, nearby T_c, of the spontaneous tensorial components, and of the associated susceptibilities, when the considered transition is an improper ferroic transition.

As usual in Landau's theory, this behaviour has to be derived from the free-energy associated to the transition. Since in a improper ferroic the relevant tensorial components $S_{ij\ldots l}$ constitute a set of secondary order-parameters, the free-energy F will contain three kinds of invariant polynomial terms, as already shown by eq. (5.2) in chapter II. The first ones are polynomials of the primary OP-components η_r, the second ones are polynomials of the relevant tensorial components, and the third ones are "mixed" polynomials of the η_r, and the $S_{ij\ldots l}$. As shown in chapter II §5, the latter terms of interest are linear in the $S_{ij\ldots l}$, and of degree $f \geqslant 2$, in the η_r. Also, in the vicinity of T_c one can restrict the expansion to the second degree terms in the $S_{ij\ldots l}$ and to the "mixed" term involving the smallest value of f. As pointed out in the preceding paragraph, the possible values of f comply with $2 \leqslant f \leqslant 4$, for structural transitions.

Equation (5.14) in chap. II, shows that, below T_c, the spontaneous tensorial components are related to the equilibrium value of the primary-OP by :

$$S \propto (\eta_r^o)^f \quad [T \leqslant T_c] \qquad (6.2)$$

This relation means that the components of $S_{ij\ldots l}$ are small quantities as compared to the η_r^o. The larger f, the smaller the $S_{ij\ldots l}$. This situation justifies the terminology introduced by Aizu for f [34] : f is the faintness index of the considered ferroic tensor S.

For the sake of simplicity let us restrict to a single S component, and denote E the macroscopic field thermodynamically conjugated to S (if S is the polarization, E is the electric field ; if S is a strain component, E is a stress component). Two types of susceptibilities can be defined for S, depending if the η_r, or their conjugated quantities ξ_r are maintained constant. In usual experiments, η_r is free to vary, and no conjugated field ξ_r is applied (this is, in particular impossible when $\vec{k}_o \neq o$). In this case, the susceptibility of interest is :

$$\chi_s = \frac{\partial S}{\partial E}\bigg|_{\xi_r = o,\ E = o} \qquad (6.3)$$

On the other hand, following eqs. (5.2)-(5.6) in chapter II, and putting $\rho^2 = (\sum \eta_r^2)$, and $\eta_r = \rho\cdot\gamma_r$, we can write the Landau free-energy in the presence of a non-zero field E as :

$$G = \frac{\alpha}{2}\rho^2 + \frac{\rho^4}{4}.f_4(\gamma_r) + \frac{S^2}{2\chi_o} + \delta.S.\rho^f.\phi(\gamma_r) + \frac{\lambda}{2}S^2.\rho^2 - S.E \qquad (6.4)$$

As will be shown hereunder, the term with λ-coefficient must be considered when $f \geqslant 3$. Its presence is always allowed by symmetry, in the free-energy of any transition, since it is the product of two totally-symmetric terms (cf. chap. II, eq. 5.1).

The equilibrium state of the system corresponds to the set of necessary conditions :

$$\frac{\partial G}{\partial \rho} \equiv \rho\left[\alpha + \rho^2.f_4(\gamma_r) + \delta f S\rho^{f-2}.\phi(\gamma_r) + \lambda S^2\right] = o \qquad (6.5)$$

$$\frac{\partial G}{\partial \gamma_r} = \frac{\rho^4}{4}.\frac{\partial f_4}{\partial \gamma_r} + \delta S\rho^f.\frac{\partial \phi}{\partial \gamma_r} = o \qquad (6.5')$$

$$\frac{\partial G}{\partial S} = o : E = \frac{S}{\chi_o} + \delta\rho^f.\phi(\gamma_r) + \lambda S\rho^2 \qquad (6.5'')$$

On the other hand, the susceptibility χ_S is obtained by derivation of (6.5") with respect to E (at constant ξ) :

$$1 = \chi_S\left(\frac{1}{\chi_o} + \lambda\rho^2\right) + \frac{\partial\rho}{\partial E}\left(\delta f\rho^{f-1}.\phi(\gamma_r) + 2\lambda S\rho\right) + \delta\rho^f\sum\frac{\partial\phi}{\partial\gamma_r}.\frac{\partial\gamma_r}{\partial E} \qquad (6.5''')$$

In many cases, equation (6.5') will determine symmetry directions in the OP-space (cf. chap. II, eq. 4.4.6) which do not depend on the field E. We can restrict to this case and put $\partial\gamma_r/\partial E = o$ in eq. (6.5'''). The susceptibility χ_S can then be derived from (6.5''') by expressing $(\partial\rho/\partial E)$ with help of (6.5) and (6.5").

a) behaviour above T_c

Above T_c, the relevant solution of (6.5) is $\rho = o$, independent of E. Reporting this value into (6.5'''), we find :

$$\chi_S = \chi_o \qquad (6.6)$$

χ_o, being the coefficient of S^2, is strictly positive nearby T_c, and has no anomalous behaviour related to the transition (unlike the coefficient α associated to the primary-OP). Hence, above T_c, the <u>susceptibility χ_S is not affected by the transition</u>. It has the same temperature dependence as χ_o which is temperature independent, or smoothly and weakly temperature dependent across T_c. This behaviour contrasts with the susceptibility relative to the η_r which diverges at T_c.

b) behaviour below T_c.

Below T_c eq. (6.5) reduces to the cancellation of the expression between brackets. Taking into account the constancy of the γ_r, and differentiating this expression with respect to E we are led to distinguish two cases.
If f = 2, $\rho^{f-2} = 1$, and the differentiation yields :

$$\frac{\partial\rho}{\partial E} = -\frac{\delta f\phi + 2\lambda S}{2\rho . f_4} \cdot \frac{\partial S}{\partial E} \qquad (6.7)$$

Reporting (6.7') into (6.5''') we obtain, to lowest order in ρ and S :

$$\chi_S = \frac{\chi_o}{1 - \chi_o\, 2\delta^2\phi^2/2f_4} \qquad (6.7')$$

Note that, below $T_c(\alpha < o)$, the set of eqs. (6.5) (6.5") requires

$$f_4\,(\gamma_r) > 2\delta^2\phi^2 > o \qquad (6.7")$$

Hence, in the vicinity of T_c, χ_S is temperature independent or smoothly varying with temperature. Since $\chi_S > \chi_o$, the anomaly of the susceptibility at T_c consists in a discontinuous upward step, on cooling through T_c.

If f $\geqslant$ 3, ρf^{-2} is not a constant. Differentiating the bracket in (6.5), and taking into account (6.5") for E = o, we obtain (neglecting the higher order terms $2\lambda S \ll \rho^{f-2}$ and $\rho^{f-3}.S \ll \rho$) :

$$\frac{\partial\rho}{\partial E} \approx -\frac{\delta f \rho^{f-3}.\phi(\gamma_r)}{2f_4\,(\gamma_r)} \qquad (6.8)$$

Reporting (6.8) into (6.5''') yields, to lowest order :

$$\chi_S = \frac{\chi_o}{1 - \frac{\chi_o\delta^2 f^2}{2f_4}\rho^{2f-4} + \lambda\rho^2} \qquad (6.8')$$

For f = 3, $\lambda\rho^2$ is of the same order of magnitude as ρ^{2f-4}, while for f = 4 $\lambda\rho^2$ is the dominant term of the two. For these values of f there is no discontinuity of χ_S at T_c, in contrast with the case f = 2, since $\chi_S(T<T_c) \to \chi_o$ $[\rho \to o$ when $T \to T_c]$. The anomaly at T_c consists in a upward or downward break of the slope of $\chi_S(T)$, depending on the coefficients of the free-energy.

The different variations obtained are not essentially modified by the consideration of terms of higher degrees in the expansion [34]. The results are summarized on fig.11 .

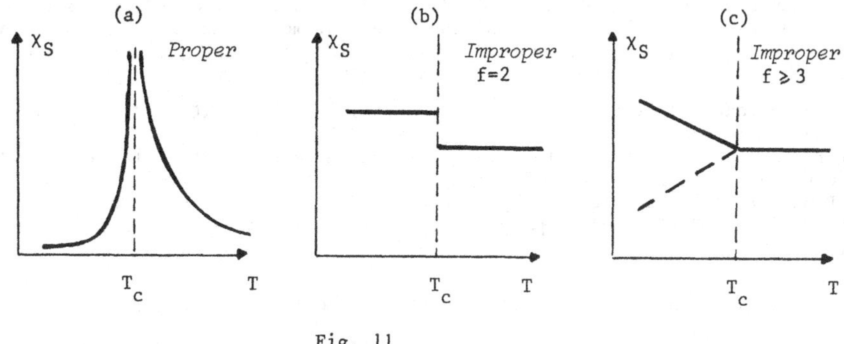

Fig. 11

The preceding results have been obtained on the basis of the simplified free-energy (6.4). The use of the correct form of the expansion as a function of the $S_{ij..l}$ components would not modify the qualitative temperature dependence obtained [36]. We have explained in chapter II § 5 the group theoretical method which can be used to obtain this correct expression of the free-energy, and, in particular, the form of the mixed invariant polynomials.

6.3. Pseudo-proper ferroics

In chapter II, § 5, we had discarded the possibility of having, in the Landau free-energy, a "mixed" invariant term, bilinear in the components of the primary OP, η_r, and in the secondary OP ζ_s. The basis for rejecting such a term is that the procedure leading to the definition of the primary OP (chap. II, § 3.5) implies that the second degree contribution to the free-energy is taken in the form of a sum of squares associated to the various irreducible degrees of freedom of the system.

Nevertheless, the existence of a bilinear mixed invariant has often been considered [30, 37] in the application of Landau's theory, in order to account for the properties of systems in which two distinct irreducible degrees of freedom are of interest.

The physical relevance of such a method can be understood in the following way. Assume, for instance, that ζ_s is a set of strain-components (e.g. tensorial components in a ferroelastic transition), and that these components are "clamped" (i.e. ζ_s is maintained to the value ζ_s = o) during the variation of T. In this case, even if the symmetries of ζ_s and η_r allow the existence of a bilinear term ($\sum \zeta_r . \eta_r$) in the free-energy, the effective second-degree term will be in the canonical form:

$$\frac{\alpha}{2} (\sum \eta_r^2) + \frac{b}{2} (\sum \zeta_s^2) \qquad (6.9)$$

In this system, it can happen that a continuous transition exists, defined by the linear vanishing of α at T_c.

Assume now that the clamping is released. The bilinear mixed invariant must be taken into account. It has the effect of modifying the characteristics of the transition. It is this modification which is the subject of the application of Landau's theory mentioned above [30,37].

Note that a bilinear term can only exist if (η_r) and (ζ_s) transform according to the same IR of G_o (Cf. chap.II, § 2.2.3). If the (ζ_s) coincide with ferroic tensorial components, the corresponding ferroic transition is said to be pseudo-proper: the tensorial components S are distinct from the primary OP but transform according to the same IR of G_o. Since S corresponds to $\vec{k}_o=0$, the primary OP of a pseudo-proper ferroic transition necessarily corresponds to an IR with $\vec{k}_o=0$.

7. EXAMPLE: THE STRUCTURAL TRANSITION IN GADOLINIUM MOLYBDATE.

The phenomenological theory of the 160°C transition in gadolinium molybdate $Gd_2(MoO_4)_3$ (GMO), is a good illustration of the considerations developped in this chapter, as well as of their limitations in accounting for the detailed behaviour of a real system.

7.1. Symmetry of the primary-order parameter.

Crystallographic data show [39] that the transition in GMO occurs between two phases with symmetries $G_o=P\bar{4}2_1m$ and $G=Pba2$. Thus, the transition involves a lowering of the point symmetry ($\hat{G}_o=\bar{4}2m \rightarrow \hat{G}=mm2$), as well as a lowering of the translational symmetry: the group $\mathcal{T}_o$ is generated by the three basic translations $\vec{a}_1, \vec{a}_2, \vec{a}_3$, of the P-tetragonal Bravais lattice ($\vec{a}_3$ directed along the 4-fold axis of G_o, and $|\vec{a}_1|=|\vec{a}_2|$), while $\mathcal{T}$ is generated by:

$$\vec{a}'_1 = \vec{a}_1 - \vec{a}_2 \qquad \vec{a}'_2 = \vec{a}_1 + \vec{a}_2 \qquad \vec{a}'_3 = \vec{a}_3 \tag{7.1}$$

The index of $\mathcal{T}$ in $\mathcal{T}_o$ is 2. Since the index of $\hat{G}$ in $\hat{G}_o$ is also 2, the index of the space group G in G_o is 4.

The relation (7.1) allows to identify unambiguously the $\vec{k}_o$ vector of the IR of G_o associated to the primary OP. Indeed, if we put $\vec{k}_o = \Sigma\, q_i^o . \vec{a}_i^*$, equations (3.19) must be satisfied by $\vec{k}_o$ and the translations $\vec{a}'_i$ in (7.1), and by no other set of primitive translations of $\mathcal{T}_o$ corresponding to a unit-cell smaller than (7.1). Expressing eqs. (3.19) and the latter condition, we obtain, up to a primitive translation of the reciprocal lattice ($\vec{a}_i^*$):

$$\vec{k}_o = \frac{1}{2}\,(\vec{a}_1^* + \vec{a}_2^*) \tag{7.2}$$

We will see, in §8, how $\vec{k}_o$ can be deduced directly from the diffraction spectra of the crystal on either sides of T_c.

The star k_o^* generated by applying $\hat{G}_o$ to $\vec{k}_o$ consists of the vectors $\pm(\vec{a}_1^* \pm \vec{a}_2^*)/2$ which are equivalent to $\vec{k}_o$. This star is therefore reduced to a single vector. Consequently, the IR's of G_o, $\Gamma_i(\vec{k}_o)$, coincide with the small representations τ_i. Consultation of standard tables [5,6] allows to specify the τ_i, and shows that, for the vector (7.2), the space-group G_o has 4 one-dimensional complex IR's (τ_1, $\bar{\tau}_1$, τ_2, $\bar{\tau}_2$) and one two-dimensional real IR τ_5. The possible OP-symmetries are represented by τ_5, and by the two physically irreducible" representations (Cf.chap.II §2) constructed from the sets of complex conjugate IR's $(\tau_1+\bar{\tau}_1)$ and $(\tau_2+\bar{\tau}_2)$. The matrices of the three preceding representations are reproduced on table 9.

Table 9.

G_o	$\{E\|0\}$, $\vec{a}_3$	$\{S_4\|0\}$	$\{U_z\|0\}$	$\{S_4^3\|0\}$	$\{\sigma_{xy}\|\vec{t}_o\}$	$\{\sigma_{\bar{x}y}\|\vec{t}_o\}$	$\{U_x\|\vec{t}_o\}$	$\{U_y\|\vec{t}_o\}$	$\vec{a}_1$, $\vec{a}_2$
						$\vec{t}_o=(\vec{a}_1+\vec{a}_2)/2$			
Γ_5	$\begin{vmatrix}1&0\\0&1\end{vmatrix}$ (E)	$\begin{vmatrix}1&0\\0&-1\end{vmatrix}$ (σ_y)	$\begin{vmatrix}1&0\\0&1\end{vmatrix}$ (E)	$\begin{vmatrix}1&0\\0&-1\end{vmatrix}$ (σ_y)	$\begin{vmatrix}0&1\\1&0\end{vmatrix}$ $(\sigma_{\bar{x}y})$	$\begin{vmatrix}0&1\\1&0\end{vmatrix}$ $(\sigma_{\bar{x}y})$	$\begin{vmatrix}0&-1\\1&0\end{vmatrix}$ (C_4)	$\begin{vmatrix}0&-1\\1&0\end{vmatrix}$ (C_4)	$\begin{vmatrix}-1&0\\0&-1\end{vmatrix}$ (C_2)
$\left.\begin{matrix}\Gamma_1\\\Gamma_2\end{matrix}\right\}$	$\begin{vmatrix}1&0\\0&1\end{vmatrix}$ (E)	$\begin{vmatrix}0&1\\-1&0\end{vmatrix}$ (C_4^3)	$\begin{vmatrix}-1&0\\0&-1\end{vmatrix}$ (C_2)	$\begin{vmatrix}0&-1\\1&0\end{vmatrix}$ (C_4)	$\pm\begin{vmatrix}1&0\\0&1\end{vmatrix}$ (E,C_2)	$\pm\begin{vmatrix}-1&0\\0&-1\end{vmatrix}$ (C_2,E)	$\pm\begin{vmatrix}0&1\\-1&0\end{vmatrix}$ (C_4^3,C_4)	$\pm\begin{vmatrix}0&-1\\1&0\end{vmatrix}$ (C_4,C_4^3)	$\begin{vmatrix}-1&0\\0&-1\end{vmatrix}$ (C_2)

In principle,the IR related to the OP of the considered transition can be directly deduced from experimental data:e.g. from neutron diffraction measurements performed above T_c (Cf. ref. 40 and §8 hereafter).In the present example,this identification can also be performed by working out the symmetries G compatible with each of the 3 IR's in table 9,and comparing the results to the experimentally determined group G=Pba2 [39]. In this view,in the line of the method described in §5, we determine,first,the images of G_o in each representation,as well as the kernels of the corresponding homomorphisms.

7.1.1 Images,kernels,and isotropy groups.

Inspection of table 9 reveals that the set of distinct matrices of Γ_5 contains 8 elements forming a group I_o isomorphous to $\mathbb{C}_{4v}$ (Cf.§4.2.2 and chap.II §4.5.3). The correspondance $I_o \leftrightarrow \mathbb{C}_{4v}$ has been specified in table 9 under each matrix.

The image I_o' of G_o in the two other representations is the same:it consists of 4 distinct matrices isomorphous to the group $\mathbb{C}_4$.

The kernels K_5,K_1 and K_2 all contain,as subgroup, the translation group $\mathcal{T}$ of the low symmetry phase, generated by the $\vec{a}_i'$ in eq. (7.1).Let us write each kernel K as an expansion into cosets with respect to $\mathcal{T}$:

$$K=\mathcal{T}+\{R|\vec{t}\}\mathcal{T}+\ldots \tag{7.3}$$

where $\{R|\vec{t}\}$ is an element of G_o not belonging to $\mathcal{T}$,whose matrix is the identity matrix. Hence it is either an element appearing explicitly on table 9,or the product of such an element by the translations $\vec{a}_1$ or $\vec{a}_2$ (it is sufficient to consider $\vec{a}_1$ since $(\vec{a}_1+\vec{a}_2)$ belongs to $\mathcal{T}$).Inspection of table 9 shows that the elements generating the cosets of K_5, K_1 and K_2 are, respectively:

$$
\begin{array}{lll}
K_5 & \longrightarrow & \{E|0\}\ \{U_z|0\} \\
K_1 & \longrightarrow & \{E|0\}\ \{U_z|\vec{a}_1\}\{\sigma_{xy}|\vec{t}_o\}\{\sigma_{\bar{x}y}|\vec{t}_o+\vec{a}_1\} \\
K_2 & \longrightarrow & \{E|0\}\ \{U_z|\vec{a}_1\}\{\sigma_{xy}|\vec{t}_o+\vec{a}_1\}\{\sigma_{\bar{x}y}\ |\vec{t}_o\ \}
\end{array}
\qquad (7.4)
$$

The isotropy groups $I \subset C_{4v}$ have been determined in chapter II §4.6.2 and fig.7). There are three types of orbits respectively represented by the directions (1,0), (1,1), and ($x,y \neq x$) in the OP-space. Their isotropy groups are respectively the subgroups (E,σ_y), $(E,\sigma_{\bar{x}y})$ and (E) of the image I_o.

A simpler result holds for the isotropy groups of the second type of image $I_o' = C_4$. In the two-dimensional space of the OP, no direction is invariant by the rotations of this group. Hence, there is a single type of orbit characterized by the trivial isotropy group (E).

7.1.2. Low-symmetry space groups.

As stated in §5, the possible low symmetry groups G compatible with the preceding representations are the groups G=IxK. Using the isotropy groups I determined above, and the kernels (7.4) we find the generators of the groups G associated to each of the three considered IR's (table 10). Note that in the case of Γ_1 and Γ_2, G is identical to the kernel.

Table 10.

	Generators		Space group
Γ_5	$\{E\|0\}\{U_z\|0\}\ \ \{S_4\|0\}\ \ \{S_4^3\|0\}$	X $\mathcal{T}$	$P\bar{4}$
	$\{E\|0\}\{U_z\|0\}\ \ \{\sigma_{xy}\|\vec{t}_o\}\ \ \{\sigma_{\bar{x}y}\|\vec{t}_o\}$		Pma2
	$\{E\|0\}\{U_z\|0\}$		P2
Γ_1	$[\{E\|0\}\{U_z\|\vec{a}_1\}\ \{\sigma_{xy}\|\vec{t}_o\}\ \{\sigma_{\bar{x}y}\|\vec{t}_o+\vec{a}_1\}]$	X $\mathcal{T}$	Pmm2
Γ_2	$[\{E\|0\}\{U_z\|\vec{a}_1\}\ \{\sigma_{xy}\|\vec{t}_o+\vec{a}_1\}\ \{\sigma_{\bar{x}y}\|\vec{t}_o\}]$	X $\mathcal{T}$	Pba2

The symbols of the space groups, indicated on the last column of table 10, have been identified as in § 5.2, and transformed into a standard origin and orientation [4].

We can see that one, only, of these groups coincides with the experimentally observed space-group of GMO, below T_c. This allows the identification of the OP-symmetry of the considered transition as $\Gamma_2(\vec{k}_o)$.

7.1.3. Compliance with the Landau and Lifschitz conditions.

Let us show that the OP-representation Γ_2 is active (§4.1).

Note first that $3\vec{k}_o = \vec{k}_o \neq 0$: the sum of any three vectors belonging to the star k_o^* (which is reduced to the single vector $\vec{k}_o$) is different from zero. This result ensures the fulfilment of the Landau condition (Cf.§4.1).

$\vec{k}_o$ being one of the vectors compatible with the Lifschitz condition (table 6), a further probing of the fulfilment of this condition by $\Gamma_2(\vec{k}_o)$ is necessary. As $\vec{k}_o$ is equivalent to $(-\vec{k}_o)$, the appropriate expression of the Lifschitz criterion is provided by eq. (3.13'). Using the content of table 9, and applying eq.(3.13') in the same manner as in table 5, one can check straightforwardly that the Lifschitz criterion is satisfied.

Hence, Γ_2 is an active IR of G_o : the considered OP is apt to describe a continuous transition between two crystalline phases. Experimentally, the transition has a small discontinuity. In the subsequent discussion we will neglect this feature, and discuss its relevance at the end of the paragraph.

7.2. Secondary OP and free-energy expansion.

7.2.1. Improper ferroelectricity and ferroelasticity in GMO.

The point symmetry-change observed at the transition of GMO is from a non-polar group $\hat{G}_o = \bar{4}2m$ of the tetragonal system, to a polar group $\hat{G}$=mm2 of the orthorhombic system. As emphasized in § 6.1.b), such a symmetry change identifies the considered transition as simultaneously ferroelectric and ferroelastic. Consultation of appropriate tables [14,22,25-29] shows that the polarization component associated to the "symmetry breaking" is P_3 (the coordinate axes are parallel to the translations $\vec{a}_i$): this component is not preserved by the group $\hat{G}_o$, while a non-zero value for it is compatible with the group $\hat{G}$. Likewise, one finds that the symmetry breaking component of the strain tensor is the shear e_{12}. The magnitude of e_{12} is a measure of the angle between $\vec{a}_1$ and $\vec{a}_2$ below T_c, while the symmetry $\hat{G}_o$, existing above T_c, requires $\vec{a}_1$ and $\vec{a}_2$ to be perpendicular.

The two quantities P_3 and e_{12} carry, separately, the same one-dimensional IR (denoted γ) of $\hat{G}_o$. γ is defined by the characters (+1) for the elements (S_4, S_4^3, U_x, U_y) of $\hat{G}_o$, and the characters (-1) for the remaining elements. As this representation is distinct from the primary OP-representation $\Gamma_2(\vec{k}_o)$, identified above, the transition in GMO must be classified as improper ferroelectric and improper ferroelastic (Cf. § 6.2.1): P_3 and e_{12} are secondary order parameters of the transition.

7.2.2. Form of the Landau free-energy.

In agreement with § 6.2.2, the free-energy F has to contain 3 parts, respectively corresponding to the primary-OP, to the secondary-OP, and to the expression of the coupling between the two sets of degrees of freedom.

a) Primary-OP contribution.

We have pointed out in § 7.1.1, that the image of G_o in the OP representation is $\mathbb{C}_4$. For this image, the 4-th degree expansion has been indicated in chapter II eq.4.19. Using this result, and making explicit the dependence of the free-energy on the expansion's coefficients β_i, we write:

$$F_1(\beta_1,\beta_2,\beta_3 | \rho,\theta) = \frac{a(T-T_c)}{2}.\rho^2 + \frac{\beta_1}{4}\rho^4 + \frac{\beta_2}{4}\rho^4\cos 4\theta + \frac{\beta_3}{4}\rho^4\sin 4\theta \qquad (7.5)$$

where $\eta_1 = \rho\cos\theta$ and $\eta_2 = \rho\sin\theta$,are the primary OP components whose transformation properties are defined by the matrices of Γ_2 in table 9.

b) **Secondary order parameter contribution.**

This contribution can be restricted to quadratic terms (Cf. chap.II, §5). It has the form :

$$F_2 = \frac{P_3^2}{2\chi_o} + \frac{C}{2} e_{12}^2 + \delta P_3 . e_{12} \tag{7.6}$$

The existence of the last term in (7.6) results from the identical symmetries of P_3 and e_{12}: their product is invariant by G_o. The quadratic form (7.6) can be transformed into a sum of squares. We have chosen to keep it in this form in order to preserve the physical meaning of the quantities considered (polarization ,and strain). The linear coupling term then represents the existence of a piezoelectric coupling [22-24] between the polarization and the strain, which is compatible with the symmetry G_o.

c) **Coupling- term.**

As stressed in Chap.II § 5.2, the construction of the "mixed-invariant" contribution to the free-energy, can be achieved by searching the smallest number $f \geq 2$,for which the symmetrized f-th power of the representation of (η_1, η_2), contains the representation of P_3 or e_{12}. The representation $|\Gamma_2^2|$ is spanned by $(\eta_1^2 + \eta_2^2 = \rho^2 ;\ \eta_1\eta_2 = \rho^2\sin 2\theta ; \eta_1^2 - \eta_2^2 = \rho^2\cos 2\theta)$. The first of these polynomials is totally symmetric. As easily checked on table 9,each one of the two others carries the representation γ . Hence, the mixed invariant contribution to the free-energy has the form:

$$F_3 = P_3 . \rho^2 (\delta_1' . \sin 2\theta + \delta_2' . \cos 2\theta) + e_{12}\rho^2 (\delta_1'' \sin 2\theta + \delta_2'' \cos 2\theta) \tag{7.7}$$

7.3. Phenomenological theory of GMO.

The free-energy $F = (F_1 + F_2 + F_3)$ provides the basis for a discussion of the physical properties of GMO related to its structural transition:symmetry characteristics of the phase stable below T_c , temperature dependences of the primary and secondary order parameters , and of the susceptibilities associated to these degrees of freedom.

7.3.1. Low symmetry phase.

The image $\mathbb{C}_4$ of G_o, does not contain any plane-symmetries,but only rotations about an axis perpendicular to the OP-space. Hence, this space has no symmetry-directions . The isotropy groups of all directions in the OP-space are identical,and reduced to the identity transformation. This situation has the consequence that it is not necessary to determine the minimum of F if we want to specify the symmetry G below T_c:whatever the "equilibrium" direction in the OP-space, the symmetry G is the same, in agreement with the indications of table 10. The need of minimizing effectively F is restricted to two aspects: specify the range of coefficients $(\beta_i ,\delta_i' ,\delta_i'' ...)$ corresponding to the positivity of F for infinite values of the variables; determine the temperature dependence of the physical quantities of interest.

a) **Temperature dependence of θ.**

The equilibrium state of the system, at each temperature, is determined by the absolute minimum of F with respect to ρ, θ, P_3, and e_{12}. We can eliminate, in the first place the variables P_3 and e_{12}, by expressing the minimum of (F_2+F_3) with respect to these variables. As pointed out in Chap.II §5, such an elimination leads to a free-energy $F=F_1(\beta_1', \beta_2', \beta_3' | \rho, \theta)$, formally identical to (7.5), but with "renormalized" coefficients $\beta_i' = \beta_i'(\beta_j, \delta_j, \delta_j', \delta_j'', C, \chi_o)$. The minimum of F with respect to θ then yields:

$$\tan(4\theta_o) = \frac{\beta_3'}{\beta_2'} \qquad (7.8)$$

$$\beta_2' \cos(4\theta_o) < 0 \qquad (7.8')$$

Unlike the coefficient $\alpha \propto a(T-T_c)$, the coefficients $(\beta_i, \delta_j', \ldots)$ are assumed to be temperature independent, or to possess a weak, "background", temperature dependence (i.e. not induced by the existence of the transition at T_c). This property also holds for the β_i', and consequently, the angle θ_o, determined by eq. (7.8), is either constant, or weakly temperature dependent. This possible temperature dependence is consistent with the fact that the equilibrium direction in the OP-space, which is specified by θ_o, is not a symmetry direction.

In the view of estimating a possible temperature dependence of θ_o, let us rely on experimental data, and consider the "mechanical" degrees of freedom of the system.

Up to now, we have only considered the symmetry breaking strain component e_{12}. We can also study the behaviour of strain components which preserve the point-symmetry of the system (this was the case of the strain component considered in chap.I § 2.9). In the present system, such a component is provided, for instance by e_{33}, which represents the elongation of the crystal along the z-axis. In the same manner as the strain component investigated in chap.I, e_{33} carries the totally symmetric IR of G_o. Accordingly, its contribution to the free-energy of the system has the form:

$$\frac{C'}{2} e_{33}^2 + \delta_3 . e_{33} . \rho^2 \qquad (7.9)$$

(The linear term in e_{33}, which is symmetry-permitted, is assumed to have been eliminated through the procedure used in chap.I §2.9). The derivative of (7.9) with respect to e_{33} yields the equilibrium value e_{33}^o of this strain component:

$$e_{33}^o \propto \rho_o^2(T) \qquad (7.9')$$

On the other hand, the derivative of (F_2+F_3) with respect to P_3 and e_{12} (eqs. (7.6) and (7.7)), yield the equilibrium value of e_{12}:

$$e_{12}^o = \rho_o^2(T) . f(\sin 2\theta_o \; ; \cos 2\theta_o) \qquad (7.10)$$

where f is a linear combination of sin(2θ) and cos(2θ).A comparison of (7.9') and of (7.10) shows that the ratio of the two strain components e^o_{33} and e^o_{12} will disclose the possible temperature dependence of θ_o.

Actually,from the available data [42] ,one infers that this ratio is constant ,within experimental accuracy, in the entire temperature range investigated (20°C < T < 160°C)

We can conclude that,scaled to the accuracy of the measurements, [42,43] ,the angle $\theta_o(T)$ can be considered independent of the temperature. As stressed above, this constancy is not imposed by the symmetry.

b) Rotation of the reference frame in the OP -space.

Since θ_o is constant, it is of interest to use eq.(7.8) and substitute in F the coefficient β'_3 by its expression as a function of β'_2 and θ_o. Reporting this expression in the free-energy $F=F_1\ (\beta'_1,\beta'_2,\beta'_3\ |\rho,\theta)$,we obtain:

$$F = \frac{a(T-T_c)}{2}\rho^2 + \frac{\beta'_1}{4}\rho^4 + \frac{\beta'_2}{4\cos\theta_o}.\rho^4.\cos4(\theta-\theta_o) \qquad (7.11)$$

This transformation brings F to a simpler form,only containing 2 independent fourth degree terms ,and identical to the free-energy pertaining to the image $\mathbb{C}_{4v}$ (Cf. Chap.II,eq. 4.19',and Chap.I §2.7.2). In agreement with results derived in previous chapters,the condition of stability of the low-symmetry phase consists in $\beta'_1 > 0$,in addition to (7.8').

In order to complete the identification of eq.(7.11) with eq.(4.19') in chap.II,it is sufficient to rotate,by the angle θ_o,the reference frame in the OP-space,thus bringing the η_1-axis in this space,in coincidence with the direction $\theta = \theta_o$. The inequality (7.8') then imposes,consistently, that the minimum of F corresponds to $\theta_o=0$ (modulo $\pi/2$).

c) Ferroic and antiphase domains in GMO.

The $\mathbb{C}_4$ symmetry of the OP imposes a 4-fold degeneracy of the absolute minimum of F. The 4 equivalent equilibrium states are defined by the same value of $\rho_o(T)$ and by the 4 angles $\theta_o=(0,\pi/2,\pi,3\pi/2)$.Table 11,hereunder, shows the correspondance between these angles and the equilibrium values of the polarization P_3 and of the strain e_{12} . This correspondance stems from the linearity of the function f in eq.(7.10) in $\sin2\theta_o$ and $\cos2\theta_o$.

Hence,there are only two orientation states,in the sense defined in §6.These states corresponding to pairs of degenerate states ($\theta_o=0,\pi$) and ($\theta_o=\pi/2,3\pi/2$),are distinguished by the opposite values taken by both P_3 and e_{12}. If simultaneously present in a sample,the 2 orientation states constitute ferroelectric and ferroelastic domains.

Table 11

o	0	$\frac{\pi}{2}$	π	$\frac{3\pi}{2}$
P_3	P	-P	P	-P
e_{12}	e	-e	e	-e
η_1	η_1	$-\eta_2$	$-\eta_1$	η_2
η_2	η_2	η_1	$-\eta_2$	$-\eta_1$

The last lines of table 11 shows the values of the order parameter components (η_1,η_2) in the 4 states. We can see that two states which have the same ferroic tensors (e.g. $\theta_o=0,\pi$) have opposite values of (η_1,η_2). Inspection of table 9 shows that the only matrix of G_o which achieves such an inversion of both η_1 and η_2 is the matrix representing the primitive translations $\vec{a}_1$ and $\vec{a}_2$. Hence the states $\theta=0$, and $\theta=\pi$ correspond to crystal structures "shifted" with respect to eachother by a translation $\vec{a}_1$ or $\vec{a}_2$. If simultaneously present in a sample, these states constitute translations domains, or, according to an accepted terminology, [44], antiphase domains. Fig. 12 shows the expected domain pattern of GMO, involving orientation and antiphase states. It also shows the operations of G_o which establish the geometrical correspondance between these states.

Fig. 12

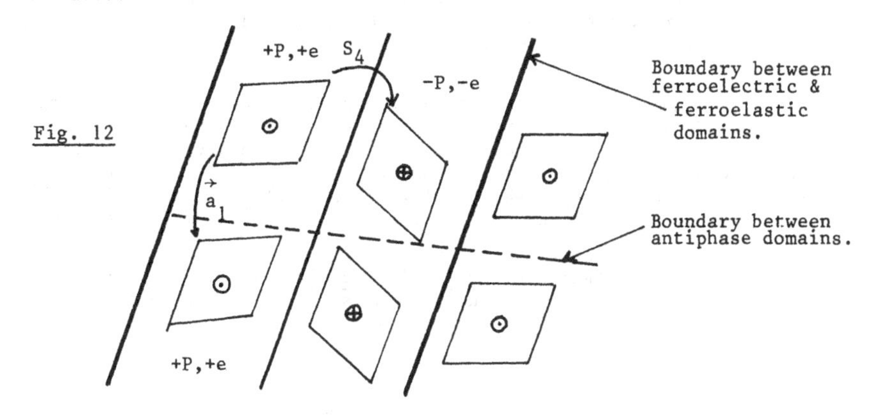

7.3.2 Temperature dependence of the physical quantities.

The form (7.11) of F being identical to the free-energy studied in chapter I § 2.7.2, we can use the results already established in this case. The temperature dependence of ρ_o is given by:

$$\rho_o = \sqrt{\frac{a(T_c-T)}{\beta_1' - |\beta_2' \cos 4\theta_o|}} \qquad (7.12)$$

Likewise, the "radial" and "angular" susceptibilities respectively associated to ρ and θ, both diverge at T_c (Cf. chap.I, eq.2.19).

The general results derived in §6.2.2 for improper ferroics also apply to GMO. As apparent in eqs.(7.6) and (7.10), the faintness indices f relative to P_3, e_{12}, and e_{33} are all equal to 2. Hence, these quantities being proportional to ρ_o^2, should vary linearly below T_c, as a function of the temperature. Their associated susceptibilities, which are respectively a dielectric susceptibility χ_{33} and elastic "compliances" |22-24| S_{12} and S_{33}, should display the behaviour indicated on fig. 10, for f=2: no temperature dependence on either sides of T_c (or a "background" temperature dependence), and an upward step variation on cooling through T_c.

However, we expect this description to be qualitative only. Indeed, the calculation of the susceptibilities performed in § 6.2.2 has assumed no variation of the angle θ, under application of the fields conjugated to the secondary order parameters. This is not necessarily valid in GMO since P_3 and e_{12} are coupled to θ -dependent terms. The influence of a variation of θ has been examined in the theory developped by Dvorak [39]. The behaviour deduced is in qualitative agreement with fig. 10.

Fig.13 shows schematically the calculated variations of $P_3(T)$, $\chi_{33}(T)$, and C(T) (which is the inverse susceptibility associated to e_{12}), as well as the results of the experimental observations of these quantities. The figure discloses the successes and the limitations of the above theoretical treatment. The absence of divergence of the two susceptibilities, and the occurence of plateaux on either sides of T_c is correctly accounted for by the theory. On the other hand, the observed variation of $P_3(T)$ is not linear, and C(T) has an unexpected temperature dependence in the vicinity of T_c.

Two arguments can be invoked to explain these differences. One is the consideration of the discontinuous character of the transition in GMO. This character requires expanding F up to a degree higher than 4. The general consequences of such an expansion will be discussed in chapter IV. In the present case, it allows to account satisfactorily for the behaviour of $P_3(T)$ [45], and partly for the variations of C(T) below T_c [46]

The second argument, which explains the variation of C(T) above T_c, [46] is the consideration of critical fluctuations. The latter phenmenon, which constitutes a general limitation to the validity of Landau's theory will be discussed in chapter VIII.

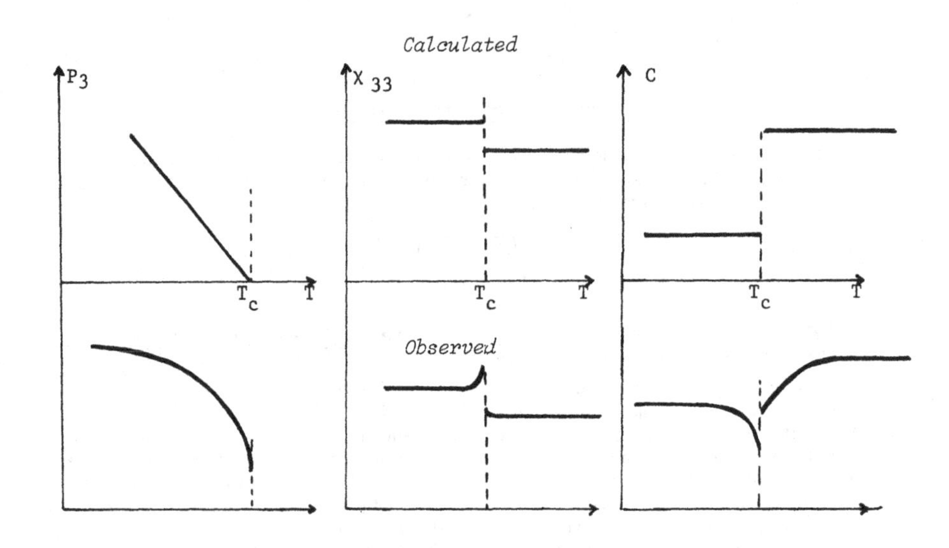

Fig.13

8. CONNECTIONS TO THE EXPERIMENTAL SITUATION.

In this paragraph we examine various types of connections between the theoretical aspects developped in the present chapter, and the structural transitions observed in real systems.

8.1. Physical nature of the order-parameter.

In the framework of Landau's theory, the general physical meaning of the OP is rooted in the relationship between the OP and the density increment (eq.(3.9) in chap. II):

$$\delta\rho(\vec{r}) = \Sigma\ \eta_r \cdot \phi_r(\vec{r}) \qquad (8.1)$$

In the spirit of the convention made in chap.II,§3.5, we can assume that the normalized functions ϕ_r are also dimensionless. The OP-components must then be considered to possess the same physical nature as the density increment. In agreement with the definition of $\rho(\vec{r})$ (chap.II,§ 3.1), $\delta\rho(\vec{r})$ is a variation of the probability density of the particles in the system. In the case of structural transitions, the equilibrium probability densities associated to the various types of particles of the system define the structure of the crystal (Cf.§1). In this case, the set of OP-components η_r represents a certain type of structural deviation, the structure of reference being that of the high-symmetry phase. This structural deviation has, in addition, well defined symmetry properties (i.e. its irreducibility).

Note that the difference between the structures of the low-symmetry phase, stable below T_c, and the high-symmetry one, is not only only determined by the equilibrium value η_r^o of the OP. This difference actually corresponds to the total density increment, $\delta\rho_{eq.}(\vec{r})$ (chap.II, eq.(5.15)) which incorporates the structural deviations determined by the primary OP, (η_r^o), and by the secondary order parameters (ζ_s^o).

8.1.1 . Different types of order-parameters.

It is of interest to distinguish two types of structural transitions on the basis of the physical nature of their order-parameter.

a) Displacive type.

In displacive structural transitions, the OP represents a collective displacement of the atoms, ions,.. of the structure, with respect to the average positions occupied, at equilibrium, in the high-symmetry phase.

The model-example investigated in chap.I provides a simple illustration of such a transition. In this model, the OP coincides with the displacement of a single ion in each cell of the crystal. Generally speaking, the collective displacement will involve several atoms. This is, for instance, the case of GMO (Cf. §7). The OP of its transition coincides with a complicated set of rotations of the tetrahedra (MoO_4) forming the skeletton of the structure [41]. Fig. 14 shows, schematically, the tetrahedra concerned by the rotations, for two values of the angular component of the OP ($\theta=0$, and $\theta=\pi/2$) respectively corresponding to the two ferroelectric-ferroelastic states ($\pm P_3$, $\pm e_{12}$). It can be checked that the rotations of the tetrahedra break the high-symmetry ($P\bar{4}2_1m$), and establish the polar-orthorhombic point-symmetry in the low symmetry phase (mm2).

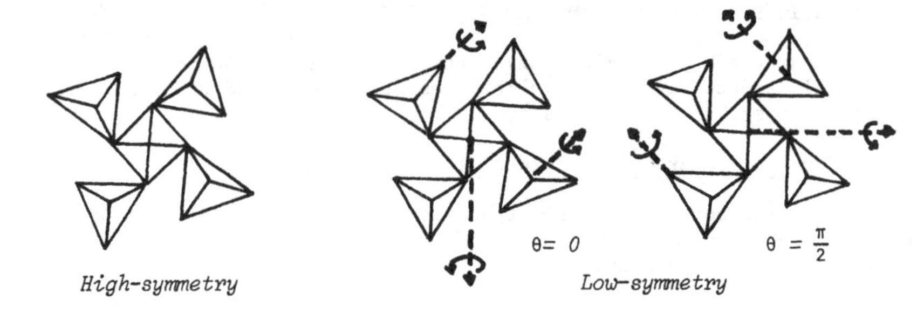

Fig. 14

However, though they determine a point-symmetry compatible with the existence of a polarization P_3 , these rotations do not generate directly the occurence of an electric dipole in each cell, in contrast to the situation in the model-example of chap.I . This structural characteristics of GMO is a general feature of improper ferroics : the onset of the macroscopic tensorial quantities constituting the secondary order parameters is a small effect at the structural level; it is not readily apparent in the main structural displacements taking place across the transition.

It is worth recalling that collective displacements of atoms of the type involved in the order-parameter are also considered in the study of the collective oscillations of the atoms of the high-temperature crystalline structure, around their equilibrium positions [47].In this case, each irreducible set of collective displacements constitutes the eigenvector of a normal-mode of the lattice. One is therefore entitled to establish a correspondance between the OP of a displacive structural transition adjacent to a phase of "high-symmetry" G_0 , and one of the normal modes of this phase. However,we must keep in mind that the OP is a variational set of displacements (some components of which become a static set below T_c), while the normal mode pertains to a time dependent set of dynamic variables.

The correspondance between normal mode and OP imposes a specific constraint to the symmetry of the OP of a displacive structural transition. This constraint derives from the fact that the considered irreducible set of displacements is part of the (3N-3)-dimensional reducible set of displacements constituting the so-called "mechanical-representation" of the crystal [47,48]. This reducible representation of G_0 is carried by the set of arbitrary displacements in space,of each of the N atoms of the crystal about their equilibrium position (excluding the overall translation of the crystal).

Consequently, the OP-representation $\Gamma_i(\vec{k}_0)$ must be contained in the decomposition of the mechanical representation into IR's of G_0.

The above correspondance also leads to another important inference. Indeed, the potential energy $\mathcal{V}$ of the crystal is a G_0-invariant function of the eigenvector-components relative to the normal modes [47]. This property has the consequence that the polynomial expansion of $\mathcal{V}$,is formally identical to the Landau free-energy F. In particular, the coefficient α of the quadratic term in F can be put in correspondance with the coefficient $(m\omega^2/2)$ of the quadratic term in $\mathcal{V}$,where $\nu=(\omega/2\pi)$ is the frequency of the relevant mode.

From the vanishing of α at T_c, it is then infered that $\nu(T_c)$ vanishes. This inference is consistent with the fact that a static set of displacements η_I^o onsets at T_c: the restoring force f ($f \propto \nu^2$) relative to this type of deviation from equilibrium has therefore vanished at T_c.

The vanishing of ν at T_c, and the precursor decrease of its value on approaching T_c from above, justify the term "soft-mode" used to qualify the normal mode associated to the OP of the transition [48,49].

b) Order-disorder type.

In an order-disorder transition,certain atoms of the crystal can occupy several structural sites with equal probability,in the high-symmetry phase, while the occupancy probability of the various considered sites is different below T_c. The order parameter of such a transition consists in the difference between the probabilities assigned to each site. This definition encompasses a variety of distinct physical situations. In the copper-zinc alloy CuZn , each of the copper and of the zinc atoms have equal probabilities to occupy sites belonging to distinct sublattices above T_c. Below T_c copper occupies preferentially one sublattice and zinc the other sublattice (Fig.15a). In the insulating crystal KH_2PO_4 , protons H^+ can occupy two distinct sites nearby a $(PO_4)^-$ tetrahedron. One of these sites is preferentially chosen below T_c (Fig. 15b). The two sites are geometrically related by an operation of G_o which does not belong to G. For order-disorder transitions, the structure of the high-symmetry phase can thus be considered as the statistical average of the various structures of the "degenerate states" which are individually stable below T_c.

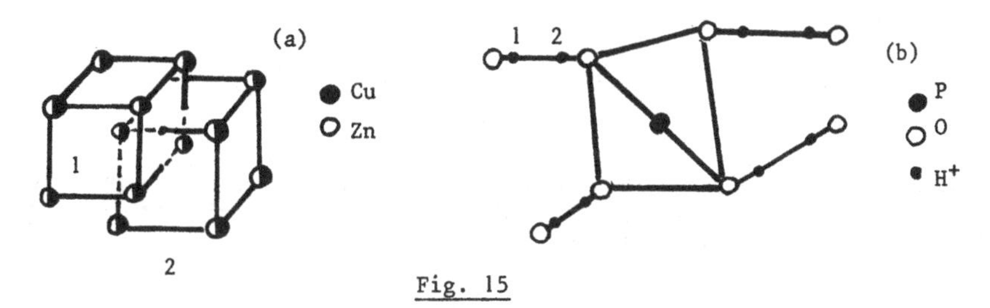

Fig. 15

For obvious reasons, order-disorder transitions have no "soft-mode". The dynamical precursor effect of these transitions is a slowing down of the rate of switching of the relevant structural element between its various possible states (e.g. of a proton between its two sites in KH_2PO_4).

Experimental investigations have shown that transitions in real systems seldom belong,in a clearcut way, to either of the preceding types of structural transitions. For instance, in KH_2PO_4 ,the ordering of the protons on one type of sites is accompanied by a displacive rotation and distortion of the $(PO_4)^-$ tetrahedra [50]. More generally, whenever the primary-OP, is of the order-disorder type, secondary OP of the displacive type will often exist (e.g. a symmetry-breaking strain).

8.1.2. $\vec{k}_o$-vector of the OP-representation and superlattice reflections below T_c

Let us show that the modification of the diffraction spectrum of a crystal which takes place across a structural transition is directly related to the $\vec{k}_o$ vector of the OP.

This modification is the Fourier transform of the structural deviation $\delta\rho_{eq}(\vec{r})$ characterizing the low symmetry phase. Consider first the the Fourier transform of the primary density increment:

$$\delta\rho_1(\vec{q}) = \int \delta\rho_{1\,eq}(\vec{r}).e^{i\vec{q}.\vec{r}}.d\vec{r} = \sum_{\vec{k},r} \eta^o_{\vec{k},r} \int \phi_{\vec{k}r}(\vec{r})e^{i\vec{q}.\vec{r}}.d\vec{r} \quad , \qquad (8.2)$$

where the discrete sum runs over the vectors $\vec{k}$ of the star of $\vec{k}_o$,and over the basis of the small representation of the OP $\eta_{\vec{k}r}$.

As defined in chap. II § 2.2.1, any function $f(\vec{r})$ transforms under a translational shift $\vec{a}$ as:

$$f(\vec{r}) \rightarrow g(\vec{r})=f(\vec{r}+\vec{a}) \qquad (8.3)$$

Since the functions $\phi_{\vec{k}r}(\vec{r})$ in (8.2) carry an IR of the translation group $\mathcal{T}_o$ of the crystal, we can write (Cf. table 3) :

$$\phi_{\vec{k}r}(\vec{r}+\vec{T}) = e^{-i\vec{k}.\vec{T}}.\ \phi_{\vec{k}r}(\vec{r}) \qquad (\vec{T}\in\mathcal{T}_o) \qquad (8.4)$$

If we make the change of variables $\vec{r}\rightarrow\vec{r}'+\vec{T}$ in the integrals of (8.2) ,and use eq. (8.4), we obtain:

$$\delta\rho_1(\vec{k}) = \sum_{\vec{k},r} \eta^o_{\vec{k}r}\ .e^{i(\vec{q}-\vec{k}).\vec{T}} \int \phi_{\vec{k}r}(\vec{r}').e^{i\vec{q}.\vec{r}'}\ .d\vec{r}' \qquad (8.5)$$

Comparison of (8.2) and (8.5) shows that $(\vec{q}-\vec{k}).T=0$ for any $\vec{T}\in\mathcal{T}_o$. This condition implies that $(\vec{q}-\vec{k})$ is a vector of the reciprocal lattice of $\mathcal{T}_o$ (eq. 2.7). Thus , $\delta\rho_1(\vec{q})$ will be non-zero only if:

$$\vec{q} = m_1\vec{a}_1^* + m_2\vec{a}_2^* + m_3\vec{a}_3^* + \vec{k} \quad , \qquad (8.6)$$

where $\vec{k}$ is a vector of the star k_o^* ,associated to a non-zero value for , at least, one of the OP-components $\eta^o_{\vec{k}r}$.

It can easily be shown ,by means of the same argument, that the Fourier transform of the <u>total</u> density increment $\delta\rho(\vec{q})$ is non-zero only if:

$$\vec{q} = \vec{K} + \Sigma\ m_p.\vec{k}_p \qquad (8.7)$$

where $\vec{K}$ is a reciprocal lattice vector of the high-symmetry phase, and where the $\vec{k}_p$ belong to k_o^*

Since the diffraction spectrum of the crystal, above T_c, already contains reflections (i.e. non-zero components of $\rho_o(\vec{q})$) corresponding to $\vec{q}=\vec{K}$, the considered phase transition will induce, below T_c, the onset of new reflections, associated to the vectors of the star k_o^* , and to their combinations with integer coefficients. Referring to (8.2) or to (8.5), we can see that the amplitudes of these reflections depend linearly of the $\eta^o_{\vec{k}i}$ and are therefore small. These reflections onsetting below T_c with a vanishingly small amplitude are called "superlattice reflections". Their experimental detection allows a straightforward specification of the star of $\vec{k}$-vectors of the OP-representation. Fig. 16 hereunder shows the location of these reflections for GMO (Cf. §7) ,in agreement with the value determined in eq. (7.2).

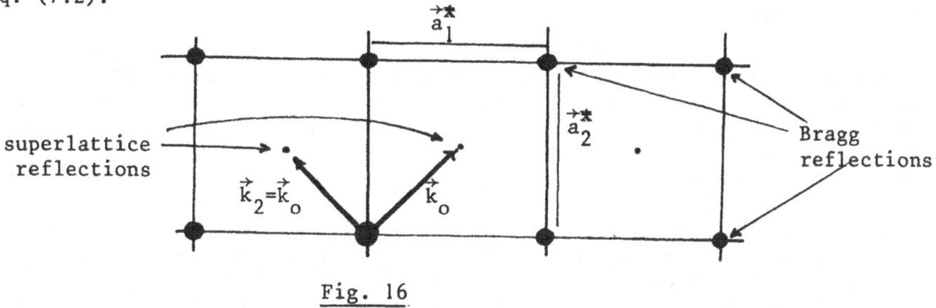

Fig. 16

Clearly, if $\vec{k}_o=0$,there are no superlattice reflections.

8.1.3. OP-symmetry and modifications of the vibrational spectrum across T_c.

The collective oscillations of the atoms of a crystal can be studied by means of a number of experimental techniques. In particular these oscillations can give rise to components ("lines") in the infrared and Raman spectra of a crystal [47]. The number of these components depend on the number of atoms in the primitive unit cell, as well as on the point-symmetry of the crystal. Since both these features are apt to change across a structural transition, the spectra will differ on either sides of the transition. Namely, new lines can appear below T_c , because the number of atoms in each unit cell changes (this number is multiplied by the index of $\mathcal{T}$ in $\mathcal{T}_o$). On the other hand, lines corresponding to "degenerate" collective oscillations (i.e. oscillations associated to a multidimensional IR of G_o), can be splitted into several components, due to the lower-symmetry existing below T_c. Shifts of existing lines can also take place.

A qualitative information on these effects can be deduced from a comparison of the symmetries G and G_o: number of new lines expected, existence of a splitting. On the other hand, since the order parameter of the transition is apt to describe the structural change, one expects a quantitative account of the preceding effects (magnitude of the splitting, intensities of the new lines,and their temperature dependence) on the basis of an evaluation of the coupling between the OP and the relevant degrees of freedom. Such a task has been carried out in detail by Petzelt and Dvorak [51].

We restrict here to an illustration of the method,and account for the onset of a new line in the Raman spectrum of gadolinium molybdate (Cf. §7), below T_c.

In the framework of the simplest mechanism [47] a collective oscillation of a crystal will couple to light waves ,and give rise to a line in the Raman spectrum of the crystal if the corresponding degrees of freedom carries an IR $\Gamma_i(\vec{k}_o)$ of the crystal's space-group, complying with 2 conditions:

i) $\vec{k}_o = 0$. ii) $\Gamma_i(0)$ is equivalent to one of the IR's carried by the components of a symmetric second rank tensor (e.g. the strain tensor).

a) Qualitative considerations.

Consider in GMO the collective oscillation associated to the OP,above T_c (i.e. the "soft" normal mentioned in § 8.1.1). Since the OP carries the representation $\Gamma_2(\vec{k}_o)$, with $\vec{k}_o \neq 0$, this mode does not give rise to a line in the Raman spectrum: it is "Raman inactive" above T_c. Below T_c, the same degree of freedom (η_1,η_2) carries a representation of the low symmetry group G. As stressed in § 7.1 , η_1 and η_2 are left invariant by all the elements of G (G being equal to the kernel of the homomorphism $G_o \rightarrow I_o$). Hence, each of these components,as well as any linear combination $(a\eta_1 + b\eta_2)$ carries the totally symmetric IR of G. Since any symmetric second rank tensor always has totally symmetric components (if only its trace), we can conclude that,below T_c , the collective oscillations associated to (η_1,η_2) will be Raman active. Besides,as (η_1,η_2) decomposes into 2 irreducible degrees of freedom (both of Γ_1 symmetry) with respect to G, the soft mode splits into 2 normal modes below T_c.

b) Quantitative considerations.

Let us examine the temperature dependence of the intensities of the lines associated to the splitting of the soft mode ,below T_c. These intensities must vanish at T_c , in agreement with the "inactivity" of the mode above T_c.

Let Q be the collective atomic coordinate relative to a totally symmetric lattice mode of the high-temperature phase of GMO . For the reason stated above, this mode is Raman active:the atomic displacements Q couple to the light, the intensity of the corresponding Raman line being $I_Q \propto Q^2$ [51]. Due to the Γ_1 symmetry of Q, one can form a G_o invariant polynomial of the form $Q(\eta_1^2+\eta_2^2)$. If Q and the η_i are considered as variational parameters, such a term is a contribution to the free-energy of the system. Referring to § 7.2,we can see that its existence warrants the onset ,below T_c, of a non-zero equilibrium value Q_o of the secondary- order parameter Q. If,on the other hand, Q and the η_i are considered as normal mode coordinates, the former term is an anharmonic contribution to the potential energy of the crystal above T_c. This potential energy, restricted to the Q and η_i degrees of freedom is:

$$\mathcal{V} = \frac{m\omega^2}{2}\cdot(\eta_1^2+\eta_2^2) + \frac{m_Q\cdot\omega_Q^2}{2}\cdot Q^2 + \delta Q\cdot(\eta_1^2+\eta_2^2) + \dots \qquad (8.8)$$

Below T_c, both the η_i and Q acquire non-zero equilibrium values. Assume,for instance, that $\eta_1^o = \eta_o \neq 0$, and $Q_o \neq 0$. The potential energy relative to the oscillation about the new equilibrium atomic positions can be obtained by putting $\eta_i=(\eta_i^o+\eta_i')$ and $Q=(Q_o+Q')$ in eq.(8.8) ,we obtain the following quadratic form:

$$\mathcal{V}' = (\frac{m\omega^2}{2}+\delta Q_o)(\eta_1'^2+\eta_2'^2)+\frac{m_Q\omega_Q^2}{2}.\ Q'^2 + 2.\delta\eta_o Q'\eta_1' \tag{8.9}$$

This expression discloses that Q' and η_1' are not normal coordinates below T_c since they appear as bilinearly coupled in $\mathcal{V}'$. Diagonalization of (8.9) in the case where ω and ω_Q are not too close, and where the bilinear term is a small correction yields:

$$\mathcal{V}' = (\frac{m\omega^2}{2}+\delta Q_o)\eta'^2 + \frac{M\Omega^2}{2}.\zeta^2 + \frac{M_Q\Omega_Q^2}{2}\ q^2 \tag{8.10}$$

with,

$$\zeta = \eta_1' + \frac{2\delta\eta_o Q'}{\Delta} \tag{8.10'}$$

and,

$$M\Omega^2 = m\omega^2 + \frac{4\delta^2\eta_o^2}{\Delta} + 2\delta Q_o \tag{8.10''}$$

ζ and η' are the two normal coordinates resulting from the decomposition of the soft mode below T_c . Eq.(8.10') shows that ζ contains a component proportional to the "Raman active" coordinate Q'. Through this component,light waves can couple to the atomic displacements ζ which become Raman active. The intensity of the corresponding line is $I\propto\delta^2.\eta_o^2.I_Q$,where I_Q is the intensity of the line associated to Q. Since $\delta^2\eta_o^2$ is small near T_c, the ζ-line will be weak. Besides, the temperature dependence of its intensity is specified by the relation $\eta_o^2\propto(T_c-T)$. This intensity decreases linearly, and eventually vanishes at T_c. On the other hand, (8.10) and (8.10") show that the two components ζ,and η' of the soft mode have different frequencies below T_c , the difference between the squared frequencies being proportional to $\eta_o^2(T)$.

In summary,the mechanism of Raman activation of a normal mode comes from the existence of anharmonic couplings between the considered degree of freedom, and Raman active degrees of freedom. Below T_c,the anharmonic coupling gives rise to a bilinear coupling between modes which induces the mixing of Raman active atomic displacements into the inactive one.

Other mechanisms have also been considered [51],such as the conversion below T_c of "higher order" Raman scattering into ordinary Raman scattering. These mechanisms have in common to rely on the onset,below T_c, of a non-zero equilibrium value of the considered atomic degree of freedom. This has a "biasing" effect on the interaction terms, either between distinct degrees of freedom,or between a given degree of freedom and electromagnetic waves. In the understanding of this mechanism, the Landau theory provides the tool for working out the quantities which acquire non-zero values below T_c, and for specifying the temperature dependence of these quantities.

8.2 Physical realization of different OP symmetries.

The example of GMO, investigated in §7 is the concrete realization of one of the situations analyzed theoretically in this chapter : 2-dimensional irreducible OP complying with the Landau and Lifschitz criteria, "intrinsic" symmetry of the OP characterized by the image $\mathbb{C}_4$, existence of secondary order parameters related to ferroelectric and ferroelastic properties, improper nature of the ferroelectricity and of the ferroelasticity.

In this paragraph we indicate other examples of real systems undergoing structural transitions,which constitute concrete realizations of the various theoretical situtations pointed out in this chapter. In the first place,table 12 summarizes the characterictics of a selected set of transitions in real substances : nature of the space-group change, OP-dimension,intrinsic OP-symmetry (i.e. image I_o of G_o in the OP-representation), physical nature of the OP.

Table 12

Substance	G_o	→ G	OP dimension	Image I_o	Continuous?	Physical nature	Ref.
LaP_5O_{14}	Pmna	$P2_1/b$	1	$\bar{\mathbb{C}}_1$	yes	displacive (D)	37
CuZn	Im3m	Pm3m	1	$\bar{\mathbb{C}}_1$	yes	order disorder(O)	53
Hg_2Cl_2	I4/mmm	Cmcm	2	$\mathbb{C}_{4v}$	yes	D	54
GMO	$P\bar{4}2_1m$	Pba2	2	$\mathbb{C}_4$	no	O-D	41
$C_4O_4H_2$	I4/m	$P2_1/m$	2	$\mathbb{C}_4$	no	D	55
$SrTiO_3$	Pm3m	I4/mcm	3	O_h	yes	O-D	56
CuAu	Fm3m	P4/mmm	3	O_h	no	O-D	57
Sb_5O_7I	$P6_3/m$	$P2_1/b$	3	T_h	no	D	58
VO_2	$P4_2/mnm$	$P2_1/b$	4	$(D_2/D_2;D_2/D_2)^*$	no	D	59

Some of these transitions show a more or less pronounced discontinuity of the OP at T_c, while the others appear continuous within experimental accuracy. For the latter category of transitions,the conditions set by the Landau theory should be strictly obeyed. In the first place, there should be a consistency of the crystallographic data with the results of the theory: the low-symmetry group G must be one of the isotropy groups in the carrier space of an irreducible representation Γ of G_o. Moreover, if F is the free-energy expansion of fourth degree (or of lowest relevant degree) associated to the Γ-symmetry,the direction of the OP-space which is invariant by G must correspond to the absolute minimum of F for some range of the expansions coefficients.

Finally, the degeneracy of the low-symmetry phase, disclosed by the domain pattern should be equal to the index of G in G_o.

All the examples listed in table 12 comply with these conditions. A systematic comparison, performed in ref.14 shows ,more generally, that no exception is found among continuous transitions in real systems to this applicability of Landau's theory. Besides, a number of discontinuous transitions (e.g. the ones listed on table 12) also satisfy the above conditions: in particular, their observed symmetry-change is satisfactorily described by an irreducible OP (Cf. also chap. IV).

In the second place,for continuous transitions, the OP-components should carry an active IR of G_O. This is the case of the examples in table 12. A few cases,departing from this situation have been pointed out[14].Thus ,the structural transition in NbO_2 [52] has an OP which does not comply with the Lifschitz criterion though this transition appears continuous and surrounded by perfectly periodic phases. However,it has also been pointed out[14] that the experimental data relative to it are somewhat less accurate than for the transitions enumerated on table 12, and the possibility is still open to detect in this substance, either a discontinuity of the transition, or an intermediate,non-strictly crystalline (i.e. incommensurate) phase below T_c.

As for discontinuous transitions, the active character of the OP-representations involved in the description of the transition will be discussed in chapter IV.

Table 13 focuses on the non-ferroic or ferroic nature of a selected set of structural transitions observed in real substances. This table also shows the relationship between the relevant macroscopic tensors pertaining to the ferroic transitions,and the order parameter. A systematic examination of the available experimental observations[14] has revealed that ferroelastic transitions (with or without the simultaneous occurence of ferroelectricity) constitute the largest class of transitions observed in real systems. In this class one encounters as many examples of proper ferroelasticity as of improper ferroelasticity. In the latter case,the faintness index relative to the strain components is almost invariably equal to 2. The example of GMO studied in §7 therefore illustrates a situation representative of a large class of the transitions in real systems.Conversely, the examples of ferroelectric non-ferroelastic transitions all possess proper ferroelectric properties. Conjectured examples of improper ferroelctricity (without the simultaneous occurence of ferroelasticity) have been eventually proven to involve the existence of an incommensurate phase (Cf. chap. V fig.1). Examples of non-ferroic transitions are essentially found in organic substances and in metallic alloys [14] ,while few examples of secondary ferroics are available.

9. CONCLUSION.

The specific character of structural transitions among continuous phase transitions stems from the applicability of the Lifschitz criterion.

Table 13

Substance	Ferroic class $\hat{G}_o \rightarrow \hat{G}$	Proper(P) or Improper(I)	Faintness index	Continuous(?)	ref.
Triglycine sulphate	ferroelectric 2/m 2	P	-	yes	50
$LaNbO_4$	ferroelastic 4/m 2/m	P	-	yes	60
Hg_2Cl_2	ferroelastic 4/mmm mmm	I	2	yes	54
KH_2PO_4	ferroelectric &ferroelastic $\bar{4}2m$ mm2	P	-	no	50
GMO	"	I	2	no	41
$LaCoO_3$	secondary ferroic $\bar{3}m$ 3	P	-	yes	61
NH_4Cl	secondary ferroic m3m $\bar{4}3m$	P	-	yes	62
CuZn	non-ferroic m3m m3m	-	-	yes	53
$C_6Cl_4O_2$	non-ferroic 2/m 2/m	-	-	yes	63

This criterion imposes a "limited symmetry breaking" of the translation-group the index of $\mathcal{T}$ in $\mathcal{T}_o$ is in most cases equal to 1 or 2 and it restricts considerably the number of distinct intrinsic OP-symmetries and the number of distinct free-energy expansions of interest (Cf. §4). Another specific feature, is the importance of the secondary order parameters. These quantities coincide with macroscopic tensors most directly accessible to the experimental investigation while the primary order parameter concerns quantities more difficult to probe experimentally :its complete identification requires a refined determination of the structures of the phases surrounding the transition. An important step in the understanding of a structural transition is therefore, the clarification of the coupling scheme between the primary OP and the secondary order parameters.

REFERENCES

|1| G.Ya. Lyubarskii *The application of group theory in physics.* (Pergamon London 1960).

|2| G.F. Koster *Space groups and their representations. In*Solid state physics Ed. F.Seitz and D. Turnbull (Academic Press .NY.Y 1957).

|3| C.J.Bradley and A.P. Cracknell. *The mathematical theory of symmetry in solids.* (Clarendon Press. Oxford 1972).

|4| *International tables for X-ray crystallography.* Vol 1 Kynoch press (Birmingham 1952).

|5| O.V. Kovalev. *The irreducible representations of space groups* (Gordon and Breach N.Y. 1965).

|6| J. Zak, A. Casher, H. Glück, and Y. Gur .*The irreducible representations of space groups.* (Benjamin N.Y. 1969).

|7| S.C. Miller, and N.V. Love. *Tables of irreducible representations of space groups and magnetic corepresentations of magnetic space groups.* Pruett. Boulder (Colorado 1967).

|8| E.M. Lifschitz Zh. Eksp. Teor. Fiziki 11,255 (1941).

|9| I.E. Dzyaloshinskii Sov. Phys. JETP 19,960 (1964).

|10| C. Haas Phys. Rev. 140,863 (1965).

|11| S.Goshen ,D. Mukamel, ans S. Shtrikman. Int. J. Magnetism 6,221 (1974).

|12| Y.Onodera,and Y. Tanabe, J.Phys. Soc. Jpn 45,1111 (1978).

|13| V. Janovec,V. Dvorak, and J. Petzelt Czech. Journ. Physics B25,1362 (1975).

|14| (a) P. Tolédano,and J.C. Tolédano Phys.Rev. B14,3097 (1976).
(b) " " " B16,386 (1977)
(c) J.C. Tolédano and P. Tolédano " B21,1139 (1980)
(d) P. Tolédano,and J.C. Tolédano " B25,1946 (1982)

|15| H.T. Stokes and D.M. Hatch Phys. Rev. B30,4962 (1984).

|16| V.N. Syromiatnikov,Dissertation. Institute of Metal Physics. Sverdlovsk (USSR 1977). V.E. Naish and V.N. Syromiatnikov Sov. Phys. crystallogr. 21,627 (1976). V.E.Naish and V.N. Syromiatnikov Sov. Phys. Crystallogr. 22, 2 (1977).

|17| Yu.M. Gufan,and V.P. Sakhnenko. Sov Phys. JETP 36,1009 (1973).

|18| Yu.M. Gufan and V.P. Popov. Sov. Physics Crystallogr. 25,527 (1980). Yu.M. Gufan, V.P. Dmitriev,V.P. Popov,and G.M. Chechin. Sov. Physics Solid state 21,327 (1979).

|19| J.S. Kim ,H.T. Stokes and D.M. Hatch. Phys. Rev. B33,6210 (1986).

|20| E.Meimarakis and P. Tolédano in Phonons Physics Ed. J. Kollar (World Scientific 1985). J.S. Kim,D.M. Hatch and H.T. Stokes. Phys. Rev. B33, 1774 (1986).

|21| D.C. Wallace. *Thermodynamics of crystals.* (John Wiley and sons N.Y. 1972).

|22| J.F. Nye *Physical Properties of Crystals*. (Clarendon Press Oxford 1960).

|23| S. Baghavantam . *Crystal Symmetries and Physical Properties*.(Academic Press .N.Y. 1966).

|24| J. Sapriel *L'Acoustooptique*.(Masson ,Paris 1976).

|25| K. Aizu; J.Phys. Soc. Jpn 27,387 (1969),and 34,121 (1973)

|26| J.C. Tolédano. Ann. Télécommunications. 29,249 (1974).

|27| L.A. Shuvalov J. Phys. Soc Jpn Suppl. 28, 38 (1970).

|28| R.E. Newnham and L.E. Cross Mat. Res. Bull. 9,927 (1974)

|29| R.E. Newnham American Mineralogist 59,906 (1974)

|30| V. Dvorak Journ. Physique 33,Suppl.4 ,C-289 (1972)

|31| A.P. Levanyuk and D.G. Sannikov Sov. Phys. Uspekhii 17,199 (1975)

|32| N. Boccara. Ann. Physics 47,40 (1968)

|33| V.L. Indenbom Sov Phys. Crystallogr. 5 ,106 (1960)

|34| K. Aizu J. Phys Soc Jpn 33,629 (1972)

|35| P.Tolédano Thesis (Faculté des Sciences Amiens 1979)

|36| V. Dvorak Ferroelectrics 7,1 (1974)

|37| G. errandonéa Phys. Rev. B21, 5221 (1980)

|38| V.Dvorak Phys. stat. solidi 45 (b),147 (1971)

|39| V. Dvorak Phys. Stat. solidi 46(b),763 (1971)

|40| P. Bastie,J.Bornarel,J.Lajzerowicz,M.Vallade Nucl. Inst.Methods 166,53(1979)

|41| W. Jeitshko Acta Crystallogr. B28, 60 (1972)

|42| J. Kobayashi, Y. Sato, and T. Nakamura Phys. Stat. Solidi 14(a),259(1972)
E. Du Trémolet de Lacheisserie,J.M. Courdille and J. Dumas, Journ. de Physique. 38,65 (1977)

|43| B. Dorner, J.D. Axe, and G. Shirane Phys. Rev B6,1950 (1972)

|44| J.R. Barkley and W. Jeitshko J. Appl. Physics 44,938 (1973)

|45| L. Sawaguchi and L.E. Cross J.Appl. Physics 44,2541 (1973);L.E. Cross, A. Fouskova, and S.E. Cummins. Phys. Rev. Letters 21,812 (1968)

|46| W. Yao,H.Z. Cummins and R.H.Bruce Phys. Rev B24,426 (1981)

|47| H. Poulet and J.P. Mathieu *Spectres de vibration et dynamique cristalline* (Gordon and Breach Paris 1970).

|48| W. Cochran *The dynamics of Atoms in Crystals* (Edward Arnold London 1973)

|49| J.F. Scott Reviews of modern Physics 46,83 (1974)

|50| F. Jona and G. Shirane *Ferroelectric crystals* (Pergamon Press NY 1962)

|51| J. Petzelt and V. Dvorak J. Physics C9,1571 (1976);C9,1587 (1976)

|52| S.M. Shapiro,J.D. Axe ,G. Shirane,and P.M. Raccah Solid State Commun. 15,377 (1974)

|53| W.P.Pearson.*A handbook of Lattice Spacings and Structures of Metals* (Pergamon London 1958).

|54| Ch. Barta,A.A. Kaplyanskii,V.V. Kulakov,and Yu.M. Maskov Sov. Physics Solid State 17,717 (1975).

|55| E.J. Samuelsen and D.Semmingsen Solid State Commun. 17,217 (1975)

|56| E. Pytte,and J. Feder Phys. Rev. 187,1077 (1969)

|57| C.S. Barrett,and T.B. Massalski *The Structure of Metals* (Mc Graw Hill NY 1966)

|58| W. Prettl and K.H. Rieder Phys. Rev. B14,2171 (1976)

|59| J.R. Brews. Phys. Rev. B1,2557 (1970)

|60| L.H.Brixner,J.F. Whitney, F.C. Zumsteg,and G.A. Jones Mat. Res. Bull. 12,17 (1977).

|61| P.M. Raccah,and J.P. Goodenough Phys. Rev. B155 ,932 (1967)

|62| C.H. Wang and R.B. Wright J. Chem Phys. 60,849 (1974)

|63| H. Chihara and N. Nakamura J. Phys. Soc Jpn 44,1567 (1978)

CHAPTER IV

FIRST ORDER PHASE TRANSITIONS

1. INTRODUCTION

The basic assumption of the Landau theory is that the transition between the high and low-temperature phases takes place continuously at a given temperature T_c. From this assumption are deduced the main symmetry and thermodynamic features of the phase diagram, namely: 1) the involvement of a single active irreducible degree of freedom which determines the symmetry breaking, 2) the weak value of the order-parameter modulus in the vicinity of T_c, which justifies the form of the thermodynamic potential as a Taylor expansion of the order-parameter components, 3) the non-coexistence of the stability regions of the two phases, and their group-subgroup relationship, 4) the absence of latent heat, and a discontinuity at T_c for the physical quantities related to second order derivatives of the Landau potential.

Is the Landau approach still valid when the assumption of continuity at T_c, is not verified experimentally ? In other words, to what extent the theory can be used to interpretate the phenomenological properties of first order transitions ? This question is not of secondary interest, as most of the structural transitions, and a number of magnetic transitions (see Chapter VI) are found to be discontinuous. In this way, there exist a comprehensive set of experimental data allowing to check the validity of the Landau theory for first order transition. These data will be discussed in this chapter, restricting for reasons of coherency to structural transitions in solids. However, our conclusions will apply also to magnetic and liquid crystal systems.

Even when considering only structural systems, first-order transitions constitute a very heterogeneous family. Obviously, the thermodynamic classification of phase transitions which distinguishes only between second and first-order transitions is insufficient as it denominates similarly weakly discontinuous transitions and reconstructive ones. In spite of the fact that the various types of first-order transitions possess a number of common characteristics (latent heat, coexistence of the phases, discontinuity of the order-parameter and of other quantities related to the first derivatives of the Landau expansion) it is possible to differentiate them qualitatively and quantitatively by a complete set of measurements. Of course, such an experimental differentiation entails to take into account the peculiarities of each transition, which include the effect of imperfections (defects, impurities, inhomogeneities) existing in any real system. However, we will see in this chapter, that there exist some general properties, connected with the symmetry of the systems under consideration, which allow to discriminate a number of classes of first order transitions.

Actually, the Landau theory provides implicitely the basis for a qualitative classification of first-order transitions. Two main categories of transitions can be distinguished. A first category contains the transitions which can be predicted to be first-order from symmetry considerations. For these transitions, general results can be obtained independently of any particular microscopical mechanism and accounting only on the symmetry properties of the system. In this respect, four classes of symmetry predicted first-order transitions can be inferred from the standard Landau theory exposed in the preceding chapters.

1) Transitions induced by a single irreducible representation (IR) which does not fulfil the Landau symmetric cube criterion.

2) Transitions between strictly periodic structures induced by a single IR, which does not satisfy the Lifshitz criterion. The validity of these two conditions, as being predictive for the first-order character of transitions, has been discussed by a number of authors [1-5], and will be examined in §.3 and §.4 respectively.

3) Transitions which are related to more than one order-parameter

4) Transitions for which the high and low-temperature phases are not related by a group-subgroup relationship.

Although the discontinuous character of the transitions belonging to these two classes is not questioned, few studies have precised their features. In §.5 we examine some examples of transitions, the properties of which can be described by several order-parameters. For transitions of class 4 a semi-phenomenological theory generalizing the Landau approach is presented in §.6.

A second category of first order phase transitions is constituted by the transitions whose discontinuous character depends on the values of the coefficients in the order-parameter expansion, the obtained low symmetry phases being unstable relatively to a fourth degree expansion*, but stable when taking into account higher degree terms. These transitions, which are examined in §.2, are related to a single active IR of the high-temperature phase, and it is well known that the line of transitions in their phase diagram, may stop or not at a tricritical point, depending on the variation of the coefficients as a function of the external parameters. Thus, the order of this category of transitions is essentially dependent on the particular mechanism expressed by the order-parameter, and it cannot be predicted, in principle, on a symmetry basis.

In the following section, we restrict ourselves to discuss the standard models of first-order transitions which necessitate higher degree terms in the order-parameter expansion. The more elaborated treatments which take into account the role played by critical fluctuations, defects, or local strains, are reviewed in Chapter VIII.

2. FIRST-ORDER TRANSITIONS ASSOCIATED WITH HIGH-DEGREE EXPANSIONS OF THE ORDER-PARAMETER

In this section different types of phase diagrams containing first-order transitions lines are considered, assuming that the transitions

* When there is a single quartic term in the Landau expansion, higher-order invariants (of degree higher than four) have also to be considered for continuous transitions induced by multidimensional representations, as indicated in Chapter II.

are induced by a single active IR, i.e. an IR fulfilling the Landau and Lifshitz conditions. Accordingly, only even degree invariants appear in the order-parameter expansions. Denoting n the number of components of the order-parameter and d the degree of the higher order term in the thermodynamic potential, four examples will be successively considered corresponding to different values of n and d. A complete picture of the phase diagrams corresponding to thermodynamic potentials with high-degree invariants, can be found in the book of Gufan [6] and in a number of papers by Gufan and Coauthors [7-29].

2.1. Case with n = 1, and d = 6 : general features of first-order transitions.

Let us start by the simplest case of a one component order-parameter associated with the expansion :

$$F(\eta) = \frac{\alpha}{2}\eta^2 + \frac{\beta}{4}\eta^4 + \frac{\gamma}{6}\eta^6 \tag{2.1}$$

which describes, when assuming $\beta < 0$ and $\gamma > 0$, a first-order transition between two phases denoted I and II. As usually we take $\alpha = a(T - T_c)$ with $a > 0$.

The equations minimizing $F(\eta)$ are :

$$\frac{\partial F}{\partial \eta} = \eta(\alpha + \beta\eta^2 + \gamma\eta^4) = 0 \tag{2.2}$$

$$\frac{\partial^2 F}{\partial \eta^2} = \alpha + 3\beta\eta^2 + 5\gamma\eta^4 \geqslant 0 \tag{2.3}$$

The discussion of the conditions (2.2) and (2.3) is visualized in Figures 1a and 1b. It can be summarized as follows : For $T > T_2 = T_c + \frac{\beta^2}{4\gamma a}$ only one phase is stable, namely the parent phase I corresponding to $\eta = 0$. At $T = T_2$ phase II with :

$$\eta = \pm \left[\frac{-\beta + (\beta^2 - 4\alpha\gamma)^{1/2}}{2\gamma} \right]^{1/2} \tag{2.4}$$

appears as a metastable state (i.e. corresponding to secondary minima in the non-equilibrium curve $F(\eta)$). As the temperature is lowered, the stability of phase II increases. Phases I and II are equally stable when the additional condition $F(\eta) = 0$ is fulfilled, i.e. at :

$$T_1 = T_c + \frac{3\beta^2}{16\gamma a} \tag{2.5}$$

Below T_1, phase I becomes less stable than phase II, and remains as a metastable state until α changes sign for $T = T_c$.

From the preceding discussion, the distinctive properties of first-order transitions can be inferred. Thus, these transitions are characterized :

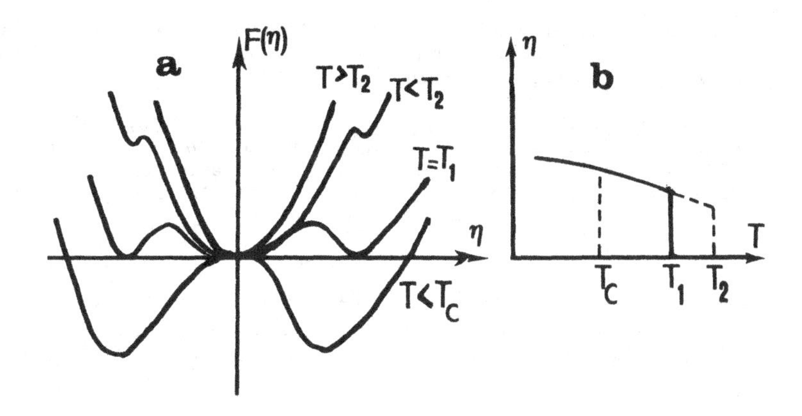

Figure 1 : a) Variation of the non-equilibrium potential $F(\eta)$ as a function of temperature. b) Temperature dependence of the order-parameter η .

a) by a region of coexistence of the two phases I and II in the interval :

$$\Delta T = T_2 - T_c = \frac{\beta^2}{4\gamma a} \tag{2.6}$$

The two phases are alternatively metastable for $T > T_1$ and $T < T_1$. Although the transition may in reality take place at any temperature within this interval, the more probable transition temperature is T_1, which corresponds to an equal stability for the two phases. As a consequence of the region of metastability (2.6) for the two phases, a thermal hysteresis should generally be observed, i.e. the measured transition temperature will not be the same on heating or cooling.

b) by a discontinuous jump of the order-parameter at the transition, which is given by :

$$\eta(T_1) = \left(-\frac{\beta}{4\gamma}\right)^{1/2} \tag{2.7}$$

Below T_1, $\eta(T)$ increases following the law (2.4)

c) by a latent heat, which results from the discontinuity of $\eta(T)$. One has :

$$\Delta Q = \frac{a}{2} T_1 \eta^2(T_1) = -\frac{a\beta T_1}{8\gamma} \tag{2.8}$$

Analogously a discontinuous variation should be found at T_1 for all the physical quantities proportional to the first-derivatives of the thermodynamic potential (e.g. volume, lattice constants, induced polarization or strain).

d) by specific discontinuities at T_1 for the physical quantities related to second-order derivatives of the thermodynamic potential. If ξ is the field conjugated to η , minimization of $F(\eta) - \eta\xi$ with respect to η and ξ provides the following form for the susceptibility:

$$= \lim_{\xi \to o} \left(\frac{\partial \eta}{\partial \xi}\right) = \begin{cases} 1/a\,(T - T_c) & \text{for } T > T_1 \\ 16\gamma\,/3\beta^2 & \text{for } T \doteq T_1 \\ 1/\alpha + 3\beta\eta^2 + 5\gamma\eta^4 & \text{for } T < T_1 \end{cases} \quad (2.9)$$

χ is represented in Fig. 2.a, it differs from the standard Curie-Weiss law obtained for second-order transition (see chapter I, figure 5), by

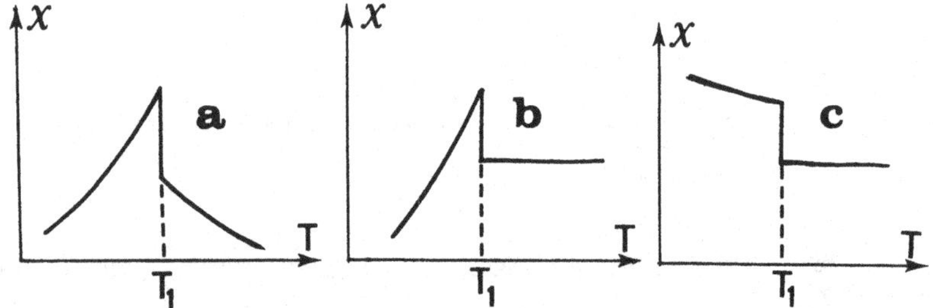

Figure 2 : Temperature dependence of the susceptibility χ for first-order transitions : a) proper transition ; b) improper transition with $\nu = 2$ c) improper transition with $\nu > 2$ and a negative lower degree coupling coefficient.

a finite jump : $\chi\,(T_1) = 16\,\gamma/3\,\beta^2$ which takes place at the transition. Fig. 2.a) holds for proper first-order transitions. As shown for GMO in chapter III the working out of the dielectric (or elastic) susceptibility for an improper first-order transition requires to take into account higher degree coupling terms between the primary and secondary order-parameters. Thus depending on the value of the faintness index ν (see chapter III) one may get typical forms for the susceptibilities, as shown in Figures 2b and 2c.

Although at T_1, the specific heat curve $C(T)$ undergoes an upward step on cooling proportional to $\frac{a^2}{\beta}$ as for a second-order transition (Fig. 4 of chap. I), the influence of higher degree terms in the thermodynamic potential will manifest itself in the temperature dependence of $C(T)$ below T_1. In this region $C(T)$ varies as :

$$C(T) = C_o + a^2(\beta^2 - 4\alpha\gamma)^{-1/2} \quad (T < T_1) \qquad (2.10)$$

and thus the decrease of $C(T)$ below T_1 should be sharper for a first-order transition than for a second-order one.

In the preceding discussion only the temperature was considered as an external variable. Assuming a variation of the coefficients α and β in (2.1) as a function of two external variables (e.g. the pressure P and the temperature T), and a constant value for γ, one gets the $\alpha(\beta)$ phase diagram shown in Figure 3. In this diagram, one can see that the line of first-order transitions defined by :

$$\alpha = \frac{3\beta^2}{16\gamma} , \qquad \beta < 0 \tag{2.11}$$

goes over the line of second-order transitions :

$$\alpha = 0 \quad , \quad \beta > 0 \tag{2.12}$$

at the tricritical point $\alpha_{tc} = \beta_{tc} = 0$. Close from this point, for $\beta < 0$, β can be neglected in Eq.(2.2), and one has : $\eta \simeq (-\frac{\alpha}{\gamma})^{1/4}$.

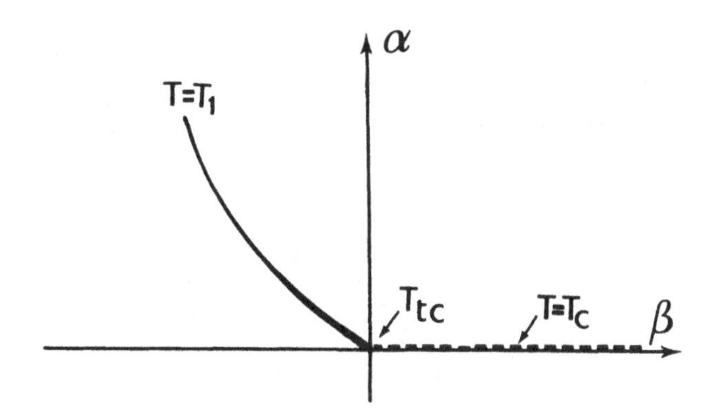

Figure 3 : phase diagram $\alpha(\beta)$ corresponding to the potential (2.1) $T = T_c$ is a second-order transition line ; $T = T_1$ is a first-order transition line.

For $\beta > 0$ close from the tricritical point, one can assume the power law $\beta = A^u$ where A is a positive constant and $0 \leqslant u \leqslant 1/2$. Thus the order-parameter near from the second-order transition line $T = T_c$ can be written :

$$\eta = (-\frac{\alpha}{\beta})^{1/2} \simeq (-\frac{\alpha}{A})^{\frac{1}{2} - \frac{u}{2}} \tag{2.13}$$

Accordingly, when approaching the point ($\alpha_{tc} = \beta_{tc} = 0$) the critical exponent associated with the order-parameter varies in the interval $[\frac{1}{4}, \frac{1}{2}]$.

2.2. General features of phase diagrams associated with one-component order-parameter expansions.

The discussion performed in the preceding section for a thermodynamic potential with $n = 1$, and $d = 6$, can be generalized to any even value of d. As shown by Gufan and Larin [6,13,15], one-component order-parameter expansions with $d \geqslant 8$ may describe succession of isomorphous (isostructural) phases separated by first-order transitions. In this subsection we summarize the main results obtained by these authors.

a) Let us consider the general one component order-parameter potential :

$$F(\eta) = a_1\eta^2 + a_2\eta^4 + a_3\eta^6 + a_4\eta^8 + \ldots \tag{2.14}$$

The corresponding equation of state is :

$$a_1 + 2a_2\eta^2 + 3a_3\eta^4 + 4a_4\eta^6 + \ldots = 0 \tag{2.15}$$

If $d = 2m$ is the higher degree assumed in (2.14), i.e. if m is the maximal number of non-zero coefficients a_i, Eq.(2.15) possesses at the utmost $(m - 1)$ positive roots for the variable η^2, which are associated with $(m - 1)$ stable low-temperature phases possessing identical symmetries. If m is even the phase diagram will contain a succession of $\frac{m}{2}$ transitions, whereas the number of transitions is $\frac{m-1}{2}$ for m odd. In Fig. 4, one can see that for m odd all the transitions are first-order, while for even m, one has a second order transition followed by $\frac{m}{2} - 1$ first-order transitions. These results have been illustrated in chapter I for the case $m = 2$ $(d = 4)$ and in this chapter (§2.1) for $m = 3$ $(d = 6)$. We will now discuss in more details the case $m = 4$ $(d = 8)$, which has been studied in refs. 6, 13 and 15.

The equation of state and the limit stability equation can be written in the case $m = 4$ (Fig. 4b) :

$$\eta(a_1 + 2a_2\eta^2 + 2a_3\eta^4 + 4a_4\eta^6) = 0 \tag{2.16}$$

and

$$a_1 + 6a_2\eta^2 + 15a_3\eta^4 + 28a_4\,\eta^6 = 0 \tag{2.17}$$

In addition to the solution $\eta = 0$, which corresponds to the high temperature phase (phase 0), Eq.(2.16) may possess one or three positive solutions for η^2, depending on the sign of a_3. Thus for $a_3 > 0$ one has only one low-temperature stable phase, as for the case $m = 3$, while for $a_3 < 0$ (and $a_4 > 0$) we find two stable low-temperature phases of identical symmetries denoted I and II. Assuming this latter condition for the sign of the coefficients a_3 and a_4, one can represent the complete phase diagram corresponding to $m = 4$, using as variables the coefficients a_1 and a_2, or more likely the dimensionless variables :

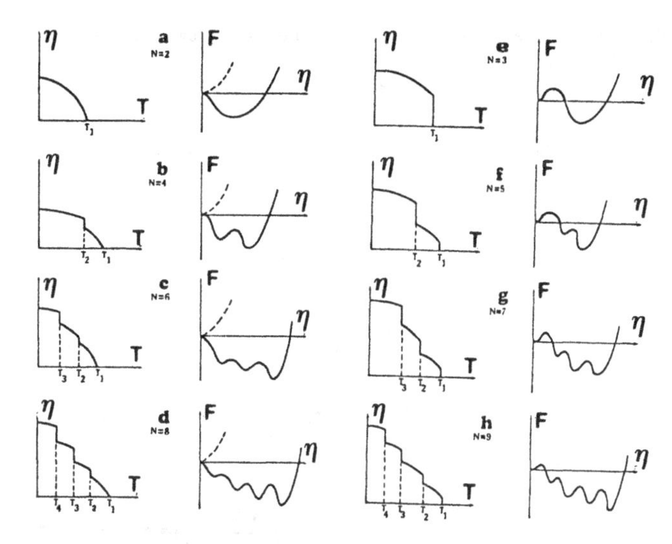

<u>Fig. 4</u> : Temperature dependence of the order-parameter $\eta(T)$ and order-parameter dependence of the non-equilibrium thermodynamic potential $F(\eta)$, for different maximal degrees of the order-parameter expansion,on the left are represented the cases where $d = 2m$ with m even, whereas figures on the right represent the cases with m odd.

$$\tilde{a}_1 = a_1 a_4^2/|a_3^3| \qquad \text{and} \qquad \tilde{a}_2 = a_2 a_4/a_3^2 \tag{2.18}$$

The phase diagram representing the solutions of Eqs.(2.16) and (2.17) in the plane $(\tilde{a}_1, \tilde{a}_2)$ is schematized in Figure 5. The stability line separating phases I and II has the equation :

$$\tilde{a}_1 = -\frac{1}{2}\tilde{a}_2 + \frac{1}{8} \tag{2.19}$$

It crosses the $\tilde{a}_2$ axis at a triple point Q with coordinates $(\tilde{a}_1^Q = 0, \tilde{a}_2^Q = \frac{1}{4})$ where converge the line of second-order transitions 0-I, the line of first-order transitions 0-II, and the line of isostructural transitions (2.19). This latter line terminates at a critical point K with coordinates $(\tilde{a}_1^K = -\frac{1}{6}, \tilde{a}_2^K = \frac{3}{8})$, at which the values of the order-parameter η in phases I and II coincide :

$$\eta^2_{K,I} = \eta^2_{K,II} = \frac{|a_3|}{4a_4} , \tag{2.20}$$

The jump of the order-parameter at the 0-II transition is equal to

$$(\Delta\eta^2)_{0-II} = \frac{|a_3|}{3a_4}\left[1 + (1 - 3\tilde{a}_2)^{1/2}\right] \tag{2.21}$$

whereas on the I-II transition line, the jump is :

$$(\Delta\eta^2)_{I-II} = \frac{|a_3|}{2a_4} (3 - 8\,\tilde{a}_2)^{1/2} \qquad (2.22)$$

The jumps (2.21) and (2.22) coïncide at the three phase point Q .

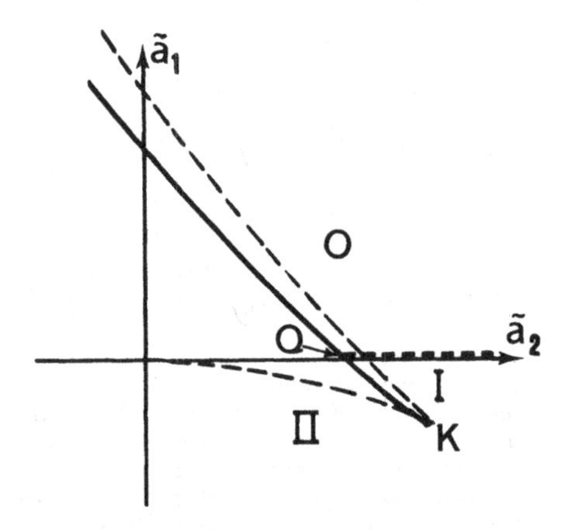

Figure 5 : Phase diagram associated with a one-component order-parameter expansion limited to the eight degree, for $a_3 < 0$. Full lines are first-order transition lines. Dashed lines indicate the limit of stability of each phase. Doted lines represent second-order transition lines.

b) Following refs. 6, 15, let us examine the behaviour of the susceptibilities, the thermal effects, and the effect of an external field for a system associated with the preceding eight-degree expansion. As the O-I transition is an ordinary second-order transition, the anomalies will coïncide with those for a standard second-order transition, i.e. the Curie-Weiss law will be satisfied if η is the polarization. On the first-order transition line O-II the value of the inverse susceptibility is :

$$(\chi^{-1})_o = \tilde{a}_1\,(T_c)\,\frac{|a_3|^2}{a_4^2} \qquad (2.23)$$

on the phase O side, whereas on the phase II side it is :

$$(\chi^{-1})_{II} = 8\,\eta_{II}^{2}\,\frac{a_3^2}{a_4}\,(-\tilde{a}_2 + \frac{a_4}{|a_3|}\,\eta_{II}^{2}) \qquad (2.24)$$

where $\tilde{a}_1(T_c)$ is the value of $\tilde{a}_1$ on the line O-II, and η_{II}^2 is given by (2.21). On the line of isostructural transitions I-II, there should be a positive jump in the inverse susceptibility, given by :

$$\Delta\chi_{I-II}^{-1} = 16\sqrt{2}\,\frac{|a_3|^3}{a_4^2}\,(\frac{3}{8} - \tilde{a}_2)^{3/2} \qquad (2.25)$$

Assuming $\tilde{a}_2$ constant, and a linear temperature dependence $a_1 = \alpha(T - T_1)^2(\alpha > 0)$ where T_1 is the transition temperature I-II the different forms of the function $\chi^{-1}(\tilde{a}_1)$ for different values of a_2 are shown in Figure 6.

The jump in entropy at the isostructural transition I-II is

$$\Delta S_{I-II} = \frac{1}{8}\,(1 + \lambda\,\frac{a_4}{|a_3|})(3 - 8\,\tilde{a}_2)^{1/2}\,\frac{\partial\tilde{a}_2}{\partial T} \qquad (2.26)$$

where $\tilde{a}_1 = \lambda\,\tilde{a}_2 + b$ is the equation of the transition line I-II Actually, due to a region of coexistence of phases I and II, thermal hysteresis effects should be observed for all physical quantities when cycling through the I-II transition line. Analogously hysteresis loops should be observed under the application of an external field. If ξ is the external field conjugate to η, the non-equilibrium potential takes the form :

$$F = a_1\eta^2 + a_2\eta^4 + a_3\eta^6 + a_4\eta^8 - \eta\xi \qquad (2.27)$$

The corresponding equation of state is :

$$2a_1\eta + 4a_2\eta^3 + 6a\eta^5 + 8a_4\eta^7 = \xi \qquad (2.28)$$

Let us restrict ourselves to the region of the phase diagram $(0 < \tilde{a}_2 < \frac{3}{8})$ where the isostructural transition I-II is possible for $\xi = 0$. In the corresponding temperature interval, the equation of state (2.28) may admit seven real solutions associated with four minima and three maxima. Depending on the value of $\tilde{a}_1(T)$, the form of the functions $\eta(\xi)$ will display different aspects which are represented in Figure 7. Thus, two minima correspond to stable phases I and II, between which occurs a first order transition in the field ξ, along a line given by the equation :

$$\tilde{a}_1 - a_o = \frac{1}{2\tilde{\eta}_1}\,(\tilde{\xi} - \xi_o) \qquad (2.29)$$

where $\tilde{\eta}_1^2 = \frac{5}{28}\,[1 - (1 - \frac{56}{25}\,\tilde{a}_2)^{1/2}]$, $a_o = -5\tilde{\eta}_1^4[1-(1-\frac{56}{25}\,\tilde{a}_2)^{1/2}]$, $\tilde{\xi}_1 = \xi\sqrt{a_4}/|a_3|$ and $\xi_o = \frac{16}{7}\,\tilde{\eta}_1^2\,[2 + (1 - \frac{56}{25}\,\tilde{a}_2)^{1/2}]$. If we denote a_1^- and a_1^+ the values of a_1 at which there occurs a loss of stability

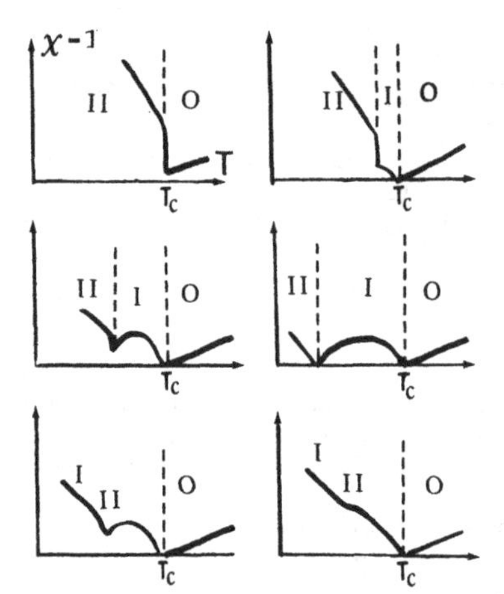

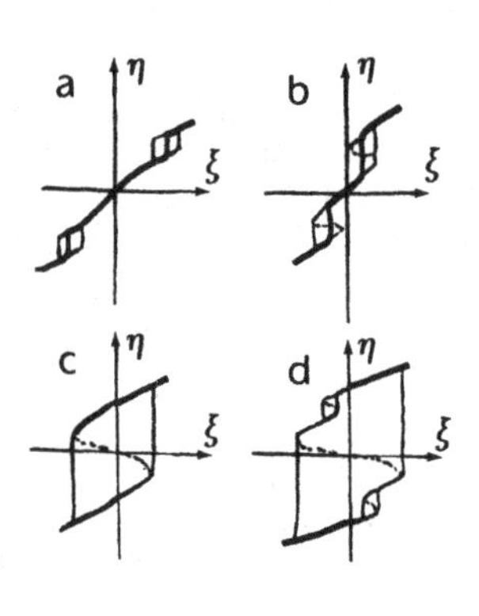

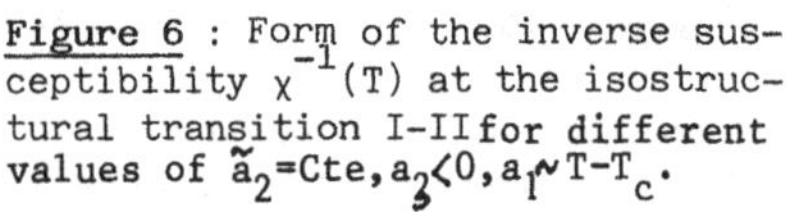

Figure 6 : Form of the inverse susceptibility $\chi^{-1}(T)$ at the isostructural transition I-II for different values of $\tilde{a}_2$=Cte, $a_3<0$, $a_1 \sim T-T_c$.

Figure 7 : $\eta(\xi)$ curves for various values of $\tilde{a}_1$ = constant

of phases I and II respectively, one can distinguish four types of hysteresis curves : 1) for $a_1^1 \lesssim \tilde{a}_1 < a_o$, double hysteresis loops appear at the ends of the hysteresis loop usually observed for second order transitions (Fig. 7a), 2) for $\tilde{a}_1 \lesssim a_1^2$ the double loops are superposed on the usual hysteresis loop (figures 7b and 7c), 3) For $\tilde{a}_1 < a_1^2$ the shape of the hysteresis loop takes the form shown in Fig. 7d.

The detailed working out of the preceding results, as well as a more complete picture of the potential (2.27) can be found in refs. 6, 15.

2.3. Phase diagrams with two-component order-parameters

Let us examine the phase diagram associated with a two-component order-parameter which transforms as the image C_{4v} (see chapters II and III). Restricting to the fourth degree terms, the thermodynamic potential can be written in polar coordinates ($\eta_1 = \rho \cos \phi$, $\eta_2 = \rho \sin\phi$)

$$F = \frac{\alpha}{2} \rho^2 + \frac{\beta_1}{4} \rho^4 + \frac{\beta_2}{4} \rho^4 \cos 4\phi \qquad (2.30)$$

It admits two distinct low-temperature phases, which may be stable across a second-order transition from the high temperature phase (0). These phases, denoted I and II, correspond to the equilibrium values :

$$\begin{aligned} &\text{phase I} : \eta_1 \neq 0,\ \eta_2 = 0 \quad \text{i.e. } \rho \neq 0 \ ,\ \sin\phi = 0 \\ &\text{phase II}: \eta_1 = \eta_2 \neq 0 \quad \text{i.e. } \rho \neq 0 \ ,\ \sin\phi = \cos\phi \end{aligned} \qquad (2.31)$$

Following ref. 6, we will show that the more complete phase diagram associated with an order-parameter symmetry C_{4v}, is obtained using an eight degree expansion F. This diagram contains an additional phase (III) of symmetry different from I and II, corresponding to the condition :

$$\text{phase III :} \quad \eta_1 \neq \eta_2 \quad \text{i.e.} \quad \rho \neq 0, \quad \cos\phi \neq 0 \qquad (2.32)$$

Besides, the 0-III transition can only be of the first-order. Following the method introduced by Gufan [7], we introduce the full rational basis of invariants for the image C_{4v}. It is constituted by the two invariants :

$$I_1 = \eta_1^2 + \eta_2^2 = \rho^2 \quad , \quad I_2 = (\eta_1^2 - \eta_2^2)^2 - 4\eta_1^2\eta_2^2 = \rho^4 \cos 4\phi \qquad (2.33)$$

As the higher degree of the I_i is n = 4, the minimum degree of the potential which accounts for all possible symmetries is [6] 2n = 8. F can be written as :

$$F = a_1I_1 + a_2I_1^2 + a_3I_1^3 + a_4I_1^4 + b_1I_2 + b_2I_2^2 + C_{12}I_1I_2 + C_{112}I_1^2I_2 \qquad (2.34)$$

Accordingly the equations of states of the system are :

$$\frac{\partial F}{\partial \rho} = 2\rho(F_1 + 2\phi^2 F_2 \cos 4\phi) = 0$$

$$\frac{\partial F}{\partial \phi} = -4\rho^4 F_2 \sin 4\phi = 0 \qquad (2.35)$$

where $F_1 = \frac{\partial F}{\partial I_1}$, $F_2 = \frac{\partial F}{\partial I_2}$

Equations (2.35) possess four solutions :

1) $\rho = 0$ (i.e. : the high-temperature phase 0)
2) $\sin 4\phi = 0$, which can be decomposed in :

$$\sin 2\phi = 0 \quad \text{and} \quad 2\rho\,(F_1 + 2\rho^2 F_2) = 0 \qquad \text{phase I}$$

or

$$\cos 2\phi = 0 \quad \text{and} \quad 2\rho\,(F_1 - 2\rho^2 F_2) = 0 \qquad \text{phase II}$$

3) $F_2 = 0$ and $F_1 = 0$

The determination of the stability regions in the phase diagram for the preceding phases, require to check the conditions :

$$\frac{\partial^2 F}{\partial \rho^2} \geqslant 0 \quad , \quad \left(\frac{\partial^2 F}{\partial \rho^2}\right)\left(\frac{\partial^2 F}{\partial \rho^2}\right) - \left(\frac{\partial^2 F}{\partial \rho \partial \phi}\right)^2 \geqslant 0 \qquad (2.37)$$

where :

$$\frac{\partial^2 F}{\partial \rho^2} = 2\,(F_1 + 2\rho^2.F_2 \cos 4\phi) + 4\rho^2[F_{11} + 4\rho^2 \cos 4\phi(F_{12} + F_{22})]$$

$$\frac{\partial^2 F}{\partial\rho\partial\phi} = -16\,\rho^3 \sin 4\phi\, F_2 + \rho^5 \sin 4\phi(-8F_{21} - 16F_{22}\rho^2 \cos 4\phi) \qquad (2.38)$$

$$\frac{\partial^2 F}{\partial\phi^2} = -16\,\rho^4\, F_2 \cos 4\phi + 16\,\rho^8\, F_{22} \sin^2 4\,\phi$$

Figure 8 shows the regions where conditions (2.37) are fulfilled for phases I, II and III, in the plane (a_1, b_1). Accordingly, we can verify that phase III can be reached from phase 0 either across a line of first-order transition (Figure 8a) or at a singular point of second order phase transition (Figures 8b and 8c). One can also obtain phase III through first-order transition lines from phase I or II. Let us note that the symmetry of phase III is always lower than the symmetries of phases I and II. For example, if we take for phase 0 the point-group C_{4v} (or the space-groups C^i_{4v} (i = 1-12)

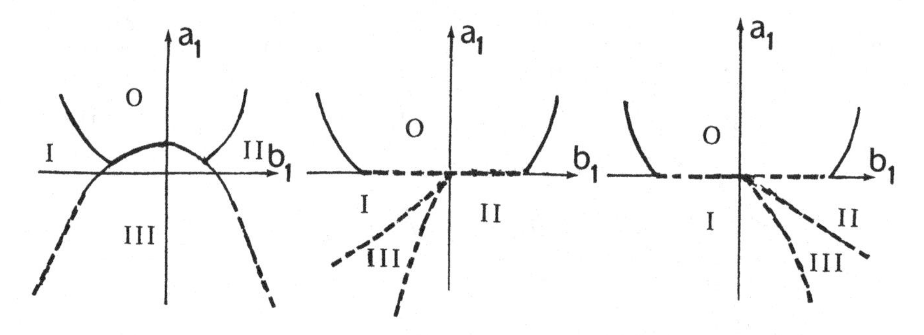

Figure 8 : Schematic phase diagrams for the two-component order-parameter (2.34)a) $a_2 < 0$, $\Delta = 4a_2b_2 - c^2_{12} < 0$
b) $a_2 > 0$, $\Delta > 0$, $c_{12} > 0$ c) $a_2 > 0$, $\Delta > 0$, $C_{12} < 0$
Dashed and continuous lines are respectively lines of second and first order transitions.

at k = 0) phases I and II will respectively coïncide with the monoclinic groups $C_s(x)$ and $C_s(xy)$, while phase III corresponds to the triclinic symmetry C_1.

The preceding results can be generalized to all images associated with a two-component order-parameter. Thus, for the images C_{nv} the full basis of integrity is formed by the invariants : $I_1 = \rho^2$ and $I_2 = \rho^n \cos n\,\phi$, whereas for the images C_n, one has to include the additional invariant $I_3 = \rho^n \sin n\,\phi$. For structural transitions (see chapter III), where n takes the values n = 3,4,6,8,12, in order to account for the full set of low-temperature phases, including the phases which can be obtained as the result of a first-order transition, one has to consider potentials of degrees 6,8, 12,16 and 24 respectively. The case with n = 3, in which a third degree invariant is involved, will be discussed in §. 3.

2.4 Phase diagrams with multicomponent order-parameter

When the number of components of the order-parameter becomes larger than two, the number of low-temperature phases which are stabilized by considering higher degree terms in the thermodynamic potential increases drastically. The enumeration of such phases was given by Gufan and Popov [23] and Gufan and Chechin [17], respectively for four and six dimensional order-parameters. However, the equations allowing to work-out the detailed features of the corresponding phase diagrams, are not solvable using the standard methods indicated in the preceding chapters, and require use of more powerful mathematical tools, which can be deduced from the Catastrophe theory[6].

As stressed by Gufan and Sakhnenko [10] phase diagrams associated with a number of components $n \geqslant 3$, differ essentially from the diagrams with $n = 1,2$, by the fact that the lines bounding the region of existence of the phases may intersect at a finite angle, at a N-phase point, i.e. a point at which a number N of phases undergo a continuous transition into each other. As noted in ref.10, the existence of points where more than three phases meet in the plane of the thermodynamic parameters is not forbidden by the Gibbs rule. This rule forbids the coexistence of more than three phases only when these phases are separated by lines of first-order transitions. This restriction results from the fact that for first-order transitions, the thermodynamic potentials of the phases are assumed to be independent. Accordingly, the requirement that the potentials are equal at the transition point leads to independent equations for the thermodynamic parameters. For second-order transitions, the number of phases which may be stable near a meeting point can be larger than three, as the equilibrium potentials of the different phases are extrema of the same non equilibrium potential, and thus related to each other, since they are expressed in terms of the same phenomenological coefficients.

As an illustrative example of phase diagram containing a N phase point, we will consider the example of the diagram associated with the three-component order-parameter with image O_H, that was discussed in ref. 10 . The three invariants forming the full rational basis are :

$$I_1 = \eta_1^2+\eta_2^2+\eta_3^2 \quad , \quad I_2 = \eta_1^4+\eta_2^4+\eta_3^4, \qquad I_3 = \eta_1^2\eta_2^2\eta_3^2 \tag{2.39}$$

When restricting to the fourth degree expansion, the corresponding thermodynamic potential admits two possible low-temperature phases associated with the equilibrium values :

$$\eta_1 = \eta_2 = \eta_3 \neq 0 \qquad \text{and} \qquad \eta_1 \neq 0 \ , \ \eta_2 = \eta_3 = 0 \tag{2.40}$$

Up to the eighth degree, the thermodynamic potential can be written :

$$\begin{aligned} F = &\alpha_1 I_1 + \alpha_2 I_1^2 + \alpha_3 I_1^3 + \alpha_4 I_1^4 + \beta_1 I_2 + \beta_2 I_2^2 \\ &+ \gamma I_3 + C_{12} I_1 I_2 + d I_1^2 I_2 + f I_1 I_3 \end{aligned} \tag{2.41}$$

The non-linear equations determining the equilibrium values of the order-parameter are :

$$\frac{\partial F}{\partial \eta_1} = 2\,\eta_1\,(F_1 + 2\,\eta_1^2 F_2 + \eta_2^2\eta_3^2\,F_3) = 0$$
$$\frac{\partial F}{\partial \eta_2} = 2\,\eta_2\,(F_1 + 2\,\eta_2^2 F_2 + \eta_1^2\eta_3^2\,F_3) = 0 \qquad (2.42)$$
$$\frac{\partial F}{\partial \eta_3} = 2\,\eta_3\,(F_1 + 2\,\eta_3^2 F_2 + \eta_1^2\eta_2^2\,F_3) = 0$$

where $F_k = \frac{\partial F}{\partial I_k}$ $(k = 1, 2, 3)$. Eqs.(2.42) admit <u>six</u> types of solutions describing low-symmetry phases with different symmetries, namely :

$$\eta_1 \neq 0, \eta_2 \neq 0,\ \eta_3 \neq 0,\quad F_1 = F_2 = F_3 = 0 \qquad \text{(phase I)}$$
$$\eta_1 \neq 0,\ \eta_2 \neq 0,\ \eta_3 = 0,\quad F_1 = 0,\ F_2 = 0 \qquad \text{(phase II)}$$
$$\eta_1 = \eta_2 \neq 0,\ \eta_3 \neq 0,\ 2F_2 - \eta_1^2 F_3 = 0,\ F_1 + (2\eta_1^2 + \eta_3^2)F_2 = 0 \qquad \text{(phase III)}$$
$$\eta_1 = \eta_2 \neq 0,\ \eta_3 = 0,\quad F_1 + 2\,\eta_1^2\,F_2 = 0 \qquad \text{(phase IV)} \qquad (2.43)$$
$$\eta_1 = \eta_2 = 0,\ \eta_3 \neq 0,\quad F_1 + 2\,\eta_3^2\,F_2 = 0 \qquad \text{(phase V)}$$
$$\eta_1 = \eta_2 = \eta_3 \neq 0,\quad F_1 + 2\eta_1^2\,F_2 + \eta_1^4 F_3 = 0 \qquad \text{(phase VI)}$$

one can see that the solutions (2.40) correspond to phase V and VI. Thus, higher degree terms in the thermodynamic potential may stabilize <u>four</u> additional phases (I-IV) which take place as the result of <u>first-order transitions.</u> Phase I has the lowest possible symmetry for the order-parameter under consideration. Phase II and III have twice as many elements as the symmetry group of phaseI. Using the notation $F_{ij} = \frac{\partial^2 F}{\partial I_i\,\partial I_j}$, one can write the stability conditions for phases II and III respectively :

$$\text{II} : F_{11} > 0,\quad F_{11}F_{22} - F_{12}^2 > 0,\ F_3 > 0, \quad \text{III} : F_{11} > 0, (\eta_1^2 - \eta_3^2)F_3 > 0,$$
$$(\eta_3^2 - \eta_1^2)F_3 > 0 \qquad (2.44)$$

For phase IV, V and VI, the stability conditions can be written :

$$\text{IV} : F_{11} > 0,\quad F_2 > 0,\quad -2F_2 + \eta_1^2\,F_3 > 0$$
$$\text{V} : 2\,F_2 + F_{11} > 0,\quad F_2 < 0 \qquad (2.45)$$
$$\text{VI} : 2\,F_2 + F_{11} > 0,\ 4F_2^2 + 4\,F_2F_{11} - 2\eta^2 F_3F_{11} > 0,\ 2F_2 + 3F_{11} > 0$$

The discussion of the stability conditions (2.44), (2.45)and of the respective boundaries of the phases [10] lead to the following conclusions :

1) In a plane of two thermodynamic variables (X_1, X_2) five phases (0, II, IV, V and VI) can be contiguous at one point defined by: $F_1(X_1, X_2) = 0$, $F_2(X_1, X_2) = 0$

2) Near the preceding five-phase point, the sequence of low-symmetry phases is the following 0 → V → II → IV → IV → VI → 0

3) Phase IV is stable only up to the sixth degree for F, while phase II requires eight degree invariants to be stabilized.

The possible types of phase diagrams corresponding to the potential (2.41) are shown in Figure 9. Their features depend on the respective sign of the discriminant :

$$\Delta = 4\,\alpha_2\beta_2 - C_{12}^2 \tag{2.46}$$

and of the coefficient γ of the sixth degree-invariant I_3. A detailed description of the diagrams can be found in ref. 10

The two other images corresponding to a three component order-parameter which fulfil the Landau condition, are T_H and 0 (see chapter III). The phase diagram associated with T_H require introduction of an additional basic invariant : $I_4 = \eta_1^4(\eta_2^2 - \eta_3^2) + \eta_2^4(\eta_3^2 - \eta_1^2) + (\eta_3^4(\eta_1^2 - \eta_2^2)$, while for the image 0 one has to introduce the invariant of ninth degree $I_4 \cdot \eta_1\eta_2\eta_3$. The images T_d and T allow existence of a cubic invariant $\eta_1\eta_2\eta_3$ and will be discussed in §.3.

For a number of components n ⩾ 4, a detailed discussion of the thermodynamic potentials was performed only for the lower degree terms allowing <u>second order</u> phase transitions [30-37], for n = 4, 6 and 8. However, in refs. 17 and 23, the number of stable phases

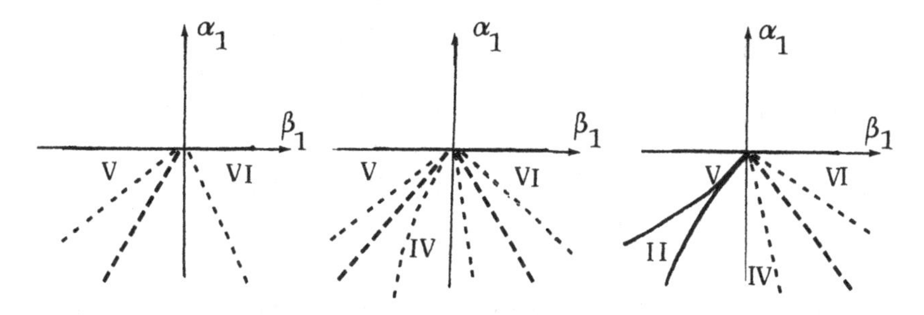

Figure 9 : Possible types of phase diagrams near a five phase point for a three-component order-parameter with image O_H. Solid lines are second order-transition lines. Dashed lines are fist-order transition lines. Dotted lines indicate the stability boundaries.

which may take place for any degree of the potentials, are enumerated for n = 4 and n = 6. As an example of the drastic increase in the number of phases which may be stabilized across a first-order transition, when considering higher degree terms in the thermodynamic potential, we list in Table I the equilibrium values corresponding to all possible stable phases for the six-dimensional image L_7 [32] (denoted E 192 b in ref. 17). In this table the symmetries of the low-temperature phases correspond to the case where the high-temperature phase is O_H^5 and the six-dimensional IR τ_{10} $(\vec{k}_{10})$ in Kovalev's notation [38]. As one can see in Table I, among the 26 phases induced by the preceding IR, only seven can be obtained as the result of a second-order transition. The remaining nineteen phases are stable only through a first-order transition, i.e. when taking into account higher degree terms in the thermodynamic potential. In this respect, there exist fiveteen independent invariants [39] forming a full rational basis, for the image L_7. The highest degree invariant is of the 12th degree, and thus, one has to expand the thermodynamic potential up to the 24th degree, in order to obtain the complete set of phases given in Table I

No.							Symmetry	
1.	η	0	0	0	0	0	D_{2h}^{17}	(2)
2.	η	η	0	0	0	0	D_{2h}^{13}	(2)
3.	0	0	η	0	0	η	D_{4h}^{7}	(4)
4.	0	0	0	η	η	0	D_{4h}^{14}	(4)
5.	η	η	η	η	η	η	C_{3v}^{5}	(4)
6.	η	$\bar{\eta}$	$\bar{\eta}$	0	$\bar{\eta}$	η	D_{3}^{7}	(4)
7.	η	0	$\bar{\eta}$	0	$\bar{\eta}$	0	T^{4}	(4)
8.	η_1	η_1	η_3	0	0	η_3	C_{2v}^{14}	(4)
9.	η_1	η_1	0	η_4	η_4	0	C_{2v}^{14}	(4)
10.	0	0	η_3	η_4	η_4	η_3	C_{2h}^{3}	(4)
11.	η_1	0	η_3	0	0	0	D_{2h}^{16}	(4)
12.	0	η_2	η_3	0	0	0	D_{2h}^{12}	(4)
13.	η_1	0	0	η_4	0	0	D_{2h}^{13}	(4)
14.	η_1	η_2	$\bar{\eta}_1$	$\bar{\eta}_2$	$\bar{\eta}_1$	$\bar{\eta}_2$	C_{3}^{4}	(4)
15.	0	0	0	0	η_5	η_6	C_{2h}^{2}	(2)
16.	η_1	η_2	η_3	η_4	η_4	η_3	C_{s}^{3}	(4)
17.	η_1	η_2	η_2	η_1	η_5	$\bar{\eta}_5$	C_{2}^{3}	(4)
18.	η_1	η_2	η_3	η_4	0	0	C_{i}^{1}	(4)
19.	η_1	η_2	η_3	η_4	η_5	η_6	C_{i}^{1}	(4)
20.	0	η_2	η_3	0	η_5	η_6	C_{s}^{1}	(4)
21.	η_1	0	0	η_4	η_5	η_6	C_{s}^{2}	(4)
22.	0	η_2	0	η_4	η_5	η_6	C_{2}^{2}	(4)
23.	η_1	0	η_3	η_4	0	0	C_{2h}^{2}	(4)
24.	0	η_2	η_3	η_4	0	0	C_{2h}^{5}	(4)
25.	0	η_2	η_3	0	η_5	0	C_{2v}^{7}	(4)
26.	0	η_2	0	η_4	0	η_6	D_{2}^{4}	(4)

Table I : Equilibrium values of the order-parameter and corresponding symmetries, for the phases induced by the six-dimensional IR τ_{10} $(\vec{k}_{10})$ [38] of the O_H^5 space-group. Only phases 1 to 7 may take place across a second-order transition. As indicated into the brackets the multiplication of the elementary unit-cell can be 2 or 4.

2.5. Experimental examples

In table II, have been listed a number of materials in which the first-order structural transitions occurs between strictly crystalline structures, and is connected to a single active IR of the high-temperature space-group. The materials are classified following the dimensionality of their order-parameter. One can verify that most of the listed systems correspond to a one-two or three-dimensional order-parameter, and we have found only one confirmed example of transitions associated with a four or six-dimensional order-parameter (i.e. VO_2 and Cd Sn As_2). Let us briefly examine the experimental confirmations of the Landau theory predictions, concerning this class of first-order transitions.

1) Five examples of isomorphous transitions are given in Table I, namely $Ni(C_5H_5)_2$, $(V_{1-x}\ Cr_x)_2O_3$, SmS, Ce and EuO. Such transitions can be interpreted, as was shown in §.2.2., by considering active IR's with high-degree expansions (of at least eight degree for one-dimensional IR's). Another possible interpretation, is to relate isomorphous transitions to the identity representation of one of the space-groups, which allows a cubic invariant in the corresponding potential [40]. In both cases, the transition between the isostructural phases should be first-order. This result is confirmed by the five cases listed in Table I where an abrupt discontinuity of the lattice parameters was observed at the Curie point. A more striking prediction of the Landau theory for isomorphous transitions, is that one may find in the pressure-temperature (or concentration-temperature) phase diagram, a critical point terminating the equilibrium line between the phases, as for the liquid-vapor transition. Such a point has been effectively observed in SmS (at 923 K and 6.8 kbar) [41], in Ce (at 580 K and 17.5 Kbar) [42] and in $(V_{1-x}\ Cr_x)_2O_3$ (for x = 0.037) [43].

2) For the remaining transitions indicated in Table I a change in the space-group occurs at the transition. When the transition is slightly first-order, it may be possible, by varying another external parameter to reach the tricritical point predicted by the Landau theory. This was realized in NH_4Cl [44] and in a number of ferroelectrics (KH_2PO_4 [45], SbSI [46], TGSe [47], $BaTiO_3$ [48] by applying an external hydrostatic pressure, or in solid solutions, such as $Sn_2P_2(Se_xS_{1-x})_6$ [49] and $SbSe_xS_{1-x}I$ [50], by varying the concentration. A decrease of the critical exponent for the order-parameter is verified when approaching the tricritical point, although it dos not follow in general[$\frac{1}{2}$, $\frac{1}{4}$] interval predicted by the theory [51].

3) Another clearcut series of results confirming the predictions of the Landau theory of first-order transitions, is the stabilization of additional phases, when including higher degree terms in the thermodynamic potential. The perovskites $BaTiO_3$, $KNbO_3$ and $KTaO_3$, constitute well-known illustrative examples of such a prediction. In these compounds, an orthorhombic phase of symmetry

	(a)	(b)	(c)		(d)
n = 1	$Ni(C_5H_5)_2$	423	$P2_1/c \to P2_1/C$	(1)	Γ
	$(V_{1-x}Cr_x)_2O_3$	197	$R\bar{3}c \to R3c$	(1)	Γ
	SmS	6.5 kbar			
	Ce	8 kbar	Fm3m → Fm3m	(1)	Γ
	EuO	300 kbar			
	$CsH_3(SeO_3)_2$	145	$P\bar{1} \to P\bar{1}$	(2)	Z
	$Cu(HCOO)_2.4H_2O$	234	$P2_1/b \to P2_1/b$	(2)	A
	$C_{10}F_8$	266			
	$(CH_2ClCOO)_2HNH_4$	128	B2/b → Bb	(1)	Γ
	$KTiF_4$	488	Cmcm → Pmmn	(2)	Y
	$(C_2H_5NH_3)_2MnCl_4$	225	Cmca → Pbca	(2)	Y
	$(C_3H_7NH_3)_2MnCl_4$	180	Cmca → Pbam	(2)	Y
	$(C_{10}H_{21}NH_3)_2CdCl_4$	234	Cccm → Pnnm	(2)	Y
	KCN	63	Immm → Pmmn	(2)	X
	$(NH_4)_2SO_4$	223	$Pnma \to Pna2_1$	(1)	Γ
	SbSI	293	$Pnma \to Pna2_1$	(1)	Γ
	SbSr	93	$Pnma \to Pna2_1$	(1)	Γ
	KH_2PO_4	396	$I\bar{4}2d \to Fdd2$	(1)	Γ
	$(NH_4)_2H_3IO_6$	252	$R3 \to R\bar{3}$	(2)	Z
	NH_4Cl	243	$Pm3m \to P\bar{4}3m$	(1)	Γ
n = 2	$C_4O_4H_2$	470	$I\bar{4}/m \to P2_1/m$	(2)	Z
	ADP	148	$I\bar{4}2d \to C222_1$	(2)	Z
	$Gd_2(MoO_4)_3$	432	$P\bar{4}2_1m \to Pba2$	(2)	M
	$CH_3NH_3CdCl_4$	279	$I4/mmm \to P4_2/ncm$	(4)	X
	$PbNb_2O_6$	833	P4/mbm → Amm2	(1)	Γ
	$C_{18}H_{24}$	303	Immm → Fddd	(4)	R
n = 3	$PbZr_{0.9}Ti_{0.1}O_3$	373	$R\bar{3}m \to R3c$	(8)	A
	$Pb_3(PO_4)_2$	453	$R\bar{3}m \to B2/b$	(2)	L
	Sb_5O_7I	481	$P6_3/m \to P2_1/b$	(2)	M
	$NH_4Fe(SO_4)_2.12H_2O$	361	$Pa3 \to Pca2_1$	(1)	Γ
	$BaTiO_3$	393			
	$PbTiO_3$	663	Pm3m → P4mm	(1)	Γ
	$KNbO_3$	608			
	$CuAu_3$	653	Fm3m → P4/mmm	(2)	X
	EuO	400 Kbar	Fm3m → Pm3m	(4)	X
	SmTe	110 Kbar			
	Cu_3AuI	667			
n = 4	VO_2	343	$P4_2/mnm \to P2_1/b$	(2)	R
n = 6	$CdSnAs_2$	840	$F\bar{4}3m \to I\bar{4}2d$	(4)	W

Table II : Experimental examples of systems undergoing first-order structural transitions induced by a single active irreducible representations.(b) Transition temperatures or pressures in Kelvin or kbar (c) Symmetry change (cell multiplication) (d) Brillouin zone point.

Amm2 takes phase, which can be shown to be stable only when sixth degree terms are included in the three-component order-parameter expansion [52]. In another perovskite $PrAlO_3$ [53], the following sequence of phases is observed : Pm3m $\xrightarrow{1320\ K}$ $R\bar{3}c$ $\xrightarrow{205\ K}$ Imma $\xrightarrow{146\ K}$ B2/m $\xrightarrow{92\ K}$ I4/mcm. Gufan has shown [6] that the orthorhombic phase is stable only for a sixth degree expansion of the three-dimensional order-parameter, whereas the monoclinic phase requires an eighth degree expansion. In VO_2 the same author has proved [6] that a complete description of the phase diagram, which includes two monoclinic and one triclinic phases [54] , requires a tenth degree expansion for the corresponding four-component order-parameter. As will be shown in Chapter V, the stabilization of observed lock-in phases in incommensurate systems necessitates even higher degree terms in the thermodynamic potential.

3. TRANSITIONS PREDICTED TO BE DISCONTINUOUS BY THE LANDAU CONDITION

When a cubic invariant of the order-parameter is allowed by symmetry in the Landau potential, the transition cannot be continuous, except for isolated points in the phase diagram, where the coefficient of the cubic term vanishes accidentally. This is the mere formulation of the so-called Landau condition introduced by Landau in its original paper [55], and whose group-theoretical expression was given in chapter II. In this section we examine the phenomenological implications of the Landau condition (§. 3.1) and show, using the available experimental data (§.3.2) that it is indeed a very reliable sufficient condition for the corresponding transitions to be first-order.

3.1. Role of third order invariants of the order-parameter in the phenomenological description of structural transitions

1. Let us consider, as an introductory example, the one-component order-parameter expansion :

$$F = \frac{\alpha}{2}\eta^2 + \frac{B}{3}\eta^3 + \frac{\beta}{4}\eta^4 \tag{3.1}$$

which admits a third order invariant. The minima equations are:

$$\begin{aligned} \eta\,[\,\alpha + B\,\eta + \beta\,\eta^2\,] &= 0 \\ \alpha + 2\,B\,\eta + 3\,\beta\eta^2 &\geqslant 0 \end{aligned} \tag{3.2}$$

Assuming $B < 0$, $\beta > 0$ and $\alpha = a(T-T_c)$, the discussion of the conditions (3.4) is visualized in Figure 10. Thus, for :

$$T_1 = T_c + \frac{B^2}{4\beta a} \tag{3.3}$$

the low-temperature phase appears as a metastable state. The high and low temperature phases are equally stable for :

$$T_2 = T_c + \frac{2}{9}\,\frac{B^2}{\beta a} \tag{3.4}$$

T_2 coïncides with the actual first-order transition temperature, at which the order-parameter undergoes the jump :

$$\eta(T_2) = -\frac{B}{3\beta} \qquad (3.5)$$

then increases following the law :

$$\eta(T) = \frac{-B + (B^2 - 4\alpha\beta)^{1/2}}{2\beta} \qquad (3.6)$$

which is represented in Figure 11. The preceding discussion resembles closely to the one performed in §.2.1 for the potential (2.1) and most of the features characterizing first-order transitions associated with active IR's can also be found for the first-order transitions corresponding to an (inactive) IR which does not fulfil the Landau condition : region of coexistence for the high and low temperature phases, discontinuous jump for the order-parameter, latent heat etc... However the phase diagram for the two classes of first order transitions can be distinguished by the property

that no tricritical point can be found for transitions predicted to be first-order by the Landau condition. Besides, for these transitions a special critical point should be found (so-called Landau critical point [56, 57]) which is defined by the conditions:

$$\alpha(T_2, P) = 0 \quad , \quad B(T_2, P) = 0 \qquad (3.7)$$

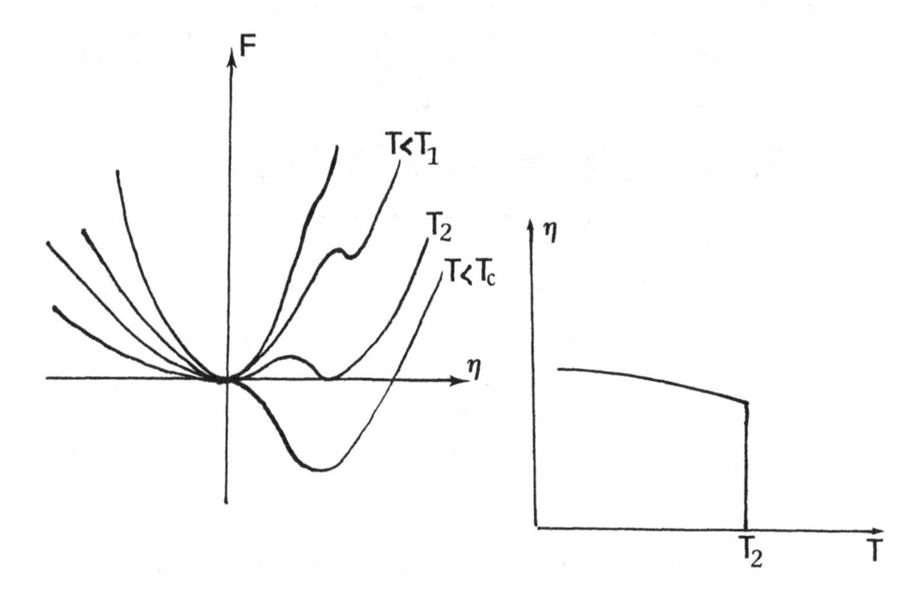

Figure 10 : variation of the potential (3.1) as a function of temperature.

Figure 11 : Temperature dependence of the order-parameter (3.6)

It corresponds to an isolated point of continuous transition on the first-order transition line $\alpha\,(T_2,P) = 0$. At this point (denoted L in Figure 12) terminates the first-order transition line B (T,P) = 0 separating two phases of identical symmetries (I and II) associated with the opposite values of the order-parameter:

$$\eta_{I,II} = \pm\left(-\frac{\alpha}{\beta}\right)^{1/2} \qquad (3.8)$$

At the L point all three phases become identical. As was shown by Landau [55] the latent heat for the isomorphous I-II transition is proportional to η_o^3, where $\eta_o = \eta_I - \eta_{II} = 2\left(-\frac{\alpha}{\beta}\right)^{1/2}$ is the jump of the order-parameter across the $B(T,P) = 0$ line.

Thus, the distinctive feature of the class of first-order transitions associated with a thermodynamic potential containing a cubic order-parameter invariant is the existence of a Landau critical point, of the type shown in Figure 12.

2. In addition to the identity representation of the 230 space-groups (wich allow a cubic and a linear invariant of the corresponding one-component order parameter), IR's which are inactive because of the sole Landau condition (i.e. which fulfil the Lifshitz condition) can be found exclusively in the rhombohedral, hexagonal and cubic systems. Less than three hundred IR's

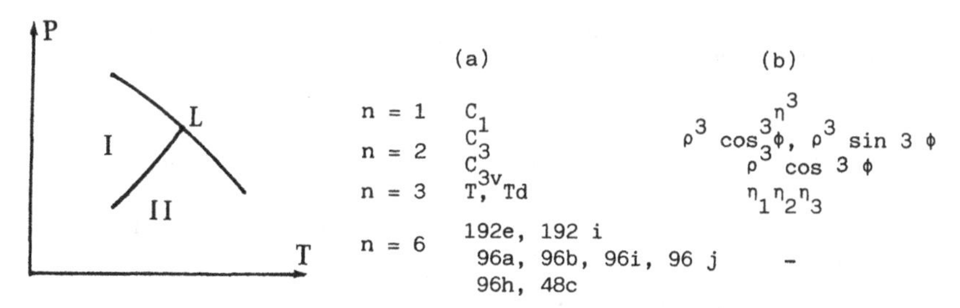

	(a)	(b)
n = 1	C_1	η^3
n = 2	C_3	$\rho^3\cos 3\phi$, $\rho^3 \sin 3\phi$
	C_{3v}	$\rho^3 \cos 3\phi$
n = 3	T, Td	$\eta_1\eta_2\eta_3$
n = 6	192e, 192 i 96a, 96b, 96i, 96 j 96h, 48c	-

Figure 12 : Neighborhood of a Landau critical point.

Table III (a) Images associated with IR's violating the Landau condition (b) form of the cubic terms. For n = 6, the notation of the images refers to [17].

are concerned [58] , which belong to the Brillouin zone centers, and to only six points of Brillouin zone boundaries, namely X (rhombohedral), K and M (hexagonal), M (cubic P), X (cubic F) and N (cubic I). This reveals that, on a theoretical ground, the Landau condition is not very eliminative. The images corresponding to the preceding IR's are listed in Table III. They are associated with one, two, three and six-dimensional IR's. Let us note that some images (e.g. C_3 and some of the six-dimensional images) allow existence of two distinct cubic invariants. Accordingly the corresponding phase diagram will not display a Landau critical point. It must also be stressed that the preceding results are valid only for the IR's fulfilling the Lifshitz condition. For IR's violating this latter condition, cubic invariants can be found also for four dimensional IR's, at specific points located inside the Brillouin zone [59].

3. The one component order-parameter expansion (3.1) coincides with the image C_1, providing the linear invariant (which is allowed by the identity IR) is eliminated by a suitable change in variable. It corresponds to the simplest phase diagram (Fig. 12) which can be obtained when a cubic invariant figures in the thermodynamic potential. Let us examine the more complex diagram associated with the two-dimensional image C_{3v}, that was discussed by Gufan [6] and by Sakhnenko and Talanov [60]. The full rational basis of invariants is in this case :

$$I_1 = \rho^2 = \eta_1^2+\eta_2^2 \,, \quad I_2 = \rho^3 \cos 3\phi = \eta_1^3 - 3\,\eta_1\eta_2^2 \tag{3.9}$$

The more complete phase diagram is obtained by considering the sixth degree expansion :

$$F = \alpha_1 I_1 + \alpha_2 I_1^2 + \alpha_3 I_1^3 + \beta_1 I_2 + \beta_2 I_2^2 + \gamma_1 I_1 I_2 \tag{3.10}$$

The minimization of F, leads to three possible stable phases:

1) Phases I and II corresponding to $\eta_1 = \eta$, $\eta_2 = 0$, associated with the extremum condition :

$$2\,\alpha_1 + 3\beta_1\eta_1 + 4\alpha_2\eta_1^2 + 5\gamma\eta_1^3 + 6\alpha_3\eta_1^4 = 0 \tag{3.11}$$

2) phase III, with $\eta_1 \neq 0$, $\eta_2 \neq 0$, associated with the conditions :

$$\begin{aligned} \alpha_1 + 2\alpha_2 I_1 + 3\alpha_3 I_1^2 + \gamma I_2 &= 0 \\ \beta_1 + 2\beta_2 I_2 + \gamma I_1^3 &= 0 \end{aligned} \tag{3.12}$$

Phases I and II are isostructural and differ only by the sign of η. They are characterized by a group of symmetry of index 3 with respect to the high-symmetry phase (0). As shown in Figure 13, they result from a first-order transition 0-I or 0-II. Using the notation $F_k = \frac{\partial F}{\partial I_k}$, $F_{ik} = \frac{\partial F}{\partial I_i \partial I_k}$, one can easely verify that in the limit $\eta \to 0$, the stability conditions (2.37) for these phases lead to the contradictory inequalities $F_2 \leqslant 0$, $F_2 \geqslant 0$, which expresses the fact that the transitions 0-I and 0-II can be second order only at the Landau critical point defined by

$$\alpha_1 = F_1\,(\eta = 0) = 0 \qquad \underline{\text{and}} \qquad \beta_1 = F_2\,(\eta = 0) = 0 \tag{3.13}$$

The stability boundary of phases I and II is provided by the inequality $F_{11} \geqslant 0$ which is explicited by :

$$-\,9\,\beta_1^2 - 32\,\alpha_2\alpha_1 = 0 \tag{3.14}$$

represented by line 3 in Figure 13.

The solution of the system (3.12) can be obtained near the Landau critical point by the approximative expressions :

$$I_1 = \frac{1}{\delta}(\gamma\beta_1 - 2\,\beta_2\alpha_1) \quad , \quad I_2 = \frac{1}{\delta}(\gamma\alpha_1 - 2\alpha_2\beta_1) \qquad (3.15)$$

where $\delta = 4\,\alpha_2\beta_2 - \gamma^2$. Since I_1 and I_2 are linear with respect to α_1 and β_1, but are of different order with respect to η, phase III may be stable only inside the region of the phase diagram bounded by the lines denoted 2 and 2' in Figure 13, whose equations are:

$$2\,\alpha_2\beta_1 - \gamma\alpha_1 \pm \delta\left(-\frac{\alpha_1}{2\alpha_2}\right)^{3/2} = 0 \qquad (3.16)$$

The stability conditions (2.37) for phase III can be written:

$$\delta > 0 \quad , \quad \alpha_2 > 0 \qquad (3.17)$$

Let us note that the Landau critical point, denoted L in Figure 13, is here a four phases point, at which merge two first order transition lines (1 and 1') and two second order transition lines (2 and 2'). Finally, we can give the approximative expression

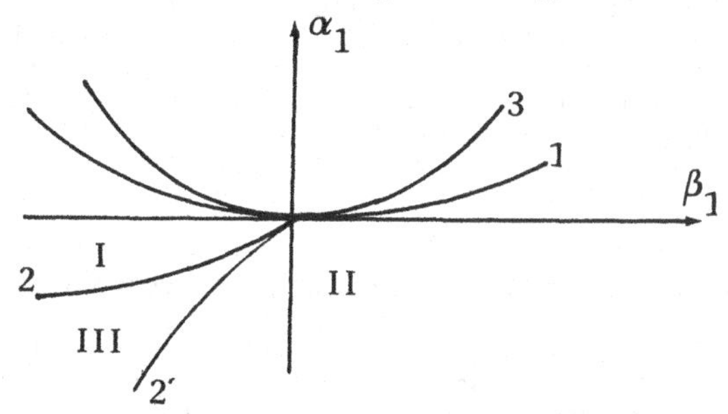

Figure 13 : Phase diagram corresponding to the image C_{3v}. Line 1 is a first-order transition line. Lines 2 and 2' are second-order transition lines. The stability line 3 is given by eq. (3.14).

for the order-parameter components in phase I and II :

$$\eta = \frac{1}{8\alpha_2}\left[-3\beta_1 \pm (9\,\beta_1^2 - 32\,\alpha_2\alpha_1)^{1/2}\right] \qquad (3.18)$$

and in phase III :

$$\eta = \pm\left(-\frac{\alpha_1}{2\alpha_2}\right)^{1/2} + \frac{1}{8\alpha_2}\left[3\,\beta_1 + 5\,\delta\left(-\frac{\alpha_1}{2\alpha_2}\right)\right] \qquad (3.19)$$

As shown in Ref. 58, the image C_{3v} can be associated to a large number of symmetry changes corresponding for example to the following point group modifications: 3m → (1,m), $\bar{3}$m → ($\bar{1}$,2/m), 6 → (3,6), 6mm → (mm2, 2), 622 → (622, 312), 6/mmm → (mmm, 2/m), 432 → (422, 222), $\bar{4}$3m → ($\bar{4}$2m, 222), m3m → (4/mmm, mmm).

3.2. Experimental examples

A systematic investigation was performed for the IR's which do not fulfil the Landau condition [58] and the corresponding experimental data were discussed by Tolédano and Pascoli [40]. Table IV summarizes some of the experimental examples of transitions associated with a thermodynamic potential containing a cubic invariant. All of them possess a first-order character, including V_3Si which was given for a long-time as a counter example to the validity of the Landau condition [61, 62], until the discontinuous nature of its 21 K transition was proved [63]. Accordingly, the argument raised by some authors [1-3], who suggested that under the effect of thermal fluctuations the coefficient of the cubic term may vanish, allowing existence for a second-order transition, has not yet received any experimental confirmation. Along the same line, no experimental example of Landau critical point was reported up to now [64].

Most of the examples listed in Table IV correspond to ferro-elastic transition (proper or pseudoproper) for which the order-parameter is proportional to a combination of spontaneous strain tensor components [60,65]. Accordingly, the coefficients in the corresponding thermodynamic potentials, coïncide with second and higher-order elastic constants [65] . In particular, as noted in refs. 60,65, for a cubic to tetragonal symmetry modification (e.g. in V_3Si, Nb_3Sn, InTl or NH_4CN), the sign of the combination of third order elastic constants coïnciding with the third degree coefficient B, indicates whether the ratio of the c and a tetragonal parameters is greater or smaller than unity. If $B < 0$, one has $\frac{c}{a} -1>0$, so that $c > 0$.

4. TRANSITIONS PREDICTED TO BE DISCONTINUOUS BY THE LIFSHITZ CONDITION

As indicated in chapter III, the Lifshitz condition requires that the antisymmetric square of an IR, according to which the transition order-parameter transforms, has no representation in common with the vector representation of the high symmetry group. Lifshitz has shown [66] that this condition is equivalent to the absence in the thermodynamic potential associated with the transition, of invariants (Lifshitz invariants) which couple bilinearly the components of the order-parameter with their spatial derivatives other than those which are full differentials. When an IR violates the Lifshitz criterion, it can still be associated with :

1) a continuous or discontinuous transition towards an inhomogeneous (incommensurate) phase, or

2) a discontinuous transition towards a normal crystalline (commensurate) phase.

The respective stability of the commensurate and incommensurate phases cannot be predicted by symmetry considerations as it depends on the relative magnitude of the anisotropic homogeneous terms contained in the potential and of the Lifshitz invariant. As

Substances	T_c(K)	Order	Symmetry change	order parameter dimension	Brillouin zone-point IR	Image
NaN_3	293	1	$R\bar{3}m \rightarrow B2/m(1)$	2	$\Gamma(\tau_3)$	C_{3v}
$C_3H_3N_3$	210	1	$R\bar{3}c \rightarrow B2/b(1)$	2		
Mg_3Cd	523	1	$P6_3/mmc \rightarrow P6_3/mmc(4)$	3	$M(\tau_1)$	Td
Ti_3Al	1423	1				
$KMn_2(SO_4)_3$	200	1	$P2_13 \rightarrow P2_12_12_1(1)$	2	$\Gamma(\tau_2+\tau_3)$	C_3
NH_4CN	–	1	$Pm3m \rightarrow P4/mmm(1)$	2	$\Gamma(\tau_3)$	C_{3v}
V_3Si	21	1	$Pm3n \rightarrow P4_2/mmc(1)$	2	$\Gamma(\tau_3)$	C_{3v}
Nb_3Sn	43	1				
KCN	170	1	$Fm3m \rightarrow Immm$ (1)	3	$\Gamma(\tau_4)$	Td
NaCN	288	1				
InTl	320	1	$Fm3m \rightarrow I4/mmm$ (1)	2	$\Gamma(\tau_3)$	Td
KNO_2	–	1	$Fm3m \rightarrow R\bar{3}m$ (1)	3	$\Gamma(\tau_4)$	Td
EuO^2	400 kbar	1	$Fm3m \rightarrow Pm3m$ (4)	3	$X(\tau_1)$	Td
SmTe	110 kbar	1				
Cu_3AuI	671	1				
$Fe_{1-x}Al_x$	$x \sim 0.5$	1				
CrN	273	1	$Fm3m \rightarrow Pnma$ (2)	6	$X(\tau_5)$	96b
$NiCr_2O_4$	274	1	$Fd3M \rightarrow I4_1/amd$ (1)	2	$\Gamma(\tau_3)$	Td
$Ca_3Mn_2GeO_{12}$	–	1	$Ia3d \rightarrow I4_1/acd$ (1)	2	$\Gamma(\tau_3)$	Td
$Mg_3B_7O_{13}Cl^2$	615	1	$F\bar{4}3c \rightarrow Pca2_1$ (2)	6	$X(\tau_5)$	96b

Table IV : Examples of materials undergoing a structural transition induced by an IR which does not fulfil the Landau condition.

will be shown in chapter V, the first terms favour a phase where the wave k-vector locks-in at a special rational value, corresponding to a commensurate phase, whereas a prevalence of the inhomogeneous part of the potential stabilizes the incommensurate structure. When the incommensurate structure occurs first, it becomes generally (but not always) unstable at lower temperature, as the quartic or higher order homogeneous terms increase more rapidly than the inhomogeneous ones, and the commensurate phase arises.

The various types of phase diagrams where commensurate and incommensurate phases possess a respective region of stability separated by lines of first or second order transitions, will be discussed in chapter V. In this section we confine ourselves to a brief discussion of the available experimental data with regard to the order of the observed transitions.

On table V are listed materials in which the commensurate phase takes place directly. According to the theoretical prediction the corresponding transition <u>should be first-order</u>. This is verified except for niobium dioxide and rubidium silver iodide. Neutron scattering studies report a continuous transition for NbO_2 with a temperature independent k-vector located on the surface of the Brillouin zone [67,68] . The high-temperature transition in this compound has been described by a four-component order-parameter which violates the Lifshitz condition [69]. However alternate interpretations are suggested by earlier data, which could preserve the predictive value of the Lifshitz condition for first-order commensurate-commensurate transitions:

a) DTA measurements [70] reveal a small thermal hysteresis and an enthalpy change characterizing a first-order transition
b) X-ray results [71] allege in favour of a two steps transition: one in the range 850-900° C and the other near 800° C with a two dimensional OP [30].

In $RbAg_4I_5$ the continuous character of the 208 K transition [72] needs further confirmation, more especially as the corresponding IR violates also the Landau condition. Otherwise an additional first-order transition found at -152° C in this compound suggests that an incommensurate intermediate has possibly failed to be observed.

Among the materials possessing a structural incommensurate phase, the high-symmetry commensurate-incommensurate phase transition is generally reported as second order. However there exist a little number compounds for which the preceding transition is reported to be first-order. The more clearcut example is constituted

<u>Table V</u> : Examples of materials undergoing a phase transition associated with an IR which does not fulfil the Lifshitz condition.

Substances	T_c(K)	order	Symmetry change	Order Parameter dimension	Brillouin point
$NaH_3(SeO_3)_2$	194	1	$P2_1/c \rightarrow P\bar{1}(4)$	2	Z
$RbH_3(SeO_3)_2$	153	1	$P2_12_12_1 \rightarrow P2_1(2)$	2	X
$NaNH_4C_4H_4O_6.4H_2O$	109	1	$P2_12_12 \rightarrow P2_1(2)$	2	X
$LiNH_4SO_4$	283	1	$Pna2_1 \rightarrow P2_1(2)$	2	S
NbO_2	1073	(2)	$P4_2/mnm \rightarrow I4_1/a(8)$	4	1/4 1/4 1/2
FeS	410	1	$P6_3/mmc \rightarrow P\bar{6}2c$ (6)	4	H
C_5CuCl_3	423	1	$P6_3/mmc \rightarrow P6_122(3)$	4	0 0 1/3
$RbAg_4I_5$	433	(2)	$P4_132 \rightarrow R32$ (1)	3	Γ
$Cd(NO_3)_2$	-	1	$Pa3 \rightarrow Pca2_1$ (2)	6	X

by the pervoskite Pb_2CoWO_6 [73] which undergoes a strongly first-order transition from the Fm3m phase to an incommensurate phase, which remains stable in an interval of more than two hundred degrees. For $BaMnF_4$ [74] and Rb_2ZnBr_4 [75] the transition is considered as weakly first-order. However the evidence in $Ba_2NaNb_5O_{15}$ of an incommensurate transition [76] possessing a number of second-order characteristics but a strong thermal hysteresis (see chapter V) indicates that a non-standard behaviour must be expected for the order of the transitions to an incommensurate phase.

In summary, the violation of the Lifshitz condition leads in most cases to first-order transitions when the lock-in (commensurate) phase appears directly below the high-symmetry phase. Such a situation cannot be predicted only on a symmetry basis, as the respective magnitude of the anisotropic and Lifshitz invariants must be taken into account. On the other hand, first-order transitions to incommensurate phases constitute a small minority in incommensurate systems and seem to take place only when the incommensurate structure remains stable within a wide interval of temperature.

5. TRANSITIONS ASSOCIATED WITH MORE THAN ONE ORDER-PARAMETER

The transitions examined in the three preceding sections were described by a single order-parameter, in agreement with the essential statement of the Landau theory which is that to take place a continuous transition should be associated with a single irreducible degree of freedom. This single critical degree of freedom, which determines the symmetry of the low-temperature phase, can be coupled to secondary non-critical degrees of freedom (secondary order-parameters) which modify in a non-essential manner the equilibrium and non equilibrium properties of the system. There exist however a number of transitions reported experimentally, where the entire phase diagram, including the symmetry of the phases, must be connected to more than one order-parameter. This means that the number of phenomenological variables which have to be introduced in the description of the system is not minimal, as for a single order-parameter, and consequently that the theory will generally lead to less conclusive predictions for the transition properties.

As each order-parameter expresses a set of atomic displacements associated to a certain normal mode of the system, the fact that several order-parameters are involved signifies that the displacements related to one order-parameter brings about an instability in parameters inducing other displacements. Due to the non linear interaction between the various parameters, the modifications accompanying a transition (change of symmetry, thermal, dielectric and elastic anomalies, etc...) are more complex, making more difficult a phenomenological description. However, such a description is still realizable on the basis of the Landau theory, and we will see, at least when not more than two order-parameters are involved, that very specific consequences can be deduced for the phase diagram, depending on the nature of the coupling between the parameters.

We will first examine the symmetry aspects related to the existence of more than one order-parameter, then examine typical phase diagrams associated with the various types of coupling between two order-parameters. As the first-order character of the corresponding transitions is not questionable, it will not be discussed. Besides the large family of reconstructive transitions which are generally connected with more than one order-parameter, will be examined in §. 6.

5.1. General properties of phase transitions associated with several order-parameters

Let G_o be a space-group describing a given high-symmetry phase and $\tau_1, \tau_2, \ldots, \tau_j$ a set of IR's of G_o (active or inactive) associated with different k-vectors of its Brillouin-zone. The sum $\tau = \sum_{i=1}^{j} \tau_i$ forms a reducible representation of G_o which is liable to induce a transition towards a low-symmetry phase of space-group G. This transition has the following characterics:

1) If the spontaneous values of the order-parameters, which transform like the τ_i, arise simultaneously at a single transition temperature, G is the intersection $\bigcap_{i=1}^{j} G_i$ where G_i is a space-group induced by τ_i. This situation occurs for a sufficiently strong coupling of the order-parameters and corresponds to the genuine case of a transition associated with a reducible representation. Actually, three types of situations are realized experimentally which corresponds to the coupling of two order-parameters : the transition, which occurs from the high to the low-symmetry phase is connected with two different IR's associated (1) with the same wave-vector, as it is found in ZrO_2 [77] or $Bi_4Ti_3O_{12}$ [78], 2) with two different wave-vectors, as reported for Ag_2HgI_4 [79], $ZnSnAs_2$ [80], N_3AlF_6 [81], $(C_6H_5CO)_2$ [82], and in a number of perovskites such as $SmAlO_3$ [83]. The detailed features of these transitions are given in Table VI.

A third case is illustrated experimentally by a number of reconstructive transitions, as in Co, Fe or C (see § 6) : here the observed symmetry modification can be interpreted as a transition between <u>two low-symmetry phases</u>, derived from a <u>latent</u> structure which is not realized experimentally, and can be defined as the maximal substructure common to the two observed phases.

<u>Table VI</u> : Examples of transitions associated with two order-parameters

Substances	T_c(K)	Order	Symmetry change	Order-parameter dimension	IR's and Brillouin-zone points
$Bi_4Ti_3O_{12}$	948	1	I4/mmm → Aba2 (2)	4	$\tau_3(X) + \tau_5(X)$
$(C_6H_5CO)_2$	83	1	$P3_121$ → B2/m (4)	5	$\tau_2(M) + \tau_3(\Gamma)$
Ag_2HgI_4	596	1	$F\bar{4}3m$ → $I\bar{4}$ (4)	9	$\tau_4(\Gamma) + \tau_1(W)$
$ZnSnAs_2$	1193	1	$F\bar{4}3m$ → $I\bar{4}$ (4)	9	$\tau_4(\Gamma) + \tau_1(W)$
$SmAlO_3$	1100	1	Pm3m → Pnma (4)	4	$\tau_8(R) + \tau_5(M)$
N_3AlF_6	1103	1	Fm3m → $P2_1/c$(2)		$\Gamma + X$
Li_2SO_4	-	1	Fm3m → $P2_1/c$(4)	3	$X + L$
ZrO_2	-	1	Fm3m → $P2_1/c$	3	$\tau_{10}(X)+ \tau_7(X)$

The indication that more than one order-parameter is involved at a given transition can be revealed either by the fact that the superstructure is related to several wave-vectors, by a drastic lowering of the rotational symmetry, or by the absence of group-subgroup relationship between the phases (see §.6). The method for determining the relevant IR's related to the symmetry change starts in any case by the consideration of the concrete modifications observed in each phase (see below). Janovec et al [84] have calculated the symmetry changes induces by the reducible representations of the point groups, and Pascoli [58] has determined the possible Bravais-lattice modifications corresponding to sets of distinct $\vec{k}$-vectors allowed by the Lifshitz criterion. It must be stressed, that most of the symmetry changes obtained by single IR's can also be obtained by a reducible representation. In this case, one has to consider the whole set of physical properties, other than symmetry, in order to decide if the transition can be interpreted with a single IR or with a reducible representation. In particular, considering a reducible representation may explain, in a number of cases, the first-order character of transitions which have been improperly related to a single active IR.

2) The thermodynamic potential associated with a reducible representation contains invariants of each order-parameter plus a mixed invariant transforming as the products of the IR's corresponding to each order-parameter. The case of two order-parameters η and ξ has been considered by a number of authors. Lifshitz [85] has examined an order-parameter expansion restricted to the fourth degree terms :

$$F = \frac{\alpha_1}{2}\eta^2 + \frac{\beta_1}{4}\eta^4 + \frac{\alpha_2}{2}\xi^2 + \frac{\beta_2}{4}\xi^4 + \frac{\delta}{2}\eta^2\xi^2 \tag{5.1}$$

which describes a succession of one second-order transition, followed by a first order transition respectively associated with the equilibrium values ($\eta \neq 0$, $\xi = 0$) and ($\eta = 0$, $\xi \neq 0$), or two second-order transitions via the intermediate state ($\eta \neq 0$, $\xi \neq 0$). As was shown by Holakovsky [86], to represent the simultaneous onset of η and ξ(i.e. the solution $\eta \neq 0$, $\xi \neq 0$) at a first order transition line, a sixth order term in ξ has to be added to (5.1). In the corresponding asymmetric expansion a large value of the "primary" order-parameter η triggers an instability with respect to ξ. The more comprehensive work concerning transitions induced by two order parameters is due to Gufan et al [6, 18, 19, 20]. In the following section we review some of the phase diagrams studied by the preceding authors.

5.2. Examples of phase diagrams associated with two order-parameters

5.2.1. Case ofa biquadratic coupling between the order-parameters

Let us first discuss in a complete manner the case of two order-parameters connected by a biquadratic coupling of the form $\delta\eta^2\xi^2$, as in (5.1), that was discussed in the book of Gufan [6] and in a serie of papers by Gufan and Larin [20], Larin [28], and Torgashev et al [87]. We will start by the simplest form (5.1) which depends on five phenomenological parameter α_1, β_1, α_2, β_2 and δ. These parameters should obey the conditions $\beta_1 > 0$, $\beta_2 > 0$ and $\delta > -\sqrt{\beta_1\beta_2}$, i.e. F must be positive definite at high values of η and ξ. Assuming α_1, α_2 and δ to be linear functions of temperature T and pressure P, the T-P phase diagram is a plane, in the three dimensional space of the coefficients α_1, α_2, δ. Minimization of (5.1) corresponds to the equations :

$$\begin{aligned} \eta(\alpha_1 + \beta_1\eta^2 + \delta\xi^2) &= 0 \\ \xi(\alpha_2 + \beta_2\xi^2 + \delta\eta^2) &= 0 \end{aligned} \tag{5.2}$$

In addition to the high-symmetry phase ($\eta = 0$, $\xi = 0$) denoted 0 one finds three low-symmetry phases :

$$\text{I)}\ \eta^2 = -\frac{\alpha_1}{\beta_1}, \ \xi = 0 \ ; \qquad \text{II)}\ \eta = 0, \ \xi^2 = -\frac{\alpha_2}{\beta_2}$$

$$\text{III)}\ \eta^2 = \frac{\delta\alpha_2 - \alpha_1\beta_2}{\Delta}, \ \xi^2 = \frac{\delta\alpha_1 - \beta_1\alpha_2}{\Delta} \quad \text{with } \Delta = 16\,\beta_1\beta_2 - \delta^2 \tag{5.3}$$

As already mentioned here above the sequence of stable phases can be either 1) one second order-transition 0-I followed by a first-order transition I-II, or 2) a sequence of three second-order transition 0-I, I-III, III-II. Accordingly the model (5.1) can be considered as irrealistic as the transitions between states I and III, and II and III are found experimentally to be first-order. Furthermore, the transition 0-III appears to be possible (figure 14) only at a second-order transition point in the phase diagram. One can note that in figure 14, one finds a three-phase point where converge two second-order transition lines I-III and II-III, and a first-order transition line I-II. Besides the potential (5.1) does not allow a negative interaction $\delta < -\sqrt{\beta_1\beta_2}$ between the order-parameters.

Thus, one can conclude that the model (5.1) is incomplete and consider the following potential :

$$F = \frac{\alpha_1}{2}\eta^2 + \frac{\beta_1}{4}\eta^4 + \frac{\gamma_1}{6}\eta^6 + \frac{\alpha_2}{2}\xi^2 + \frac{\beta_2}{4}\xi^4 + \frac{\delta}{2}\eta^2\xi^2 \tag{5.4}$$

which differs from (5.1) by the sixth degree term in η . The equations of state for (5.4) are :

$$\begin{aligned} \eta(\alpha_1 + \beta_1\eta^2 + \gamma_1\eta^4 + \delta\xi^2) &= 0 \\ \xi(\alpha_2 + \beta_2\xi^2 + \delta\eta^2) &= 0 \end{aligned} \tag{5.5}$$

As for the potential (5.1) we obtain four possible stable phases, namely :

$$\text{0)}\ \eta = 0, \ \xi = 0 \ ; \qquad \text{I)}\ \eta^2 = \frac{-\beta_1 + (\beta_1^2 - 4\alpha_1\gamma_1)^{1/2}}{2\gamma_1}, \ \xi = 0$$

$$\text{II)}\ \eta = 0, \ \xi^2 = -\frac{\alpha_2}{\beta_2} \qquad \text{III)}\ \eta^2 = \frac{(\delta^2/\beta_2)-\beta_1 + [(\beta_1 - \delta^2/\beta_2)^2 - 4\gamma_1(\alpha_1 - \delta\alpha_2/\beta_2)]^{1/2}}{2\gamma_1}$$

$$\xi^2 = \frac{-\delta\eta^2 - \alpha_2}{\beta_2} \tag{5.6}$$

The discussion of the stability conditions [6,20] leads to the various phase diagrams represented in Figure 15, which display the following features :

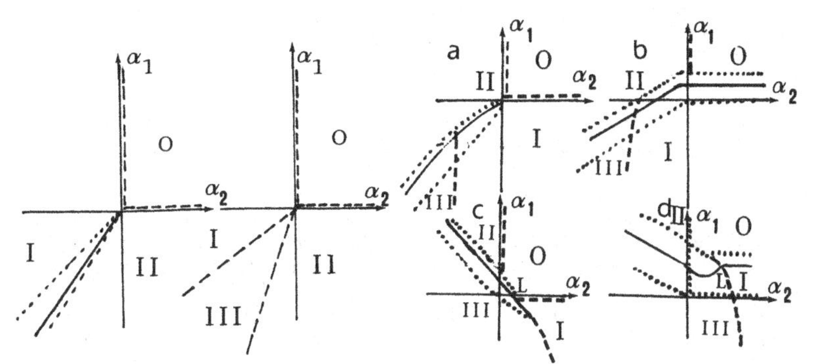

Figure 14 : Phase diagram corresponding to potential (5.1). Same conventions as in Fig.15

Figure 15 : Phase diagram in the (α_1,α_2) plane for potential (5.4) $\delta>2\sqrt{\beta_1\beta_2}, \beta_1>0$ b) $\delta>2\sqrt{|\beta_1|\beta_2}$, $\beta_1<0$ c) $\delta<-2\sqrt{\beta_1\beta_2}$, $\beta_1>0$ d) $\delta<-2\sqrt{|\beta_1|\beta_2}$, $\beta_1<0$. The continuous, dashed and dotted lines represent respectively first-order, second-order and stability lines.

a) If $0 < \delta < \sqrt{\beta_1\beta_2}$ and $\beta_1 > 0$ one has the diagram corresponding to Fig. 15a. In this diagram the region of coexistence of phases II and III is bounded by the loss of stability lines of phase II and III given respectively by :

$$\alpha_1 = \frac{2\delta\beta_1}{\beta_2} \quad \text{and} \quad \left(\alpha_1 = \frac{\delta\alpha_2}{\beta_2} + \frac{\Delta^2}{\gamma_1\beta_2^2}\ ,\ \alpha_1 = \frac{4\beta_1\alpha_2}{\delta} - \frac{\gamma_1\alpha_2^2}{2\delta}\right) \tag{5.7}$$

while the line of first-order transition between phases II and III has the equation :

$$\alpha_1 = \frac{\delta}{\beta_2}\alpha_2 + \frac{3\Delta^2}{4\gamma_1\beta_2^2} \tag{5.8}$$

At the three-phase point Q whose coordinates are :

$$\alpha_1^Q = \frac{2\Delta\delta}{\gamma_1\beta_2}\ ,\quad \alpha_2^Q = \frac{4\Delta\delta^2}{\gamma_1\beta_2^2} \tag{5.9}$$

the line (5.8) becomes a line of first-order transitions between the phases I and II. If $\beta_2 > 0$, the line O-I is always of the second order (Figure 15a) whereas for $\beta_2 < 0$, the line of equality of the potentials I-II goes over continuously to the line $\alpha_1 = \frac{3}{16}\frac{\beta_2^2}{\gamma_1}$ which is the first-order transition line O-I (figure 15b)

b) If $\delta < -\sqrt{\beta_1\beta_2}$, then for $\beta_1 > 0$ and $\beta_1 < 0$, there is a O-III boundary phase at which $\Delta < 0$. The stability of the phases II and III are given by the same Eqs. (5.7). There exist a first-order transition line O-III (Figures 15c) and 15d)) and a tricritical point L with coordinates :

$$\alpha_1^L = \frac{\beta_1^2\beta_2^2 - \delta^4}{4\,\gamma_1\beta_2^2} \quad , \quad \alpha_2^L = \frac{2\,\Delta\,\delta}{\gamma_1\beta_2} \qquad (5.10)$$

at which the line of second-order transitions between phases I and III changes to a line of first-order transition (Fig. 15c)

c) Two types of three phase points are found in the phase diagrams. The origin point O (α_1^o, α_2^o) is a tetracritical point (see § 7) since it is the point of convergence of the stability lines of phase O, I, II and III. Near to the three phase points Q and L , which lie on a line of first order transitions (Fig. 15a to 15e), second-order transitions between metastable phases should be possible.

A still more complete phase diagram can be obtained when adding to (5.4) a sixth degree term $\frac{\gamma_2}{6}\xi^6$. This diagram was briefly discussed in refs 6, 20, and in more details in ref. 28. It differs from the diagrams shown in Figure 15, by : 1) the fact that first-order transition lines may occur between phases I and III, and II and III , 2) the existence of tricritical points on both the lines I-III and II-III, 3) reentrant sequences of phases I-III-I and II-III-II, 4) a three phase point where converge three lines of first-order transitions between phases I, II and III, the sequence O-I-III-II being a succession of first-order transitions.

The coupled order-parameter model developped in this section was applied by Torgashev et al [87] to the sequence of transitions observed in $LiNH_4SO_4$. In this model the two order-parameter coincide with the component P_y of the spontaneous polarization, and with a two-component order-parameter (η_1, η_2) transforming as in IR associated with the X-point of the primitive orthorhombic Brillouin zone. Larin [28] used the more complete model with two sixth-degree terms, to interpretate the Pm3m → P4/mbm → Cmcm → $P2_1/m$ sequence of phases observed in $CsPbCl_3$ (where the two order-parameters are three-dimensional). This latter model may possibly explain the reentrant behaviour reported in a number of materials, such as Rochelle salt or deuterated sodium trihydrogen selenite.

5.2.2. Case of linear-quadratic coupling between the order-parameters.

Another type of typical coupling between two order-parameters is of the form $\delta\eta^2\xi$. The corresponding phase diagram has been worked out by Gufan and Sakhnenko [8], Gufan and Torgashev [19,88] and Latush et al [29]. As in §.5.2.1, let us first examine the simplest model described by the thermodynamic potential:

$$F = \frac{\alpha_1}{2}\eta^2 + \frac{\beta_1}{4}\eta^4 + \frac{\alpha_2}{2}\xi^2 + \frac{\beta_2}{4}\xi^4 + \delta\eta^2\xi \qquad (5.11)$$

In (5.11) we assume that F is positive definite for large η and ξ, i.e. $\beta_1 > 0$ and $\beta_2 > 0$, and that α_1, α_2 vary linearly with the temperature*. In the absence of interaction ($\delta = 0$) between η and ξ, the minimization of F yield four different stable phases :

$$\text{I) } \eta = \xi = 0, \quad \text{II) } \eta \neq 0, \xi = 0, \quad \text{III) } \eta = 0, \xi \neq 0, \quad \text{IV) } \eta \neq 0, \xi \neq 0 \quad (5.12)$$

with two domains for phases II and III ($\pm\eta$, or $\pm\xi$) and four domains for phase IV ($\pm\eta$ and $\pm\xi$). All these phases meet along second order-transition lines $\alpha_1 = 0$, and $\alpha_2 = 0$.

The nonlinear interaction due to $\gamma \neq 0$ has the resultthat there are only three different phases, namely :

$$\text{I) } \eta = \xi = 0, \quad \text{II') } \eta = 0, \xi \neq 0, \quad \text{III') } \eta \neq 0, \xi \neq 0 \quad (5.13)$$

phase III' has <u>two</u> domains, with $\xi < 0$ when $\delta > 0$. It corresponds to phase II with an <u>induced</u> (improper) value of ξ. Thus, phase IV becomes metastable when $\delta \neq 0$. Figure 16 shows the phase diagram corresponding to the potential (5.11) in the (α_1, α_2) plane. Phase I is stable for $\alpha_1 > 0$, $\alpha_2 > 0$ (whatever is the value of δ). When $\alpha_2 > \frac{\delta^2}{2}$, there is a second-order transition line between phase I and III' along $\alpha_1 = 0$. At the tricritical point T_1 with coordinates $(\frac{\delta^2}{\beta_1}, 0)$ the preceding line transforms into a first-order transition line between phases I and III', and at

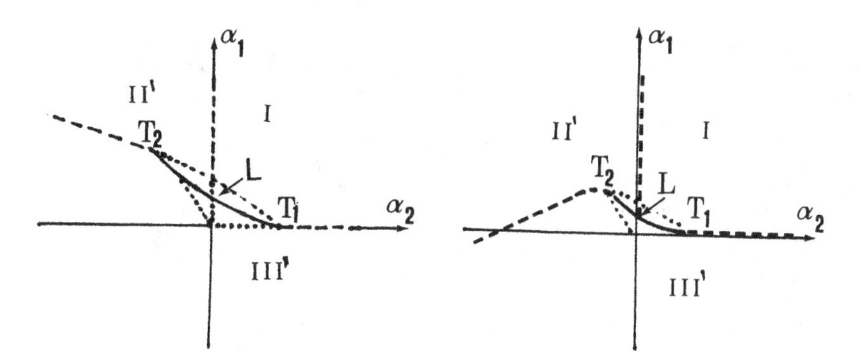

Figure 16 : Phase diagram corresponding to potential (5.11) with $\beta_1 > 0, \beta_2 > 0, \delta > 0$. Dashed lines : second order transitions. Continuous lines: first-order transition. dotted lines: phase stability.

Figure 17 : Phase diagram corresponding to potential (5.11) plus an additional biquadratic invariant $\gamma\eta^2\xi^2$, with $\gamma > 0$. Lines have the same meaning as in Fig. 16.

* A <u>non-linear</u> variation for α_1 and α_2 can be shown to modify the phase diagram in the same way then taking into account higher degree terms in the thermodynamic potential (and a <u>linear</u> variation for α_1 and α_2).

the triple point L of coordinates $(0, \frac{\delta^2}{2(\beta_1\beta_2)^{1/2}})$ it becomes a straight first-order line between phases II' and III'. At a second tricrital point T_2 defined by $(-\frac{\delta^2}{2\beta_1}, \frac{\delta^2}{(\beta_1\beta_2)^{1/2}})$ the first-order transition line II'-III' turns into the second-order transition line defined by the equation : $\alpha_1 = \delta(2|\alpha_2|/\beta_2)^{1/2}$

It must be emphasized that the I-III' transition is improper with respect to the ξ parameter, and the T_1 L line is a line of triggered transitions [86] as the symmetry of the resulting phase is induced by the η parameter (and not ξ). The discontinuity of the order-parameters η across the line T_1L is :

$$(\Delta\eta)^2_{I-II'} = 2(\alpha_1 + \delta\Delta\xi_{I-II'})/\beta_1 \text{ with } \Delta\xi_{I-II'} = \frac{\alpha_1[3\delta + (4\alpha_1\beta_1 + \delta^2)^{1/2}]}{\frac{\beta_1\beta_2}{2} - 2\delta^2} \quad (5.14)$$

and across the line LT_2 :

$$(\Delta\eta)_{II'-III'} = [\frac{2\delta^2(\beta_1\beta_2)^{-1/2} - \alpha_1}{\beta_1}], \Delta\xi_{II'-III'} = (-\frac{\alpha_2}{\beta_2})^{1/2} - \delta(\beta_1\beta_2)^{-1/2} \quad (5.15)$$

the discontinuities (5.14) and (5.15) vanishing at T_1 and T_2. A more complete model consists in taking into account in (5.11), the additional biquadratic invariant $\frac{\gamma}{2}\eta^2\xi^2$, whose phase diagram (Figure 17) differs qualitatively from the phase diagram corresponding to (5.11), by the fact that it allows the following sequences of phases : $I \overset{2}{\leftrightarrow} II' \overset{1,2}{\leftrightarrow} III' \overset{2}{\leftrightarrow} II'$, and $I \overset{1}{\leftrightarrow} III' \overset{2}{\leftrightarrow} II'$, where the order of the transitions is indicated above the arrows.

The preceding results apply, as pointed out by Gufan and Torgashev [19], to improper ferroelectric or ferroelastic transitions in which the polarization, or strain are triggered by a primary order-parameter (but are not secondary order parameters in the sense of §.II.5, as they are symmetry breaking quantities). This is for example the situation occuring in $SrTeO_3$ [89], or in HCL [90], but not in boracites [91], as suggested in ref. 92. In the case of a two-component primary order-parameter, and a one component triggered polarization, the thermodynamic potential, may take the explicit form :

$$F = \frac{\alpha_1}{2}(\eta_1^2 + \eta_2^2) + \frac{\beta_1}{4}\eta_1^2\eta_2^2 + \frac{\beta_2}{4}(\eta_1^4 + \eta_2^4) + \frac{\alpha_2}{2}P^2 + \frac{\beta_3}{4}P^4$$
$$+ \delta\eta_1\eta_2 P + \frac{\gamma}{2}(\eta_1^2 + \eta_2^2)P^2$$

Neglecting the non linear interaction between the two order-parameters ($\delta = 0$) would lead to the possible stabilization of five low-symmetry phases, namely : I) $\eta_1 = 0$, $\eta_2 \neq 0$, $P = 0$, II) $\eta_1 = \eta_2 \neq 0$, $P = 0$, III) $\eta_1 = \eta_2 = 0$, $P \neq 0$ IV) $\eta_1 \neq 0$, $\eta_2 = 0$, $P \neq 0$, V) $\eta_1 \neq \eta_2 \neq 0$, $P \neq 0$, which exhibit respectively four domains (I and II), two domains (III) and eight domains (IV and V). When taking into account the δ term, one finds that phases II and V coincide, i.e. $\delta\eta_1\eta_2$ plays the role of an internal electric field which induces a non-zero value for the spontaneous polarization. The form of the dielectric susceptibilities corresponding to the potential (5.16) can be found in ref. 19. A three order-parameter model is considered in ref. 29, where two triggered parameters (e.g. the polarization and the strain) are coupled via a $\eta^2\xi$ type invariant to the primary (η_1, η_2) order-parameter. Gufan and Torgashev [87] also considered the case where the two-component primary order-parameter allows existence for a Lifshitz invariant, and when a cubic term is allowed by the symmetry of the triggered parameter. The thermodynamic potential, which includes a ξ^3 term :

$$F = \frac{\alpha_1}{2}\eta^2 + \frac{\beta_1}{4}\eta^4 + \frac{\alpha_2}{2}\xi^2 + \frac{B}{3}\xi^3 + \frac{\beta_2}{4}\xi^4 + \delta\eta^2\xi \qquad (5.17)$$

is of physical interest as it allows interpretation of the liquid-nematic transition (see chapter VII). The minimization equations are :

$$\begin{aligned} &\eta(\alpha_1 + \beta_1\eta^2 + 2\,\delta\eta\xi) = 0 \\ &\xi(\alpha_2 + B\xi + \beta_2\,\xi^2\,) + \delta\eta^2 = 0 \end{aligned} \qquad (5.18)$$

They lead to the phase diagram shown in Figure 18, which contains two possible low-symmetry phases, namely: I) $\eta \neq 0$, $\xi = 0$, and II) $\eta \neq 0$, $\xi \neq 0$. These phases are separated from the high-temperature phase by lines of first-order transitions.

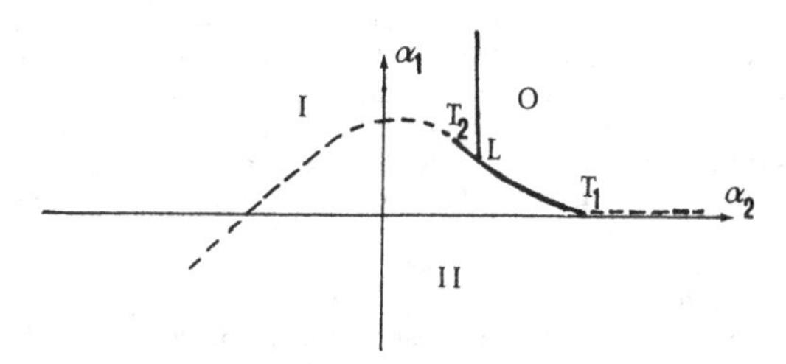

Figure 18 : Phase diagram corresponding to the potential (5.17).

As for te phase diagrams represented in Figs.16 and 17 one finds in Fig.18 two tricritical points and one triple point.

An experimental example was discussed by Toledano [93] which both allows existence for a Lifshitz invariant and a cubic term for the triggered order-parameter. It is the case of benzil $(C_6H_5CO)_2$ where the primary order-parameter has three components, and the triggered parameter has two components, the coupling term being of the form : $\xi_1(2\eta_1^2 - \eta_2^2 - \eta_3^2)+\xi_2\sqrt{3}(\eta_2^2-\eta_3^2)$

5.2.3. Case of a linear-cubic coupling between the order-parameters

Let us finally briefly mention the situation where the thermodynamic potential associated with the two order-parameters is of the form :

$$F = \frac{\alpha_1}{2}\eta^2 + \frac{\beta_1}{4}\eta^4 + \frac{\alpha_2}{2}\xi^2 + \frac{\beta_2}{4}\xi^4 + \delta\eta\xi^3 \qquad (5.19)$$

This case was partially discussed for the improper ferroelectric transitions in $PbZrO_3$ [19] and boracites [91] where (5.19) has to be supplemented by additional invariants (e.g. $\gamma\eta\xi^2$ and $\frac{B}{3}\eta^3$ in the case of boracites). Assuming ξ to be the polarization P, one finds the remarkable property that P varies as $(T_c-T)^{3/2}$, and the corresponding dielectric susceptibility is proportional to (T_c-T) in the ferroelectric phase. The detailed phase diagram corresponding to (5.19) has not yet been worked out.

6. PHENOMENOLOGICAL THEORY OF PHASE TRANSITIONS WHICH HAVE NO GROUP-SUBGROUP RELATIONSHIP BETWEEN THE PHASES

The possibility of defining an order-parameter is a preliminary condition for applying the symmetry and thermodynamic concepts which underly the Landau theory. This definition presupposes that a group-subgroup relationship is realized between the high-and low-symmetry phases which take place on each side of the transition. This is always verified for second-order phase transitions, and for the first order-transitions considered in the preceding sections. These transitions are in most cases characterized by small displacements of the atoms from one equilibrium site to another, and no rupture of the atomic bonds occur. By contrast, for reconstructive transitions [94] , in which large shifts and drastic reordering are found, a group-subgroup relationship is generally absent and no order-parameter is currently defined. For this category of transitions, to which belongs most of the elements [95] , a number of alloys [95] and some insulators [96] , it is usually admitted that the Landau theory cannot be used, and only crystallographic and atomistic models have been proposed [94-96] , in order to describe their mechanism.

In contrast to the preceding ideas, we show in this section that an order-parameter can be defined for reconstructive transitions. The content of this section corresponds to still unpublished results [97].

6.1. Reconstructive transitions for which the order-parameter is a sinusoïdal function of the atomic shifts.

Let us first focus on the $\beta - \omega$ transition reported for Ti, Zr and Hf [98]. The displacement of the Ti atoms in the ω phase with respect to the β phase are shown in Figure 19. They are deduced from the superstructure observed by electron diffraction pattern [99]. One can see on this figure that the shifts take

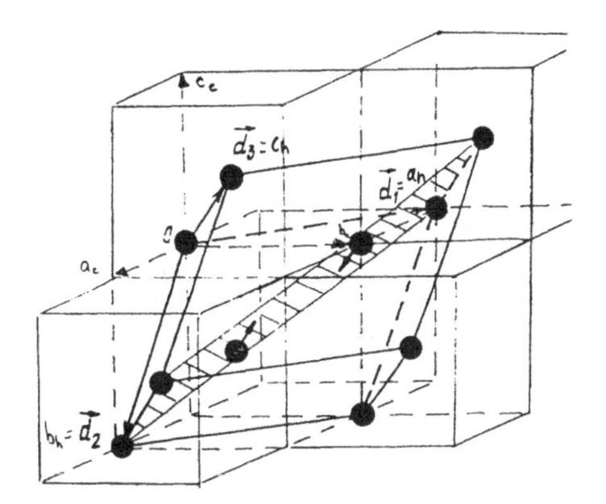

Figure 19 : Average shifting of the Ti atoms from the β to the ω phase. The rhombohedral unit-cell is drawn in thick lines. The hatched plane is the cubic $[0\bar{1}\bar{1}]$ plane.

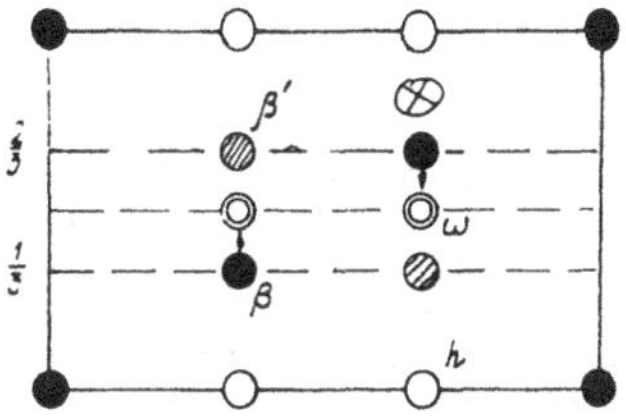

Figure 20 : Shifts of the Ti atoms in the $[0\bar{1}\bar{1}]$ plane. The unshifted atoms are symbolized by black dots (β phase), whereas white, hatched and circled dots correspond to the h, β' and ω phases respectively.

place along the [III] direction in such a manner that in the ω phase, the atoms lie in the planes $(0\bar{1}\bar{1})_{cubic}$. As stressed by de Fontaine [100] the displacement of the Ti atoms can be described by the three independent vectors :

$$\vec{a}_h = \vec{a}_c - \vec{b}_c, \quad \vec{b}_h = \vec{b}_c - \vec{c}_c, \quad \vec{c}_h = \frac{1}{2}(\vec{a}_c + \vec{b}_c + \vec{c}_c) \qquad (6.1)$$

corresponding respectively to the $[\bar{2},2,0]$, $[2,0,\bar{2}]$ and [III] directions, in cubic coordinates. These displacements can be viewed in the $(0\bar{1}\bar{1})$ plane wich is shown in Figure 20. Depending on the magnitude of the shifts along [III], five phases associated with different structures can be singled out : 1) the β phase which corresponds to unshifted atoms (black dots in Fig. 20). It has the "parent" Im3m symmetry with one atom per unit cell.

2) the ω phase which appears as the result of a shift of $\sqrt{3}/12$ along the [III] direction. It possesses the P6/mmm symmetry with 3 atoms per unit-cell.

3) the β' phase corresponding to a shift of a $\sqrt{3}/6$ along [III] . This phase, denoted by hatched circles in Figure 2, has the same symmetry as the β phase, within an exchange of coordinate system, i.e. β' can be deduced from β by a rotation about [III]of 60°.

4) the h phase (white circles in Fig. 20) which takes place for a shift of a $\sqrt{3}/3$ while the space-group of this phase is P6/mmm, as the ω phase, the two phases differ by their translational symmetry, the h phase having only one atom per unit-cell.

5) the non-symmetric ω phase (crossed circles in Fig. 2) which is induced by any general shift along [III]. It possesses the rhombohedral symmetry $P\bar{3}m1$, with Z = 3.

The wave-vector describing the breaking of translational symmetry given by (6.1) is $\vec{k}_1 = \frac{1}{3}(a^*_\beta + b^*_\beta + c^*_\beta)$. Its invariance group is 3m, and its star possesses eight branches. Accordingly the vector representation discribing the $\beta - \omega$ shifts is composed by an eight dimensional IR, denoted $\tau_1(k_1)$ which expresses the shifts along [III] and by a sixteen dimensional IR which accounts for the displacements in the plane perpendicular to [III]. Decomposing the basis vectors $\vec{\phi}_i$ which span the order-parameter space in terms of symmetric coordinates associated with the shifts along [III], one finds that all basis vectors are zero, except $\vec{\phi}_1$ and $\vec{\phi}_2$ which correspond to $\pm \vec{k}_1$. Two space-groups may be associated with the superstructure (6.1), namely P3m1 (Z = 3) and $P\bar{3}m1$ (Z = 3). One can easely verify that the first of these groups describes non-symmetric displacements of the atoms 2 and 3 in Fig. 1, whereas for symmetric shifts, assumed in Fig. 20, we obtain the group $P\bar{3}m1$. For a shift of magnitude a $\sqrt{3}/12$ the group-subgroup relationship $Im3m \rightarrow P\bar{3}m1$ disappears as the rhombohedral symmetry increases to P6/mmm (Z = 3), which is the actual symmetry of the ω phase.

Because of the magnitude of the shifts (denoted ξ), it should be unphysical to assume that the order-parameter (denoted η) is proportional to ξ. A good understanding of the β-ω transition mechanism can be obtained by introducing a non-linear (periodic) dependence of η as a function of ξ. We will take here :

$$\eta = \eta_o \left[\sin\left(\frac{4\pi\xi}{a\sqrt{3}} + \frac{\pi}{6}\right) - \frac{1}{2}\right] \qquad (6.2)$$

The form (6.2) is justified by Figure 21, which shows that for successive values of ξ along [III], one gets periodically the sequence β, ω, β' and h phases. More precisely, for ξ = 0, 1, 2, 3,... (in $\frac{a\sqrt{3}}{2}$ units) one has $\eta = 0$ (β phase) ; for $\xi = \frac{1}{6}, \frac{7}{6}, \frac{13}{6}, \frac{19}{6}, \ldots$, $\eta = \frac{1}{3}$ (ω phase ; for $\xi = \frac{1}{3}, \frac{4}{3}, \frac{7}{3}, \frac{10}{3}, \ldots$, $\eta = 0$ again (β' phase), and for $\xi = \frac{2}{3}, \frac{5}{3}, \frac{8}{3}, \frac{11}{3}, \ldots \eta = -1$ (h phase). The preceding numbers demonstrate that Eq. (6.2) holds for infinite shifts. Taking into account the equilibrium values ($\eta_1 = \eta_2 = \eta \neq 0$; $\eta_{i \neq 1,2} = 0$) one can write the effective potential associated with $\tau_1(\vec{k}_1)$:

$$F = F_o + a_1\eta^2 + b\eta^3 + a_2\eta^4 \qquad (6.3)$$

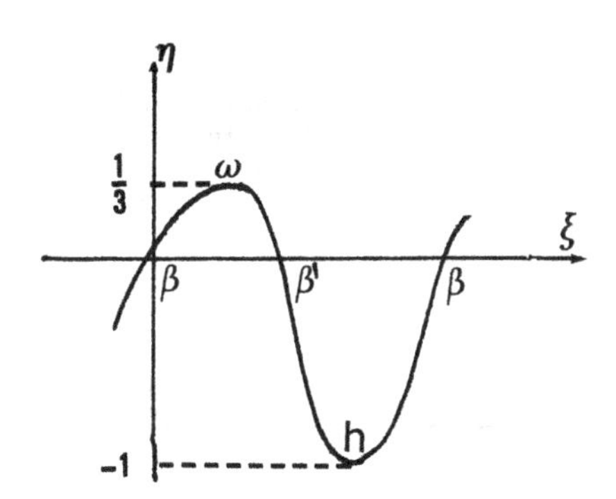

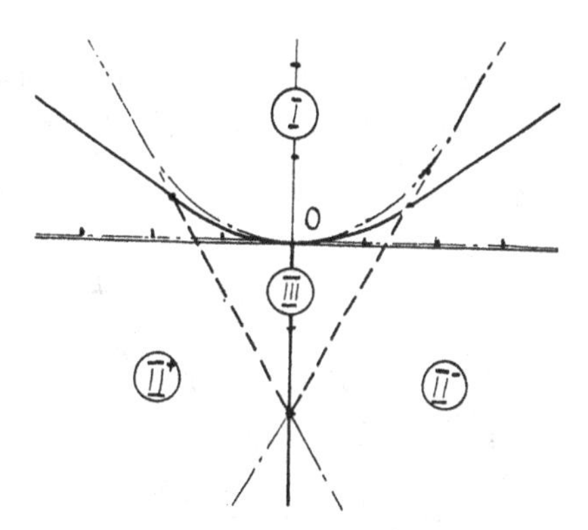

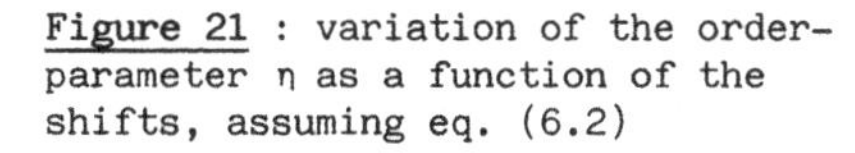

Figure 21 : variation of the order-parameter η as a function of the shifts, assuming eq. (6.2)

Figure 22 : Phase diagram corresponding to the potential (6.3). Full lines represent first-order transition lines. Dotted lines are second-order transitions lines. Stability lines are dashed lines.

Replacing (6.2) in (6.3), and minimizing with respects to the shifts, yields the equation of state :

$$[\sin f(\xi)-\tfrac{1}{2}]\cos f(\xi)\{2a_1+3b[\sin f(\xi)-\tfrac{1}{2}]+4a_2[\sin f(\xi)-\tfrac{1}{2}]^2\} = 0 \quad (6.4)$$

with $f(\xi) = \frac{4\pi\xi}{a\sqrt{3}} + \frac{\pi}{6}$. Eq.(6.4) leads to three possible states : the β and ω phases for respectively $\sin f(\xi) = 1$ (e.g. $\xi = 0,1,2,3,\ldots$) and $\cos f(\xi) = 0$ (e.g. for $\xi = \frac{1}{6}, \frac{7}{6}, \frac{13}{6}, \frac{19}{6},\ldots$), namely for constant values of the order-parameter. The vanishing of the expression into brackets in (3) corresponds to the non-symmetric ω phase. It is characterized by a temperature dependent order-parameter which identifies to the Landau solution, obtained by minimizing F with respect to η , and not with respect to the shifts. The h phase which is also obtained for $\cos f(\xi) = 0$, represents an unstable configuration of the atoms with respect to the α structure, that will be discussed below. The phase diagram corresponding to the thermodynamic potential (6.3) is represented in Figure 22.

A similar scheme applies to the $\beta-\alpha$ transition also reported in Ti, Zr and Hf. In the hexagonal closed-packed α modification the angle between the three fold axes, which is about 72° in the β phase, becomes 60°. The basis vectors of the hexagonal α unit-cell can be expressed in function of the primitive cubic lattice vectors as :

$$\vec{a}_\alpha = \vec{a}_\beta + \vec{b}_\beta + \vec{c}_\beta \ , \quad \vec{b}'_\alpha = -\vec{c}_\beta \ , \quad \vec{c}_\alpha = \vec{a}_\beta - \vec{b}_\beta \qquad (6.5)$$

The directions corresponding to the three vectors (6.5) are represented in Figure 23, as well as the shifts of the atoms from the β to the α phase. From (6.5) one can deduce the wave-vector $\vec{k}_2 = \frac{1}{2}(\vec{a}^*\beta + \vec{b}^*\beta)$ associated with the translational ordering, which is located at the N point of the bcc Brillouin-zone. $\vec{k}_2$* has six branches, and the shifts of the atoms which take place

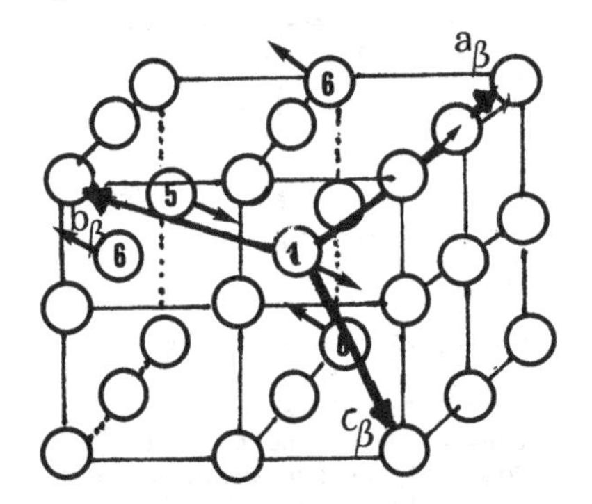

Figure 23 : Average shifting of the atoms at the β-α transition

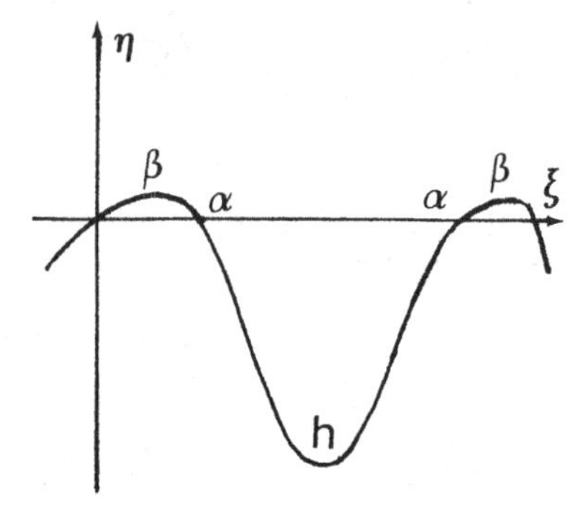

Figure 24 : Variation of the order-parameter ζ given by (6.6) as a function of the shifts .

along the [IIO] direction are connected with a single six-dimensional IR, denoted $\tau_4(k_9)$ in Kovalev's tables [38]. From the symmetries induced by the preceding IR, one can deduce the following picture for the $\beta - \alpha$ transition : when the atoms are shifted along [IIO] the system undergoes the overall symmetries

1) Cmcm (Z = 2) for displacements corresponding to a general position.

2) $P6_3/mmc$ (Z $\neq$ 2) for the high-symmetric shifts of magnitude $\frac{a\sqrt{2}}{12}$, $\frac{7a\sqrt{2}}{12}$,...., <u>which corresponds to the α phase</u>.

3) P6/mmm (Z = 1), for shifts $\frac{a\sqrt{2}}{4}$, $\frac{3a\sqrt{2}}{4}$,..., which corresponds to <u>the h phase</u> precedingly discussed for the $\beta - \omega$ transition. This unstable phase provides the link between the $\beta - \omega$, and $\beta - \alpha$ mechanisms, the $\alpha - \omega$ transitions appearing as a transition between two low-symmetry phases.

The periodicity of the order-parameter ζ as a function of the shifts ξ , is accounted, for the $\beta - \alpha$ transition, by the function

$$\zeta = \zeta_o \ \{\sin\left(\frac{4\pi\xi}{a\sqrt{2}} + \frac{\pi}{2}\right) - \frac{1}{2}\} \qquad (6.6)$$

the variation of which is shown in Figure 24.

In summary, the order-parameters for the $\beta - \omega$ and $\beta - \alpha$ transition have been defined as a sinusoïdal function of the shifts, and not as a linear function as in the standard Landau theory. For specific high-symmetric values of the shifts, the symmetry group of the lower-symmetric phase has been shown to increase, and the group-subgroup relationship between the phases was subsequently destroyed. In this approach, the equilibrium states of the system, have been obtained by minimizing the thermodynamic potential of the transition with respect to the shifts, and not with respect to the order-parameter components, in contrast to the traditional procedure, but in agreement with the statistical mechanics methods [101]

The preceding model applies analogously [97] to the BCC-HCP and BCC-FCC modifications which are found for example in Ba, Te, Fe and Yb. However, it does not apply mechanically to all reconstructive transitions, but only to transitions where the variational parameter coincides with an atomic shift. As will be shown in the following section, a different formulation of the phenomenological theory must be used when the variational parameter is the degree of occupation of a latent unit-cell. .

6.2. Reconstructive transitions for which the variational parameter is the degree of occupation of a latent unit-cell.

Let us consider the FCC to HCP transition which is observed in ^{4}He, ^{3}He, Co, Tl, Pb, Fe, La, Ce, Pr, Nd, Sm, Yb and Am. We will first show that the maximal substructure common to the FCC and HCP unit-cells possesses the P6/mmm symmetry. The two cells are represented in Figures 25 and 26. One can see that the FCC cell is formed by three stacked hexagonal layers, each layer being shifted with respect to another by $a_h/\sqrt{3}$ in the direction $[120]_h$ where a_h is the basic vector of the rhombohedral FCC cell, in hexagonal coordinates. The HCP phase is analogously formed by two stacked hexagonal layers which are also shifted, one with respect to another by $a_h/\sqrt{3}$. The basis vectors of the HCP phase can be expressed in function of the basis vectors of the FCC phases :

$$\vec{a}_{1h} = \vec{a}_{2c}-\vec{a}_{1c}\ ,\quad \vec{a}_{2h} = \vec{a}_{1c}-\vec{a}_{3c},\quad \vec{a}_{3h} = \frac{2}{3}\,(\vec{a}_{1c} + \vec{a}_{2c} + \vec{a}_{3c}) \qquad (6.7)$$

One can demonstrate [97] using the preceding relationship, that the maximal substructure common to the FCC and HCP unit-cells has a unique unit-cell of volume $V_l = V_h/6 = V_c/3$. It is composed by a one-layer structure shown in Figures 25 and 26, which corresponds to the hexagonal space-group P6/mmm. It can be stressed that, as the FCC and HCP structures have respectively one and two atoms per unit-cell, the P6/mmm structure is filled by only 1/3 atom per unit-cell and is thus a disordered phase.

Let us now show that the FCC-HCP transition can be interpreted as a transition between two low-symmetry phases, i.e. as the result of two independent transitions from the preceding latent parent phase, of symmetry P6/mmm, namely P6/mmm $\rightarrow$ P6$_3$/mmc and P6/mmm $\rightarrow$ Fm3m.

The relationship between the basis vectors of the monolayer latent phase and the two-layered HCP phase can be written :

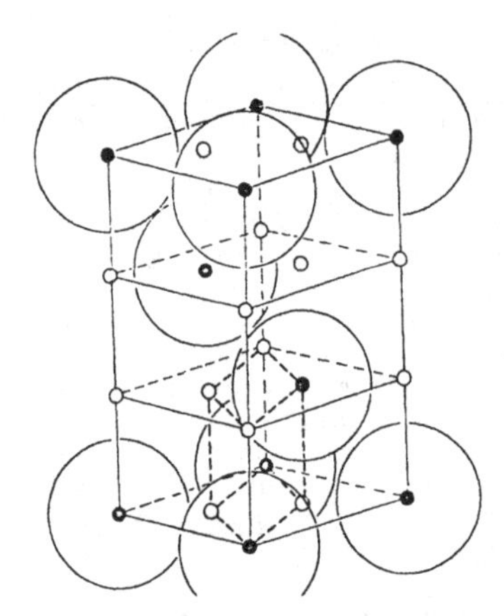

Figure 25 : Structure of the FCC unit cell. Atoms are symbolized by circles. The black and white dots correspond respectively to atoms which are initially in equivalent positions in the latent phase (dashed lines).

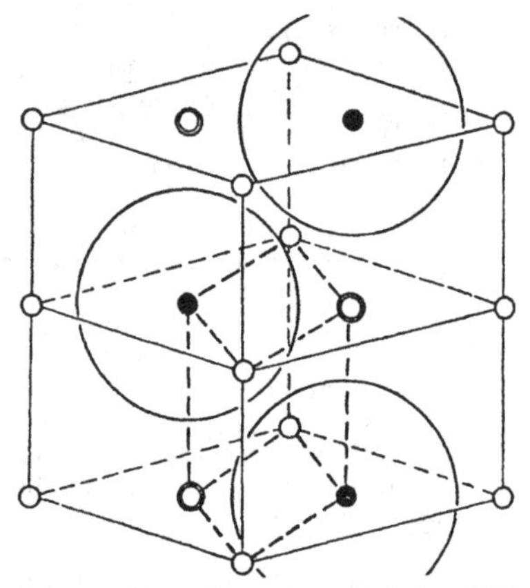

Figure 26 : Structure of the HCP unit cell. Same convention as in Fig. 25. The latent phase is represented in dashed lines.

$$\vec{a}_{1h} = 2\vec{a}_{11}+\vec{a}_{21}\ ,\quad \vec{a}_{2h} = \vec{a}_{11} + 2\vec{a}_{21},\quad \vec{a}_{3h} = 2\vec{a}_{31} \qquad (6.8)$$

It corresponds to the wave-vector $\vec{k}_1 = \frac{1}{3}(\vec{a}_1^* + \vec{a}_2^*) + \frac{1}{2}\vec{a}_3^*$ where $\vec{a}_i^*$ (i = 1-3) are the reciprocal lattice vectors in the hexagonal system. $\vec{k}_1$ has two branches and corresponds to the H point of the hexagonal Brillouin-zone surface [38]. One can deduce from the standard Landau procedure that the IR associated with the P6/mmm → $P6_3$/mmc (V x 6) symmetry change has two dimensions [38]. The thermodynamic potential associated with the two-component order-parameter, can be written :

$$F = \frac{\alpha}{2}\rho^2 + \frac{\beta}{4}\rho^4 + \frac{\gamma_1}{6}\rho^6 + \frac{\gamma_2}{6}\rho^6 \cos 6\phi \qquad (6.9)$$

Along the same line, the relationship between the FCC and latent phase basis vectors is :

$$\vec{a}_{1c} = 2\vec{a}_{11} + \vec{a}_{21},\quad \vec{a}_{2c} = \vec{a}_{11} + 2\vec{a}_{21},\quad \vec{a}_{3c} = 3\vec{a}_{31} \qquad (6.10)$$

It corresponds to the wave-vector $\vec{k}_2 = \frac{1}{3}(\vec{a}_1^* + \vec{a}_2^* + \vec{a}_3^*)$ which lies inside the hexagonal Brillouin zone. The star of $\vec{k}_2$ has four branches [38]. Taking into account the disordered nature of the latent phase, and the fact that it possesses one atom per unit-cell, one can identify the IR of P6/mmm associated with $\vec{k}_2$ as a four-dimensional representation. Among the possible symmetries induced by this representation, figures the rhombohedral group R3m (V x 3). <u>This group can be identified as the FCC (Fm3m) structure, due to the specific geometrical and crystallographic conditions fulfilled by the lattice parameters</u>. These conditions are : 1) a ratio $a_{31}/a_{11} = \sqrt{2}$ which corresponds to a closed packing structure.

2) an angle of 60° between a_{11}, a_{21} and a_{31}.

3) a $\frac{1}{3}$ occupation of the latent structure by the atoms.

The identification of the $R\bar{3}m$ structure as a cubic Fm3m structure when the preceding requirements are fulfilled is illustrated in Figures 27 and 28. From these figures, one can see that the connection between the four-fold axis of the FCC structure, and the six-fold axis of the hexagonal structure, is given by :

$$(C_4^z|000)_c \longrightarrow (C_6^z \mid \vec{a}_{31})_1 \tag{6.11}$$

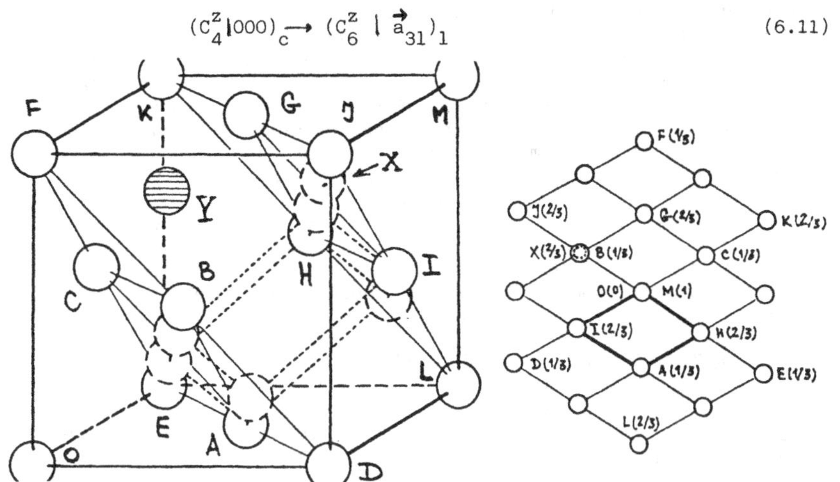

Figure 27 : Identification of the R3m → Fm3m symmetry for a 1/3 occupation of the latent unit-cell. The lettered atoms refer to Fig.28

Figure 28 : Projection of the FCC cell on the (III) plane. Eq.(6.11) is illustrated by the transformation of the I atom into the B position under a four-fold rotation.

In summary, the FCC-HCP transition can be understood as a transition between two low-symmetry phases which are induced by two different IR's of the P6/mmm latent phase, corresponding to two-and four-component order-parameters. This model, which is schematized in Figure 29, does not differ fundamentally from the model proposed for the $\beta - \omega$, and $\beta - \alpha$ transition (see §.6.1). The difference is here that the parent phase is not observed. Besides, the variational parameter ξ appears to be <u>the degree of occupation of the latent unit-cell</u> (i.e. the concentration) : for a non fractional occupation of the latent cell, the symmetry of the low-temperature phase is lowered to $R\bar{3}m$, while for the symmetric value $\xi = 1/3$ the symmetry increase to Fm3m. However, the dependence on ξ of the four-dimensional order-parameter associated with the P6/mmm → $R\bar{3}m$ → Fm3m sequence of phases is less obvious. Thus for $\xi = 1/3$ but for a 90° angle between the lattice vectors, the symmetry increases to Pm3m. For $\zeta = 2/3$, one obtains the diamond type structure, providing an additional compression of the lattice along z [97].

Although the approach presented in this section, applies in a general manner to reconstructive transitions in metals and insulators [97], we have chosen to illustrate its applicability to the restricted case of elements, because the simplicity of their structures allows concrete consideration on the shift and ordering of the atoms. The existence of a small number of sublattices is indeed an important necessary condition for one of the essential mechanism assumed in the approach : an increase of symmetry of one phase due to high-symmetric shifts of the atoms, or to specific conditions fulfilled by the latent unit-cell. The existence of a complex set of sublattices, would be defavourable for the preceding mechanism to take place, as the increase of symmetry associated with one type of sublattice would be compensated by the modifications involved in the other sublattices, and as a consequence, the group-subgroup relationship between the phases should be preserved. In this respect, let us stress that the group-subgroup relationship, which is generally interpreted as typifying transitions with small shifts and little ordering, may also be the signature of complex sublattice structures.

7. SINGULAR POINTS IN PHASE DIAGRAMS

In the phase diagrams discussed in this chapter a number of singular points have been predicted to take place. Different sorts of singular points can be distinguished :

1) Multicritical points which can be defined phenomenologically as points of sudden change of behaviour on a line of critical points (i.e. a second-order transition line). Thus at a tricritical point a continuous critical transitions becomes abruptly first-order. A bicritical point [102] may similarly be characterized as the meeting of two separate critical lines corresponding to two distinct order-parameters. This latter definition can be extended to polycritical points, and in particular to tetracritical points where four critical lines meet. If one of the ordered phase is spatially modulated one may have the so-called Lifshitz points.

2) Isolated critical points, as the Landau critical point, which corresponds to a second-order transition point lying on a first order transition line at the end of another first-order transition line separating two isomorphous phases.

3) End points. This class of singular points include the critical point which terminates a line of first-order transitions between two isomorphous phases, but also the so-called critical end point which has been predicted to end some lines of second order transitions.

4) N phase points, at which merge N phases and the corresponding separation lines of second and first-order transitions.

As shown in the case of the tricritical point, in §.2.1, for most of the preceding singular points, we may predict specific values for the critical exponents characterizing the variation of the order-parameter, susceptibility, specific heat,etc... The discussion of the universal classes of critical exponents characterizing singular point has been the subject of abundant theoretical works (see for example refs. 102-105) the review of which are beyond the scope of this book. Let us only note, that the little number of accurate experiments, which have been performed to verify the theoretical predictions, do not give in general a satisfactory agreement (see chapter VIII).

REFERENCES

1. S. Alexander, Solid State Commun. 14, 1069 (1974)
2. A.L. Korzhnevskii, and B.N.Shalaev, Sov.Phys.Solid State 21, 1311 (1979)
3. S.Alexander, and D.J.Amit, J.Phys. A8, 1988 (1975)
4. M.Hosoya, J.Phys.Soc. Japan 42 399 (1977)
5. Z.Racz and T.Tél, Phys.Rev. A26, 2968 (1982)
6. Yu.M.Gufan "Structural phase transitions" (in russian), Nauka,Moscow (1982)
7. Yu.M.Gufan, Sov. Phys.Solid State 13, 175 (1971)
8. Yu.M.Gufan and V.P.Sakhnenko, Sov.Phys.Solid state 16, 1034 (1974)
9. E.B. Vinberg, Yu.M.Gufan,V.P.Sakhnenko and Yu.I.Sirotin,Sov. Phys. Crystallogr. 19, 10 (1974)
10. Yu.M.Gufan and V.P.Sakhnenko, Sov.Phys.JETP, 42, 728 (1975)
11. Yu.M.Gufan, V.P.Dmitriev, V.P.Popov and G.M.Chechin, Phys. Met. Metall. 46, 6 (1978)
12. Yu.M.Gufan and E.S.Larin, Sov.Phys.Solid State,20, 998 (1978)
13. Yu.M.Gufan and E.S.Larin, Sov.Phys.Dokl. 23, 754 (1978)
14. Yu.M.Gufan, V.P.Dmitriev,V.P.Popov, and G.M.Chechin,Sov.Phys. Solid state 21, 327 (1979)
15. Yu.M.Gufan and E.S.Larin, Izvest.Akad.Nank.SSSR 43,1567 (1979)
16. Yu.M.Gufan, V.P.Dmitriev and V.I.Torgashov, Sov.Phys.Crystallogr 24, 342 (1979)
17. Yu.M.Gufan and G.M.Chechin, Sov.phys.Crystallogr. 25, 261 (1980)
18. Yu.M.Gufan, and V.P.Dmitriev, Sov.Phys.Crystallogr.25, 6 (1980)
19. Yu.M.Gufan and V.I.Torgashev, Sov.Phys.Solid.State 22, 951 (1980)
20. Yu.M.Gufan and E.S.Larin, Sov.Phys.Solid State 22, 270 (1980)
21. V.P.Dmitriev, Yu.M.Gufan,V.P.Popov and G.M.Chechin, Phys.Stat.sol (a) 57, 59 (1980)
22. V.P. Dmitriev, Sov.Phys. Crystallogr. 25, 683 (1980)
23. Yu.M.Gufan and V.P.Popov, Sov.Phys.Crystallogr.25, 527 (1980)
24. E.V.Gorbunov, Yu.M.Gufan, N.A.Petrenko and G.M.Chechin,Sov.Phys. Crystallogr. 26, 3, 1981
25. Yu.M.Gufan and V.B.Shirokov, Sov.Phys.Solid state 23, 1992 (1981)
26. Yu.M.Gufan and V.P.Dmitriev, Phys.Met.Metall.53, 852 (1982)
27. Yu.M.Gufan and V.L. Lorman, Sov.Phys.Solid State 25, 598 (1983)
28. E.S.Larin, Sov.Phys.Solid state 26, 3019 (1984)
29. L.T.Latush,V.I.Torgashev, and F.Smutny,Ferroelectrics Letters 4, 37 (1985)
30. P.Tolédano and J.C.Tolédano, Phys.Rev. B14, 3097 (1976)
31. P.Tolédano and J.C.Tolédano, Phys.Rev. B16, 386 (1977)
32. J.C.Tolédano and P.Tolédano, Phys.Rev. B21, 1139 (1980)
33. P.Tolédano and J.C.Tolédano, Phys.Rev. B25, 1946 (1982)
34. P.Tolédano, Thesis, University of Picardie, unpublished (1979)
35. H.T.Stokes and D.M.Hatch, Phys.Rev. B31, 7462 (1985)
36. D.M.Hatch, H.T.Stokes, J.S.Kim and J.W.Felix, Phys.Rev. B32, 7624 (1985)
37. J.S.Kim, D.M.Hatch and H.T.Stokes, Phys.Rev.B33, 1774 (1986)
38. O.V.Kovalev, "Irreducible representations of the space-groups", Gordon and Breach, New-York (1965)
39. Yu.M.Gufan and V.P.Dmitriev, unpublished.
40. P.Tolédano and G.Pascoli in "Symmetry and Broken symmetry in condensed matter physics"Ed.N.Boccara (IDSET, Paris, 1981)

41.I.L.Aptekar',V.I.Rashupkin and E.Yu.Tonkov, Sov.Phys.Solid state 21, 897 (1979)
42.A.Jayaraman, Phys.Rev.137, A179, (1965)
43.A.Jayaraman, D.B.Mc Whan, J.P.Remeika and P.D.Dernier,Phys.Rev. B2, 3751 (1970)
44.D.E.Bruins and C.W.Garland,J.Chem.Phys.63, 4139 (1975)
45.M.Vallade, Phys.Rev.B12, 3755 (1975)
46.P.Bastie, M.Vallade, C.Vettier and C.M.E.Zeyen, Phys.Rev.Lett. Soc.Japan 49, 1874 (1980)
47.H.Yamashita, Y.Takeuchi and I.Tatsuzaki, J.Phys.Soc.Japan 49, 1874 (1980)
48.R.Clarke and L.Benguigui, J.Phys.C10, 1963 (1977)
49.V.Yu.Slivka, Yu.M.Vysochanshii,L.A.Salo,M.I.Gurzan and D.V.Chepur, Sov.Phys.Solid state 21, 1845 (1979)
50.E.I.Gerzanich, Sov.Phys.Solid state 23, 1896 (1981)
51.K.S.Aleksandrov and I.N.Flerov, Sov.Phys.Solid state 21, 195 (1979)
52.A.F.Devonshire, Phil.Mag.40, 1040 (1949)
53.Landölt-Börnstein, Ferro and Antiferroelectric substances,Springer-Verlag Band 3 (1969)
54.J.P.Pouget and H.Launois, J.de Physique 37, C4, 49 (1976)
55.L.D.Landau, Zh.Eksper.Teor.Fiz.7, 19 (1937)
56.P.B.Vigman,A.I.Larkin, and V.P.Filev,Sov.Phys.JETP 41,944 (1976)
57.C.Vause and J.Sak, Phys.Rev. B18, 1455 (1978)
58.G.Pascoli, Thesis (University of Picardie 1981) unpublished and P.Toledano and G.Pascoli, Ferroelectrics 25, 427 (1980)
59.One finds a cubic invariant for the four-dimensional IR associated with the wave vector $k = 1/3\ (a_1+a_2) + 1/6\ a_3$ located inside the hexagonal Brillouin zone.
60.V.P.Sakhnenko and V.M.Talanov, Sov.phys.Solid state 21,1401(1979)
61.P.W.Anderson and E.I.Blount, Phys.Rev.Letters 14, 217 (1965)
62.J.Wanagel and B.W.Batterman, J.Appl.Phys.41, 3610 (1970)
63.B.S.Chandrasekhar, H.R.Ott and B.Seeker, Solid state commun. 39, 1265 (1981)
64.Y.Ishibashi and V.Dvorak, J.Phys.Soc.Japan 44, 941 (1978)
65.P.Tolédano, M.M.Fejer and B.A.Auld, Phys.Rev.B27, 5717 (1983)
66.E.M.Lifshitz, Zh.Eksp.Teor.Fiz.11, 255 (1941)
67.S.M.Shapiro,J.D.Axe,G.Shirane,and P.M.Raccah,Solid State Commun. 15, 377 (1974)
68.R.Pynn, J.D.Axe and P.M.Raccah, Phys.Rev. B17, 2196 (1978)
69.D.Mukamel, Phys.Rev.Letters 34, 481 (1975)
70.C.N.R.Rao, G.R.Rao, and G.V.S.Rao, J.Solid State Chem.6,340 (1973)
71.T.Sakata,K.Sakata and I.Nishida, Phys.Stat.Sol.20, K155 (1967)
72.See Ref. 31 page 403
73.W.Bührer, H.Schmid, W.Brixel, F.Maaroufi and P.Tolédano, to be published in Phys.Rev.B (1988)
74.M.Regis, M.Candille, and P.St-Grégoire, J.Phys.Lett (Paris) 41, L423 (1980)
75.C.J.de Pater, J.D.Axe, and R.Currat, Phys.Rev.B19, 4684 (1979)
76.J.Schneck and F.Denoyer, Phys.Rev.B23, 383 (1981)
77.G.M.Wolton, Acta Cryst. 17, 763 (1964)
78.J.F.Dorrian, R.E.Newnham and D.K.Smith, Ferroelectrics 3,17(1971)
79.K.W.Browal,J.S.Kasper and H.Wiedemeier, J.Solid state Chem. 10, 20 (1974)

80.J.Jerphagnon, in Proceedings Ternary compounds conference, Institute of Physics, London (1977)
81.E.G.Stewart, and H.P.Rooksky, Acta cryst.6, 49 (1953)
82.P.Esherik and B.E.Kohler, J.Chem. Phys. 59, 6681 (1973)
83.S.C.Abrahams,J.L.Bernstein and J.P.Remeika, Mater.Res.Bull 9, 1613 (1974)
84.V.Janovec, V.Dvorak and J.Petzelt, Czech.J.Phys.B25, 1363 (1975)
85.E.M.Lifshitz, Zh.Eksp.Teor.Fiz.19, 353 (1944)
86.J.Holakovsky, Phys.Stat.Sol.(b) 56, 615 (1973)
87.V.I.Torgashev, V.Dvorak, and F.Smutny, preprint
88.Yu.M.Gufan, and V.I.Torgashev, Sov.Phys.Solid state 23, 1129 (1981)
89.T.Yamada, and H.Iwasaki, J.Appl.Phys.44, 3934 (1973)
90.N.Niimura, K.Shimaoka, H.Motegi and S.Hoshino, J.Phys.Soc.Japan 32, 1019 (1972)
91.P.Tolédano, H.Schmid,M.Clin and J.P.Rivera, Jap.Jour.Appl.Phys. Soc.Japan 32, 1019 (1972)
92.A.P.Levanyuk and D.G.Sannikov, Sov.Phys.Solid state 17, 327 (1975)
93.J.C.Tolédano, Phys. Rev.B20, 1147 (1979)
94.J.M.Buerger, Elementary crystallography, John Wiley, New-York (1963)
95.W.B.Pearson:"A Handbook of lattices spacings and structures of metals and alloys", Pergamon Press, Oxford (1967)
96.Z.Nishiyama", Martensitic transformation", Academic Press, New-York (1979)
97.V.P.Dmitriev, S.B.Rochal, Yu.M.Gufan and P.Tolédano, to be published
98.Y.A.Bagariatskii, T.V.Tagunova and J.C.Williams, in proceedings. of the 29th annual meeting of the Electron Microscopy Society of America, p.122, Boston, Mass. (1971)
99.N.E.Paton, D. de Fontaine, and J.C.Williams, in Proceedings of the 29th annual meeting of the Electron Microscopy Society of America, p.122, Boston, Mass. (1971)
100.D. de Fontaine, Acta Met. 18, 275 (1970)
101.L.D.Landau, and E.M.Lifshitz, statistical physics, Addison-Wesley, Reading, Mass. (1958)
102.R.B. Griffiths, Phys.Rev.Letters 24, 715 (1970)
103.Multicritical phenomena, Ed. by R.Pynn and A. Skjeltrop, NATO ASI series vol. 106, Plenum Press New-York 1983
104.N.Boccara, Symétries brisées, Hermann (1976)
105.H.E.Stanley, Introduction to phase transitions and critical phenomena, Clarendon Press, Oxford (1971)

CHAPTER V

LANDAU THEORY OF INCOMMENSURATE PHASES

1. INTRODUCTION

In the framework of Landau's theory, incommensurate (INC) phase transitions can be defined as a special case of continuous structural transitions (ST) adjacent to a high-symmetry (HS) crystalline phase. Their specificity stems from the fact that the k-vector associated to their order-parameter (OP), $\vec{k} = \Sigma\, q_i . \vec{a}_i^*$ ($\vec{a}_i^*$ being a reciprocal lattice primitive translation in the HS phase) fulfills the condition that one q_i , at least, is an irrational number. This situation was not considered in the preceding chapters because we only examined order-parameters complying with the Lifshitz symmetry criterion (chap.III §3) and that this criterion imposes to the OP wavevector to possess simple rational q_i coordinates. In itself, the "incommensurate" character of $\vec{k}$ would not require developping special methods for the construction of the free-energy associated to the order-parameter. However, this free-energy would only be useful in a limited temperature interval below the transition temperature (which will be denoted T_I in the present chapter). It would not account for the complex sequence of phenomena, displaying universal features (without a counterpart below ordinary ST) which is revealed by the idealized experimental situation of INC systems, in a wider temperature interval below T_I. To account for this situation it is necessary to elaborate a specific phenomenological model consisting in an extension of the Landau theory (LT).

1.1. Standard experimental scheme for INC systems

This standard scheme, illustrated by fig. 1, involves a sequence of three phases, separated by two transitions, the upper one being the INC transition at T_I

Just below T_I, the situation appears similar to that of ordinary ST : the structural and physical properties of the LS phase can be related to a primary OP, associated to a single irreducible representation of the HS space-group , whose k-vector, $\vec{k}_o$, has temperature independent components q_i^o . One important distinctive feature is the non-cristallinity of the low-symmetry (LS) phase. As will be shown in § 2.1 the LS phase cannot be identified by one of the 3-dimensional space-groups, due to the irrational character of certain of the q_i^o components. Nevertheless, the

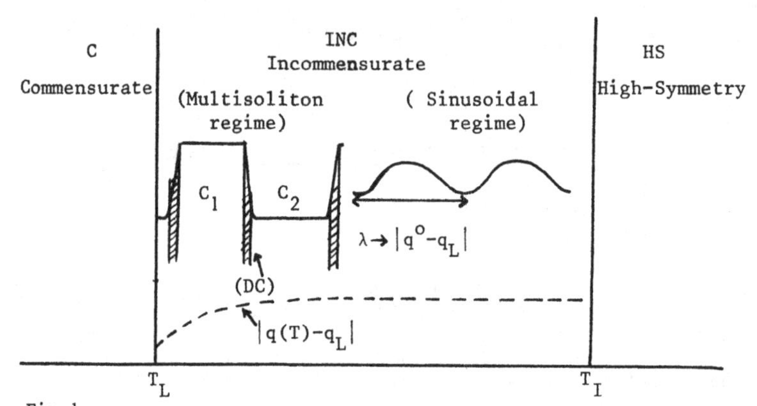

Fig.1. Schematic representation of the main specific properties of incommensurate systems: sequence of 3 phases; change of regime from a sinusoidal modulation to an array of commensurate domains separated by discommensurations; temperature dependence of the incommensurate wavevector.

distortion onsetting in the LS phase has a Fourier transform displaying, besides the 3 elementary periods $\vec{a}_i^*$, a limited number of additional elementary periods, related to the components q_i, thus denoting that the LS phase is a perfectly ordered phase. The Fourier components relative to the additional periods (which can be termed satellite components) have a small amplitude as compared to the Fourier components relative to the $\vec{a}_i^*$ periods (which can be termed principal components) : this reflects the fact that the structural modification which contains the additional periods, is a small "modulation" of the particles density in the system. The main part of the density has the 3-dimensional periodicity defined by the $\vec{a}_i^*$. Thus the LS phase can be termed an "incommensurately modulated crystalline phase".

Further away from T_I, in the LS phase, three main specific features distinguish INC systems :

a) Though at each temperature the incommensurate distortion is, as usual, described by an irreducible primary OP, the corresponding k-vector, $\vec{k} = \Sigma q_i \vec{a}_i^*$, has temperature dependent components q_i. Thus, as the temperature varies, the OP does not remain associated with a given irreducible representation of the HS space-group but with a continuous set of unequivalent irreducible representations indexed by distinct k-vectors. This situation is realized down to a temperature T_L. The temperature range between T_I and T_L is considered as the stability range of a single phase termed the INC phase, though on varying the temperature the components of $\vec{k}$ necessarily take rational as well as irrational values. The validity of this assignment is discussed in § 2.2.3.

b) Below T_L the stable phase is termed the <u>commensurate (C) phase</u>. Like the INC phase, it consists in a structural distortion of the HS phase, described by a primary OP associated to a single irreducible representation of the HS phase. However, by contrast to the INC phase, the $\vec{k}$-vector indexing the OP, $\vec{k}_L = \Sigma\, q_i^L . \vec{a}_i^*$, has temperature <u>independent</u> q_i^L components.

The values of the q_i^L are close to the q_i^o, but are all equal to "simple" rational numbers, i.e. to rational numbers with small denominators such as (1/2),(1/3),(1/4),(2/5),.... Below T_L, one recovers a situation similar to that of ordinary ST, since the C-phase can be identified by a three-dimensional space-group involving primitive translations which are simple multiples of the primitive translations $\vec{a}_i$ of the HS phase. T_L is termed the "lock-in" transition, since it induces, on cooling, a "locking" of all the elementary periods of the system on definite multiples of the "basic" $\vec{a}_i$ periods. As usual for structural transitions (chap. III §7), the lowering of translational and orientational symmetries occuring in the C-phase with respect to the HS-phase gives rise to a finite number of energetically equivalent states (denoted C_i) for the C-phase. These states, which are in geometrical correspondance with eachother by means of the space-symmetry operations lost in the HS to C transformation, can happen to coexist in a given sample in the form of "domains".

c) In its lower range of stability, nearby T_L, the INC phase has a structure which can be considered to a good approximation, as a periodic array of C_i domains separated by thin walls which are termed "spatial solitons" or "discommensurations". The periods of the array are determined by the differences $(q_i - q_i^L)$, where the q_i and the q_i^L are respectively the INC components of the OP wavevector at the considered temperature, and the C-components. These differences decrease as measure as T is lowered towards T_L (i.e. the size of the C_i-domains increase in anticipation of the onset of the C-phase).

A good illustration of the preceding standard scheme is provided , for instance, by the experimental situation in the compound with formula (Rb_2ZnCl_4), for which the data are indicated on fig. 2) [1] .

The wavevector of the OP has only one temperature dependent component within the INC phase : $\vec{k} = q_3\vec{a}_3 = |(1/3)-\delta|\, a_3^*$ In an interval of $\sim$ 50°C below T_I, q_3 keeps an almost constant value, $q_3^o = (\frac{1}{3} - \delta_o)[\delta_o \sim 0.03]$. On further cooling towards T_L, δ decreases to 0.015 before abruptly vanishing at T_L. The lock-in wavevector is $\vec{k}_L = a^*/3$. Between T_I and T_L, δ shows a characteristic concave variation. The transition at T_I is continuous, while that at T_L appears discontinuous. In agreement with the description in c) hereabove, electron microscopic observations disclose clearly, within the INC phase, the existence of a periodic array of C_i domains [2]. The C-phase is associated to 6 types of C_i domains differing either by the sense of their spontaneous polarization, or by relative translational shifts (of magnitude $\vec{a}_3$ or $2\vec{a}_3$) of the type already described for improper ferroics (cf.§ III-). Actually, the symmetry change

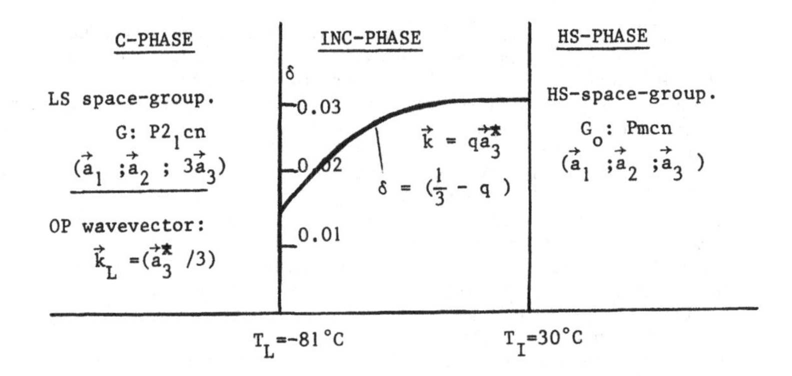

Fig.2. Characteristics of the 3 consecutive phases of Rb_2ZnCl_4.

between the HS phase and the C phase can be classificated as an improper ferroelectric transition (cf chap.III §6).

1.2. Basic ideas for the adaptation of Landau's theory

Our aim, in this chapter, is to relate the set of specific manifestations of the existence of the INC-phase : limited stability between T_I and T_L, temperature dependence of the q_i, progressive structuration of the INC phase into an array of C_i-domains. We also wish to determine, as we have done for ordinary ST, the anomalous behaviour of various relevant physical properties which is induced by the INC phase and its adjacent transitions T_I and T_L. Finally we wish to discuss the symmetry characteristics of the OP and of the LS phases.

The basic idea underlying the required extension of the LT is to assume that the set of considered phenomena is the result of the competing influences of two types of contributions to the free-energy of the system.

The first contribution, F_1, generally of degree 2 in the OP components, expresses the tendency of the system to establish an INC phase with wavevector components q_i^o, just below T_I. At the phenomenological level considered in this book, the microscopic origin of this tendency is not discussed, and the values of the q_i^o are taken as a starting point of the theory.

The second contribution F_2, of degree higher than F_1, favours the onset of a C-phase with the q_i^L wavevector components. F_2 can be termed the "lock-in" contribution. Due to its degree, higher than that of F_1, the influence of F_2 will only become significant when the amplitude of the OP has sufficiently increased, i.e. when $|T-T_I|$ has become sufficiently large. This influence is twofold. For $T < T_L$, it offsets completely the influence of F_1, and stabilizes the C-phase. Between T_I and T_L it induces the onset of regions having the structure of the C_i states in a fraction of the volume of the system. This fraction increases as measure as T approaches T_L, i.e. as measure as the magnitude of F_2 grows with respect to the magnitude of F_1. This increase is achieved in two ways i) By structuring the system into a periodic array of C_i-domains separated by walls which become thinner for $T \to T_L$; ii) by increasing the distance between the walls, thus provoking a reduction of the wavevector difference $|k - k_L|$, and determining a temperature dependence for the q_i components. These ideas can be formulated in a variety of manners.[1] The corresponding theories are related to two schemes.

a) The first scheme has in common with the general formulation of the LT the feature that the components of the irreducible OP are quantities η_i which are independent of the spatial coordinates (i.e. spatially uniform). In this scheme, F_1 is the usual quadratic invariant $\alpha(\Sigma\, \eta_i^2)$, and the wavevector $\vec{k}_0$ corresponds to the minimum of the k-dependent coefficient $\alpha(\vec{k})$. F_2 has a qualitatively different form in the INC phase and in the C-one. In the former phase it is constituted by certain of the non-linear coupling terms between the primary OP [with wavevector $\vec{k}$] and secondary order-parameters associated to multiples of $\vec{k}$ [harmonics](If $\vec{k}$ is INC, an infinite number of such secondary OP with distinct symmetries exist). In the latter phase, when $\vec{k} = \vec{k}_L$, F_2 coincides with one of the invariants of degree higher than 2, built from the components of the primary OP. This "primary" term is the limit of many INC coupling terms when $\vec{k} \to \vec{k}_L$.

In § 3 we will consider two phenomenological theories based on this scheme. These theories have the advantage of showing the qualitative origin of the specific behaviour of INC systems. In particular, they show that the structuration of the system as an array of C_i-domains reflects the fact that the spontaneous values of secondary OP, corresponding to certain definite harmonics of $\vec{k}$ acquire large amplitudes on approaching T_L. Conversely, these theories have the drawback to account only imperfectly for the temperature dependence of the wavevector components, and to predict systematically a sharply discontinuous character for the lock-in transition.

b) The second scheme considers, in the entire temperature range investigated, a primary OP associated to an irreducible representation with $\vec{k} = \vec{k}_L$, thus having a symmetry which is different from the actual OP symmetry in the INC phase. Also, unlike the situation in the standard formulation of Landau's theory, it assumes that the OP components are functions $\eta_i(\vec{r})$ of the spatial coordinates. In this framework, the system's free-energy is a functional of the OP components and of its spatial derivatives. The F_1 term is related to the lowest degree terms containing the spatial derivatives of the $\eta_i(\vec{r})$. This term vanishes for spatially uniform η_i, and is minimum for sinusoïdally modulated η_i with wavevector $\vec{k}_o$. The F_2 term is a functional of degree higher than two of the $\eta_i(\vec{r})$, which vanishes for a sinusoïdal variation of the $\eta_i(\vec{r})$, and is minimum if these components are spatially uniform. As will be shown in §4&5, various forms of this scheme are apt to account for the set of properties constituting the standard INC behaviour described above.

In the first and second schemes, the INC phase respectively appears as a homogeneously distorted HS-phase, and as an inhomogeneously distorted C-phase.

In this chapter, before outlining in §3 the two different types of Landau theories, we will examine, in §2 the construction of the free-energies which pertain to the two schemes.

In § 4 , the investigation of the physical consequences of the second scheme will be performed in the restricted framework of a two-component OP, and of standard simplifying assumptions.

In § 5 we will discuss the consequences of lifting such restrictions, and consider various types of refinements and generalizations of the second scheme. Finally, in § 6 , we will survey the pending theoretical problems and briefly evaluate the experimental situation as compared to the developped theoretical schemes.

2. SYMMETRY PROPERTIES OF THE ORDER PARAMETER AND OF THE FREE ENERGY.

The first scheme described above involves two types of Landau free-energies. One associated with an INC wavevector [$\vec{k}_o$, or $\vec{k}(T)$] and one with the lock-in wavevector $\vec{k}_L$. The second scheme corresponds to a third form of free-energy involving, locally, derivatives of the spatially non-uniform OP components. In the latter case it will be relevant to distinguish two cases according to the presence or absence of a bilinear term in the OP components and their first derivatives (the so-called Lifshitz invariant). Fig. 3 summarizes the main topics and results exposed in this section.

2.1. Order-parameter and free-energy for an incommensurate wavevector.

In this paragraph we consider the case of a physically irreducible OP, whose representative wavevector $\vec{k}_o = \Sigma\, q_i^o . \vec{a}_i^*$, has at least one irrational q_i^o component. As mentioned in the introduction, this case corresponds to the situation nearby the T_I transition. We first derive the symmetry characteristics of the OP, which are related to the irrationality of the q_i^o components. We then examine the translational properties of the INC phase, and define the number of independent modulation directions. Finally, we investigate the specific form of the INC free-energy.

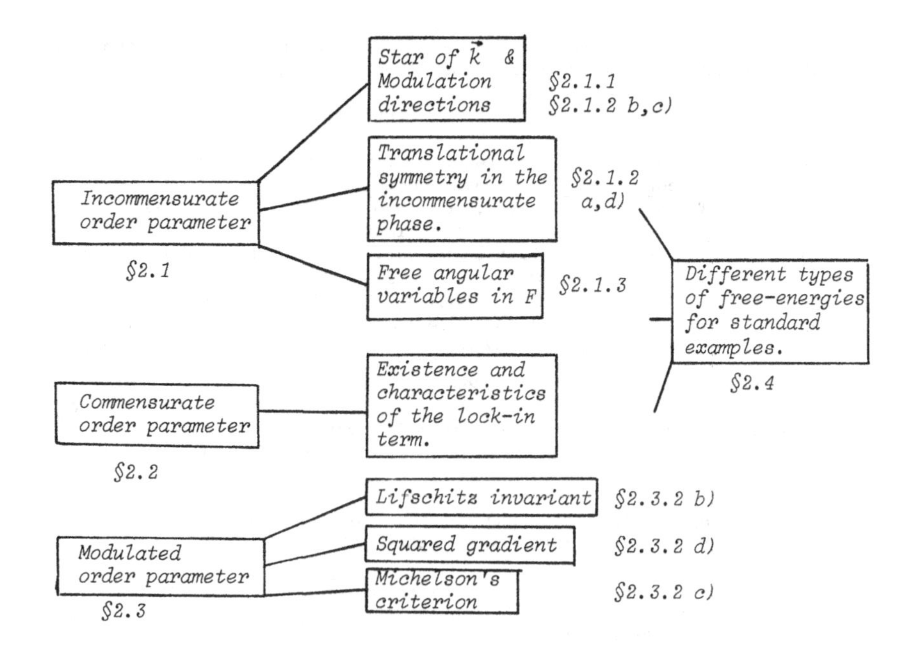

Fig. 3

2.1.1. Symmetry characteristics of the INC order-parameter.

a) INC character of all the vectors of the star of $\vec{k}_o$.

Let $\vec{k}_p$ be the vectors of the star of $\vec{k}_o$. It is easy to show that each $\vec{k}_p$ has, at least, one irrational component, referred to the $\vec{a}_i^*$. We know that any $\vec{k}_p$ is transformed from $\vec{k}_o$ by application of an element $\hat{g}_p$ of the point group $\hat{G}_o$ of the HS phase. Denote $\vec{k}_p = \sum_i q_i^p . \vec{a}_i^*$, we have :

$$\vec{k}_o = \hat{g}_p^{-1} \vec{k}_p = \sum_i q_i^p . \hat{g}_p^{-1} \vec{a}_i^* = \sum_i \sum_j q_i^p . m_{ij} . \vec{a}_j^* \qquad (2.1)$$

where the m_{ij} are integers since the reciprocal lattice $\vec{a}_i^*$ is invariant by any $\hat{g}_p^{-1}$. Thus,

$$\vec{k}_o = \sum_j q_j^o . \vec{a}_j^* = \sum_j \left(\sum_i m_{ij} q_i^p \right) . \vec{a}_j^* \qquad (2\cdot2)$$

If all the q_i^p were rational, the q_j^o would also be rational in contradiction with the starting assumption.

b) **Even dimensionality of the order-parameter**

As noted in chap.III§3 the star of wavevectors associated to a <u>physically irreducible OP</u> always contains pairs of vectors $\vec{k}_p$ and $(-\vec{k}_p)$. Whenever $\vec{k}_p$ is INC, the two vectors $\vec{k}_p$ and $(-\vec{k}_p)$ are <u>unequivalent</u>. Indeed, $2\vec{k}_p$ has irrational components and cannot coincide with a reciprocal lattice vector. Accordingly, the physically irreducible OP is associated to an <u>even number</u> of unequivalent wavevectors $(\pm \vec{k}_p)$. The dimensionality of the OP, which is the product of the number of $(\pm \vec{k}_p)$ vectors in the star, and of the dimension of the small representation τ, will itself be even.

The simplest INC order-parameter has therefore two components. As indicated in table 7, this simple situation is realized for a large fraction of the presently investigated INC systems. Other situations encountered in real systems correspond to $n = 4$, and 6.

c) **Infinite (continuous) character of the image of the HS space-group in the OP space.**

Let q_j^o be an irrational component of $\vec{k}_o$, and consider the set of diagonal matrices $[m\vec{a}_j]$ representing the translations $m\vec{a}_j$ (m integer) in the OP space. Their first diagonal element is :

$$e^{-i\vec{k}_o(m\vec{a}_j)} = e^{-2i\pi.mq_j^o} \tag{2.3}$$

As q_j^o is irrational, all the elements corresponding to different m values are distinct. The arguments $(2\pi.mq_j^o)$ of the exponential, form an infinite discrete set which, modulo 2π, "fills" the interval $[0, 2\pi]$: This infinite set is said to be <u>dense</u> in the interval $[0, 2\pi]$ The set of matrices $[m\vec{a}_j$ is, itself, an infinite discrete set of distinct matrices, dense in a continuous set of diagonal matrices whose first diagonal element is $e^{i\theta}$ with θ any number belonging to the $[0,2\pi]$ interval. The image of the space-group G_o in the OP space contains this set of matrices. Thus this image is isomorphous to a subgroup I_o of $O(n)$ dense in a continuous subgroup I_∞ of $O(n)$, n being the <u>even</u> number of components of the OP. It can be shown that the free-energy associated to the OP, which is invariant by I_o (chap.II§4.5), is also invariant by the continuous group I_∞ [3] .

2.1.2. <u>Number of modulation directions in the INC phase</u>.

a) **Lack of three dimensional translational symmetry.**

The symmetry of the LS phase stable below T_I, is represented (ch.III §5) by the group of invariance of the "primary" **density** increment $\delta\rho_1(\vec{r})$. This function, associated to the OP, represents the distortion of the LS phase with respect to the HS one . As explained in chapter II (§ 3.6), $\delta\rho_1(\vec{r})$ can be considered as a vector in the OP space whose direction is determined by the non-zero components of the OP. Consider a standard basis $\eta^l_{\pm p}$ in the OP space (the index l enumerates the components of the small representation τ, and $\pm p$ refers to the opposite arms $\pm\vec{k}_p$ of the star of the OP wavevector). Any translation $\vec{T}_i=(\Sigma\, m_{ij}.\vec{a}_j)$ does not interchange the components η^l_p , and transforms each component as:

$$|\vec{T}_i|\,\eta^l_{\pm p} = e^{\pm i\vec{k}_p.\vec{T}_i}\,.\,\eta^l_{\pm p} \qquad (2.4)$$

In order to belong to G, a translation $\vec{T}_i$ must therefore leave unchanged each of the non-zero η^l_p components. Let η^l_p be one such component. As shown in § 2.1.1 hereabove, $\vec{k}_p$ possesses at least one irrational component q_j^p along $\vec{a}^*_j$. According to eq.(2.4), the translation $\vec{T}_i$ will belong to G if:

$$\frac{1}{2\pi}\,(\Sigma\, m_{ij}\vec{a}_j)\,\vec{k}_p = \Sigma\, m_{ij}\, q_j^{\,p} = 0 \quad \text{(modulo 1)} \qquad (2.5)$$

If condition (2.5) is satisfied by 3 non-coplanar translations $\vec{T}_i$, the matrix of the m_{ij} is non-singular and conditions (2.5), for i = 1, 2, 3, imply that all the q_j^p are rational numbers contrary to the assumption. Thus (2.5) can, at most, be satisfied by 2 independent translations $\vec{T}_i$, and consequently, the symmetry of the LS phase is not represented by a 3-dimensional space-group (since there is no three-dimensional translational group of invariance of $\delta\rho_1(\vec{r})$).

For instance, if the star of $\vec{k}_0$ only contains $(\pm\vec{k}_0) = \pm q\vec{a}^*_1$ (with q irrational), the only translations of G_0 which preserve $\delta\rho_1(\vec{r})$ are of the form $(m_2\vec{a}_2 + m_3\vec{a}_3)$: the LS phase lacks translational symmetry along $\vec{a}_1$.

In the general case, the LS phase can have translational periodicity along two, one or even no direction, depending on the number of irrational components q_j^p of the $\vec{k}_p$ vectors and of the non-zero components η^l_p of the OP. In these different situations, the INC phase is said to possess, respectively, one, two, or three independent modulation directions.

Let us examine the relationship between the number of modulation directions and the characteristics of the star of $\vec{k}_0$.

b) "Rational" dependence between the vectors in the star.

Consider any p vectors belonging to the star of $\vec{k}_0$, denoted $\vec{k}_{\alpha_1}, \vec{k}_{\alpha_2}, \ldots, \vec{k}_{\alpha_p}$. These vectors are rationally dependent if one can find a set of integers $m_{\alpha_1}, \ldots, m_{\alpha_p}$, not all equal to zero, and such as :

$$m_{\alpha_1} \vec{k}_{\alpha_1} + \ldots + m_{\alpha_p} \vec{k}_{\alpha_p} = \vec{G} \qquad (2.6)$$

where $\vec{G}$ is a reciprocal lattice vector. Conversely, if (2.6) can only be realized by setting all the m_{α_i} equal to zero, the considered vectors are <u>rationally independent.</u>

Note that every $\vec{k}_p$ in the star of $\vec{k}_o$ can be written as:

$$\vec{k}_p = \hat{g}_p \vec{k}_o = \sum_j \hat{g}_p (q_j^o \cdot \vec{a}_j^*) = \sum_i (\sum_j m_{ij}^p q_j^o).\vec{a}_i^* \qquad (2.7)$$

Thus, the coordinates of all the $\vec{k}_p$ are integral combinations of, at most, three irrational numbers q_1^o, q_2^o, and q_3^o. On the basis of equations (2.6) and (2.7), one can show the following properties:

i) The maximum number of rationally independent vectors in the star of $\vec{k}_o$ is $1 < d' < 3$.

ii) If $\vec{k}_o, \ldots \vec{k}_{d'-1}$ are a set of d' rationally independent vectors in the star of $\vec{k}_o$, then all the other vectors in the star can be written as :

$$\vec{k}_p = \sum_0^{d'-1} m_i \vec{k}_i + \frac{r_1}{r_2} \vec{G} \qquad (2.8)$$

where the m_i, r_1 and r_2 are integers and $\vec{G}$ is a reciprocal lattice vector.

iii) $\vec{k}_o, \ldots \vec{k}_{d'-1}$ are linearly independent [eg. if d' = 3, the three vectors $\vec{k}_o, \vec{k}_1, \vec{k}_2$ are non-coplanar]

Let us illustrate these properties on two examples [1], [4].

-Barium manganese fluoride, $BaMnF_4$, has the following symmetry characteristics: HS space-group $G_o = A2_1am$; the star of the OP comprises 4 vectors, $\pm\vec{k}_1 = \pm(q\vec{a}_1^* + \vec{a}_3^*/2)$, and $\pm\vec{k}_2 = \pm(q\vec{a}_1^* + \vec{a}_2^*/2)$, with q irrational and close to 0.39. We note that, though the 4 vectors of the star are unequivalent, $2(\vec{k}_1 - \vec{k}_2)$ is a reciprocal lattice vector. Thus, in agreement with the definition (2.6), this star has only <u>one</u> rationally independent vector (d'= 1). The realization of equation (2.8) for this example is:

$$\vec{k}_2 = \vec{k}_1 + \frac{\vec{a}_3^* - \vec{a}_2^*}{2} \qquad (2.9)$$

-The symmetry characteristics of tantalum diselenide 2H-$TaSe_2$ are the following: $G_o = P6_3/mmc$; The star of the OP wavevector contains 6 arms; $\pm\vec{k}_1 = \pm q\vec{a}_1^*$; $\pm\vec{k}_2 = \pm q\vec{a}_2^*$; $\pm\vec{k}_3 = \mp q(\vec{a}_1^* + \vec{a}_2^*)$. In this case, d'=2. Indeed, any relation of the type (2.6) between two of the three $\vec{k}_i$ vectors requires $m_{\alpha_i}=0$, while the three vectors obey the relation $(\vec{k}_1+\vec{k}_2+\vec{k}_3)=0$.

c) Number of modulation directions.

In the LS phase, certain of the η_p^l components have non-zero values. The corresponding set of $\vec{k}_p$ vectors (whose number is generally

smaller than the number of arms in the star) contains d rationally independent vectors ($d \leqslant d'$). The property of linear independence of the d vectors ,stated hereabove in iii) , allows to show that the number of modulation directions, as defined in § 2.1.2 a), is equal to d.

Let us establish, for instance that if $d = 3$, the INC phase has no translational symmetry. Assume that this phase is invariant by a given translation $\vec{T} = \Sigma\, m_i \vec{a}_i$. For each vector $\vec{k}_p$ in the above set, one must have $\vec{T}.\vec{k}_p = 0\ (2\Pi)$. The latter condition must be satisfied in particular by the 3 rationally independent vectors $\vec{k}_1$, $\vec{k}_2$, $\vec{k}_3$. Thus:

$$\vec{T}.\vec{k}_p = \vec{T}(\hat{g}_p\, \vec{k}_1) = (\hat{g}_p^{-1}\, \vec{T}).\vec{k}_1 = 0 \quad (2\Pi) \qquad (2.10)$$

As $(\vec{k}_1, \vec{k}_2, \vec{k}_3)$ are non-coplanar, the three translations $\vec{T}_p = \hat{g}_p^{-1}\vec{T}$ are also non-coplanar. Following the argument used in (2.5), relation (2.10) cannot be satisfied by any $\vec{T}$, due to the incommensurate character of $\vec{k}_1$. For $d < 3$ the proof is more lengthy and we do not indicate it explicitely.

In $BaMnF_4$, for any set of non-zero OP components, $d=d'=1$ and, in agreement with the preceding statement, one must find a single modulation direction, i.e. a two-dimensional translation group of invariance. Indeed, if only $\eta^l_{\pm 1}$ acquire non-zero values, the two independent translations $\vec{T}_1 = 2\vec{a}_2$ and $\vec{T}_2 = \vec{a}_3$ clearly satisfy the conditions $\vec{T}_i.\vec{k}_1 = 0$. If both $\eta^l_{\pm 1}$ and $\eta^l_{\pm 2}$ have non-zero values , the translation group is composed of the two independent translations $\vec{T}_1 = 2\,\vec{a}_2$ and $\vec{T}_2 = 2\vec{a}_3$.

In $2H\text{-}TaSe_2$ we can consider the two situations ($\eta^l_{\pm 1} \neq 0$, $\eta^l_{\pm 2} = 0$, $\eta^l_{\pm 3} = 0$) , or ($\eta^l_{\pm 1} \neq 0$, $\eta^l_{\pm 2} \neq 0$, $\eta^l_{\pm 3} \neq 0$). In the first case , there is one rationally independent vector ($d=1$) ,and the incommensurate phase is invariant by $\vec{T}_1 = \vec{a}_2$ and $\vec{T}_2 = \vec{a}_3$. In the second case there are two independent vectors ($d=2$) , and the INC phase is only preserved by $\vec{T}_2 = \vec{a}_3$.

d) Fourier transform of the total density increment.

As shown in Chapter III, the spatial Fourier transform of the primary density increment $\delta\rho_1(\vec{r})$,

$$\delta\rho_1(\vec{k}) = \int \delta\rho_1(\vec{r}).e^{-i\vec{k}.\vec{r}}.\, d^3\vec{r} \qquad (2.11)$$

only contains discrete components corresponding to $\vec{k}$ vectors of the form :

$$\vec{k} = \Sigma\, m_i \vec{a}_i^* + \Sigma\, m_p.\vec{k}_p \qquad (m_p = \pm 1) \qquad (2.12)$$

where the $\vec{k}_p$ in the second member are associated to the non-zero OP components η^l_p.

The total density increment $\delta\rho(\vec{r})$ also includes the structural distortions corresponding to secondary OP which acquire spontaneous values below the transition. These spontaneous values arise (chap. II §5) from terms in the free-energy which are linear in the secondary OP and non-linear in the components of the OP. Hence the $\vec{k}$ vectors associated to the secondary OP are of the form :

$$\vec{k} = \Sigma\, m'_p\, \vec{k}_p \qquad (2.13)$$

where the $\vec{k}_p$ belong to the star of the primary OP, and where $m'_p > 0$ and $(\Sigma\, m'_p) \geqslant 2$.

As a consequence, the non-zero $\vec{k}$ Fourier components of the total density increment $\delta\rho\,(r)$ is also of the form (2.12) but with m_p taking any integral value, and not only $(m_p = \pm 1)$.

In the case of ordinary structural transitions, the $\vec{k}_p$ vectors have simple rational coordinates with respect to the $\vec{a}^*_i$, and, accordingly, the $\vec{k}$ vectors defined by eq.(2.12) generate a 3-dimensional lattice of the form :

$$\vec{k} = \Sigma\, n_i . \vec{a}'^*_i \qquad (2.14)$$

where the $\vec{a}'^*_i$ are 3 independent vectors of the form $(\vec{G}_i/n_i)$, with $\vec{G}_i$ a reciprocal lattice vectors.

Here, the set of $\vec{k}_p$ vectors contains d rationally independent vectors. One can express the $\vec{k}$ vectors in eq.(2.12) as :

$$\vec{k} = \Sigma\, n_i . \vec{a}'^*_i + \sum_1^d m_p . \vec{k}_p \qquad (2.15)$$

by replacing all the $\vec{k}_p$ as functions of the d rationally independent ones, according to eq.(2.8). Thus, the $\vec{k}$ vectors are obtained by combining, with integer coefficients, (3+d) basic vectors. As $d \geqslant 1$, this Fourier spectrum does not correspond to a crystalline phase (in agreement with §2.1.2 a) but nevertheless to a perfectly ordered phase characterized by a discrete spectrum involving a limited number (3+d) of basic spatial periodicities, d being the number of incommensurate modulation directions.

The fact that for an INC phase, $\vec{k}$ is generated by (3+d) basic vectors while for a crystalline phase it is generated by 3 basic vectors, has led to an interpretation of the nature of the symmetry of the INC phase based on the consideration that the order of the INC phase is an <u>ordinary crystalline order in a speculative space of dimension (3+d).</u> The relationship between the latter space and the observed 3-dimensional structure and its Fourier spectrum is briefly introduced in table 1.[5].

e) Single phase character of the temperature range between T_I and T_L.

In the case of an ordinary ST, one can speak of a well-defined LS phase, because this phase is identified by a specific space-group G. Alternately, the LS phase can also be identified

Table 1 . Crystallographical description of modulated phases [5].

As shown by eq.(2.15), the Fourier transform of the particle density contains (3+d) basic vectors, i.e. basic spatial periods. In direct space the system is not periodic because the (3+d) periods are superimposed in a 3-dimensional space. As the d-additional periods are incommensurate with the others the superimposition destroys the periodicity at least in one direction. In order to restore a crystallographical order the idea is to reject the d-additional periods in d additional spatial dimensions. The figure hereunder shows the principle of such a construction for (1+1) dimensions.

The one-dimensional incommensurate system is the intersection of the (1+1) periodic structure with the real axis, while the real Bragg spectrum can be obtained as the projection of the Fourier transform of the (1+1) structure. The primitive translations in the (1+1) space are:

$$\vec{a}'_1 = \vec{a}_1 - q.\vec{e} \qquad \vec{a}'_2 = \vec{e}$$

where the incommensurate vector is $\vec{k}=(q\vec{a}_1^{*})$ in the "real" reciprocal space, and where $\vec{e}$ is a unit vector perpendicular to the real axis. Thus none of these basic translation vectors is contained in the real space in conformity with the lack of periodicity of the real structure. By contrast the real reciprocal axis contains a reciprocal lattice translation of the (1+1) structure:

$$\vec{a}'^{*}_1 = \vec{a}_1^{*} \qquad \vec{a}'^{*}_2 = q\vec{a}_1^{*} + 2\pi\vec{e}$$

More generally for a (3+d) dimensional space, the (3+d) basic translations in direct space are:

$$\vec{a}'_i = \vec{a}_i - \Sigma\, q_{i\ell}.\vec{e}_\ell \quad ; \quad \vec{a}'_{3+\ell} = \vec{e}_\ell \quad (i=1,3\ ;\ell=1,d)$$

where the vectors of the modulation are

$$\vec{k}_\ell = \Sigma\, q_{i\ell}.\vec{a}_i^{*}$$

In the reciprocal space, the basic vectors are:

$$\vec{a}'^{*}_i = \vec{a}_i^{*} \quad ; \quad \vec{a}'_{3+\ell} = 2\pi\,\vec{e}_\ell + \Sigma\, q_{i\ell}.\vec{a}_i^{*}$$

As the (3+d) dimensional structure has perfect periadicity it will be invariant by a crystallographical space-group in (3+d) dimensions. However this space group cannot coincide with any of the space-groups enumerated for this dimension. Indeed the nodes of the reciprocal lattice cannot be freely interchanged by the point symmetry operations in the (3+d) space. Only are acceptable the operations which interchange nodes generated by vectors of the basic lattice (i=1,3) or nodes generated by the modulation vectors (ℓ=1,d).

by the symmetry of the primary OP and by the non-zero components of the OP, since the latter features determine entirely G.

For an INC phase, the same type of identification is not possible. There is no 3-dimensional space-group, and in addition (cf. § 1.1a), the primary OP has a symmetry which changes in the range between T_I and T_L (due to the variation of $\vec{k}$).

Two methods can be used to overcome this difficulty. One, which we only mention, is to identify the INC phase by the symbol of its (3+d) dimensional space-group (cf. table 1). The other is to note that though they belong to unequivalent representations, the various considered OP possess two common symmetry features one is the symmetry group of the associated wavevector $\vec{k}$, and the other is the nature of the "weighted representation" (cf. chap.III) corresponding to the small representation τ. In addition, the INC phase is specified, as any LS phase, by the fact that certain definite OP components are non-zero.

Indeed, the wavevector $\vec{k}_o$ (i.e. the $\vec{k}$-vector of the OP nearby the T_I transition), having irrational components, does not correspond to any of the high-symmetry points of the Brillouin-zone (BZ) of the HS phase (cf.chap.III §3). In agreement with their definition, the latter points are the only ones which are isolated from the symmetry point of view (i.e. the group of their $\vec{k}$-vector $\hat{G}(\vec{k})$ is larger than the group of neighbouring vectors. Conversely, $\vec{k}_o$ belongs to a continuous set of points, all having the same group $\hat{G}(\vec{k}_o)$.This set can be a volume (if $\vec{k}_o$ is at a general point of the BZ ,and if $\hat{G}(k_o)$ is reduced to the identity),or a portion of plane, or a portion of line (if $\vec{k}_o$ lyes on a symmetry element of the reciprocal lattice). This is also valid for all vectors of the star of $\vec{k}_o$ since their symmetries are conjugated to $\hat{G}(\vec{k}_o)$ with respect to the point group $\hat{G}_o$ of the HS phase.

If we consider two vectors $\vec{k}_o$ and $\vec{k}$ belonging to the same continuous range of BZ points, there is a one to one correspondance between the irreducible representations of G_o, which they partly specify. Let $\tau'(r)$ be the matrix of the point-group element R for a weighted representation of $\hat{G}(\vec{k}_o)$,the matrices of the small representations at $\vec{k}_o$ and $\vec{k}$,based on τ' ,are respectively:

$$\tau(\vec{k}_o|\{R|\vec{t}\}) = e^{-i\vec{k}_o.\vec{t}}.\tau'(R) \quad \text{and} \quad \tau(\vec{k}\,|\{R|\vec{t}\}) = e^{-i\vec{k}.\vec{t}}.\tau'(R) \qquad (2.16)$$

The irreducible representations of G_o, based on $[\tau'(R),\vec{k}_o]$ and $[\tau'(R), \vec{k}]$ have the same dimensions (same number of arms in the stars of $\vec{k}_o$ and $\vec{k}$ and same dimension for the small representations)

We can therefore identify the INC phase by a part of the symmetry properties of the primary OP. It will be specified i) by the manifold (line or portion of plane etc...) of BZ points possessing a common symmetry group $\hat{G}(\vec{k}_o)$. ii) by the nature of the weighted representation $\tau'(R)$ of $\hat{G}(\vec{k}_o)$. iii) by the nature of the non-zero components of the OP.

The drawback of such an identification is that, in certain cases, the C-phase and the INC phase will have in common the same preceding symmetry characteristics, while their physical properties are different. This difficulty is also inherent to the (3+d) crystallographic identification of the INC phase, and expresses the present lack of understanding of the symmetry charateristics of the T_L transition : in 3-dimensions this transition induces an increase of the translational symmetry on cooling; in (3+d) dimensions it can preserve the symmetry · in the OP space it corresponds to a lowering of the symmetry of the image of the space-group, as stated hereunder (§2.2.1).

2.1.3. Partial independence of the INC free-energy on the "angular" components (phases) of the OP.

a) Examples.

For n=2, the star of $\vec{k}_o$ will just contain ($\pm\vec{k}_o$). We can take the two components $\eta_\pm$ of the OP in correspondence with the two vectors ($\pm\vec{k}_o$). The radial and angular components are defined by $\eta_\pm = \rho e^{\pm i\theta}$. Any term in the Landau free-energy is a sum of monomials of the form $\eta_+^p . \eta_-^{(p'-p)} = \rho^p e^{i\theta(2p-p')}$. Consider the translation matrix $[\vec{a}_j]$. It does not interchange η_+ and η_-, and, consequently, it must leave each of the preceding monomials invariant. Its action on $\eta_\pm$ is :

$$[\vec{a}_j]\ \eta_\mp = e^{\pm i\vec{k}_o.\vec{a}_j}\ \eta_\mp = e^{\pm 2i\pi .\ q_j^o}\ \eta_\mp \qquad (2.17)$$

Thus, the invariance of a monomial imposes the condition :

$$e^{-2i\pi\, q_j^o\ [2p - p']} = 1 \qquad (2.18)$$

Due to the irrational character of q_j^o , this can only be achieved if p' = 2p. Hence, any monomial contributing to the free-energy is independent of the angular variable θ . The Landau free-energy has the simple following form :

$$F = \frac{\alpha}{2} . \rho^2 + \sum_{p=2}^{n} \frac{\beta_{2p}}{2^{2p}} . \rho^{2p} \qquad (2.19)$$

This independence on the phase angle θ is a consequence of the continuous character of the invariance group I_∞ of F . (In the case n = 2, I_∞ coincides with $\mathbb{C}_{\infty v}$ = O(2)). Indeed, since I_∞ is a continuous group, it necessarily contains rotations by an arbitrary angle ψ , acting in the plane (η_+, η_-) of the OP components. As θ is the angular coordinate in this plane, a rotation will change θ into the arbitrary angle ($\theta + \psi$). Thus θ must be absent from the expression of the free-energy F which is invariant by the ψ rotations.

For OP dimensions larger than 2, the same type of argument based on the occurence of a continuous set of rotations in I_∞

suggests that F will be independent from certain angular variables. However, the missing angles are not necessarily associated to pairs of wavevectors $\vec{k}_{\pm p}$,but ,more generally to combinations of the angular variables θ^{l}_{p} . Let us illustrate this property on the two examples already considered in § 2.1.2.b).

-In $BaMnF_4$ the star of $\vec{k}$ has 4 arms,and the small representation is one dimensional [1] . The standard procedure,described in chapter III, for constructing the matrices of the OP representation and the Landau free-energy (which will be again illustrated in §2.4) leads to the following fourth degree expansion:

$$F= \frac{\alpha}{2} (\rho_1^2+\rho_2^2) + \frac{\beta_1}{4}(\rho_1^4+\rho_2^4) + \frac{\beta_2}{2}\rho_1^2\rho_2^2 + \frac{\beta_3}{2}\rho_1^2\rho_2^2.\cos 2(\theta_1-\theta_2) \qquad (2.20)$$

where $\eta_{\pm 1}= \rho_1 e^{\pm i\theta_1}$,and $\eta_{\pm 2}= \rho_2 e^{\pm i\theta_2}$.
We can see that this truncated free-energy only depends on the difference $(\theta_1 -\theta_2)$. Any change of the values of the two angles by the same amount leaves F invariant. The geometrical interpretation of this angular independence is the following:the invariance group of the free-energy is a subgroup of O(4) (labelled $D_2 x D_\infty$) containing one continuous set of rotations in the 4-dimensional space of the OP. This set consists in simultaneous rotations,by the same arbitrary angle ψ , in the two hyperplanes $(\eta_1;\eta_{-1})$ and $(\eta_2;\eta_{-2})$. Thus the only angular combination allowed in the free energy is the difference $(\theta_1- \theta_2)$.

- In $2H\text{-}TaSe_2$, the star of $\vec{k}_1$ has 6 arms,and the small representation is one dimensional [4] . We define 3 radial and angular variables through $\eta_{\pm j} =\rho_j.e^{\pm i\theta_j}$. The associated free-energy is:

$$F = \frac{\alpha}{2} (\rho_1^2 + \rho_2^2 + \rho_3^2) + \frac{\beta}{3}\rho_1\rho_2\rho_3\cos [\theta_1+\theta_2+\theta_3]+ \frac{\gamma_1}{4}(\rho_1^4 + \rho_2^4 + \rho_3^4] + \frac{\gamma_2}{2}[\rho_1^2\rho_2^2 +\rho_1^2\rho_3^2 +\rho_2^2\rho_3^2] \qquad (2.21)$$

The 3 angular variables θ_j appear in F through the combination $(\theta_1 +\theta_2 +\theta_3)$. F is unchanged by any change of the θ_j which preserves this sum. The matrix associated to the translation $\vec{a}_1$ respectively changes θ_1, θ_2, θ_3 into $(\theta_1 - 2 \pi q)$, θ_2, and $(\theta_3 + 2\pi q)$. Likewise $\vec{a}_2$ changes θ_1, θ_2, θ_3 into θ_1,$(\theta_2-2\pi q)$,and $(\theta_3 + 2 \pi q)$. The geometrical operations acting in the OP space, and associated to the translations $m\vec{a}_i$ consist in simultaneous rotations, by angles ψ and $-\psi$, in two planes associated to 2 pairs of opposite wavevectors.One has two free angular variables.The minimization of F only determines the value of the sum $(\theta_1 +\theta_2 +\theta_3)$.

As will be shown below in §2.4, cases exist with $n > 2$, for which the free-energy is independent of all the angular variables θ_j. However as illustrated by the two preceding examples, the full independence on the θ_j is not general. Let us examine the relationship between the number of free-angular variables and the number of modulation directions defined in § 2.1.2.

b) Angular dependence of F and number of modulation directions.

As stressed in 2.1.3.a) terms in F are composed of translationally invariant monomials. Expressed as functions of the angular variables θ_p^l defined by $\eta_{\pm p}^l = \rho_{pl}.e^{\pm i\theta_p^l}$, a given monomial can be written, omitting the dependence on the moduli ρ_{pl} , as:

$$(\eta_p^l)^{m_{pl}}.(\eta_{-p}^l)^{m'_{pl}} \propto e^{i\Sigma(m_{pl}-m'_{pl}).\theta_p^l} \qquad (2.22)$$

Invariance of the former monomial by any of the primitive $\vec{a}_i$ translations leads to the set of conditions :

$$\vec{a}_i.(\sum_{p,l}(m_{pl} - m'_{pl}).\vec{k}_p) = 0 \quad (2\pi) \qquad (2.23)$$

which requires :

$$\sum_{p=0}^{s-1}(\sum_{l=1}^{r}(m_{pl} - m'_{pl}).\vec{k}_p) = \sum_{0}^{s} M_p.\vec{k}_p = \vec{G} \qquad (2.24)$$

Refering to (2 .8) and to the existence of d' rationally independent vectors in the star of $\vec{k}_o$, we can see that there are (s-d') independent conditions of the form (2.24) with the M_p not all equal to zero. There are, in addition, (r-1)s independent combinations of the form $(\theta_p^l - \theta_p^{l'})$ (same p, $l \neq l'$) which are clearly related to translationally invariant monomials. Each of these monomials will generate invariant terms in the free-energy (Cf.chap III § 4). F contains (rs - d') independent contributions with integer coefficients of the angular variables θ_p^l . Hence there are d' free angular variables, i.e. as many as the maximum number of modulation directions compatible with the symmetry of the OP.

2.2. Order-parameter and free-energy for the commensurate wave-vector $\vec{k}_L$

As mentioned in the introduction (§.1.1.b) the C-phase is described by a primary OP associated to an irreducible representation of the HS space-group G_o. Its representative vector $\vec{k}_L = (\Sigma\ q_j^L\ \vec{a}_j^*)$ has "simple" rational coordinates q_j^L .

Accordingly, the C-phase is invariant by a 3-dimensional translation group, the Fourier components of $\delta\rho(\vec{r})$ being of the form (2.12). On the other hand, the image of G_o in the OP space is finite as in the case of ordinary ST. Finally the free-energy has a different form than in the INC phase.

In this paragraph, we first state a distinction between 3 different situations in respect of the symmetry relationship between the OP at the wavevectors $\vec{k}_o$ and $\vec{k}_L$ (fig.4). In each of the relevant cases, we illustrate ,by examples, the existence of a "lock-in" term in the C-free-energy, i.e. of a term absent from the INC free-energy, and which depends on the angular variables θ_i . We then discuss,on more general grounds, the degree of lock-in terms. This discussion provides, on the one hand, a justification for the assignment of an INC character to the phase existing between T_I and T_L. On the other hand, it allows to work out systematically the possible commensurate wavevectors $\vec{k}_L$, on the basis of symmetry considerations.

2.2.1. Relationship between the OP at $\vec{k}_o$ and at $\vec{k}_L$.

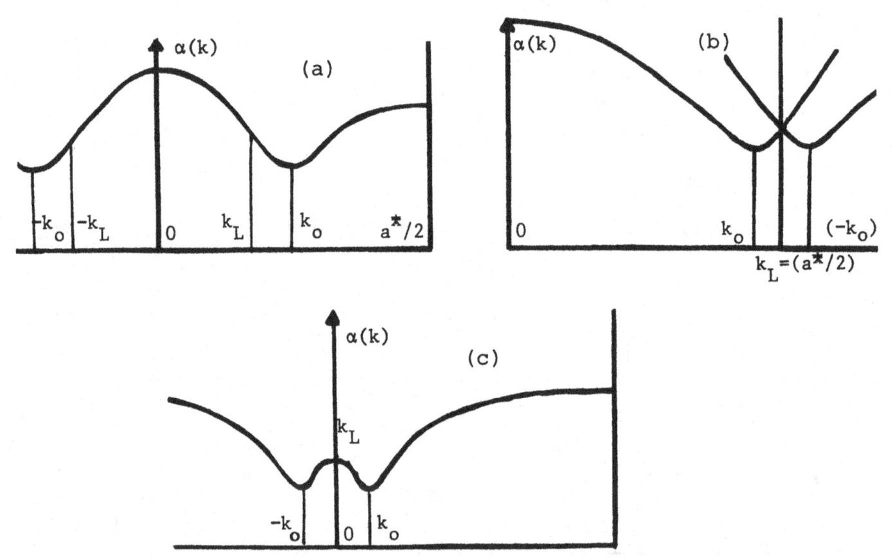

Fig 4. Three distinct situations for the locations of $\vec{k}_o$ and $\vec{k}_L$.

a) Same symmetry for $\vec{k}_o$ and $\vec{k}_L$.

This situation is realized,for instance, in Rb_2ZnCl_4 (Cf. § 1.1.c) and in 2H-$TaSe_2$ (Cf.§ 2.1.2.b).

In Rb_2ZnCl_4, $\vec{k}_o = q\vec{a}_3^*$, and $\vec{k}_L = (\vec{a}_3^*/3)$ are along the same line of HS points. They possess the same little group $\hat{G}(k) = \mathbf{C}_{2v}$. The C- and INC-order parameters are also associated to the same weighted representation $\tau'(R)$, and , accordingly, both order- parameters have two components $\eta_{\pm} = \rho e^{\pm i\theta}$. However, for the INC wavevector, the translation matrices $[m\vec{a}_3]$ constitute an infinite set, with diagonal elements $e^{\pm i\theta}$ (θ taking any value int the interval $[0,2\pi]$), while for $\vec{k}_L$ the set of distinct matrices $[m\vec{a}_3]$ is finite and generated by the diagonal matrix $e^{\pm 2mi.\pi/3}$. In the spaces of the C and INC order-parameters, the space-group G_o has respectively a finite image I_c isomorphous to C_{6v}, and an infinite image $I_\infty = C_{\infty v}$ (cf. 2.1.1). The C-free-energy, expanded up to degree 6 terms is :

$$F_L = \frac{\alpha}{2}\rho^2 + \frac{\beta_4}{4}\rho^4 + \frac{\beta_6}{6}\rho^6 + \frac{\beta'_6}{6}\rho^6\cos(6\theta) \qquad (2.25)$$

It contains one additional θ-dependent term as compared to the free-energy (2.19) of the same degree, relative to the INC order-parameter.

A similar, though more complex, case is $2H\text{-}TaSe_2$.
The vectors $\vec{k}_o = q\vec{a}_1^*$ and $\vec{k}_L = (\vec{a}_1^* / 3)$ have the same little group. The C- and INC order parameters have the same number of components, equal to the number of arms in the stars of $\vec{k}_o$ and $\vec{k}_L$. They also have in common the same "weighted" representation. However, while the INC star contains two rationally independent vectors (cf. 2.1.2.b), the C-star contains none. As compared to the INC-free-energy (2.21), the commensurate one contains additional terms which depend on the angular variables θ_i :

$$\begin{aligned} F_L = F_{INC} &+ \frac{\beta'}{3}[\rho_1^3\cos(3\theta_1) + \rho_2^3\cos(3\theta_2) + \rho_3^3\cos(3\theta_3)] \\ &+ \frac{\gamma'}{3}\rho_1\rho_2\rho_3[\rho_1\cos(2\theta_1-\theta_2-\theta_3)+\rho_2\cos(2\theta_2-\theta_1-\theta_3) \\ &+ \rho_3\cos(2\theta_3-\theta_1-\theta_2)] \end{aligned} \qquad (2.26)$$

where F_{INC} is expression (2.21).

b) $\hat{G}(\vec{k}_L) \supset \hat{G}(\vec{k}_o)$, and equal dimensions of the OP at $\vec{k}_L$ and $\vec{k}_o$.

An illustration of this second situation is provided by ammonium fluoberyllate $(NH_4)_2BeF_4$ [1]. The symmetry characteristics of the INC phase are identical to those discussed for Rb_2ZnCl_4 (Fig.2): The HS space-group is G_o=(Pmcn), and $\vec{k}_o = q.\vec{a}_3^*$. However, $\vec{k}_L = \frac{\vec{a}_3^*}{2}$ has a little group $\hat{G}(\vec{k}_L) = (D_{2h})$, which is larger than the one of $\vec{k}_o(C_{2v})$. The location of $\vec{k}_L$, on the surface of the BZ, corresponds to an end point on the symmetry line containing $\vec{k}_o$. Though the star of $\vec{k}_L$ has a single arm ($\vec{k}_L = -\vec{k}_L$), the C and INC order-parameters have the same dimension (n=2). This is due to the fact that the small representation at $\vec{k}_L$ is 2-dimensional. The symmetries of the two OP are "compatible" : the two components ($\eta_\pm$) of the INC order-parameter tend towards the two distinct components of the OP representation at $\vec{k}_L$, when $\vec{k} \to \vec{k}_L$. The image $I_c = (\mathbb{C}_{4v})$ of G_o in the space of the commensurate OP is a finite subgroup of the image $I_\infty = (\mathbb{C}_{\infty v})$ relative to the INC OP. The C-free-energy can be written, up to the fourth degree, as :

$$F_L = F_{INC} + \frac{\beta'}{4} \rho^4 . \cos (4\theta) \qquad (2.27)$$

where F_{INC} is equal to (2.19) and where $\eta_\pm = \rho . e^{\pm i\theta}$.

c) $\hat{G}(\vec{k}_L) \supset \hat{G}(\vec{k}_o)$,and smaller dimension of the OP at $\vec{k}_L$.

This last situation occurs, for instance, in Thiourea $SC(NH_2)_2$, [1] or in a more complex way, in quartz SiO_2 [1] , and biphenyl [1].

Thiourea has the same HS space-group Pmcn as those considered for Rb_2ZnCl_4 and $(NH_4)_2BeF_4$ (§. 2.2.a and 2.2.b), and an INC wavevector $\vec{k}_o = q\vec{a}_1^*$, with the same little point group $\hat{G}(\vec{k}_o) = \mathbb{C}_{2v}$ as in the two former examples. The little group $\hat{G}(\vec{k}_L) = \mathbb{D}_{2h}$ of the C-wavevector $\vec{k}_L = 0$ is ,similarly to the case examined in the preceding paragraph b), of higher-symmetry than $G(k_o)$ However ,the C- and INC order-parameter dimensions are different , respectively one, and two. This is due to the fact that when $\vec{k} \to \vec{k}_L$, the two INC OP components do not tend towards components of an irreducible representation at $\vec{k}_L = 0$, like in case b).When $\vec{k} \to \vec{k}_L$ they span a two-dimensional reducible representation which contains the one-dimensional representation relative to the OP of the C-phase.

Similarly, the dimension of the OP decreases from 6 in the INC phase of quartz to 1 in its C-phase. In biphenyl this decrease is from 4 to 2.

Cases a) and b) constitute the standard symmetry scheme for INC systems. Case c) is non-standard and will be further discussed in §5. The selected examples indicate that the standard scheme involves three main features:

i) same number of OP components in the INC-and C-phase, and "compatibility" of the OP symmetries.

ii) Lower symmetry for the image I_L of G_o ($I_L \subset I_\infty$).

iii) Presence of additional terms, functions of the angular variables θ_p^l , in the free-energy of the C-phase. These terms have the effect of leaving no "free-angular-variables". They are the "lock-in" terms mentioned in § 1.2 b).

2.2.2. Degrees of the lock-in terms.

We assume that $\vec{k}_L$ belongs to the same symmetry manifold as $\vec{k}_o$,defined by $\hat{G}(k_o)$. It corresponds either to a general point of this manifold (as in § 2.1.a), or to a point of its boundary (as in examples 2.1.b). For an OP with dimension n = 2sr (s pairs of arms in the star of $\vec{k}$ and an r-dimensional small representation), we have seen in § 2.1.5.a), that (sr - d') independent combinations of the θ_p^l angles are present in the INC free-energy. The occurence of d' free-angular variables is related to the fact that the star of $\vec{k}_o$ contains d' rationally independent vectors. When $\vec{k} = \vec{k}_L$, all the components of the vectors of the star of $\vec{k}_L$ are rational numbers, and any such vector fulfills the condition :

$$m_p \vec{k}_L^p = \vec{G}_p \tag{2.28}$$

Thus, there is no rationally independent vector in the star of $\vec{k}_L$. The argument §2.1.3 shows that there are <u>no free-angular variables</u>. Let us examine the conditions which determine the degrees of the lock-in terms, i.e. of the terms which fix, in the C-phase, the values of the "free-angular variables" found in the INC phase.

Note, first, that all the independent angular combinations which are present in the INC free-energy are also present in the C-free-energy. This is due to the fact that $\vec{k}_L$ belongs to the same symmetry manifold (plane or line...) as $\vec{k}_o$. Accordingly, the star of $\vec{k}_L$ will comply with the (s-d') conditions (2.24) . Likewise, the angular differences $(\theta_p^l - \theta_p^{l'})$ will be associated to translationnally invariant monomials, in accordance with §.2.1.3.

On the other hand, new terms will arise in the C-free-energy related to the fulfillment by the star of $\vec{k}_L$ of <u>d'</u> additional conditions of the type (2.28), not satisfied by the star of $\vec{k}_o$.

More precisely, there will be <u>d'</u> conditions, independent from the (sr - d') conditions discussed in § 2.1.3 , which can be expressed in the form :

$$\sum_p m_p . \vec{k}_L^p = \vec{G} \tag{2.29}$$

According to equations (2.22 - 2.24), each such condition is related to a set of translationally invariant monomials of a certain degree m . The degree m of these monomials obeys the inequality ($m \geqslant \Sigma_p |m_p|$) . The degrees of the terms in the free-energy which are generated by these monomials (i.e. the <u>lock-in terms of F_L</u>) are equal to, or greater than m.

Let us examine the relationship between the sums ($\Sigma_p|m_p|$)and the values of the denominators in the rational components of the commensurate wavevectors $\vec{k}_L^p$.

For instance, in the case of Rb_2ZnCl_4 (cf. Fig. 2 and §2.2.1.a), for which there is a single pair of arms in the star of $\vec{k}_o(\pm\vec{k}_o=\pm q\vec{a}^*)$, one has $k_L=(\vec{a}_3^*/3)$. Eq (2.29) reduces to (2.28), and the minimum value of m_p is 3. Consistently, eq (2.25), shows that the lock-in term is of degree 6. Likewise in $\{N(CH_4)_2\}ZnCl_4$ [1] , a structural isomorph of Rb_2ZnCl_4, there is a C-phase associated to $\vec{k}_L = \frac{2}{5} \cdot \vec{a}_3^*$. Here, $m_p = 5$, while the lock-in term of lowest degree can be shown to be the tenth degree term $\rho^{10} . \cos(10\theta)$.

In these two examples the degree of the lock-in term of lowest degree is larger than the value of the denominators in the rational components of $\vec{k}_L$. Clearly, on the basis of eq (2.28), we can see that whenever the star of $\vec{k}_L$ is reduced to $(\pm \vec{k}_L)$, the degree of the lock-in term will be at least equal to the largest denominator of the rational components of $\vec{k}_L$ (in the reduced form of these rational numbers).

In these cases we can conclude that "high order commensurate vectors", i.e. of $\vec{k}_L$ vectors having rational components with large denominators, will be associated to lock-in terms of high-degree.

Such a result does not hold necessarily if the star of $\vec{k}_L$ contains several pairs of vectors. As shown in § 2.2.4. hereunder, in this case, a high- order commensurate vector can satisfy eq.(2.29) with small m_p^i numbers. Accordingly, the degree of the lock-in term can be smaller than the denominators involved in the components of the $\vec{k}_L{}^p$ vectors (for instance this degree can be 4, while the largest denominators are 7 or 13).

However, considering the form of eq (2.29) we can expect that the preceding situation will only occur for exceptional sets of rational components in $\vec{k}_L$ Generally speaking, "high-order commensurate" vectors will give rise to lock-in terms of high-degree as in the examples considered above.

2.2.3. Similarity of INC phases and "high-order" commensurate phases.

The relationship stated in the preceding paragraph, allows to clarify certain ambiguities existing in the definition of an INC phase given in § 1.1 : i) As pointed out in § 1.1.a), in the temperature range assigned to the INC phase, the temperature dependent components of $\vec{k}$ take rational as well as irrational values.
However, in general, the rational values will not possess a small denominator and the corresponding lock-in terms will be of a high-degree and will not influence the properties of the system. The free-energy truncated to the relevant terms of low degree will have an identical form for the commensurate and the INC q-values taken as a function of temperature. The single phase character of the INC phase is therefore justified, besides the symmetry arguments stated in 2.1. 2.e), by the absence of a lock-in term of "low' degree, and the subsequent occurence of "free-angular variables" in the whole temperature range between T_I and T_L.

ii) The same argument allows to define the experimental accuracy needed to check the irrationality of the components of $\vec{k}$: an INC vector, or a "high-order" commensurate one, will determine the same truncated free-energy, and consequently the same physical properties characteristic of INC systems. In order to assert the INC character of $\vec{k}$, it is sufficient to distinguish $\vec{k}$ from a "low-order" commensurate vector associated to a lock-in term of low degree.

iii) Finally, this argument provides a physical basis for classifying the commensurate vectors according to the "simplicity" of their rational coordinates. $\vec{k}_L$ will be considered to possess "simple" rational coordinates if, for $\vec{k} = \vec{k}_L$, the free-energy contains lock-in terms of low degree (e.g. $\leqslant 6$).

2.2.4. Enumeration of possible lock-in wavevectors on the basis of symmetry considerations.

The preceding paragraph has emphasized the fact that few commensurate vectors $\vec{k}_L$ will determine low-degree lock-in terms in the C-free-energy. If we focus our attention on lock-in terms of a certain degree m (e.g. 3,4 or 6), the relevant C-vectors $\vec{k}_L^p$ must satisfy as necessary conditions equation (2.29), with $\Sigma|m_p| \leqslant m$. More precisely, fourth degree lock-in terms will be generated from fourth degree invariant monomials obeying the condition $\Sigma|m_p| = 4$. Noting that the $\vec{k}_L^p$ vectors in eq (2.29) belong to the star of $\vec{k}_L$, and are therefore related to $\vec{k}_L$ by symmetry operations, we can infer that an enumeration of possible lock-in vectors $\vec{k}_L$ compatible with a given degree of the lock-in terms can be performed by use of geometrical considerations.

Let $\pm\vec{k}_L^p$ be the set of vectors in the star of $\vec{k}_L$, with $\vec{k}_L^p = \hat{g}_p\vec{k}_L$. If we label m_p and m'_p the respective coefficients of the vectors $(+k_L^p)$ and $(-k_L^p)$ in eq.(2.29),we obtain:

$$\sum_p (m_p - m'_p)\, \hat{g}_p\, \vec{k}_L = \Sigma\, n_i . \vec{a}_i^* \tag{2.30}$$

m_p and m'_p being integers complying with the conditions: i) $m_p \geqslant 0; m'_p \geqslant 0$. ii)if m_p or m'_p is different from zero $m_p \neq m'_p$. iii) $\Sigma(m_p+m'_p) = m$.

Let us illustrate such an enumeration for m = 4 (fourth degree lock-in terms) in the case of an INC system whose HS phase possesses a simple tetragonal lattice P.

Let $\vec{k}_L = (k_x, k_y, k_z)$ be a general vector in the Brillouin zone. The projection of eq (2.30) on the k_z axis will determine the possible values of k_z independently from those of k_x and k_y, for each set of acceptable (m_p, m'_p) values. Hence we can restrict, in the first place, to the plane (k_x, k_y). In this plane, the projection of the star of $\vec{k}_L$ is generated by the elements of the point-group C_{4v}^z. Half of these elements transform (k_x, k_y) into $(-k_x, -k_y)$ and can be disregarded since the transformation $\vec{k} \to -\vec{k}$ is already considered in eq.(2.30).We retain the elements $(E, C_4^z, \sigma_x, \sigma_{xy})$, and note that, in eq (2.30) we can write $M_p = (m_p - m'_p)$. The projections of (2.30) on $\vec{a}_1^*$ and $\vec{a}_2^*$ yield :

$$\begin{aligned} (M_1 - M_3)k_x + (M_4 - M_2)k_y &= n_1\, |a_1^*| \\ (M_2 + M_4)k_x + (M_1 + M_3)k_y &= n_2\, |a_2^*| = n_2 |\vec{a}_1^*| \end{aligned} \tag{2.31}$$

The M_p have to comply with conditions derived from (2.31) with m = 4. The enumeration of the possible values can be restricted to the numbers m_p and m'_p satisfying $m_1 > m'_1$, and $m_1 > m_p$ since the other sets represent solutions which are equivalent by symmetry : they correspond to the substitution of $\vec{k}_L$ by another vector of its star. Table 2 shows the 40 unequivalent sets of M_p numbers complying with conditions (2.31)

Table 2

N°	M_1	M_2	M_3	M_4	N°	M_1	M_2	M_3	M_4	N°	M_1	M_2	M_3	M_4
1	4				9,10	2	±2			19,20	1			±1
2	2				11,12	2		±2		21-24	2	±1	±1	
3,4	3	±1			13,14	2			±2	25-28	2	±1		±1
5,6	3		±1		15,16	1	±1			29-32	2		±1	±1
7,8	3			±1	17,18	1		±1		33-40	1	1	±1	±1

Examine, for instance, set 3 in table 2 [$M_1 = 3$; $M_2 = +1$; $M_3 = M_4 = 0$]. Eq (2.31) yields, up to a reciprocal lattice vector,

$$k_x = \frac{n}{10} |a_1^*| \quad ; \quad k_y = \frac{3n}{10} |a_1^*| \qquad (2\text{-}32)$$

with n>0, an integer . For this $\vec{k}_L$ vector, one has a translationally invariant monomial of 4th degree of the form $(\eta_1)^3 . (\eta_2)$. Thus , in agreement with the property stated in §2.2.2, we can see that a fourth degree term, function of the θ_p^l , can arise for a $\vec{k}_L$ vector having components with a "large" denominator (e.g. $k_x = \frac{1}{10}$; $k_y = \frac{3}{10}$)

Solving (2 .31) for each combination of M_p numbers in table 2, one finds two types of solutions (k_x, k_y) :

i) isolated points with rational coordinates, having as denominators the numbers 1, 2, 4, 5, 8, 10. We have represented these points on fig.5.

ii) continuous manifolds of points, e.g. $k_x = k_y$.

Category i) represents possible lock-in wavevectors k_L, since at these wavevectors the free-energy can contain θ_p^l -dependent fourth degree terms which are absent from the free- energy associated to neighbouring $\vec{k}$-vectors.

By contrast, category ii) corresponds to fourth degree terms which cannot be considered as lock-in terms, though they depend on the θ_p^l angles. Indeed these terms exist for a continuous set of $\vec{k}$ vectors having rational or irrational components. They are of the same nature as the ones encountered in the examples of $BaMnF_4$ and 2H-Ta-Se_2 (eqs. 2.20 and 2.21).

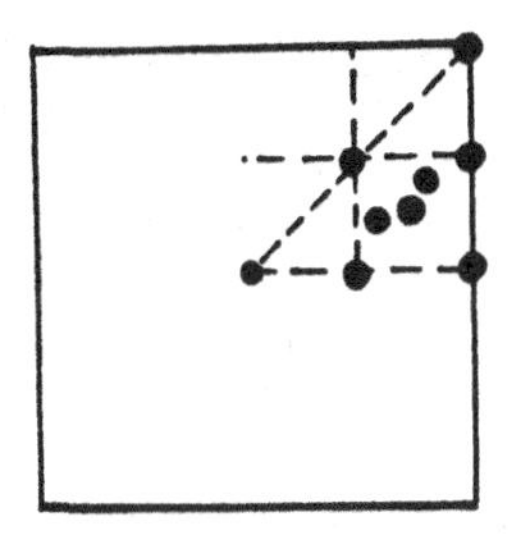

Fig.5. Possible "lock-in points" determined by 4th degree terms in the Landau free-energy.Dashed lines represent continuous sets of such points.

Along k_z, equation (2.30) also provides discrete points at $k_z = n.|\vec{a}_3^*|/4$, or continuous sets of points. Some of these sets also correspond to lines in the (k_x, k_y) plane, thus generating planes. A given θ_p^l -dependent term is allowed to occur in the free-energy, for all the points of each plane.

Similar "maps" of possible lock-in points can be worked out in other types of Brillouin zones and for lock-in terms of different degrees [6,7]. For instance, in simple cubic lattices [6] one finds possible lock-in points associated to fourth degree terms, at $na^*(\frac{1}{13}, \frac{3}{13}, \frac{9}{13})$, $na^*(\frac{2}{13}\ \frac{6}{13}\ \frac{15}{13})$, and $na^*(\frac{1}{7}, \frac{2}{7}, -\frac{3}{7})$. In addition, one also finds, in the same way as in the above example, a continuous set of points, forming a plane $[k_x + k_y + k_z] = 0$, and associated with a third degree-term.

2.3. Free-energy for a modulated order-parameter ; Lifshitz invariant

As explained in §.1.2.2., the second phenomenological scheme used to account for the properties of INC systems implies that at any temperature below T_I (i.e. within both the INC phase and the C-phase) one can use a free-energy associated to an irreducible OP with wavevector $\vec{k}_L$ (the commensurate wavevector) provided the $\eta_{p,L}^l$ components of the OP are functions of the spatial coordinates. Let us examine the justification of this method, and derive the form of the corresponding free energy.

2.3.1. Symmetry properties of the modulated order-parameter.

We assume that the primary density increment $\delta\rho_1(\vec{r})$ which describes the structural distortion occuring in the INC phase can be written in the form :

$$\delta\rho_1(\vec{r}) = \sum_{p,l} \eta_{p,L}^l \cdot \phi_{p,L}^l(\vec{r}) \qquad (2.33)$$

where the $\phi_{p,L}^l(\vec{r})$ transform according to the same irreducible representation as the OP of the commensurate phase. The sum is over the arms $\vec{k}_L^p$ of the star of the C-wavevector $\vec{k}_L$, and over the basis l of the small representation. On the other hand, $\delta\rho_1(\vec{r})$ can be expressed as a similar sum in terms of the actual INC order-parameter whose wavevector at the considered temperature is $\vec{k}$:

$$\delta\rho_1(\vec{r}) = \sum_{p',l'} \eta_{p'}^{l'} \cdot \phi_{p'}^{l'}(\vec{r}) \quad , \tag{2.34}$$

the $\eta_{p'}^{l'}$ being components independent of the spatial coordinates. Comparison of (2.33) and (2.34) shows that the $\eta_{p,L}^{l}$ are functions of $\vec{r}$.

The basic assumption of the model is that these components are slowly varying function of the spatial coordinates at the scale of the crystal's periods $\vec{a}_i$. This implies $|\vec{k}_o-\vec{k}_L| \ll |\vec{a}_i^*|$. In this framework one can consider that the $\eta_{p,L}^{l}(\vec{r})$ are approximately uniform in a volume of size intermediate between $|\vec{a}_i|$ and $(1/|\vec{k}_o-\vec{k}_L|)$. In such a volume, it is justified to make use of the standard "trick" of Landau's theory (chapII §3), and interchange the symmetry properties of the $\eta_{p,L}^{l}(\vec{r})$ and of the $\phi_{p,L}^{l}(\vec{r})$; i.e., we assume that the latter functions do not change under the symmetry operations of G_o, while the $\eta_{p,L}^{l}(\vec{r})$ transform according to the same representation as the commensurate OP (However one must consider in G_o the sole translations $[m\vec{a}_i]$ which are smaller than the volume $d\Omega$ in which the $\eta_{p,L}^{l}(\vec{r})$ are almost uniform). The $\eta_{p,L}^{l}(\vec{r})$ constitute a set of modulated OP components whose local values span an irreducible representation of G_o. We are then in a position to construct a local free-energy density $f(\vec{r})$ as an invariant function of the $\eta_{p,L}^{l}(\vec{r})$, the latter quantities being spatially smooth quantities freed from the small scale variations due to the crystal structure . These small scale variations are entirely rejected, as usual,(Chapter II) in the $\phi(\vec{r})$ functions.

The density $f(\vec{r})$ has a more complicated expression than the free-energies encountered up to now. It has to be expanded as a function of the OP components, as in the usual situation, but also as a function of the derivatives of these components. Let $f_1(\vec{r})$ be the part of $f(\vec{r})$ which only contains the $\eta_{p,L}^{l}(\vec{r})$. Since $f_1(\vec{r})$ is a sum of invariant polynomials whose form is determined by the symmetry of the $\eta_{p,L}^{l}$, the expression of $f_1(\vec{r})$ as a function of the $\eta_{p,L}^{l}$ is identical to that of the C-free-energy (§2.2). We denote $f_2(\vec{r})$ the remaining terms of $f(\vec{r})$ which depend both of the $\eta_{p,L}^{l}$ and of their spatial derivatives. The free-energy F will be an integral of $f(\vec{r})$ over the volume Ω of the system :

$$F = \int_\Omega [f_1(\vec{r}) + f_2(\vec{r})] \cdot d\Omega \tag{2.35}$$

and thus a functional of the $\eta_{p,L}^{l}(\vec{r})$

2.3.2. Contribution of the OP derivatives to $f(\vec{r})$. Lifschitz invariant.

The assumed slowness of the spatial variations of the OP

components ensures that, like the OP components, their derivatives are small quantities. It is therefore justified to expand $f_2(r)$ as a sum of polynomials of the $\eta^{l}_{p,L}(\vec{r})$ and of their derivatives of lowest order. Taking only into account the first derivatives and expanding up to second degree-terms in the $\eta^{l}_{p,L}(r)$, we have :

$$f_2(r) = \sum_{ll'} a_{ll'} \frac{\partial\eta_l}{\partial x_{l'}} + \sum_{kll'} b_{kll'}\eta_k \frac{\partial\eta_l}{\partial x_{l'}} + \sum_{kk'll'} c_{kk'll'} \frac{\partial\eta_k}{\partial x_{k'}} \cdot \frac{\partial\eta_l}{\partial x_{l'}} \qquad (2.36)$$

with the simplified notation $\eta_m = \eta^{l}_{p,L}$. Certain of the terms of this expansion can be neglected. They only contribute to the total free-energy F as surface terms, and do not fulfill the requirement of extensivity of this function. They will play no role at the limit of a large system. This is the case of the terms in (2.36) which are total derivatives. For instance,

$$\Sigma a_{ll'} \int_\Omega \frac{\partial\eta_l}{\partial x_{l'}} . d\Omega = \Sigma a_{ll'} [\Delta\eta] . \int d^2\vec{r} \qquad (2.37)$$

where $[\Delta\eta]$ is the variation of η_l across the system along l'.

Likewise, part of the second term in the second member of (2.36) is a total derivative :

$$\Sigma \left(\frac{b_{kll'} + b_{lkl'}}{2}\right)\left(\eta_k \frac{\partial\eta_l}{x_{l'}} + \eta_l \frac{\partial\eta_k}{x_{l'}}\right) = \Sigma \left(\frac{b_{kll'} + b_{lkl'}}{2}\right) \frac{\partial(\eta_k\eta_l)}{\partial x_{l'}} \qquad (2.38)$$

Thus, we only have to keep in (2.36) the remaining extensive terms. Denoting $B_{kll'} = (b_{kll'} - b_{lkl'})/2$, we obtain :

$$f_2(\vec{r}) = \sum_{k<l;l'} B_{kll'}\left(\eta_k \frac{\partial\eta_l}{x_{l'}} - \eta_l \frac{\partial\eta_k}{x_{l'}}\right) + \sum_{kk'll'} c_{kk'll'} \frac{\partial\eta_k}{\partial x_{k'}} \frac{\partial\eta_l}{\partial x_{l'}} \qquad (2.39)$$

The precise form of $f_2(\vec{r})$ is determined by the conditions of reality, and of invariance of $f_2(\vec{r})$ under the transformations of G_o.

a) Symmetry conditions imposed to the terms in $f_2(\vec{r})$.

Depending on the symmetry properties of the OP, the term in (2.39) with coefficient $B_{kll'}$ does not necessarily exist. By contrast, the other term, a quadratic polynomial of the OP derivatives, exists for any OP symmetry.

Let us first note that the symmetry conditions forbidding the presence of the first term are expressed by the Lifshitz criterion considered in chap III §3. Indeed, it is clear from (2.39) that the symmetry properties of this term are described by the tensorial product of the antisymmetized square $\{\eta_k^2\}$ of the OP representation (spanned by $\{(\eta_k,\eta_l)-(\eta_l,\eta_k)\}$), and of the vector representation V (spanned by the coordinates x_l,). For such terms to be invariant, this product must contain the totally symmetric representation Γ_1 of G_o. The expression of this condition,

$$\{\eta_m\}^2 \times V \supset \Gamma_1 \tag{2.40}$$

is the converse of the Lifschitz condition. For this reason the considered term in (2.39), is called the <u>Lifshitz invariant</u>. In order to understand the relevance of the Lifschitz criterion, which was derived in another context (chap III §3), to the expression of $f_2(\vec{r})$, it is worth recalling its initial meaning.

Let $\alpha(\vec{k}_L+\vec{q})$ be the coefficient of the quadratic term in the usual Landau free-energy, for an OP with wavevector $(\vec{k}_L+\vec{q})$ close to $\vec{k}_L$. The Lifshitz criterion expresses the absence of a <u>linear term</u> in the $\vec{q}$ expansion of $\alpha(\vec{k}_L+\vec{q})$ in the vicinity of k_L. Let us, on the other hand, define $\eta'_m(\vec{q})$, as the Fourier transform of the modulated OP component $\eta_m(\vec{r})$:

$$\eta_m(\vec{r}) = \int e^{+i\vec{q}.\vec{r}} \cdot \eta'_m(\vec{q}).d^3\vec{q} \tag{2.41}$$

The components $\eta_m(\vec{r})$ being slowly modulated, $\eta'_m(\vec{q})$ is only significant for small $|\vec{q}|$ values ($|\vec{q}| << |\vec{a}_i^*|$). We can replace the $\eta_m(\vec{r})$ by their Fourier transform in (2.35). The free-energy is then an integral over $\vec{q}$ of a free-energy density in $\vec{q}$ space. This density can be interpreted as a usual Landau free-energy associated to an OP with wavevector $(\vec{k}_L+\vec{q})$. Clearly the Fourier transform of the Lifshitz invariant is a contribution to the quadratic term of the free-energy density in q-space :

$$\int_\Omega \left(\eta_k \frac{\partial \eta_l}{\partial x_{l'}} - \eta_l \frac{\partial \eta_k}{\partial x_{l'}} \right).d\Omega = \int_{\vec{q}} q_{l'} \eta'_k(\vec{q}) \cdot \eta'_l(-\vec{q}).d^3\vec{q} \tag{2.42}$$

Its coefficient, proportional to $q_{l'}$, contains a term <u>linear in the components of $\vec{q}$</u> thus establishing the required correspondance.

As shown in chapter III (§ 3), the Lifshitz criterion is only satisfied for k_L corresponding to certain HS points of the Brillouin-zone. Conversely one or several independent <u>Lifschitz invariants exist for any OP symmetry except for certain OP symmetries corresponding to HS points of the Brillouin zone</u>.

Let us now examine the second term in eq. (2.39). Its symmetry properties are described by the symmetrized square of the representation $(\eta_m) \times V$. As this representation is real, its symmetrized square contains at least once the totally symmetric representation Γ_1. Consequently $f_2(\vec{r})$ will always contain at least one term depending quadratically of the derivatives of the OP components.

b) Number and form of the Lifshitz invariants (LI).

When condition (2.40) is satisfied, the most general form of the LI compatible with the considered OP is :

$$\sum_{\nu} B_{\nu} . L_{\nu} \tag{2.43}$$

where each L_{ν} is an independent LI, separately invariant by the operations of G_o. The number n_{ν} of independent LI equals the number of times Γ_1 is contained in the product $\{\eta_m\}^2 \times V$. Let us examine the construction of the L_{ν}.

We first note that each antisymmetric term in (2.39) must be translationally invariant. As the coordinates $x_{l'}$ fulfill this condition, each term can be written in the form :

$$\eta_p^k . \frac{\partial \eta_{-p}^l}{\partial x_{l'}} - \eta_{-p}^l . \frac{\partial \eta_p^k}{\partial x_{l'}} = \{\eta_p^k ; \eta_{-p}^l\}_{x_{l'}} \tag{2.44}$$

where we have returned to a fully explicit notation for the OP components, with (± p) representing the arms ($\pm \vec{k}_L^p$) of the star of $\vec{k}_L$. Eq.(2.44) defines, on the other hand a contracted notation for the antisymmetric combinations involved in the LI. Consider the antisymmetric products of ordered pairs of OP components :

$$(\eta_p^k;\eta_{-p}^l) - (\eta_{-p}^l;\eta_p^k) \qquad (p=1,s;k,l=1,r) \tag{2.45}$$

They span the translationally invariant subspace $\mathcal{E}^{(A)}$ of $\{\eta_m\}$, and their set is a basis for a representation $\Gamma^{(A)}$ of the point-group $\hat{G}_o$. If the small representation τ is r-dimensional, and if the star of $\vec{k}_L$ has s pairs of arms ($\vec{k}_L^p$) ,the representation $\Gamma^{(A)}$ has the dimension $r^2.s$ .Clearly,any term (2.44) belongs to the representation $\Gamma^{(A)}xV$ of $\hat{G}_o$. The L_{ν} invariants correspond to the various totally symmetric subspaces contained in $\mathcal{E}^{(A)}xV$.Thus,the number of L_{ν} is:

$$n_{\nu} = \frac{1}{r_o} (\sum_{\hat{G}_o} \chi^{(A)}(\hat{g}) . \chi^V(\hat{g})) \tag{2.46}$$

where $\chi^{(A)}(\hat{g})$ and $\chi^V(\hat{g})$ are respectively the characters of the representation $\Gamma^{(A)}$ and V for the point-group operation $\hat{g}$ of $\hat{G}_o$, and r_o the order of this group.

The effective calculation of n_{ν} does not necessarily require consideration of all the elements of G_o .It can be restricted to a smaller group . Indeed, consider the expressions (2.45) corresponding to a given index p (i.e. to a given pair of vectors $\vec{k}_L^p$. They form a basis for a subspace $\mathcal{E}_p^{(A)}$ of $\mathcal{E}^{(A)}$,and we can write :

$$\mathcal{E}^{(A)} = \sum_p \mathcal{E}_p^{(A)} \tag{2.47}$$

where all the spaces $\mathcal{E}_p^{(A)}$ have the same dimension since the combinations (2.45) associated to distinct pairs ($\pm\vec{k}_L^p$) are related by symmetry operations of $\hat{G}_o$. Let $\chi_p^{(A)}(\hat{g})$ be the contribution of the space $\mathcal{E}_p^{(A)}$ to $\chi^{(A)}(\hat{g})$ in eq.(2.46) . Due to the symmetry relationship between the $\mathcal{E}_p^{(A)}$,we can write (2.46) under the form:

$$n_{\nu} = \frac{s}{r_o} (\sum_{\hat{G}_o} \chi_p^{(A)}(\hat{g}) . \chi^V(\hat{g})) \tag{2.48}$$

Among the $\hat{g}$ operations, the only ones determining non-zero values for $\chi_p^{(A)}(\hat{g})$, are the operations of $\hat{G}_o$ which preserve the set $(\pm \vec{k}_L{}^p)$. The sum (2.48) can thus be restricted to these operations which form a subgroup $\hat{G}'_p$ of $\hat{G}_o$ of order (r_o/s). We can write (2.48) for p = 1 [i.e. $\vec{k}_L{}^p=\vec{k}_L$)

$$\eta_\nu = \frac{1}{(r_o/s)} \left(\Sigma_{\hat{G}'_1} \chi_1^{(A)}(\hat{g}) \cdot \chi^V(\hat{g}) \right) \qquad (2.49)$$

Note that $\mathcal{E}_p^{(A)}$ is invariant by $\hat{G}'_p$ and is a basis for a representation $\Gamma_p^{(A)}$ of this group. Equation (2.49) expresses that η_ν is equal to the number of times the totally symmetric representation of $\hat{G}'_1$ is contained in the product $\Gamma_1^{(A)} \times V'$, where V' is the vector representation of $\hat{G}'_1$. The group $\hat{G}'_1$ coincides with the point-group $\hat{G}(k_L)$ of $\vec{k}_L$, if $(-\vec{k}_L)$ does not belong to the mathematically-irreducible star of $\vec{k}_L$ (we have stressed in chap.III that this vector always belongs to the physically irreducible star of $\vec{k}_L$). On the other hand, if $(-\vec{k}_L)=\hat{g}'.\vec{k}_L$,with $\hat{g}' \in \hat{G}_o$,then $\hat{G}'_1=(\hat{G}(\vec{k}_L)+\hat{g}'.\hat{G}(\vec{k}_L))$.

Example 1 : Consider the case of Rb_2ZnCl_4 [G_o = Pmcn ; $\vec{k}_L{}^p = \pm\vec{k}_L = \frac{\vec{a}^*}{3}$]. The small representation τ is one dimensional (§2.2.a). We have a single term of the type (2.45) ,$(\eta_1;\eta_{-1})-(\eta_{-1};\eta_1)$.On the other hand, the set of matrices generating the OP representation are (Cf.§ 2.4.1):

$$\begin{array}{ccccc}
E,\sigma_x,\vec{a}_1,\vec{a}_2 & U_z,-\sigma_y & I,-U_x & \sigma_z,-U_y & \vec{a}_3 \\
\hline
\begin{vmatrix}1&0\\0&1\end{vmatrix} & \begin{vmatrix}\xi&0\\0&\bar{\xi}\end{vmatrix} & \begin{vmatrix}0&1\\1&0\end{vmatrix} & \begin{vmatrix}0&\xi\\\bar{\xi}&0\end{vmatrix} & \begin{vmatrix}\xi^2&0\\0&\bar{\xi}^2\end{vmatrix}
\end{array} \qquad (2.50)$$

where $\xi = e^{-i\pi/3}$. From (2.50) it is straighforward to deduce that $(\eta_1;\eta_{-1})-(\eta_{-1};\eta_1)$ transforms into itself by the operations $(E, U_z, \sigma_x, \sigma_y)$ and into its opposite by the set (U_x,U_y,σ_z, I). Thus , it transforms according to the same irreducible representation of $\hat{G}_o$ = mmm as the z-component of the vector representation. Consequently the LI will be of the form :

$$L_1 \propto \{\eta_1;\eta_{-1}\}_z \qquad (2.51)$$

Note that the set $(\eta_1 ; \eta_{-1})$ corresponds to a complex basis of the OP representation, as shown by the form of the matrices (2.50). Transforming into the real basis

$$\eta = \frac{1}{2}(\eta_1+\eta_{-1}) \qquad ; \; \zeta = \frac{i}{2}(\eta_1-\eta_{-1}) \qquad (2.52)$$

preserves the functional form of $L_1\{\eta;\zeta\}_z$. Alternately, we can make the change of variables $\eta_{\pm 1} = \rho e^{\pm i\theta}$ (cf.§2.1 and 2.2.a). We obtain :

$$L_1 \propto \rho^2 \frac{\partial\theta}{\partial z} \qquad (2.53)$$

Example 2 : Consider the case realized in Barium sodium niobate [1] $Ba_2NaNb_5O_{15}$, for which G_o = P4bm ; $\pm\vec{k}_L = \pm\{[\frac{\vec{a}^*_1 + \vec{a}^*_2}{4}] + \frac{\vec{c}^*}{2}\}$; and $\pm\vec{k}_L^{\,2} = \pm\{[\frac{\vec{a}^*_1 - \vec{a}^*_2}{4}] + \frac{\vec{c}^*}{2}\}$. The small representation τ is one-dimensional. There are two translationally invariant, antisymmetric combinations corresponding to $\pm\vec{k}_L$ and $\pm\vec{k}_L^{\,2}$:

$$A_1 = (\eta_1 \,;\, \eta_{-1}) - (\eta_{-1} \,;\, \eta_1) \text{ and } A_2 = (\eta_2;\eta_{-2}) - (\eta_{-2};\eta_2) \qquad (2.54)$$

In the basis $\eta_{\pm 1}$; $\eta_{\pm 2}$; the matrices of the OP representation can be constructed by standard methods (cf.§2.4.2). Let us write these matrices for two elements of G_o whose rotational parts C_4 and σ_x generate the point-group $\hat{G}_o$ = 4mm :

$$C_4:\ \left[\begin{array}{cc|cc} & & 0 & 1 \\ \multicolumn{2}{c|}{0} & 1 & 0 \\ \hline 1 & 0 & & \\ 0 & 1 & \multicolumn{2}{c}{0} \end{array}\right] \qquad \sigma_x:\ \left[\begin{array}{cc|cc} & & i & \\ \multicolumn{2}{c|}{0} & & -i \\ \hline 1 & 0 & & \\ 0 & 1 & \multicolumn{2}{c}{0} \end{array}\right] \qquad (2.55)$$

They allow to determine the action on A_1 and A_2 of all the point groups operations of $\hat{G}_o$ = (4mm). The representation $\Gamma^{(A)}$ spanned by A_1 and A_2 is generated by the matrices :

$$C_4:\ \begin{vmatrix} 0 & -1 \\ 1 & 0 \end{vmatrix} \qquad \sigma_x:\ \begin{vmatrix} 0 & 1 \\ 1 & 0 \end{vmatrix} \qquad (2.56)$$

One recognizes in $\Gamma^{(A)}$ the two-dimensional irreducible representation of 4mm. This representation is also associated to the (x,y) components of the vector representation. The number η_ν is equal to one, since Γ_1 is only contained once in the square of a real irreducible representation such as $\Gamma^{(A)}$. On the other hand, we can note that the two bases of $\Gamma^{(A)}$ constituted by $[A_1 ; A_2]$ and $[x,y]$ are not homologous. The form of the matrix σ_x in (2-55) indicates that $A_1;A_2$ is homologous to $[x-y ; x+y]$. Denoting X = (x-y) and Y = (x+y), we construct the LI by observing that the totally symmetric subspace of the tensorial product $[A_1,A_2] * [X,Y]$ is the "scalar product" $(A_1 * X + A_2 * Y)$.

Thus :

$$L_1 \propto \{\eta_1 ; \eta_{-1}\}_X + \{\eta_2 ; \eta_{-2}\}_Y \quad (2.57)$$

Like in the preceding example, the set $(\eta_{\pm 1}, \eta_{\pm 2})$ is a complex basis of the OP representation. We can use a real basis, either by defining $(\eta_1,\zeta_1 ; \eta_2,\zeta_2)$ as in (2.52), or by defining ρ_j, θ_j, such as $\eta_{\pm j} = \rho_j e^{\pm i\theta_j}$. Using the latter variables, the form of L_1 is:

$$L_1 \propto (\rho_1^2 \frac{\partial\theta_1}{\partial Y} + \rho_2^2 \frac{\partial\theta_2}{\partial X}) \quad (2.58)$$

Example 3. Consider a speculative case corresponding to G_o = P3m1 (C^1_{3V}), a star of wavevectors reduced to $\pm\vec{k}_L = \pm\vec{a}^*_3/3$, and a small representation τ coinciding with the two-dimensional representation of the point group 3m. The matrices generating the latter representation are, in the basis η^1_p , η^2_p (with p = ±1) :

$$C_3: \ \frac{1}{2}\begin{vmatrix} -1 & -\sqrt{3} \\ \sqrt{3} & -1 \end{vmatrix} \qquad \sigma_x: \ \begin{vmatrix} -1 & 0 \\ 0 & 1 \end{vmatrix} \quad (2.59)$$

There are 4 combinations of the type (2.45) :

$$A_l = (\eta_{1,l};\eta_{-1,l}) - (\eta_{-1,l};\eta_{1,l}) \quad A_{l+2} = (\eta_{1,l};\eta_{-1,l'}) - (\eta_{-1,l'};\eta_{1,l}) \quad (2.60)$$

with $(l, l' = 1, 2$ and $l \neq l')$. The transformation properties of the A_l derive from the matrices (2.59). Their set forms a reducible set defining $\Gamma^{(A)}$ which can be decomposed into the sum of 3 distinct irreducible representations of the 3m point group. The decomposition yields :

$(A_1 + A_2)$ transforming as γ_1 (totally symmetric)
$(A_3 + A_4)$ " " γ_2
$\left.\begin{matrix}(A_3 + A_4) \\ (A_1 - A_2)\end{matrix}\right]$ " " γ_3 (2.61)

On the other hand the vector representation V decomposes into γ_1 (z component) and γ_3(x,y components). The product $\Gamma^{(A)} \times V$ contains twice the totally symmetric representation γ_1, i.e. the "scalar products" $(A_1 + A_2) * z$, and $[(A_3 + A_4) * x + (A_1 - A_2) * y]$. Thus, the two independent LI can be written as :

$$L_1 = \{\eta_{1,1} ; \eta_{-1,1}\}_z + \{\eta_{1,2} ; \eta_{-1,2}\}_z \quad (2.62)$$

$$L_2 = \{\eta_{1,1}; \eta_{-1,2}\}_x + \{\eta_{1,2} ; \eta_{-1,1}\}_x + \{\eta_{1,1}; \eta_{-1,1}\}_y - \{\eta_{1,2};\eta_{-1,2}\}_y$$

Example 4. Consider the speculative case corresponding to $G_o = 6mm(C^1_{6v})$, a star of wavevectors reduced to $\pm \vec{k}_L = \pm(2\vec{a}^*_1 - \vec{a}^*_2)/3$ [K point of the surface of the hexagonal Brillouin zone], and a small representation coinciding, as in example 3, with the two-dimensional irreducible representation of the 3m point-group (which is the group $\hat{G}(\vec{k}_L)$). As in the preceding example, there are four combinations (2.45) having the same form as in eq. (2.60). The group 6mm is obtained by combining 3m and the twofold rotation around Z (C_2) ; the latter operation transforms $\vec{k}_L$ into $(-\vec{k}_L)$ and therefore changes each A_ℓ into its opposite. We can deduce from (2.61) the transformation properties of the A_ℓ :

$$\begin{array}{lcc} & 3m & C_2 \\ \hline (A_1 + A_2) & \gamma_1 & -1 \\ (A_3 - A_4) & \gamma_2 & -1 \\ (A_3 + A_4) & \gamma_3 & \begin{vmatrix} -1 & 0 \\ 0 & -1 \end{vmatrix} \\ (A_1 - A_2) & & \\ \hline \end{array} \qquad (2.63)$$

Of the three preceding irreducible representations of 6mm, the vector representation (V) only contains the third, spanned by the [x,y] components. Hence there is a single LI whose form is identical to that of L_2 in eq.(2.62)

c) **The Michelson restrictive symmetry criterion for k_o [8]**

In this paragraph we briefly return to the INC vector $\vec{k}_o$ pertaining to the OP symmetry in the neighborhood of T_I. Relying on the considerations of §a and b, we are in a position to discuss a symmetry criterion (also called the "weak" Lifschitz condition) restricting the possible OP symmetries of INC phase transitions.

Let $\alpha(\vec{k}_o + \vec{\kappa})$ be the coefficient of the quadratic invariant in the Landau free-energy for a wavevector $(\vec{k}_o + \vec{\kappa})$ close to $\vec{k}_o$. The assumption that the OP of the T_I transition corresponds to $\vec{k}_o$ implies that $\alpha(\vec{k}_o + \vec{\kappa})$ is minimum for $\vec{\kappa} = 0$ (chap.III §3). On the other hand, taking into account the fact that $\vec{k}_o$, being INC, does not correspond to any HS point of the Brillouin-zone, we deduce that the INC order-parameter is associated to a LI of the form (2.43) comprising n_ν independent terms L_ν. As established by eq.(2.42), each L_ν will give rise to a linear dependence of $\alpha(\vec{k}_o + \vec{\kappa})$ on the components of $\vec{\kappa}$. Such a linear dependence on the κ_ℓ is contrary to the assumption that $\alpha(\vec{k}_o + \vec{\kappa})$ is minimum at $\vec{k}_o$. Hence, this assumption requires that all the coefficients B_ν in eq. (2.43) are zero.

These coefficients $B_\nu(T,P,\vec{k})$ are functions of temperature and pressure, as well as of the considered point $\vec{k}$ in the Brillouin zone. The simultaneous cancellation of the B_ν can be expressed as :

$$B_\nu(T_I, P_I, \vec{k}_o) = 0 \qquad (2.64)$$

where T_I and P_I correspond to the transition point.

This set of equations defines a manifold $M_1(T_I,P_I)$ containing the extremity of $\vec{k}_o$. The dimension of M_1 depends on the number n_ν of coefficients $^oB_\nu$ (i.e. the number of independent L_ν). For instance if $n_\nu = 1$, (2.64) defines a plane, while if $n_\nu = 2$ it defines a line. A slight shift of T and P will displace the manifold $M_1(T,P)$ which, in general, will no longer contain the extremity of $\vec{k}_o$. Hence $\alpha(\vec{k})$ will be minimum for $\vec{k} \neq \vec{k}_o$. Since fluctuations of T and P, around (T_I, P_I), always occur, the properties of the system will not be determined by a single OP with wavevector $\vec{k}_o$, but by a continuous set of $\vec{k}$-vectors, surrounding $\vec{k}_o$, and corresponding, at each (T,P), to the minimum of $\alpha(\vec{k})$, i.e. to the cancellation of $B_\nu(T,P,\vec{k})$. If the former set of $\vec{k}$ vectors determines symmetry characteristics for the system, different from those corresponding to $\vec{k}_o$, then $\vec{k}_o$ has no physical relevance. Hence, in order to preserve this relevance, the set of $\vec{k}$ vectors must have the same symmetry characteristics as $\vec{k}_o$.

As pointed out by Michelson [8], such a situation is not always realized. Indeed, the set of $\vec{k}$ vectors defining the <u>same type of INC phase</u> as $\vec{k}_o$ belong to another manifold M_2, specified by the group of invariance $\hat{G}(\vec{k}_o)$. The physical relevance of $\vec{k}_o$ <u>imposes that M_2 and M_1</u>, which intersect at $\vec{k}_o$ for (T_I, P_I), also <u>intersect for</u> neighbouring temperatures and pressures. The fulfilment of this condition relies on the respective dimensions of the manifolds M_1 and M_2.

Consider, for instance, examples 1 and 2 hereabove (§2.3.2.b). In both cases, $\vec{k}_o$ has the same symmetry as $\vec{k}_L$ and the corresponding OP gives rise to a single term $L_\nu(n_\nu = 1)$. Accordingly, M_1 is a plane. On the other hand, in these two examples, $\hat{G}(\vec{k}_o) = G(\vec{k}_L)$ defines a line M_2 of $\vec{k}$-vectors. Hence M_1 and M_2 can intersect, in the vicinity of $\vec{k}_o$, when T, and P fluctuate.

In example 3, consider $\vec{k}_o = q\vec{a}_3^*$, on the same symmetry line as $\vec{k}_L = \vec{a}_3^*/3$. Here M_2 is again a line, while M_1, being defined by the simultaneous cancellation of two B_ν coefficients is also a line. Clearly, the intersection of these two lines at $\vec{k}_o$ for (T_I,P_I) is accidental and will not occur for neighbouring temperatures and pressures. Thus, the INC phase corresponding to an INC vector $\vec{k}_o = q\vec{a}_3^*$ and to the two-dimensional small representation defined in example 3, should not be observable.

It can be shown that n_ν is always larger than, or equal to, the dimension of the M_2 manifold. Hence the Michelson criterion is expressed by :

$$n_\nu = \dim (M_2) \qquad (2.65)$$

A detailed investigation of the various types of manifolds M_2 and of OP symmetries which can be encountered, shows [8] that the Michelson criterion excludes few OP symmetries. Condition (2.65) is generally satisfied, except for certain OP associated to <u>multidimensional</u> small representations τ. Available examples of INC phases all correspond to one-dimensional small representations, and consistently, they comply with (2.65). One has generally $n_\nu = \dim (M_2) = 1$. Quartz is an example for the situation $n_\nu = \dim (M_2) = 2$.

It is worth pointing out that the Michelson criterion is not applicable to the commensurate OP symmetry. Indeed the C-phase occurs for a non-vanishing small amplitude of the OP, and its wavevector is not controled by the $\alpha(\vec{k})$ coefficient alone. For the same reason, the Michelson criterion is not applicable to order-parameters associated to first-order transitions. This explains that one can observe commensurate phases whose OP corresponds to $n_\nu > \dim(M_2)$. Thus $\vec{k}_L$ frequently defines an isolated HS point ($\dim(M_2) = 0$), while the OP is associated to a single L_ν term ($n_\nu = 1$).

d) form of the squared gradient term

As underlined in §.2.3.2.a) the occurence of the second term in eq.(2.39) is always allowed by symmetry. To construct the precise form of this term compatible with the considered OP symmetry, we can proceed as in the case of the Lifschitz-invariant. Using the same notations as in §.2.3.2.b) we first consider all ordered products of the form $\eta_p^1 . \eta_{-p}^k$. They generate a representation Γ^s of the point-group $\hat{G}_o$. Γ^s has a dimension $2s.r^2$ if there are s pair of arms in the star of $\vec{k}_L$, and if the small representation is r-dimensional. We then form the product $\Gamma^s \times [V]^2$, where $[V]^2$ is the symmetrized square of the vector representation. Clearly all translationally invariant squared gradient terms $(\partial\eta_p^k/\partial x_{k'})(\partial\eta_{-p}^\ell/\partial x_{\ell'})$ belong to this representation. The independent terms forming the squared gradient contribution to eq.(2.36) correspond to the totally symmetric representations contained in $\Gamma^s \times [V]^2$. Let us illustrate this derivation in the case of example 2 in §.b.

Let $\mathcal{D}_1, \mathcal{D}_2, \mathcal{D}_3, \mathcal{D}_4$ be the four products $\eta_1\eta_{-1}$, $\eta_{-1}\eta_1$, $\eta_2\eta_{-2}$ and $\eta_{-2}\eta_2$. Their transformation by the operations of 4mm are specified by the two matrices :

$$C_4 : \left[\begin{array}{cc|cc} 0 & 0 & 0 & 1 \\ 0 & 0 & 1 & 0 \\ \hline 1 & 0 & 0 & 0 \\ 0 & 1 & 0 & 0 \end{array}\right] \qquad \sigma_x : \left[\begin{array}{cc|cc} 0 & 0 & 1 & 0 \\ 0 & 0 & 0 & 1 \\ \hline 1 & 0 & 0 & 0 \\ 0 & 1 & 0 & 0 \end{array}\right] \qquad (2.66)$$

deduced from (2.55). These matrices form a reducible representation Γ^s of 4mm which can be decomposed into 3 irreducible representations :

$$\begin{array}{lll} (\mathcal{D}_1 + \mathcal{D}_2 + \mathcal{D}_3 + \mathcal{D}_4) & \text{transforms as } \gamma_1 & \text{(totally symmetric)} \\ (\mathcal{D}_1 + \mathcal{D}_2 - \mathcal{D}_3 - \mathcal{D}_4) & \text{" as } \gamma_4 & \\ \mathcal{D}_1 - \mathcal{D}_2 & & \\ \mathcal{D}_3 - \mathcal{D}_4 & \text{" as } \gamma_5 & \end{array} \qquad (2.67)$$

On the other hand $[V]^2$ decomposes into $2\gamma_1$ $[(x^2+y^2) ; z^2]$; γ_3 $[x^2-y^2]$; γ_4 $[xy]$ and γ_5 $[(xz, yz)]$. Besides, for γ_5 the basis $(\mathcal{D}_1 - \mathcal{D}_2 ; \mathcal{D}_3 - \mathcal{D}_4)$ is homologous to $[Xz, Yz]$ with $X = (x-y)$ and

$Y = (x + y)$. Hence $\Gamma^s \times [V]^2$ contains four times the totally symmetric representation γ_1 : there are four independent invariants whose expression is provided by the preceding decomposition :

$$\delta_1 = \sum_p \left(\frac{\partial \eta_p}{\partial X}\cdot\frac{\partial \eta_{-p}}{\partial X} + \frac{\partial \eta_p}{\partial Y}\cdot\frac{\partial \eta_{-p}}{\partial Y}\right)$$

$$\delta_2 = \sum_p \frac{\partial \eta_p}{\partial z}\cdot\frac{\partial \eta_{-p}}{\partial z}$$

$$\delta_3 = \frac{\partial \eta_1}{\partial X}\cdot\frac{\partial \eta_{-1}}{\partial X} - \frac{\partial \eta_1}{\partial Y}\cdot\frac{\partial \eta_{-1}}{\partial Y} - \frac{\partial \eta_2}{\partial X}\cdot\frac{\partial \eta_{-2}}{\partial X} + \frac{\partial \eta_2}{\partial Y}\cdot\frac{\partial \eta_{-2}}{\partial Y} \tag{2.68}$$

$$\delta_4 = \frac{\partial \eta_1}{\partial X}\cdot\frac{\partial \eta_{-1}}{\partial z} - \frac{\partial \eta_1}{\partial z}\cdot\frac{\partial \eta_{-1}}{\partial X} + \frac{\partial \eta_2}{\partial Y}\cdot\frac{\partial \eta_{-2}}{\partial z} - \frac{\partial \eta_2}{\partial z}\cdot\frac{\partial \eta_{-2}}{\partial Y}$$

The same transformation into real variables as in (2.58) yields:

$$\delta_1 = \sum_p \left[\left(\frac{\partial \rho_p}{\partial X}\right)^2 + \left(\frac{\partial \rho_p}{\partial Y}\right)^2 + \rho_p^2 \left\{\left(\frac{\partial \theta_p}{\partial X}\right)^2 + \left(\frac{\partial \theta_p}{\partial Y}\right)^2\right\}\right]$$

$$\delta_2 = \sum_p \left[\left(\frac{\partial \rho_p}{\partial z}\right)^2 + \rho_p^2 \left(\frac{\partial \theta_p}{\partial z}\right)^2\right]$$

$$\delta_3 = \left(\frac{\partial \rho_1}{\partial X}\right)^2 - \left(\frac{\partial \rho_1}{\partial Y}\right)^2 + \left(\frac{\partial \rho_2}{\partial Y}\right)^2 - \left(\frac{\partial \rho_2}{\partial X}\right)^2 + \rho_1^2\left[\left(\frac{\partial \theta_1}{\partial X}\right)^2 - \left(\frac{\partial \theta_1}{\partial Y}\right)^2\right] + \rho_2^2\left[\left(\frac{\partial \theta_2}{\partial Y}\right)^2 - \left(\frac{\partial \theta_2}{\partial X}\right)^2\right] \tag{2.69}$$

$$\delta_4 = \rho_1\left(\frac{\partial \rho_1}{\partial z}\cdot\frac{\partial \theta_1}{\partial X} - \frac{\partial \rho_1}{\partial X}\cdot\frac{\partial \theta_1}{\partial z}\right) + \rho_2\left(\frac{\partial \rho_2}{\partial z}\cdot\frac{\partial \theta_2}{\partial Y} - \frac{\partial \rho_2}{\partial Y}\cdot\frac{\partial \theta_2}{\partial z}\right)$$

We will see in § 3 that the only coordinates of interest in such a system are the ones appearing in the Lifshitz invariant, i.e. X and Y. Thus, the two invariants δ_2 and δ_4 can be neglected in the phenomenological theory of the INC phase.

2.4. Three standard examples

In this paragraph, we determine, or complete, the determination of the free-energies considered in §. 2.1.-2.3., for three standard examples of INC phases, in order to illustrate the various methods and results discussed.

2.4.1) Rubidium-zinc-chloride (Rb_2ZnCl_4)

We have already examined the INC phase of this material. In particular, we have indicated the form of the free-energy for each INC wavevector (eq.2.19), the form of the commensurate free-energy (eq.2.25), and the form of the Lifschitz invariant involved in the expression of the "modulated" free-energy density (eqs.2.51-2.53). Let us summarize these results.

As starting elements of the theory, we rely on the experimental specification of the symmetries of the INC and C order-parameters. These symmetries consist in the locations of

$\vec{k}_o(q\vec{a}_3^*)$ and $\vec{k}_L$ $(\vec{a}_3^*/3)$, and also of the associated small representations. Consultation of standard tables (cf.§III.) show that $\vec{k}_o$ and $\vec{k}_L$ lye on the same symmetry-line of the Brillouin-zone, that both vectors possess the same point-group $\hat{G}(\vec{k}_o) = (mm2)$, and that the small representations are one-dimensional. The stars of the two vectors are reduced, respectively, to $(\pm \vec{k}_o)$ and $(\pm\vec{k}_L)$. On the other hand, experimental investigations [ ]° specify the "weighted" representation τ' of $\hat{G}(\vec{k}_o)$, which is associated to the INC and C order-parameters. This representation, as well as the small representations deduced from it at $\vec{k}_o$ and $\vec{k}_L$ are :

	$E\ ;\ \vec{a}_1\ ;\ \vec{a}_2$ $-(\sigma_x\vert\vec{a}_1/2)$	$(U_z\vert[\vec{a}_1+\vec{a}_2+\vec{a}_3]/2)$ $-(\sigma_y\vert[\vec{a}_2+\vec{a}_3]/2)$	$\vec{a}_3$
τ'	1	1	-
$\tau(\vec{k}_o);\tau(\vec{k}_L)$	1	$\xi_{\vec{k}_o,\vec{k}_L}$	$\xi^2_{\vec{k}_o,\vec{k}_L}$

(2.70)

with $\xi_{\vec{k}_o} = e^{-i\pi q}$ and $\xi_{\vec{k}_L} = e^{-i\pi/3}$. From (2.70) we deduce the set (eq.2.50) of two-dimensional matrices of the representation of the OP, for the generating operations of the space-group G_o = (Pmcn). For q irrational, it is easy to check, consistently with the general considerations of §.2.1.3., that the free-energy is of the form (2.19) involving no θ-dependent term. For q = (1/3), one recognizes in eq.(2.50), that the set of distinct matrices of the OP representation is isomorphous to the point-group $\mathbb{C}_{6v}$. Using the general results for two-components OP (chap.II §4.5) we deduce that the commensurate free-energy has the form :

$$F_L = F_L\,(\rho^2\ ;\ \rho^6 \cos(6\theta)) \tag{2.71}$$

The lowest-degree θ-dependent term is of sixth degree. Truncated to this degree F_L is expressed by eq.(2.25). As explained in §.2.3.1, eq (2.25) also corresponds to the part $f_1(\vec{r})$ of the modulated free-energy density which only contains the OP components. The remaining free-energy density $f_2(\vec{r})$ is composed of the Lifshitz invariant, already determined in eqs.(2.51-2.53), and a squared gradient term (2.39). Following the method described in §.2.3.2.d) and using the set of matrices (2.50), one straightforwardly finds that the squared gradient contains 3 independent terms of the form :

$$\delta_\nu = \frac{\partial\eta_p}{\partial x_\nu}\cdot\frac{\partial\eta_{-p}}{\partial x_\nu} = \left(\frac{\partial\rho}{\partial x_\nu}\right)^2 + \rho^2\left(\frac{\partial\theta}{\partial x_\nu}\right)^2 \tag{2.72}$$

with $x_\nu = (x,y,z)$. As will be shown in §.3 the only term of interest is the one involving the z-coordinate, which also appears in the Lifschitz invariant. Hence the free-energy density can be written, retaining the lowest-degree θ -dependent and gradient-dependent terms :

$$f(\vec{r})=\frac{\alpha}{2}\rho^2+\frac{\beta_1}{4}\rho^4+\frac{\beta_6}{6}\rho^6+\frac{\beta'_6}{6}\rho^6\cos(6\theta)+\delta\rho^2(\frac{\partial\theta}{\partial z})+\frac{\sigma}{2}[(\frac{\partial\rho}{\partial z})^2+\rho^2.(\frac{\partial\theta}{\partial z})^2] \quad (2.73)$$

Rb_2ZnCl_4 represents one of the sinplest situations encountered in INC systems, and characterized by the following features :

i) two-dimensional OP for the INC and C-phases.
ii) Existence of a (single) Lifschitz invariant.
iii) k_o-vector paralled to a basic translation ($\vec{a}_3^*$) of the reciprocal lattice. The coordinate z involved in the Lifschitz invariant corresponds to the axis defined by $\vec{k}_o$.

However, the θ-dependent term $\rho^6.\cos(6\theta)$ is not the simplest one which can be constructed for two-dimensional OP. We know that such a term is necessarily of the form $\rho^n.\cos(n\theta)$ [chap.II §4.5]. Hence, avoiding the complications introduced by third degree terms, the simplest θ-dependent term will be $\rho^4.\cos(4\theta)$. The resulting modulated free-energy density $f(\vec{r})$ can then be written, denoting $\dot{\rho}$ and $\dot{\theta}$ the spatial derivatives of ρ and θ :

$$f(\vec{r}) = \frac{\alpha}{2}\rho^2+\frac{\beta_1}{4}\rho^4+\frac{\beta_2}{4}\rho^4\cos(4\theta)+\delta\rho^2.\dot{\theta}+\frac{\sigma}{2}(\dot{\rho}^2+\rho^2.\dot{\theta}^2) \quad (2.74)$$

Eq.(2.74) constitutes the canonical free-energy density for INC systems. It corresponds to the situation realized in a structural isomorph of Rb_2ZnCl_4, Ammonium fluoberyllate [1] of formula $(NH_4)_2BeF_4$. In §4, we will examine in detail the physical consequences of expression (2.74)

2.4.2) Barium sodium niobate

This case is defined by G_o = P4bm ; $\vec{k}_o = q(\vec{a}_1^*+\vec{a}_2^*)+\vec{a}_3^*/2$ with $\vec{k}_L$ corresponding to q = (1/4). Like in the preceding example, the extremities of $\vec{k}_o$ and $\vec{k}_L$ belong to the same symmetry line. This line is situated on the surface of the Brillouin zone, and is defined by $\hat{G}(\vec{k}_o) = (m_{\bar{x}y})$.

The small representations on this line are one-dimensional, and among two possible choices for the "weighted" representation, experimental data select one. This representation, as well as the corresponding small representations at $\vec{k}_o$ and $\vec{k}_L$ are :

	E ; $(-\vec{a}_3)$	$(\sigma_{\bar{x}y}\mid[\vec{a}_1+\vec{a}_2]/2)$ $-\vec{a}_1$; $-\vec{a}_2$
τ'	1	-1
$\tau(\vec{k}_o)$, $\tau(\vec{k}_L)$	1	$e^{-2iq\pi}$

(2.75)

As $\vec{k}_o$ (or $\vec{k}_L$) are associated to a 4-arm star, $\pm\vec{k}_o$, and $\pm\vec{k}_1$ = $\pm[q(\vec{a}_1^*-\vec{a}_2^*)+a_3^*/2]$, the OP is 4-dimensional. We deduce from (2.75) the generating matrices of its representation ; following the general procedure (chap.III), and adopting the standard (complex) basis associated to the two pairs of opposite wavevectors in the star of $\vec{k}$:

$(C_4\|0)$	$(\sigma_{xy}\|[\vec{a}_1+\vec{a}_2]/2$	$\vec{a}_1$	$\vec{a}_2$	$\vec{a}_3$
$\left[\begin{array}{cc\|cc} & & 0 & 1 \\ \multicolumn{2}{c\|}{0} & 1 & 0 \\ \hline 1 & 0 & & \\ 0 & 1 & \multicolumn{2}{c}{0} \end{array}\right]$	$\left[\begin{array}{cc\|cc} \xi & 0 & & \\ 0 & \bar{\xi} & \multicolumn{2}{c}{0} \\ \hline & & 0 & 1 \\ \multicolumn{2}{c\|}{0} & 1 & 0 \end{array}\right]$	$\left[\begin{array}{cc\|cc} \xi & 0 & & \\ 0 & \bar{\xi} & \multicolumn{2}{c}{0} \\ \hline & & \bar{\xi} & 0 \\ \multicolumn{2}{c\|}{0} & 0 & \xi \end{array}\right]$	$\left[\begin{array}{cc\|cc} \xi & 0 & & \\ 0 & \bar{\xi} & \multicolumn{2}{c}{0} \\ \hline & & \xi & 0 \\ \multicolumn{2}{c\|}{0} & 0 & \bar{\xi} \end{array}\right]$	$\left[\begin{array}{cc\|cc} -1 & 0 & & \\ 0 & -1 & \multicolumn{2}{c}{0} \\ \hline & & -1 & 0 \\ \multicolumn{2}{c\|}{0} & 0 & -1 \end{array}\right]$

(2.76)

with $\xi = e^{-2iq\pi}$. Eq(2.76) contains, in particular, the result used in (2.55)

a) **INC free-energy**

For q irrational, the only independent terms invariant by the matrices (2.76), are, up to the 4th degree, and using the same notations as in eq.(2.54) :

$$(\eta_1\eta_{-1} + \eta_2\eta_{-2}) \ ; \ (\eta_1\eta_{-1} + \eta_2\eta_{-2})^2 \ ; \ \eta_1\eta_{-1}\eta_2\eta_{-2} \qquad (2.77)$$

Expressed as a function of the "polar" coordinates in the OP space, the INC free-energy is :

$$F = \frac{\alpha}{2}(\rho_1^2 + \rho_2^2) + \frac{\beta_1}{4}(\rho_1^2 + \rho_2^2)^2 + \frac{\beta_2}{2}\rho_1^2\rho_2^2 \qquad (2.78)$$

which, as in the case of Rb_2ZnCl_4, is independent of the angular variables (θ_j). As pointed out in §2.1.3, this independence on <u>two phase angles</u> is not general for 4-dimensional OP. However, the present result is consistent with the fact that the star of $\vec{k}_o$ contains two rationally independent vectors (cf. §.2.1.2.b), and accordingly, two modulation directions.

b) **Commensurate free-energy**

For q = (1/4), besides the invariants (2.77), there is one additional fourth degree invariant which can either be written in the form $(\eta_1^4 + \eta_{-1}^4 + \eta_2^4 + \eta_{-2}^4)$ or in the form $(\rho_1^2 \cdot \cos(4\theta_1) + \rho_2^2 \cdot \cos(4\theta_2))$. As expected from the general considerations in §.2.2.2), this additional term depends on the angular variables θ_j.

c) **Modulated free-energy density**

As we have already determined in §.2.3.2.b and 2.3.2.d) the Lifschitz invariant and the squared gradient invariant for this case, we can express the free-energy density :

$$f(\vec{r})=\frac{\alpha}{2}(\rho_1^2+\rho_2^2)+\frac{\beta_1}{4}(\rho_1^2+\rho_2^2)^2+\frac{\beta_2}{2}\rho_1^2\rho_2^2+\frac{\beta_3}{4}(\rho_1^4.\cos4\theta_1+\rho_2^4.\cos4\theta_2)$$

$$+\ \delta.\mathcal{L}_1+\frac{\sigma_1}{2}\cdot\mathcal{S}_1+\frac{\sigma_2}{2}\cdot\mathcal{S}_3 \qquad (2.79)$$

where $\mathcal{L}_1$, $\mathcal{S}_1$ and $\mathcal{S}_3$ are provided by eqs. (2.58), and (2.69). This free-energy density generalizes to the case of a four-component OP, and of two spatial coordinates (x,y) the canonical form (2.74)

<u>2.4.3) Thiourea, $SC(NH_2)_2$</u>

We have stressed in §.2.2.1.c, that thiourea has the same space-group G_o as Rb_2ZnCl_4, and an INC wavevector $\vec{k}_o$ possessing the same little group $G(k_o)$ as in the latter substance. Hence, the INC order-parameter is also two-dimensional, and the INC free-energy is identical to the one (eq.2.19) found for Rb_2ZnCl_4 (in thiourea $\vec{k}_o = qa_1^*$ instead of qa_3^* in Rb_2ZnCl_4).

However, as already noted in §.2.2.1.c) $\vec{k}_L$ has higher symmetry than $\vec{k}_o$ [$\vec{k}_L = 0$] , and the C-order-parameter has only <u>one-dimension</u>. The C-free energy takes the simple form $(\frac{\alpha}{2}\eta^2+\frac{\beta}{4}\eta^4)$ valid for all real one-dimensional OP

Besides, as pointed out in chap.III§3), the latter type of OP always satisfies the Lifschitz criterion. Consequently, the modulated free-energy density corresponding to k_L ,does not contain a Lifschitz invariant. We will see,in §5, that,in this case,a successful theory of the INC phase requires taking into account gradient terms of degree 4. The simplicity of the commensurate order parameter symmetry allows a straightforward derivation of $f(\vec{r})$:

$$f(\vec{r})=\frac{\alpha}{2}\eta^2+\frac{\beta}{4}\eta^4+\frac{\sigma}{2}\left(\frac{\partial\eta}{\partial x}\right)^2+\frac{\gamma}{2}\left(\frac{\partial^2\eta}{\partial x^2}\right)^2 \qquad (2.80)$$

The only relevant spatial variable being x, since $\vec{k}_o = qa_1^*$ (cf.§2.3.1) Eq.(2.80) is the simplest free-energy density for a "secondary" theoretical scheme for INC systems.

3. QUALITATIVE INTERPRETATION OF THE EXPERIMENTAL SITUATION

In this paragraph, we show how the "standard" experimental situation of INC systems described in §.1.1. can be accounted for qualitatively in the framework of each of the two theoretical schemes mentioned in the introduction of the chapter (§.1.2). In order to emphasize the basic ideas we restrict the discussion to the simple situation where both the INC and C-phases are associated to a two-dimensional OP i.e. for which the star of $\vec{k}_o$ is reduced to $\pm\vec{k}_o$, and $\vec{k}_L$ is related to cases a) or b) in fig.4 . Besides we take $\vec{k}_o$ and $\vec{k}_L$ along a coordinate axis (x) in order to replace these vectors by scalar quantities.

3.1. Simplified formulation of the first theoretical scheme

3.1.1.) INC free-energy

In this scheme (§.1.2) the INC phase is described by a spatially uniform OP associated to the INC wavevector k, and to the free-energy expansion (2.19). In this free-energy, the coefficients α and β_i are functions of k and of the intensive quantities T and P. The requirement imposed by the observation of a transition at T_I towards an INC phase with wavevector k_o only imposes that $\alpha(k)$ is minimum for $k = k_o$ and that this minimum vanishes at T_I. Thus, as usual in Landau's theory (chapII §3) we assume the β_i independent of T and k, in the first approximation. The free-energy (2.19) can then be written , up to 4th degree terms :

$$F = \frac{\alpha_o}{2} [T - T_I + a(k - k_o)^2].\rho^2 + \frac{\beta_1}{4} \cdot \rho^4 \tag{3.1}$$

with $\beta_1 > 0$ to ensure the stability of the INC phase and $\alpha_o > 0$, $a > 0$. We assume, in addition, that $(k_L - k_o)$ is not too large, and that the parabolic dispersion of $\alpha(k)$ remains valid for $k = k_L$ (Cf. Fig.4). The equilibrium values of ρ and F deduced from (3.1) are, for $\alpha(k) < 0$:

$$\rho^2 = \frac{-\alpha(k)}{\beta_1} \quad ; \quad F(k) = \frac{-\alpha^2(k)}{4 \cdot \beta_1} \tag{3.2}$$

Consistently with the above assumptions, the value of k which minimizes F(k) is $k = k_o$, below T_I. However, in this simplified model <u>k remains equal to k_o in the whole INC range, and it has no temperature dependence.</u>

3.1.2.) Commensurate free-energy

For $k = k_L$ the free-energy contains in addition to (3.1) a θ-dependent "lock-in" term. As pointed out in §.2.4, the simplest form of this term is $\rho^4 \cdot \cos(4\theta)$. F_L then coincides with the "homogeneous" part of eq.(2.74). Since k_L and k_o are close to eachother the coefficients $\alpha(k)$ and β_1 in the latter equation will be the same as in eq.(3.1) :

$$F_L = \frac{\alpha_o}{2} [T - T_I + a(k_L - k_o)^2].\rho^2 + \frac{\beta_1}{4} \cdot \rho^4 + \frac{\beta_2}{4} \cdot \rho^4 \cdot \cos(4\theta) \tag{3.3}$$

For $\alpha(k_L) < 0$, the equilibrium values ρ_c, θ_c and F determined by eq(3.3) are :

$$\rho_c^2 = -\frac{\alpha(k_L)}{\beta_1 - |\beta_2|} \quad ; \quad \beta_2 \cos(4\theta_c) = -|\beta_2| ; \quad F_L = -\frac{\alpha^2(k_L)}{4(\beta_1 - |\beta_2|)} \tag{3.4}$$

where we assume $(\beta_1 - |\beta_2|) > 0$ to ensure the positivity of (3.3). Depending on the sign of β_2, two types of C-phases are possible. For $\beta_2 < 0$, this phase can exist in 4 energetically equivalent state C_i; corresponding to $\theta_i = (0 ; \frac{\pi}{2}, \pi ; \frac{3\pi}{2})$. For $\beta_2 > 0$, the four C_i states correspond to $\theta_i = (\frac{\pi}{4}, \frac{3\pi}{4}, \frac{5\pi}{4}, \frac{7\pi}{4})$.

3.1.3.) Lock-in transition

If $\beta_2 = 0$, we can see that the INC free-energy (3.2) is always more negative than F_L due to the fact that $|\alpha(k_L)| > |\alpha(k_o)$. However the presence of the term with β_2 coefficient lowers F_L (eq. 3.4) and stabilizes the C-phase below the transition T_L. The condition of stability of the C-phase, as compared to the INC phase is :

$$F_L = \frac{-\tilde{\alpha}^2(k_L)}{4(\beta_1 - |\beta_2|)} \leqslant \frac{-\alpha^2(k_o)}{4 \cdot \beta_1} = F \qquad (3.5)$$

which yields :

$$(T_I - T_L) = \frac{a(k_L - k_o)^2}{1 - \sqrt{1 - |\beta_2| / \beta_1}} \qquad (3.6)$$

Thus, the larger the coefficient $|\beta_2|$ of the lock-in term, the narrower the range of the INC phase. On the other hand, if the minimum of $\alpha(k)$ at k_o is very pronounced (i.e. the coefficient a is large), the INC phase is stable in a larger temperature range.

Note that the transition at T_L is discontinuous : k jumps from k_o to k_L, and ρ^2 undergoes a discontinuous variation from $[\rho_+^2 = \alpha_o (T_I - T_L)/\beta_1]$ in the INC phase, to $[\rho_-^2 = \rho_+^2 \cdot \sqrt{\beta_1/(\beta_1 - |\beta_2|)}$ in the C-phase.

3.1.4) Free-sliding of the INC distortion

Within the preceding framework, the physical anomalies at T_I are determined by (3.1), which for $k = k_o$ = cte has the usual Landau form. T_I will therefore be associated to the standard thermal anomalies (no latent heat, upward jump of the specific heat). As for T_L, it will induce discontinuous variations of the physical quantities, these discontinuities being determined by the distinct free-energies (3.1) and (3.3) which are relevant on either sides of T_L. The latter transition involves, in particular, a non-zero latent heat. We do not examine here the behaviour of other physical quantities (such as the susceptibility relative to macroscopic quantities) as this problem will be dealt with in the quantitative theory described in §.4.

The free-energy (3.1) controls the properties of the INC phase. We have noticed that this free-energy has a specific feature (§.2.1.4) which consists in the absence of θ -dependence of F, for any degree of the expansion. In order to point out the important consequence of this peculiarity, let us consider the density increment $\delta\rho_1(\vec{r})$ in eq. (2.33) which represents the structural distortion associated to the onset of the INC phase. For the sake of simplicity, we can choose the $\Phi^o_{pl}(\vec{r})$ functions to be plane waves $e^{\pm ik_o x}$. Expressing $\delta\rho_1(x)$ as a function of real combinations of these waves and real OP components we obtain :

$$\delta\rho_1(x) = \rho[\cos\theta .\cos(k_o x) + \sin\theta \sin(k_o x)] = \rho\cos(k_o x - \theta) \quad (3.7)$$

Since F does not depend on θ (eq. 3.1), an arbitrary variation of θ, by $\Delta\theta$ in eq.(3.7) will transform $\delta\rho_1(x)$ into a different structural distortion $\delta\rho_1'(x)$ possessing the same free-energy. We can write :

$$\delta\rho_1'(x) = \rho\cos\left[k_o\left(x - \frac{\Delta\theta}{k_o}\right) - \theta\right] = \delta\rho_1\left(x - \frac{\Delta\theta}{k_o}\right) \quad (3.8)$$

Thus, the correspondance between $\delta\rho_1'(x)$ and $\delta\rho_1(x)$ is an arbitrary translation parallel to the direction of $\vec{k}_o$: the INC distortion can "slide" freely. (Note that since the atomic structure is discrete along x, the correspondance (3.8) between atomic positions must be understood to be modulo 2 π for $\Delta\theta$, i.e. modulo $(2\pi/k_o)$ for x)

3.2) Refinement of the first theoretical scheme

The model in §.3.1. has the drawbacks of implying no temperature dependence for k, and no precursor effects on approaching T_I. In particular, it does not account for the spatial structuration into C domains in the range of temperatures close to T_L. It is a model well adapted to the behaviour of the INC phase just below T_I.

Let us show that it can be completed in order to account, at least qualitatively, for the behaviour expected in the remaining range of stability of the INC phase.

This refinement is based on two ideas. The first one is that the coefficient β_1 in eq (3.1) can be renormalized if one takes into account the coupling, which is allowed by symmetry, between the primary OP, and secondary order-parameters associated to wavevectors different from k_o. The second idea is that a renormalization will lead to a pronounced k dependence for β_1, if the considered secondary OP correspond to wavevectors k_s close to k_o (and to k_L), up to a reciprocal lattice vector $\vec{G}$. Such a k-dependence of β_1^L will determine a temperature dependence for the OP wavevector k. On the other hand, we will show that the condition of proximity imposed to k_s and k_o is related to the structuration of the INC phase into C domains.

We have seen in the study of improper ferroics (chap.III §6) that the onset of spontaneous quantities with symmetries different from that of the OP, results from coupling terms of the form $(\delta_s . \eta' . \eta^f)$, where

η' is the secondary OP, η the primary OP and f the faintness index. If, on the other hand, the quadratic contribution of η' to the free-energy is $(\frac{\alpha_s}{2}|\eta'|^2)$, the elimination of η' leads to a renormalization of the coefficient β of the term of degree (2f) in the primary OP :

$$\beta \to \beta' = \beta - \frac{\delta_s^2}{2\,\alpha_s} \tag{3.9}$$

If α_s or δ_s are k-dependent, a k-dependence will be induced in β'. Consider now the set of order parameters belonging to the same branch as the components of the primary OP at $\pm\vec{k}_o$.

Their wavevector k_s ends on the same HS lines as $\pm\vec{k}_o$ or $\pm\vec{k}_L$. For each of these OP, the coefficient α_s in eq.(3.9) is equal to $\alpha(k_s)$, where $\alpha(k)$ is the coefficient in (3.1), provided k_s is not too distant from k_o and k_L. Besides, if k_s is close to $\pm k_o$, $\alpha_s = \alpha(k)$ possesses a strong dispersion due to the existence of a minimum at k_o. Hence, it is possible to induce a pronounced k-dependence in β' (eq.3.9) if one can find a coupling between the primary OP and a secondary OP, belonging to one of the branches of $\pm\vec{k}_o$, and with k_s close to $\pm k_o$. The translational invariance of the term $(\eta'.\eta^f)$ imposes a first condition to k_s :

$$k_s = \pm f k_o + m a^* \tag{3.10}$$

where a^* is a reciprocal lattice basic vector parallel to k_o. The commensurate wavevector can be expressed as $k_L = (\frac{r_L}{s_L})a^*$, where r_L and s_L are integers. As explained above, a pronounced dispersion of α_s imposes that k_s is close to k_o, up to reciprocal lattice translation. We can write (3.10) in the form :

$$|(k_s \pm k_o)| = |\pm(f \pm 1)k_L \quad \pm(f \pm 1)(k_o - k_L) + m'a^*| \tag{3.11}$$

The required proximity of k_s and $\pm k_o$ imposes that the first member of (3.11) is small as compared to $|a^*|$ (the allowance for the reciprocal lattice vector in this difference has been transfered into the second member). Since $(k_o - k_L)$ is small, we are left with the condition imposed to f : $[(f \pm 1)k_L = \pm m'a^*]$:

$$f = \frac{m s_L}{r_L} \pm 1 \tag{3.12}$$

where f is a positive integer satisfying $f \geqslant 2$ (f = 1 would lead to the renormalization of $\alpha(k)$ and no temperature dependence of k would be induced).

For instance, if $k_L = \frac{a^*}{2}$, the minimum value of f is f = 3, which leads to a renormalization of the coefficients of sixth degree-terms [9]. If $k_L = \frac{a^*}{3}$, f = 2 and a k-dependence is induced in the coefficient of fourth degree terms [10].

Let us detail the calculations for the latter situation. With $k_L = \frac{a^*}{3}$ and f = 2, we can see that $k_s = -2k$ will be close to (+k) up to a translation a*, while $(-k_s) = +2k$ will be close to $(-k)$. If we set $\eta'(\pm k_s) = \rho' e^{\pm i\theta'}$, and determine the real part of the coupling term $\eta'.\eta^f$, we obtain for the modified free-energy containing both the primary and secondary OP :

$$F = \frac{\alpha(k)}{2}.\rho^2 + \frac{\beta_1}{4}.\rho^4 + \frac{\alpha(k_s)}{2}.\rho'^2 + \delta\, \rho.\rho'^2 . \cos(\theta' - 2\theta) \qquad (3.13)$$

where (cf. Fig. 6a) :

$$\alpha(k) = [T - T_I + a(k-k_o)^2]\text{, and } \alpha(k_s) = \{T - T_I + a.[3(k_o - k_L) + 2(k-k_o)]^2\} \qquad (3.14)$$

The minimization of F with respect to ρ' and θ' permits to eliminate from (3.13) these variables and leads to :

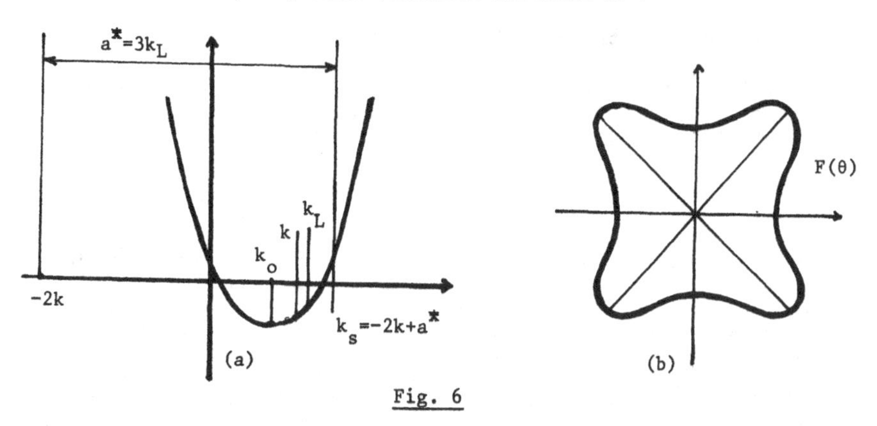

Fig. 6

$$F = \frac{\alpha(k)}{2}.\rho^2 + \frac{\beta_1'}{4}.\rho^4 \qquad (3.15)$$

with

$$\beta_1' = \beta_1 - \frac{2\,\delta_s^2}{\alpha(k_s)} \qquad (3.16)$$

a) **Temperature dependence of k**

The equilibrium free-energy is $F(k) = (-\alpha^2(k)/4\beta_1')$. Minimizing F(k) with respect to k determines the equilibrium value of k. Using (3.16), we obtain first :

$$\beta_1' . \frac{\partial\alpha(k)}{\partial k} = \frac{\delta_s^2 . \alpha(k)}{\alpha(k_s)} . \frac{\partial\alpha(k_s)}{\partial k} \qquad (3.17)$$

and, expressing $\alpha(k)$ and $\alpha(k_s)$ through (3.14) :

$$(k - k_o) = \frac{2\ \delta_s^2 \cdot \alpha(k)}{\beta_1' \cdot \alpha^2\ (k_s)} \cdot (k_o - 3\ k_L + 2k) \qquad (3.18)$$

The initial variation of k(T) nearby T_I can be obtained from eq.(3.18) in the approximation $a[k-k_o]^2 << |T - T_I| << a\ [k_o - k_L]^2$. Hence :

$$k(T) - k_L = [k_o - k_L]\ [1 - \frac{2 \cdot \delta_s^2 \cdot (T_I - T)}{27 \cdot \beta_1 \cdot a^2 \cdot (k_o - k_L)^4}] \qquad (3.19)$$

This result shows that $|k(T) - k_L|$ decreases on cooling, in agreement with the behaviour outlined in fig. 1

b) **Structuration into C-domains**

If we generalize (3.13) and consider any secondary OP (ρ', θ') involved in a coupling of the form $[\delta_s \rho' \rho^f \cdot \cos\ (\theta' - f\theta)]$, the equilibrium amplitude ρ' will have the form :

$$\rho' = \frac{|\delta_s| \cdot \rho^f}{\alpha(k_s)} \qquad (3.20)$$

In general ρ' will be small, because the coupling coefficient δ_s is small, and because $\alpha(k_s)$ does not possess a small value. Only in the cases where f complies with condition (3.12), the amplitude ρ' will be large since $\alpha(k_s)$ is close to zero. Condition (3.12) expresses that k_s is close to k_o, and therefore $\alpha(k_s) \simeq \alpha(k_o) \simeq 0$. As a consequence, the Fourier transform of the density increment $\delta\rho\ (\vec{r})$ [cf.§.2.1.2.d] which incorporates the structural distortions induced by the primary and the secondary OP, will essentially contain components corresponding to $\pm k_o$ and to the "harmonics" k_s complying with condition (3.12). The important point to notice is that conditions (3.11) or (3.12) allow a different interpretation of the Fourier transform of $\delta\rho(\vec{r})$. Let us put $k_o = k_L + qa^*$. Taking into account (3.12), condition (3.10) can be written :

$$k_s = k_L + f.qa^* \qquad (3.21)$$

Accordingly, all the harmonics of $k \approx k_o$ which possess exceptionally large amplitudes, are the sum of k_L and of f-times the difference $(k_o - k_L) = qa^*$. These harmonics can therefore be considered as resulting from a modulated C structure (characterized by k_L), with the wavevector of the modulation being $(k_o - k_L)$, and the shape of the modulation being pronouncedly non-sinusoidal since harmonics with f >>2 possess large amplitudes.

The variation of k(T) in the entire range $(T_I - T_L)$, as well as the precise shape of the former modulation will not be further examined within this "refined first scheme", since they will be studied using a more powerful method in §.4. We will show that this pronounced non-sinusoidal modulation is indeed a succession of C-regions separated by walls.

3.3) Second theoretical scheme, in the presence of a Lifschitz invariant

The second scheme is based on the use of a modulated free-energy (cf.§.2.3). We have seen that, in the free-energy density, a Lifschitz invariant can be present or absent. We consider the former case, realized in its simplest form by the free-energy density (2.74)

3.3.1) Role of the Lisfschitz invariant and of the squared gradient

Using the variables $\eta = \rho.\cos\theta$ and $\zeta = \rho.\sin\theta$, the second degree terms in (2.74) can be written in the form :

$$\frac{\alpha}{2}(\eta^2 + \zeta^2) + \delta(\eta\frac{\partial\zeta}{\partial x} - \zeta\frac{\partial\eta}{\partial x}) + \frac{\sigma}{2}[(\frac{\partial\eta}{\partial x})^2 + (\frac{\partial\zeta}{\partial x})^2] \qquad (3.22)$$

where $\eta(x)$ and $\zeta(x)$ constitute a real basis of the OP representation. As shown in §.2.3.2.a), the Fourier transform of $f(x)$ can be considered as the free-energy $F(\kappa)$ associated to an OP with wavevector $(\pm k_L + \kappa)$ close to $(\pm k_L)$.

Let us define η_κ and ζ_κ as :

$$\eta(x) = \frac{1}{\sqrt{2}}\int e^{-i\kappa x}.(\eta_\kappa + \zeta_\kappa).d\kappa \; ; \; \zeta(x) = \frac{i}{\sqrt{2}}\int e^{-i\kappa x}.(\eta_\kappa - \zeta_\kappa)d\kappa \qquad (3.23)$$

Taking into account the reality of $\eta(x)$ and $\zeta(x)$, we can write (3.22) in the form :

$$\frac{1}{2}|\eta_\kappa|^2[\alpha - 2.\delta\kappa + \sigma\kappa^2] + \frac{1}{2}|\zeta_\kappa|^2[\alpha + 2.\delta\kappa + \sigma.\kappa^2] \qquad (3.24)$$

For $\kappa = 0$ (i.e. for $k = \pm k_L$), the two components $(\eta_\kappa, \zeta_\kappa)$ possess the same coefficient $\frac{\alpha}{2}$, consistently with the degeneracy of the OP. Note that this twofold degeneracy can either correspond to a star-degeneracy (as in fig.4a) or, if $k_L = -k_L$, to a degeneracy of the small representation (as in fig.4b). We can see on the form (3.24) that the presence of a Lifschitz invariant lifts this degeneracy for a k-vector $(\pm k_L + \kappa)$. The coefficient δ of the L.I. represents the slope of the coefficients $\alpha(k)$ in fig.4, for $k = k_L$. The existence of a non-zero slope at k_L implies that the minimum of $\alpha(k)$ will not occur at k_L.

The presence of a Lifschitz-invariant at k_L has the effect of opposing the onset of a commensurate distortion with k_L wavevector. The actual distortion will correspond to $k_o \equiv (k_L + \kappa_o)$, κ_o being the value of κ which minimizes $[\alpha - 2\delta\kappa + \sigma\kappa^2]$, i.e. $\kappa_o = (\frac{\delta}{\sigma})$. The coefficient σ of the squared gradient term defines the upward curvature of $\alpha(k)$ nearby k_o, and it must have a positive sign. The coefficient of $|\zeta_\kappa|^2$ in eq.(3.24) determines the wavevector $(-k_o) = (-k_L - \kappa_o)$.

3.3.2) Role of the lock-in term

Consider the term $(\beta_2 \cdot \rho^4/4) \cdot \cos(4\theta)$, with β_2 negative. Its contribution to the total free-energy F of the system (i.e. the integral of f(x) has a minimum value if ρ and θ are uniform quantities, and if $\theta = 0$ (modulo $\pi/2$), i.e. if the uniform OP components define in the OP space one of the 4 directions represented on fig. 6.b). In this case, the spatial average of $\beta_2 \cdot \cos(4\theta)$ is equal to $-|\beta_2|$. By contrast, if $\theta(x)$ is non-uniform, this spatial average will be less negative and will determine a higher free-energy. In particular, if $\theta(x)$ is linear ($\theta = \kappa_o x$), the contribution of this term to the total free-energy vanishes. Thus, the lock-in term $(\beta_2/4) \cdot \cos 4\theta$ favours the onset of one of the commensurate states C_i corresponding to uniform values of the OP components (i.e. $\kappa = 0$, or $k = (k_L + \kappa) = k_L$), and to one of the prominent θ values indicated hereabove. Its effect is antagonist to the Lifschitz invariant which favours a modulated OP ($\kappa \neq 0$). The result of this competition will be studied in detail in §.4. Let us point out here one qualitative aspect of this result. Near the onset of the INC phase, just below T_I, the amplitude ρ of the OP is small and the Lifschitz invariant, which is of second degree in ρ, will be predominant and will stabilize an INC phase with the wavevector ($k_o = k_L + \kappa_o$). However, on cooling, the amplitude ρ grows, and the lock-in term, which is of fourth degree becomes predominant below a certain temperature, and induces the stabilization of the C-phase. In the intermediate temperature range where both terms have the same order of magnitude, the modulated order-parameter acquires a spatial variation influenced by the two terms : it is periodic with a wavevector κ, close to κ_o, but tending towards zero ; on the other hand, $\theta(x)$ is equal to 0° (modulo $\pi/2$) in large fractions of the period $(2\pi/\kappa)$, and deviates from these prominent values in narrow region of space thus determining a structure formed by commensurate regions[$\theta = \theta_i$]separated by walls (the "discommensurations").

3.3.3.) Lifschitz invariants and modulation directions

In the general case of a multidimensional OP and of a Lifschitz invariant depending of several space-variables, there can be minima of $\alpha(\vec{k})$ in more than one direction. In order to determine these directions, let us put:

$$\eta_p^\ell(\vec{r}) = \int e^{-i.\vec{\kappa}.\vec{r}} . \eta_p^\ell(\vec{\kappa}).d\vec{\kappa}$$

where $\eta_p^\ell(\vec{r})$ are the real components of the OP associated to $(\pm\vec{k}_p)$. The Fourier transform of the quadratic part of the free-energy density $f(\vec{r})$ takes the form :

$$\frac{1}{2}\Sigma|\eta_p^{\prime\ell}|^2 \{\alpha + \sigma_{p\ell uv} \cdot \kappa_u \cdot \kappa_v\} - i\Sigma\, \delta_{p\ell p'\ell' u}(\eta_p^\ell \bar{\eta}_{p'}^{\ell'} - \bar{\eta}_p^\ell \eta_{p'}^{\ell'}) \qquad (3.25)$$

where the $\sigma_{p\ell uv}$ and $\delta_{p\ell p'\ell' u}$ are related respectively to the coefficients

$C_{kk'\ell\ell'}$ and $B_{k\ell\ell'}$ in eq.(2.39)

Eq.(3.25) is a quadratic form in the η_p^ℓ. The eigenvalues of its matrix are functions of the κ_u components. The possible directions of the modulation correspond to the wavevectors $\vec{\kappa}$ which minimize either of the eigenvalues of this matrix, similarly as in (3.24).

As an illustration, consider the free-energy (2.79) adapted to the case of barium sodium niobate with L_1 given by (2.58) and $\mathcal{S}_1$, $\mathcal{S}_3$ given by (2.69). The Fourier transform (3.25) leads to 4 distinct eigenvalues :

$$[\frac{\alpha}{2} \pm 2\delta\kappa_X + \frac{\sigma_1}{2}(\kappa_X^2 + \kappa_Y^2) + \frac{\sigma_2}{2}(\kappa_X^2 - \kappa_Y^2)]$$

and (3.26)

$$[\frac{\alpha}{2} \pm 2\ \delta\kappa_Y + \frac{\sigma_1}{2}(\kappa_X^2 + \kappa_Y^2) + \frac{\sigma_2}{2}(\kappa_Y^2 - \kappa_X^2)]$$

whose minima as a function of κ_X and κ_Y are the two possible sets:

$$\kappa_X = \frac{\pm\ 2\delta}{(\sigma_1 + \sigma_2)} \qquad \kappa_Y = 0$$

or (3.27)

$$\kappa_X = 0 \qquad \kappa_Y = \frac{\pm\ 2\ \delta}{(\sigma_1 + \sigma_2)}$$

Remarks

i) There are two directions of modulation, in accordance with the result of §.2. Generally,the number of possible directions of modulations is not always determined by the form of the Lifschitz invariant, as illustrated by the case of thiourea (§.2.4.3).

ii) As announced in §.2.3.1.d), the space-coordinates which are present in the squared gradient and not in the Lifschitz invariant can be ignored if we deal with a system for which the modulation is determined by the presence of the latter invariant. Indeed (3.26) will have no minimum in directions κ_i not present in the linear term (e.g. z in the present example).

4. PHENOMENOLOGICAL THEORY FOR A TWO-COMPONENT (OP) IN THE (PMA) APPROXIMATION [11]

Let us illustrate the quantitative derivation of the properties of an INC system within the framework of the second phenomenological scheme (§.2.3), in the simple case of a two-dimensional OP (η,ζ) with a quartic lock-in term, and a single Lifschitz invariant. Using the variables $(\rho\ ,\ \theta)$ defined in (§.2.4), the free-energy density $f(\rho,\theta)$ has the form :

$$f(\rho,\theta) = \frac{\alpha}{2}\rho^2 + \frac{\beta_1}{4}.\rho^4 + \frac{\beta_2}{4}.\rho^4 \cos 4\theta - \delta\rho^2\dot{\theta} + \frac{\sigma}{2}(\dot{\rho}^2 + \rho^2.\dot{\theta}^2) \qquad (4.1)$$

where $\rho(x)$, and $\theta(x)$ are assumed to be modulated along the single modulation direction x. In order to ensure the stability of the commensurate phase in a certain temperature interval, without expanding f to higher degree terms, we must have (§.3.1.2) $\beta_1 > |\beta_2|$. On the other hand, we have seen (§.3.3.1) that the characterization of the modulation by a positive wave number κ implies ($\delta > 0, \sigma > 0$). As usual, we set $\alpha = \alpha_o(T - T_o)$. The free-energy is then :

$$F = \int_o^L f(\rho,\theta,\dot{\rho},\dot{\theta})\, dx \qquad (4.2)$$

L being the length of the crystal along x.

The equilibrium properties at a given temperature are described by the function $\rho_o(x)$ and $\theta_o(x)$ which minimize F. In order to determine these functions, we have to write the Euler-Lagrange differential equations corresponding to (4.2). Their general form is :

$$\frac{d}{dx}\left(\frac{\partial f}{\partial \dot{\rho}}\right) = \frac{\partial f}{\partial \rho} \qquad \text{and} \qquad \frac{d}{dx}\left(\frac{\partial f}{\partial \dot{\theta}}\right) = \frac{\partial f}{\partial \theta} \qquad (4.3)$$

There are no limit conditions [12] associated to this set of equations. Indeed the form (4.2) implies that the functional form of F is the same within the volume of the sample, and at its surfaces (for $x = 0$ and $x = L$). Consistently we must assume that the surface does not impose any specific value to ρ_o, $\dot{\rho}_o$, θ_o and $\dot{\theta}_o$. As a consequence, (4.3) will determine a family of "extremal" functions $\rho_o(x, C_i)$, $\theta_o(x, C_i)$ depending on the integration constants C_i. Reporting these functions into (4.2), we obtain an expression $F(C_i)$ of the free-energy. The integration constants C_i are then determined by the absolute minimum of F as a function of the C_i.

The specific form of the Euler-Lagrange equations, for the free-energy density (4.1) is :

$$\sigma.\ddot{\rho} = \rho(\alpha + \beta_1\rho^2 + \beta_2\rho^2.\cos 4\theta - 2.\delta\dot{\theta} + \sigma\dot{\theta}^2) = 0 \qquad (4.4)$$

$$\frac{d}{dx}[\rho^2.(\sigma\dot{\theta} - \delta) = -\beta_2.\rho^4.\sin(4.\theta) \qquad (4.5)$$

For general values of the coefficients, the solutions of this set of equations can only be obtained by numerical methods. We will examine the features of the solutions in §.5. Here, we begin by working out these solutions in the framework of a <u>simplifying assumption</u>, and we show that the approximate solutions thus obtained account for the principal experimental characteristics of incommensurate phases.

The assumption, which is called the "phase-modulation-only" approximation (PMA) [11] consists in considering that the amplitude ρ is not modulated, and that only $\theta(x)$ is a function of the spatial coordinate x. Clearly, there are no-exact solutions of (4.4 ; 4.5) complying strictly with this condition, since, through the vanishing of the second member of eq. (4.4), we see that, if $\theta = \theta(x)$, then $\rho \neq 0$ also depends on x. The implementation of the PMA must therefore be understood as following : consider in eq(4.5) ρ as a parameter ; find the general solutions of (4.5) in the form $\theta_o = \theta_o(x, \rho, C_i)$; report these solution in (4.2) and compute $F[\rho, C_i]$; alternately report $\theta_o(x, \rho, C_i)$ in (4.4) and replace the x dependent terms by their spatial average ; finally, determine the C_i by minimization of F and determine ρ either by the same procedure, or by solving eq.(4.4). The last step determines simultaneously the C_i and ρ as a function of the temperature. This method is a kind of mean-field method (which could be termed the mean-amplitude approximation).

Often, this procedure has been further simplified by imposing the temperature dependence of $\rho(T)$ [11] . One is then left with the sole determination of the C_i, which, as will be shown, are related to the wavenumber of the incommensurate modulation.

In the following paragraphs we first work out the general PMA solutions $\theta_o(x, \rho, C_i)$. We then examine successively in §.4.2 and §4.4. two special situations : the forms taken by θ_o respectively nearby the normal-incommensurate transition, and nearby the lock-in transition, in the framework of the "imposed $\rho(T)$" assumption. Finally in §.4.5) we release this assumption and work out the complete PMA solutions $\theta_o(x,T)$, $\rho_o(T)$. In each case we give attention to the location of the transitions T_I and T_L with respect to T_o, to their thermodynamic order, to the nature of the thermal anomalies arising at these transitions, and, to the characteristics of the INC modulation as a function of temperature.

4.1.) The θ-equation

For ρ = cte, eq(4.5) has the form :

$$\ddot{\theta} = -(\beta_2 \rho^2/\sigma) \sin (4\theta) \tag{4.6}$$

which is the non-linear pendulum equation ($\ddot{\theta} = -(g/l) \sin \theta$), also called the time-independent sine-GORDON equation. Before proceeding to find its solution, note that in §.3, we have seen that the sign of β_2 determines the nature of the stable commensurate phase, either $\theta = 0$ (modulo $\pi/2$) for $\beta_2 < 0$, or $\theta = \pi/4$ (modulo $\pi/2$) for $\beta_2 > 0$. Thus, it is likely that two different types of solutions of eq (4.6) will be found for $\beta_2 = \pm |\beta_2|$. In order to work them out jointly, we put :

$$\phi = 2\theta \ (\beta_2 > 0) \ ; \ \phi = 2\theta + \frac{\pi}{2} \ (\beta_2 < 0) \ ; \ v = \frac{4|\beta_2|\rho^2}{\sigma} \tag{4.7}$$

Eq. (4.6) then reduces, in either cases, to the form :

$$\ddot{\phi} = -(v/2).\sin(2\phi) \tag{4.8}$$

Multiplying the two members by $\dot{\phi}$ and integrating, we get :

Table 3. Definitions and results, used in chapter V, and concerning elliptic functions and integrals.

DEFINITIONS.

For $\mu^2 < 1$, the elliptic amplitude $\phi = am(X|\mu^2)$ is defined by:

$$X = \int_0^{\phi} \frac{d\theta}{\sqrt{1-\mu^2\sin^2(\theta)}} \qquad (1)$$

The elliptic sine and cosine are defined by:

$$sn(X|\mu^2) = \sin(am(X|\mu^2)) \quad ; \quad cn(X|\mu^2) = \cos(am(X|\mu^2)) \qquad (2)$$

The complete elliptic integral of the first kind $K(\mu^2)$ and of the second kind $E(\mu^2)$ are:

$$K(\mu^2) = \int_0^{\pi/2} \frac{d\theta}{\sqrt{1-\mu^2\sin^2(\theta)}} \quad ; \quad E(\mu^2) = \int_0^{\pi/2} \sqrt{1-\mu^2\sin^2(\theta)}.d\theta \qquad (3)$$

PROPERTIES

$$am(-X|\mu^2) = -am(X|\mu^2)$$
$$am(X+4K(\mu^2)|\mu^2) = am(X|\mu^2) + 2\Pi \qquad (4)$$

Thus $sn(X|\mu^2)$ and $cn(X|\mu^2)$ are periodic functions of period $4K(\mu^2)$. The shapes of $am(X|\mu^2)$ and of $sn(X|\mu^2)$ are shown on fig.7 for $\mu^2=(1/2)$, $\mu^2= 0.992$.

APPROXIMATIONS FOR $\mu^2 \approx 0$ AND $\mu^2 \approx 1$.

i) $\mu^2 \approx 0$.

Elliptic functions:

$$am(X|\mu^2) = X - \frac{\mu^2}{4}(X-\sin X.\cos X)$$
$$sn(X|\mu^2) = \sin X - \frac{\mu^2}{4}(X-\sin X.\cos X)\cos X \qquad (5)$$
$$cn(X|\mu^2) = \cos X + \frac{\mu^2}{4}(X-\sin X.\cos X)\sin X$$

Elliptic integrals:

$$K(\mu^2) = \frac{\pi}{2}(1+\frac{\mu^2}{4}+\ldots) \quad ; \quad E(\mu^2) = \frac{\pi}{2}(1-\frac{\mu^2}{4}-\ldots) \qquad (6)$$

ii) $\mu^2 \approx 1$.

Elliptic functions:

$$am(X|\mu^2) \simeq \left\{ \begin{array}{l} \mathrm{Arcsin}(\tanh X) \\ 2\mathrm{Arctang}(e^X) - \frac{\pi}{2} \end{array} \right\} + \frac{1-\mu^2}{4}.\frac{\sinh X.\cosh X - X}{\cosh X}$$

$$sn(X|\mu^2) \simeq \tanh X + \frac{1-\mu^2}{4}\frac{\sinh X.\cosh X - X}{\cosh^2(X)} \qquad (7)$$

$$cn(X|\mu^2) \simeq \frac{1}{\cosh X} + \frac{1-\mu^2}{4}.\frac{(\sinh X.\cosh X + X).\tanh X}{\cosh X}$$

Elliptic integrals:

$$K(\mu^2) \simeq \frac{1}{2}\mathrm{Log}\left(\frac{16}{1-\mu^2}\right) \quad ; \quad E(\mu^2) \simeq 1 + \frac{1-\mu^2}{2}.K(\mu^2) - \frac{1-\mu^2}{4} \qquad (8)$$

DERIVATIVES OF THE ELLIPTIC INTEGRALS.

$$\frac{dK(\mu^2)}{d\mu} = \frac{E(\mu^2)}{(1-\mu^2)} \quad ; \quad \frac{dE(\mu^2)}{d\mu} = \frac{E(\mu^2) - K(\mu^2)}{\mu} \qquad (9)$$

$$\dot{\phi}^2 = \frac{v}{2}\cos(2\phi) + C = (C + \frac{v}{2})\,[1 - \frac{v}{C + \frac{v}{2}}\cdot\sin^2\phi\,] \qquad (4.9)$$

Examine now the solutions for which C fulfills the conditions* $(C \pm \frac{v}{2}) > 0$. These conditions allow to define $0 < \mu^2 < 1$ as :

$$\mu^2 = \frac{v}{C + \frac{v}{2}} \qquad (4.10)$$

Equation (4.9) can then be written as :

$$dx = \pm \frac{\mu}{\sqrt{v}}\cdot\frac{d\mu}{\sqrt{1 - \mu^2.\sin^2\phi}} \qquad (4.11)$$

which, by integration, yields :

$$(x - x_o) = \pm \frac{\mu}{\sqrt{v}} \int_o^{\phi} \frac{du}{\sqrt{1 - \mu^2.\sin^2 u}} \qquad (4.12)$$

Table 3 indicates that the preceding integral is related to the elliptic function $am(X|\mu^2)$:

$$\phi_o(x) = am\,(\pm \frac{\sqrt{v}}{\mu}\cdot(x - x_o)|\mu^2) \qquad (4.13)$$

with $\theta_o(x)$ related to $\phi_o(x)$ by (4.7), and v and μ^2 being yet undetermined functions of the temperature.

We note that, as $am(-X|\mu^2) = -am(X|\mu^2)$, the two possible signs in eq.(4.13) correspond to solutions only differing by the sense of rotation of the angular variable $\theta(x)$. As shown by the expression of the free-energy (4.1), these solutions are physically equivalent and correspond to eachother through a reversal of the sense of the x-axis. We can adopt hereunder the sign (+) without loss of generality. On the other hand, x_o sets the origin of the modulation which is not specified**. We can therefore choose $x_o=0$. We finally get, for $\beta_2 > 0$:

$$\theta_o(x) = \frac{1}{2}\,am\,(\frac{\sqrt{v}}{\mu}\cdot x|\mu^2) \qquad (4.14)$$

* For $(C \pm \frac{v}{2}) < 0$ we would obtain periodic solutions which have been shown [13] to be unstable with respect to the minimization of F and which are not relevant to the present physical problem.

** It would be also easy to check that F is independent of x_o. This property corresponds to the free-sliding which has been discussed in §.3 of this chapter.

As mentioned in table 3 [eq(4)], $(X + 4K) = 5X) + 2\pi$ where $K(\mu^2)$ is the complete elliptic integral of the first kind (Eq.3.table 3). Thus,

$$\theta(x + \frac{8\ \mu\ K}{\sqrt{v}}) = \theta(x) + 2\pi \tag{4.15}$$

Going back to the expressions of the components (η,ζ) of the modulation, we see that these OP components, $\eta = \rho \sin\theta$ and $\zeta = \rho\cos\theta$, are periodic functions of x with period $(8\mu K/\sqrt{v})$, this period being temperature dependent through variations, yet to be specified, of $\mu(T)$, $v(T)$ and $K(\mu^2[T])$.

As shown by Fig. 7, $am(X\,|\mu^2)$ and $sn(X\,|\mu^2) = \sin(2\theta_o)$ have quite different shapes depending on the value of μ^2. If $\mu^2 \gtrsim 0$, am(X) is almost linear, and sn(X) is almost a sine-wave. This situation is necessarily realized just below the T_I transition because $\rho^2 \gtrsim 0$, and consequently $v \gtrsim 0$ [eqs.(4.7) and (4.10)].

On the other hand, if $\mu^2 \lesssim 1$, eq (1) in table 3 shows that the functions $\phi_o(X) = 2\theta_o$, and $\sin(2\theta_o) \propto sn(X)$ display wide plateaux separated by relatively narrow transition regions. On the plateaux, $\phi_o = \frac{\pi}{2}$ (modulo π). Thus, $\theta_o(x)$ has in this regions the same set of values as

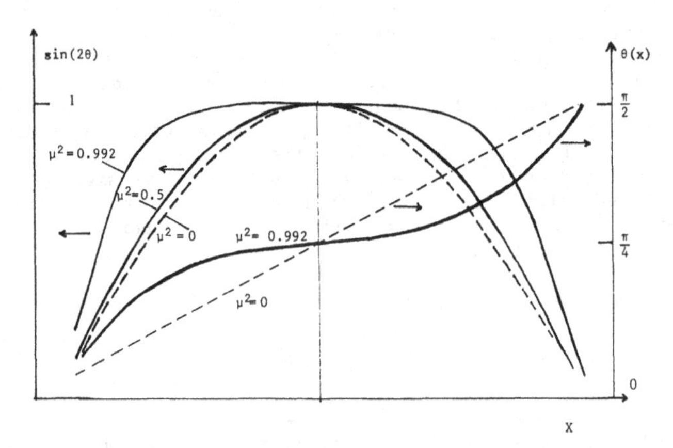

Fig. 7

in the various domains of the commensurate phase. Thus, in agreement with the standard experimental behaviour outlined in §.1.1, we find that the structure of the crystal in the interval of space-coordinates corresponding to the plateaux can be considered as isomorphous to the structure of the C-phase.

The overall structure resembles a multidomain commensurate structure where the narrow transition regions play the same role as domain walls. The specific characters of the INC structure are the periodicity of these walls and their temperature dependent width and spacing. We will see below, that the temperature range of this multidomain-like structure of the INC modulation lyes in the lower range of the INC phase, adjacent to the lock-in transition T_L. (Cf.fig.8).

4.2.) The sinusoidal limit ($\mu^2 \gtrsim 0$; $T < T_I$)

For $\mu^2 \approx 0$, one has $\phi_o(X) \gtrsim \mathrm{am}(X|0) = X$ (eq.5 in table 3). Equation (4.14) then yields :

$$(\beta_2 > 0):\ \theta_o(x) = \frac{\sqrt{V}}{2\mu}\, x = \kappa_o x, \text{ and } (\beta_2 < 0): \theta_o(x) = \kappa_o x + \frac{\pi}{4} \quad (4.16)$$

where κ_o is the unknown value of the modulation's wavenumber. To determine its value, as well as that of the amplitude ρ, let us follow the method outlined in the introduction of §4 and report (4.16) in (4.2), expressing in this way the free-energy $F(\kappa_o, \rho)$ as a function of the undetermined parameters.

If we assume $L >> \frac{2\pi}{\kappa_o}$ (the space-average of $\cos(4\theta)$ is then vanishingly small, or zero), we obtain :

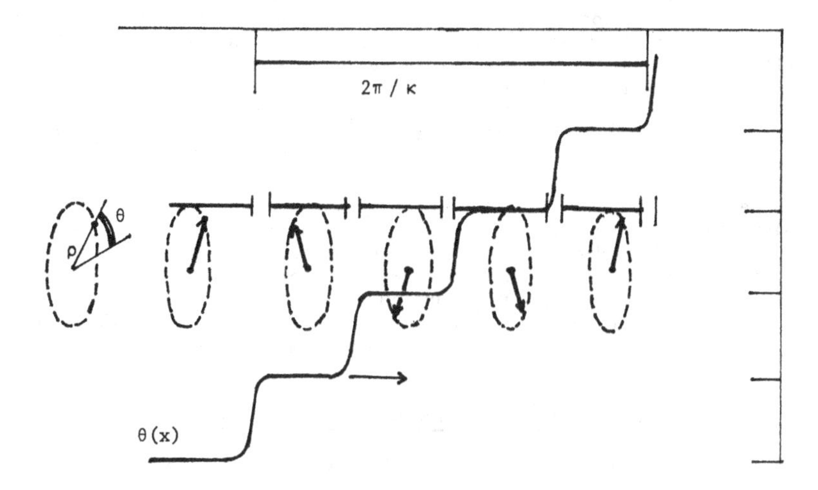

Fig. 8

$$F = \frac{1}{4}\int_0^L f(x).dx \not\approx (\frac{\alpha}{2} - \delta\kappa_o + \frac{\sigma\kappa_o^2}{2})\rho^2 + \frac{\beta_1}{4}.\rho^4 = \frac{\alpha(\kappa_o)}{2}\rho^2 + \frac{\beta_1}{4}\rho^4 \quad (4.17)$$

The values of κ_o and ρ are supplied by :

$$\frac{\partial F}{\partial \kappa_o} = \rho^2.[-\delta + \sigma\kappa_o] = 0 \quad ; \quad \frac{\partial F}{\partial \rho} = \rho[\alpha(\kappa_o) + \beta_1\rho^2] = 0 \qquad (4.18)$$

which yields :

$$\kappa_o = \frac{\delta}{\sigma} \qquad (4.19)$$

and

$$\rho_o^2 = -\frac{\alpha(\kappa_o)}{\beta_1} \qquad (4.20)$$

The transition to the HS phase occurs for $\alpha(\kappa_o)$ at a temperature :

$$T_I = T_o + \frac{\delta^2}{\alpha_o\sigma} \qquad (4.21)$$

We can then rewrite the temperature dependence of ρ_o, as :

$$\rho_o^2 = \frac{\alpha_o\ (T_I - T)}{\beta_1} \qquad (4.22)$$

In summary, nearby T_I, the INC modulation is described by components (η,ζ) which are sinusoidal functions of the coordinate x. The wavenumber of the modulation is [(eq.4.19)], determined by the coefficients δ and σ of the Lifschitz invariant and of the squared gradient in (4.1). Below T_I, the amplitude ρ_o grows with the same square-root law as usual in Landau's theory.

The lower limit of validity of this regime can be estimated by using the expression of the corrections to the approximation am(X) $\approx$ X (eq.5 ; table 3). This validity is defined by $\mu^2 \approx \frac{v}{C} \ll 1$, with $C \approx 4\kappa_o^2 = (2\delta/\sigma)^2$, and $v = (4|\beta_2|\rho_o^2/\sigma)$. Hence using (4.22), we obtain for the range of validity :

$$(T_I - T) \ll \frac{\beta_1}{|\beta_2|} \cdot \frac{\delta^2}{\sigma\alpha_o} \qquad (4.23)$$

This range is larger if the "isotropic" quartic term $(\frac{\beta_1}{4}.\rho^4)$ is large with respect to the lock-in term (with coefficient β_2). It is also larger if the Lifschitz invariant has a large coefficient δ.

As κ_o^2 is independent of the temperature, in the vinicity of T_I, it is apparent from the expression eq(4.17) of the free-energy that the thermal anomalies (and in particular that of the specific heat) have the standard shape expected for continuous transitions and already examined in chapter I (i.e. an upward jump on cooling for the specific heat).

4.3) General form of the free-energy for $\mu^2 \neq 0$

In this paragraph we compute the expression of $F(\rho,\mu^2)$ corresponding to the replacement in eq.(4.1) of $\theta(x)$ by its general solution (4.14). We perform this calculation for $\beta_2 > 0$. It would be easy to do the calculation along the same lines for $\beta_2 < 0$.

First, we write, for $\mu \neq 0$ the expression of $\dot{\theta}_o^2$ using eqs.(4.7) and (4.9) :

$$\dot{\theta}^2 = \frac{v}{4} \cdot \left[\frac{1-\mu^2}{\mu^2} + \cos^2(2\theta)\right] \tag{4.24}$$

We also replace $\cos(4\theta)$ by $(2.\cos^2\theta - 1)$, and use (4.11) to replace in (4.2) the integration over x by an integration over θ :

$$F=\int_o^L f(x)dx = L[\frac{\alpha}{2}\rho^2+\frac{(\beta_1-\beta_2)}{4}\rho^4] - \delta\rho^2[\theta(L)-\theta(o)]+\frac{\sigma v L}{8\mu^2}(1-\mu^2)\rho^2+\frac{\sigma\mu\sqrt{v}}{4}\rho^2.J \tag{4.25}$$

with ,

$$J = \int_{2\theta(o)}^{2\theta(L)} \frac{\cos^2 u.du}{\sqrt{1-\mu^2.\sin^2 u}} = \frac{(\mu^2-1)}{\mu^2}\int_{2\theta(o)}^{2\theta(L)} \frac{du}{\sqrt{1-\mu^2.\sin^2 u}} + \frac{1}{\mu^2}\int_{2\theta(o)}^{2\theta(L)} \sqrt{1-\mu^2.\sin^2 u}\,du \tag{4.26}$$

Since the integrands in J, are periodic functions of u, of period π , and that, on the other hand, $\theta_o(x)$ experiences a variation of π over half a period $(\frac{4\mu K(\mu^2)}{\sqrt{v}})$ of the modulation, we can write :

$$J = \frac{L\sqrt{v}}{2\mu^3.K(\mu^2)}\left[(\mu^2-1)\int_{-\pi/2}^{\pi/2}\frac{du}{\sqrt{1-\mu^2.\sin^2 u}} + \int_{-\pi/2}^{\pi/2}\sqrt{1-\mu^2.\sin^2 u}.du\right] =$$

$$\frac{L\sqrt{v}}{\mu^3}\cdot\left[(\mu^2-1)+\frac{E(\mu^2)}{K(\mu^2)}\right] \tag{4.27}$$

where $K(\mu^2)$ and $E(\mu^2)$ are the complete elliptic integrals of the first and second kind (see table 3).

Replacing, likewise, $[\theta(L) - \theta(o)]$ by $\frac{2\pi L\sqrt{v}}{8\mu K(\mu^2)}$, and making use of the expression (4.7) of v, we finally obtain for both β_2 signs :

$$\frac{F}{L}=[\frac{\alpha}{2}\rho^2+\frac{(\beta_1-|\beta_2|)}{4}\rho^4]-\frac{\pi}{2}\delta\sqrt{\frac{\beta_2}{\sigma}}\cdot\frac{\rho^3}{\mu K(\mu^2)}+\frac{|\beta_2|}{2\mu^2}\cdot\rho^4\cdot[\mu^2-1+\frac{2E(\mu^2)}{K(\mu^2)}] \tag{4.28}$$

4.4) Multisoliton limit ($\mu^2 \simeq 1$; $T \simeq T_L$)

4.4.1) The single soliton solution ($\mu^2 = 1$)

For $\mu^2 = 1$, the period of the modulation is infinite $[K(1) = \infty]$

Equation (5) in table 3 shows that the solution of the θ-equation has the form (for $\beta_2 > 0$) :

$$\theta(x) = \frac{1}{2} \text{Arcsin} [\tanh (x\sqrt{v})] \equiv \text{Arctang} (e^{\sqrt{v}x}) - \frac{\pi}{4} \qquad (4.29)$$

for $\beta_2 < 0$ one must add $\frac{\pi}{4}$ to the above solution. This solution describes a single wall, also called spatial SOLITON, or discommensuration, of width $\approx (1/\sqrt{v})$ [Fig. 9]

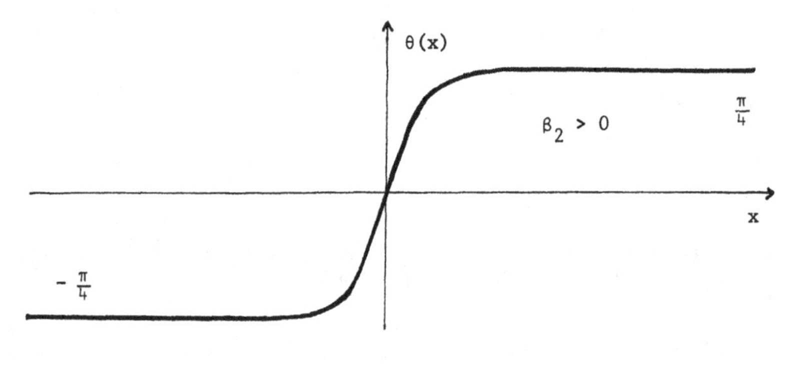

Fig. 9

As already noticed in the introduction of §.4, the regions outside the wall (for $|x| >> 1/\sqrt{v}$) have a structure close to that of the commensurate phase. The energy of the soliton can be defined as the difference between the free-energy F of the entire system, calculated for $\mu^2 = 1$, and the free-energy of the commensurate regions ($\theta = \pm\pi/4$) surrounding the soliton. Neglecting the width of the soliton, the second contribution is (eq.3), $F_c = L[(\alpha/2)\rho^2 + (\beta_1 - |\beta_2|)\rho^4/4]$. For the total free-energy we cannot use expression (4.28) which loses its validity for an infinite period of the modulation. Instead, we use (4.25) and (4.26) with $\mu^2 = 1$, and $[\theta(L) - \theta(o)] = (\pi/2)$. We obtain for the energy of the soliton :

$$E_S = F - F_c = \rho^2 . [- \frac{\delta\pi}{2} + \rho\sqrt{\sigma|\beta_2|}] \qquad (4.30)$$

This energy is negative for $\rho^2 < \frac{\delta^2\pi^2}{4\sigma\ |\beta_2|}$ and it increases (algebraically) for increasing values of ρ.

4.4.2.) Multisoliton regime, in the "imposed ρ (T)" approximation

Let us now examine the characteristics of the modulation nearby the lock-in transition T_L. As mentioned in §.4.1), in this temperature range, the shape of the modulation can be considered as a succession of commensurate regions separated by walls. Such a solution differs, strictly speaking, from a periodic array of single-solitons or "discommensurations" of the form (4.29). However, we can check that in the considered temperature range ($\mu^2 \gtrsim 1$, the difference between the actual solution (4.14) and a "multisoliton" regime is small. The exact free-energy (4.28) and the free-energy of N solitons, differ by a small quantity which can be interpreted as an interaction-energy between the solitons. Let us first determine the equilibrium value of $\mu(T)$ in the hypothesis that $\rho(T)$ is a known function of the temperature, not determined by minimization of (4.28). This assumption allows us to determine the number N of solitons in the crystal, and to write (4.28) as the free-energy of N interacting solitons. The behaviour of N(T) and the location of the lock-in transition T_L can then be derived.

a) Equilibrium value of the modulation wavenumber k(T)

The value of μ is provided by the minimum of F (eq.4.28) with respect to this parameter. Using the expressions of the derivatives of $K(\mu^2)$ and $E(\mu^2)$ (eq. 9 ; table 3), we obtain :

$$\frac{\partial F}{\partial \mu} = \frac{\rho^3}{\mu^2 . K(\mu^2)} \left[1 + \frac{E}{(1-\mu^2).K} \right] \left[\frac{\pi}{2} . \delta \sqrt{\frac{|\beta_2|}{\sigma}} - \frac{|\beta_2|}{\mu} \rho E \right] = 0 \quad (4.31)$$

As K and E are positive quantities, this equation yields :

$$\frac{E(\mu^2)}{\mu} = \frac{\pi \delta}{2\rho\sqrt{\sigma|\beta_2|}} \quad (4.32)$$

We now assume $(1 - \mu^2)$ small. The above equation provides us directly with the equilibrium value of $\mu(T)$, since $E(\mu^2) \approx 1$,

$$\mu(T) = \frac{2|\beta_2|\,\sigma}{\pi\delta} \cdot \rho(T) \quad (4.33)$$

The wavenumber of the modulation is then (eq.4.15) :

$$\kappa(T) = \frac{\pi . \sqrt{v}}{2.\mu K(\mu^2)} \approx \frac{\pi^2 . \kappa_o}{4 \operatorname{Log}\left[\dfrac{16}{1 - \dfrac{|\beta_2|\sigma}{\pi^2 . \delta^2} . \rho^2} \right]} \quad (4.33)'$$

with $\kappa_o = (\delta/\sigma)$ (eq. 4.19).

As there are 4 solitons per period, the distance between 2 solitons is $(\pi/2\kappa)$. The number of solitons per unit length of the crystal is :

$$n = \frac{2\kappa(T)}{\pi} \quad (4.34)$$

b) Form of the free-energy of N = nL interacting solitons

Equation (4.30) shows that the free-energy of the system with N non-interacting solitons is :

$$F_{NS} = L\left[\frac{\alpha}{2}\cdot\rho^2 + \frac{\beta_1 - |\beta_2|}{4}\rho^4\right] + N\rho^2\left[-\frac{\delta\pi}{2} + \rho\sqrt{\sigma|\beta_2|}\right] \quad (4.35)$$

Noting that,

$$N = \frac{L}{\mu K(\mu^2)}\cdot\sqrt{\frac{|\beta_2|}{\sigma}}\cdot\rho \quad (4.36)$$

we can rewrite (4.28) in the form :

$$F = F_{NS} + \frac{N\sqrt{|\beta_2|\sigma}}{\mu}\rho^3\cdot\left[E(\mu^2) - \frac{(1-\mu^2)K(\mu^2)}{2} - \mu\right] = F_{NS} + F_{INT} \quad (4.37)$$

(F_{INT}/N) being the interaction energy between two nearest neighbour solitons. On the basis of the values of $E(\mu^2)$ and $K(\mu^2)$, one can check that $F_{INT} > 0$: <u>the interaction is repulsive</u>.

For $\mu^2 \gtrsim 1$ the expression of (F_{INT}/N) is :

$$\frac{F_{INT}}{N} \simeq \sqrt{|\beta_2|\sigma}\,\frac{(1-\mu^2)}{4}\cdot\rho^3 \quad (4.38)$$

Drawing μ as a function of $(\frac{N}{L})$ from (4.36) for $\mu^2 \gtrsim 1$, we have:

$$(1-\mu^2) \gtrsim 16.e^{-2\cdot\sqrt{\frac{|\beta_2|}{\sigma}}\cdot\rho(\frac{L}{N})} \quad (4.39)$$

and, finally :

$$\frac{F_{INT}}{N} \simeq 4\sqrt{|\beta_2|\sigma}\rho^3\cdot e^{-2\sqrt{\frac{|\beta_2|}{\sigma}}\cdot\rho\cdot(\frac{L}{N})} \quad (4.40)$$

The repulsive interaction <u>decreases exponentially</u> with the distance $(\frac{L}{N})$ between solitons .

c) Lock-in transition at T_L

Equations (4.33) and (4.30) show that the temperature

dependent wavenumber $\kappa(T)$ and the soliton energy $E_S(T)$ vanish for the same value of $\rho(T)$. The corresponding value of the amplitude ρ is :

$$\rho^2 = \frac{\pi^2\delta^2}{4\sigma|\beta_2|} \tag{4.41}$$

Let us assume that in the vicinity of the lock-in transition $\rho(T)$ has the same temperature dependence as in the commensurate phase, i.e.:

$$\rho^2(T) = \frac{\alpha_o(T_o - T)}{(\beta_1 - |\beta_2|)} \tag{4.42}$$

Eq.(4.41) defines a characteristic temperature T_L :

$$T_L = T_o - \frac{\pi^2(\beta_1 - |\beta_2|)}{4|\beta_2|} \cdot \frac{\delta^2}{\sigma\alpha_o} = T_I - [1 - \frac{\pi^2}{4} \cdot (\frac{\beta_1}{|\beta_2|} - 1)] \frac{\delta^2}{\alpha_o\sigma} \tag{4.43}$$

This temperature corresponds to a <u>continuous transition</u> between the INC phase and the C-phase. Indeed, expressing (4.37) as a function of $\rho(T)$, we obtain the difference between the free-energies of the INC phase and of the C-phase in the vicinity of $T_L (T > T_L)$:

$$\Delta F = F_{INC} - F_C \approx \frac{A\ (T - T_L)}{\mathrm{Log}\left[\dfrac{T - T_L}{16(T_o - T_L)}\right]} < 0 \tag{4.44}$$

with :

$$A = \frac{\pi^2 \cdot \delta^2 \cdot \alpha_o}{8 \cdot \sigma(\beta_1 - |\beta_2|)} \tag{4.45}$$

Equation (4.44) shows that <u>ΔF goes to zero with zero slope</u> for $T \to T_L^+$. For $T < T_L$ we see that, in eq.(4.36), the INC phase has higher free-energy than the C-phase, as both the energy per soliton, and the interaction between solitons are positive. Our postulate (4.42) ensures the continuity of $\rho(T)$ at T_L, while the variation of $\kappa(T)$ is also continuous and given, for $T > T_L$, by :

$$\kappa(T) = \frac{\kappa_o}{4\ \mathrm{Log}\left[\dfrac{T - T_L}{16(T_o - T_L)}\right]} \tag{4.46}$$

The range of stability $[T_I - T_L]$ of the INC phase is indicated by (4.43). In agreement with the qualitative considerations of §.3.3) the INC phase is stabilized by the Lifschitz invariant (large δ), and destabilized by the lock-in term ($|\beta_2|/\beta_1$ close to 1).

d) **Latent heat and specific heat at T_L**

Using (4.33), we can put (4.37) under the form :

$$\frac{F_{INC}}{L} = \frac{F_C}{L} - \frac{\pi^2 \cdot \delta^2}{4\sigma} \cdot \frac{(1-\mu^2)}{(E(\mu))^2} \cdot \rho^2 \qquad (4.47)$$

The difference between the entropies of the two phases is, per unit length :

$$[S_{INC} - S_C] = -\frac{\partial}{\partial T}\left[\frac{F_{INC}-F_{COM}}{L}\right] = \frac{\pi^4 \cdot \delta^4}{32|\beta_2|\sigma^2} \cdot \frac{d}{dT}\left[\frac{\mu^2(1-\mu^2)}{(E(\mu))^4}\right] \qquad (4.48)$$

Using the values [eq(9), Table 3] of $\frac{dE}{dm}$ and $\frac{dK}{dm}$, and (4.33), we obtain

$$\Delta S = \frac{\pi^2 \cdot \delta^2 \cdot \alpha_o}{16(\beta_1 - |\beta_2|)\sigma}\left[\frac{2}{EK} - \frac{4 \cdot (1-\mu^2)}{E^2}\right] \qquad (4.49)$$

for $T \to T_L$, ρ is continuous and $\mu \to 1$. The leading term in ΔS is :

$$\Delta S \sim \frac{1}{EK} \to 0 \qquad (4.50)$$

Thus, the T_L transition has no latent heat, consistently with its continuous character. The specific heat in either phases is provided by : $C_p = [T.\partial S/\partial T]$. As S_c determines no anomaly, due to the continuity of $\rho(T)$, the anomalous part of C_p is given by :

$$C_p = T\,\frac{\partial(\Delta S)}{\partial T} \qquad (4.51)$$

or, within the INC phase :

$$C_p = \frac{|\beta_2| T \cdot \alpha_o^2}{4(\beta_1 - |\beta_2|)^2} \cdot \left[\frac{E^2}{\mu K^2} - \frac{3E}{\mu^2 K} + \frac{2(1-\mu^2)}{\mu^2} + \frac{E^3}{\mu^2 K^3 (1-\mu^2)}\right] \qquad (4.52)$$

for $T \to T_L^+$, we have :

$$C_p \propto \frac{\beta_1 \cdot \alpha_o^2 \cdot T_L}{4(\beta_1 - |\beta_2|)^2} \cdot \frac{1}{K^3(1-\mu^2)} \qquad (4.53)$$

which diverges for $T \to T_L^+$. The behaviour of $C_p(T)$, across the range of the INC phase, is outlined on fig. 10 .

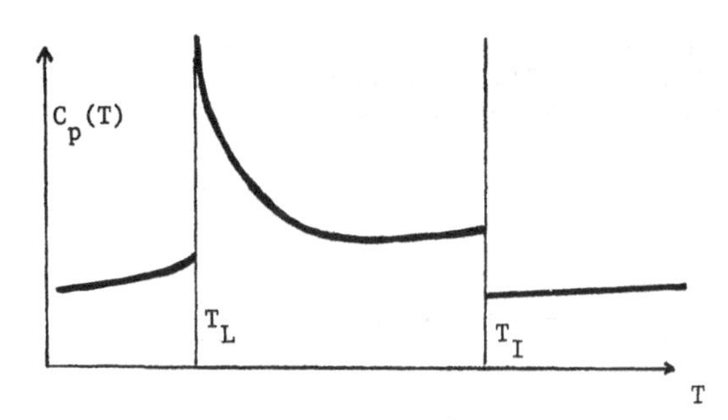

Fig. 10

4.4.3.) Lock-in transition with variational amplitude $\rho(T)$

If we release the "imposed amplitude approximation", equation (4.33) does not provide us with the temperature dependence of $\mu(T)$. The two quantities $\mu(T)$ and $\rho(T)$ are simultaneously determined by the set of equations $\partial F(\mu,\rho, T)/\partial\mu = 0$, and $\partial F(\mu,\rho, T)/\partial\rho = 0$. Deriving the second condition from eq.(4.28) and eliminating ρ by use of eq.(4.33), we obtain, on the one hand, an equation determining $\mu(T)$ and, on the other hand, the expression of the free-energy as a function of μ, and T :

$$\alpha + \frac{\delta^2\pi^2}{4\sigma}\cdot\left[\frac{1}{EK} + \frac{\mu^2(1+\beta_1/|\beta_2|)-2}{E^2}\right] = 0 \tag{4.54}$$

$$(F_{INC}/L) = \frac{\pi^2\cdot\delta^2}{16\cdot\sigma|\beta_2|}\cdot\left(\frac{\mu^2}{E^2}\right)\cdot\left(\alpha - \frac{\pi^2\cdot\delta^2}{4\cdot\sigma\cdot EK}\right) \tag{4.55}$$

Let us put $\alpha_L = \alpha(T_L)$, where T_L is the lock-in transition within the "imposed amplitude approximation" (eq.4.44). We can deduce from (4.55), (4.30) and (4.33) the difference of free-energies between the INC phase and the C-phase :

$$\Delta F = -\frac{1}{4(\beta_1 - |\beta_2|)}\left[\alpha_L^2\left(\frac{\mu}{E}\right)^2\cdot\left(\frac{\alpha}{\alpha_L} + \frac{1}{KE}\right) - \alpha^2\right] \tag{4.56}$$

Unlike the case studied in 4.4.2, the approach of the lock-in transition cannot be studied by expanding (4.56) and (4.54) in the vicinity of $\mu = 1$, because as we now point out, the transition determined by eqs.(4.54-4.56) is discontinuous, and it occurs for $T_L' < T_L$ and $\mu < 1$. Fig. 11 shows the variations $\mu(\alpha)$ and $\Delta F(\alpha)$ deduced from the above equations. We can see that for $\alpha_m < \alpha < \alpha_L$, $\mu(\alpha)$ is a two-valued function, with $\mu_-(\alpha)$, the smallest of the two $\mu(\alpha)$ values, being associated to the lowest value of the INC free-energy. It shows, on the other hand, that ΔF becomes positive, on cooling, below a temperature T_L' such as $\alpha_m < \alpha_L' < \alpha_L$, and $\mu(T_L') < 1$. As the modulation wavenumber is $\kappa = [\pi\sqrt{\beta_2}.\rho/2.\sqrt{\sigma}.\mu K(\mu^2)]$, κ is finite at the transition, and vanishes discontinuously at the onset of the C-phase.

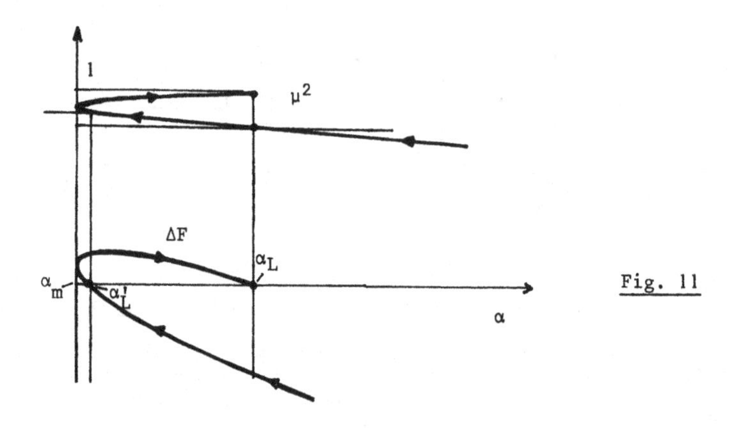

Fig. 11

Thus, in the PMA approximation with a variationally determined amplitude $\rho(T)$, the lock-in transition, which occurs at $T_L' < T_L$, appears <u>discontinuous</u>, while it is continuous in the "imposed amplitude" scheme. This discontinuity can be pronounced. Indeed if we take $\beta_1 = 2\cdot|\beta_2|$ in eq.(4.54) and eq.(4.56), we find numerically that $\mu^2(T_L') \gtrsim 0.8$. Hence the amplitude ρ will jump upwards, on cooling through T_L', by an amount of 25 % (cf.eq.4.33).

Note that the imposed amplitude $\rho^2 \propto |T-T_o|$ insures the continuity of ρ at T_L. However the validity of the imposed form clearly fails nearby T_I since the function $\rho(T)$ vanishes for $T_o < T_I$. This drawback is avoided with the variational amplitude. Indeed, eq.(4.54) yields $\mu = 0$ for $\alpha = \alpha_I = (\delta^2/\sigma)$, and therefore, (eq.4.33) $\rho(T_I) = 0$. Using eqs (4.54-4.56) one can easily show along the same lines as in §.4.4.2.d), that, consistently with its discontinuous character, the transition T_L' is associated with a latent heat and a specific heat anomaly.

4.5) Macroscopic quantities and anomalies of the susceptibilities

For an INC phase, similarly to the cases of improper ferroelectric or ferroelastic phases (chapIII), several susceptibilities are of interest : namely the susceptibilities relative to the primary OP, and those relative to certain secondary OP associated to macroscopic quantities. Let us discuss here the case of the considered two-dimensional modulated OP. Other cases will be examined in §.5 .

As stressed in §.2.3.1, the free-energy density associated to the modulated OP contains the same "homogeneous terms" as the commensurate free-energy. This property which we have used (§.2.4) to work out the form of the expansion as a function of the primary OP is valid for secondary ones. Accordingly, we can use the results worked out in chapter III for improper ferroics. If we consider, first, macroscopic quantities P which are coupled to second-degree-powers of the primary OP components (i.e. quantities corresponding to a faintness index of 2), the coupling terms can take one of 3 possible forms :

$$P_1\rho^2 \quad ; \quad P_2\rho^2\cos(2\theta) \quad ; \quad P_3\rho^2\sin(2\theta) \qquad (4.57)$$

In the first term, ρ^2 is an invariant, and P_1 must also be an invariant by the HS space-group. Such a term describes the coupling of the OP to macroscopic quantities which do not break the orientational symmetry of the system. In the framework of the PMA approximation, these quantities will also be spatially uniform.

By contrast $\rho^2 \cdot \sin(2\theta)$ and $\rho^2 \cdot \cos(2\theta)$ are not invariants, and the onset of $P_2 \propto \rho^2\cos(2\theta)$, or $P_3 \propto \rho^2 \sin 2\theta$, will induce locally (i.e. in the volume where θ is approximately uniform) a lowering of the orientational symmetry. For instance P_2 can be a component of the dielectric polarization $\vec{P}$, which will determine locally the onset of a ferroelectric structure. However, as $\cos[\theta(x)]$ and $\sin[\theta(x)]$ are periodic functions, P_2 and P_3 are also modulated quantities with a period equal to half the period of the components of the primary OP (eq.4.15) . Fig.12.a) shows the respective variations of the primary components (η,ζ), and of P_2 along the x-direction. Clearly the spatial average of P_2 or P_3 on a large distance L is zero : <u>there is no breaking of the macroscopic orientational symmetry induced by P_2 or P_3</u> within the INC phase. The symmetry breaking can only occur in the C-phase [14].

This property holds for higher degree-coupling terms. Indeed, according to eqs (2.34) and (2.35), the description of the INC phase by a modulated OP is equivalent, with respect to the symmetry of the INC distortion, to the description by a uniform OP having the actual INC wavevector. In the latter description, the only <u>translationally invariant</u> powers of the OP, are of the form ρ^2 (cf. eq.2.6), i.e. they are <u>totally symmetric.</u> Macroscopic (translationally invariant) quantities which can aquire spontaneous values in the INC phase will be present in coupling terms of the form $(P.\rho^{2p})$. These quantities will be, like ρ^{2p}, totally symmetric, and they will preserve the orientational symmetry of the HS phase, within the INC phase. Hence <u>the INC phase and the HS phase necessarily have the same orientational (macroscopic) symmetry.</u> We point out, in §.5, that such a property is not always valid for OP dimensions larger than 2.

4.5.1.) Qualitative discussion of the susceptibilities

For a two-dimensional OP, we can distinguish a "longitudinal" component of the susceptibility, describing the variations of the amplitude ρ under application of a field conjugated to the primary OP, and a "transverse" component associated to the variations of θ. Only the second will have a specific behaviour for INC systems, since the main properties of an INC system can be satisfactorily accounted for (§.4) in an approximation where the modulation of ρ is neglected. The longitudinal susceptibility will undergoe a standard divergence at T_I, in accordance with the validity of the expansion (2.19) in the vicinity of this transition. In the "imposed amplitude" approximation (§.4.4.2), this quantity will have no anomaly at T_L.

Let us now examine qualitatively the anomalies related to the modulation of θ. In this respect the main result is the existence, on approaching T_L from above, of a divergence of the susceptibilities associated to all the quantities (the primary and certain secondary OP) whose spatial average is zero in the INC phase, and non-zero in the C-phase.

Consider on fig. 12b) the effect on $\eta = \rho\cos\theta$, and on $P = \rho^2 . \cos(2\theta)$ of the application of the respective conjugated fields ξ and E, the modulation being in the multisoliton regime (§.4.4.2). The effect of these fields is to favour one η or P orientation with respect to the others, and, consequently to enlarge certain C_i domains, and to reduce others (fig. 12b). Such a variation of the distances between solitons is only opposed by the repulsive interaction (eq.4.41). The smaller this interaction, the larger the volume of the favourably oriented C_i domains for a given value of the fields, and the larger the susceptibilities which are proportional to the non-zero space-average of the (η/P) quantities induced by (ξ/E). As the interaction vanishes for $T \to T_L^+$ (eq.4.41, and 4.46), the susceptibilities related to both η and P will tend to infinity for $T \to T^+$. Fig.12c) shows the differences $[\eta(x,\xi) - \eta(x,o)]$, and $[P(x,E) - P(x,o)]$ for small values of ξ and E. Due to the existence of 4 types of C_i domains [$\theta = 0$, or $\pi/4$ modulo $\pi/2$] in the considered model, the periodic pattern induced by ξ in the variations of $\eta(x)$ is more complex than the pattern induced by E in the variations of $P(x)$. The calculation of the solution $\theta(x)$ of the Euler-Lagrange equations, in the presence of a small conjugated field, is likely to be easier if this field is conjugated to P(x). We will restrict to this case.

4.5.2.) Quantitative derivation of the susceptibility relative to $P \propto \rho^2\cos(2\theta)$ [15]

The free-energy density corresponding to a non-zero field E can be written, following Currat and Axe [15], as :

$$g(\rho,\theta,\dot{\rho},\dot{\theta}, P) = f(\rho,\theta,\dot{\rho},\dot{\theta}) + \frac{\alpha_p}{2}.P^2 - \nu P\rho^2.\cos(2\theta) - E.P \quad (4.58)$$

where $f(\rho,\theta,\dot{\rho},\dot{\theta})$ is provided by eq.(4.1) and where $\alpha_p > 0$ and $\nu > 0$. The minimum of the integrated free-energy density with respect to $P(x)$ yields :

$$P(x) = \left(\frac{\nu\rho^2}{\alpha_p}\right)\cos(2\theta) + \frac{E}{\alpha_p} \tag{4.59}$$

Reporting this expression into (4.58), we obtain :

$$g(\rho,\theta,\dot{\rho},\dot{\theta}) = \frac{\alpha}{2}\rho^2 + \frac{\beta_1'}{4}\rho^4 + \frac{\beta_2'}{4}\rho^4\cos(4\theta) - \delta\rho^2\dot{\theta} + \frac{\sigma}{2}(\dot{\rho}^2 + \rho^2\dot{\theta}^2) - \frac{\nu E\rho^2}{\alpha_p}\cos(2\theta) - \frac{E^2}{\alpha_p} \tag{4.60}$$

where $\beta_1' = (\beta_1 - \frac{\nu^2}{\alpha_p})$ and $\beta_2' = (\beta_2 - \frac{\nu^2}{\alpha_p})$. We assume $\beta_2' < 0$ in order that, in the C-phase and in the C_i-domains separating the spatial solitons, the equilibrium value θ_o of θ, in the absence of fields, is $\theta_o = 0$ (modulo $\pi/2$). In that case, (4.59) shows that a non-zero spontaneous value of $P_s(x)$ exists in the C_i-domains, in agreement with fig.6b). The field E will induce an additional polarization comprising two terms : i) a temperature independent one (E/α_p); ii) a term arising from the modification of the function $\theta(x)$, in eq. (4.59). The method we use to evaluate the susceptibility associated to $P(x)$ involves three steps [15].First, we express θ as

$$\theta(x) = \theta_o(x) + \theta_1(x) \tag{4.61}$$

and determine the small correction $\theta_1(x)$ induced by the field In the second place we express as a function of E the contribution to $P(x)$ which is induced by the field. To lowest order, this contribution is :

$$p(x) = P(x) - P_s(x) \sim \frac{E}{\alpha_p} - \left(\frac{2.\nu\rho^2}{\alpha_p}\right).\theta_1(x).\sin(2\theta_o) \tag{4.62}$$

Finally, we determine the contribution of $(\frac{p(x)}{E})$ to the macroscopic susceptibility by averaging $(p(x)/E)$ over the length of the sample (or over a period of the modulation).

a) **Determination of $\theta_1(x)$**

In the PMA approximation, the Euler-Lagrange equation for $\theta(x)$, associated to (4.60) is :

$$\ddot{\theta} = -\frac{\beta_2'\rho^2}{\sigma}\cdot\sin(4\theta) + \frac{2\nu}{\alpha p\sigma}\sin(2\theta) \tag{4.62}$$

Reporting in (4.62) a solution of the form (4.61) and keeping the terms of lowest order in $\theta_1(x)$ and E, we obtain :

$$\ddot{\theta}_1 = -\frac{4\beta_2'\cdot\rho^2}{\sigma}\cdot\cos(4\,\theta_o)\,.\,\theta_1 + \frac{2\nu\xi}{\alpha_p\sigma}\cdot\sin(2\,\theta_o) \tag{4.63}$$

A solution of (4.63) is $\theta_1(x) = a\cdot\sin(2\,\theta_o)$ [15] . Such a solution determines an induced polarization $p(x) \propto \sin^2(2\theta_o)$ which agrees with the qualitative result infered in §.4.5.1 (fig.12c). Replacing in (4.63), θ_1 by the former function, and making use of eqs.(4.6) and (4.9), for $\beta_2' < 0$, we can determine the value of the coefficient a :

$$a = \frac{-\nu\xi}{\alpha_p\cdot|\beta_2'|\cdot\rho^2}\cdot\left(\frac{\mu^2}{1-\mu^2}\right) \tag{4.64}$$

b) **Form of the induced polarization and susceptibility**

Using (4.64) and (4.62), we can write :

$$p(x) = \frac{E}{\alpha_p}\left[1 + \frac{\nu^2}{\alpha_p\cdot|\beta_2'|}\cdot\left(\frac{\mu^2}{1-\mu^2}\right)\sin^2(2\,\theta_o)\right] \tag{4.65}$$

The macroscopic susceptibility can be written as the spatial average of $(p(x)/E)$:

$$\chi = \frac{\langle p(x)\rangle}{E} = \frac{1}{L}\int_o^L p(x)\,dx \tag{4.66}$$

Using the same transformation as in §.4.3), for $\beta_2' < 0$, we obtain :

$$\langle\sin^2(2\theta_o)\rangle = \frac{1}{\mu^2}\left[\frac{E}{K} - (1-\mu^2)\right] \tag{4.67}$$

where $K(\mu^2)$ and $E(\mu^2)$ are the elliptic integrals of the first and second kind (cf. table 3). Finally :

$$\chi = \frac{1}{\alpha_p}\left\{1 + \frac{\nu^2}{\alpha_p\cdot|\beta_2'|\cdot(1-\mu^2)}\left[\frac{E}{K} - (1-\mu^2)\right]\right\} \tag{4.68}$$

For $T \rightarrow T_I^-$, in the sinusoidal limit, $\mu \approx 0$ and $(E/K) \approx (1 - \frac{\mu^2}{2})$. Thus :

$$\chi = \frac{1}{\alpha_p}\left[1 + \frac{3\,\nu^2}{2\cdot\alpha_p\cdot|\beta_2'|}\cdot\mu^2(T)\right] \tag{4.69}$$

where $\mu^2 \propto (T_I - T)$ (Cf. eq. 4.22, 4.10 and 4.7). On the other hand, above T_I, χ reduces to $\frac{1}{\alpha_p}$ (eq. 4.59). In this model, the T_I transition induces no anomaly of the susceptibility relative to the secondary OP. This result is in contrast with the step variation of the susceptibility obtained at an improper ferroic transition (chap III). The different behaviour is due to the fact that no macroscopic value of

P onsets below T_I.

We can examine the behaviour nearby T_L in the framework of the "imposed amplitude" approximation. The transition T_L corresponds to $\mu^2 \to 1$ (§4.4.2). In this limit, $E(\mu^2) \to 1$ and $K(\mu^2) \to \infty$. Using the equivalents of these functions (table 3) and eq.(4.33), one finds :

$$\chi(T \to T_L^+) \underset{\sim}{\sim} \frac{\nu^2}{2\alpha_p^2 |\beta_2'|} \cdot \frac{1}{[E(\mu^2)-1]} \quad \frac{\pi^2 \nu^2 (\beta_1 - |\beta_2'|)}{4 \cdot \alpha_p^2 \cdot \beta_2'^2} \quad \frac{T_I - T_o}{T - T_L} \qquad (4.70)$$

In agreement with the qualitative discussion in §.4.5.1, χ diverges for $T \to T_L^+$, with a Curie-Weiss law $\propto (T - T_L)^{-1}$.

Eq.(4.68) also allows to study the behaviour of χ in the case where $\rho(T)$ is determined variationally (§.4.4.3). Fig.12d summarizes the resulting variations [15]as well as those obtained (eq.4.70) in the "imposed amplitude" approximation.

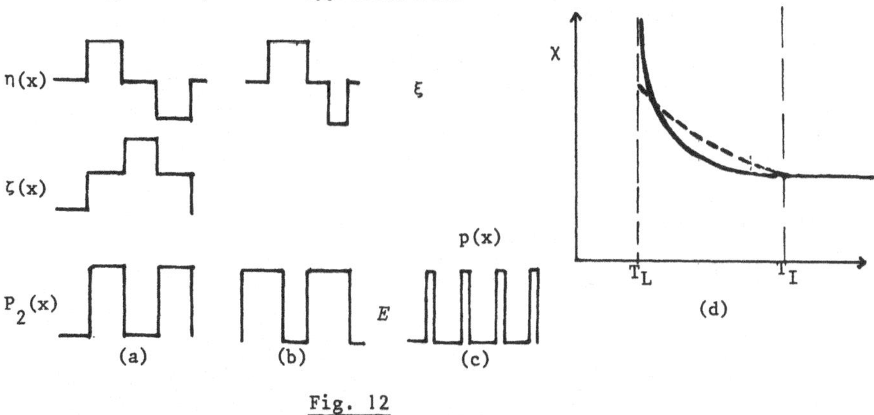

Fig. 12

It also shows the behaviour of the spatial average of the spontaneous value of P(x). This quantity undergoes a jump at T_L even when this transition is continuous. Indeed, below T_L it is proportional to ρ^2 which has a finite value at T_L, while above T_L, $\langle P(x) \rangle = 0$

5. EXTENSIONS

In this paragraph we examine first the physical implications of the free-energy density (4.1) when the restrictions imposed by the PMA are lifted. We then give our attention to the situation in which the modulated free-energy involves no Lifschitz-invariant (§.2.4.3). In §.5.3, we consider various additional contributions to the free-energy density which allow to account, differently from the manner already discussed, for the onset of the INC phase, the temperature dependence of κ, and the characteristics of the lock-in transition. Finally we examine the case of INC phases associated to OP dimensions larger than 2.

5.1) Lifting of the PMA

As stressed in the introduction of §.4, the equilibrium state of the INC system is determined by a set of equations (4.4.-4.5) which cannot be solved analytically in the general case. It is worth pointing out, however, that the PMA provides rigorous solutions in the limit of vanishing amplitudes ρ , i.e. in the sinusoidal region, just below T_I. Indeed, in this limit, eq.(4.5) is reduced to its left member which admits the solution ρ = cte, $\theta(x) = (\delta/\sigma)x$. Deviations form the PMA are expected at lower temperatures, and, in particular when $T \to T_L$. Also note that deviations from the sinusoidal regime will also vanish when $\beta_2 \to 0$ (i.e. when the lock-in term tends to zero). In agreement with the qualitative discussion in §.3, it is the term ($\beta_2 \cdot \cos 4\theta$) which is at the origin of the lock-in transition and of the precursor structuration of the modulation into spatial solitons.

To study the deviations from the PMA one has to use numerical methods to solve eqs (4.4-4.5) and determine the minimum of the free-energy (4.2). The free-energy density (4.1) depends on 5 coefficients. However, by renormalizing the various quantities, it is possible to reduce to two the number of physically relevant coefficients, and to use dimensionless quantities. For instance we can put :

$x = (\sigma/\delta)X$; $\rho = \delta/\sqrt{\sigma\beta_1}\,\rho_1$; we obtain :

$$f_1(X) = \frac{\varepsilon}{2}\rho_1^2 + \rho_1^2\frac{(\dot{\theta}-1)^2}{2} + \frac{\rho_1^4}{4} + \frac{d}{4}\rho_1^4\cos(4\theta) + \frac{\dot{\rho}_1^2}{2} \qquad (5.1)$$

with,

$$\varepsilon = \frac{\sigma\cdot\alpha_o\cdot(T - T_I)}{\delta^2} \quad ; \quad d = \frac{\beta_2}{\beta_1} \quad ; \quad F = \frac{\delta^3}{\sigma\beta_1}\int f_1(X)\,dX \qquad (5.2)$$

where d controls the relative amplitude of the lock-in term and ε is a function of temperature and of the coefficient δ of the Lifschitz invariant.

By using the complex components $\eta_1 = \rho_1 \cdot e^{i\theta}$ and $\zeta_1 = \bar{\eta}_1 = \rho_1 e^{-i\theta}$ the free-energy density (5.1) can also be written in the form :

$$f_1(X) = \frac{(\varepsilon+1)}{2}\eta_1\zeta_1 + \frac{\dot{\eta}_1\dot{\zeta}_1}{2} + \frac{i}{2}(\dot{\eta}_1\zeta_1 - \dot{\zeta}_1\eta_1) + \frac{(\eta_1\zeta_1)^2}{4} + \frac{d}{8}(\eta_1^4 + \zeta_1^4) \qquad (5.3)$$

Euler-Lagrange equations derived from free-energy densities (5.1) or (5.3) can then be solved for various values of ε and d. Up to now, two methods have been used in this view : i) finite difference methods [17] applied to the set of differential equations. ii) methods [18] using a truncated Fourier series expansion of the periodic functions $\eta_1(X)$ and $\zeta_1(X)$, with the unknown quantities being the fundamental wavenumber q and the coefficients of the expansion.

It is worth noting that the latter Fourier series will contain certain harmonics only, whose rank depend on the form of the lock-in term. Consider for instance the Euler-Lagrange equations associated to the reduced form (5.3) :

$$\frac{\ddot{\zeta}_1}{2} + i\,\dot{\zeta}_1 = \frac{(\varepsilon+1)}{2}\,\zeta_1 + \frac{\eta_1\zeta_1^2}{2} + \frac{d}{2}\cdot\eta_1^3$$
$$\frac{\ddot{\eta}_1}{2} - i\,\dot{\eta}_1 = \frac{(\varepsilon+1)}{2}\,\eta_1 + \frac{\eta_1^2\zeta_1}{2} + \frac{d}{2}\cdot\zeta_1^3 \qquad (5.4)$$

This set of equations cannot be satisfied with a single term $\eta_1 = \bar{\zeta}_1 \propto e^{iqx}$, since the lock-in term with coefficients (d/2) generates a third harmonic e^{-3iqx}. Likewise, if we put $\eta_1 = (a_1 e^{iqx} + a_3\, e^{-3iqx})$ the various terms in (5.4) will generate harmonics of (+5) (-7) (+9) etc... Finally, a truncated Fourier series must be written as [18]:

$$\eta_1 = \bar{\zeta}_1 = \sum_{n=o} a_{4n+1}\cdot e^{i(4n+1)qx} + \sum_{n=1} a_{4n-1}\cdot e^{-i(4n-1)qx} \qquad (5.5)$$

Reporting the solution (5.5) into the free-energy density (5.3) and calculating the spatial average (5.2) of $f_1(X)$, one obtains, as a q-dependent contribution to F :

$$-\frac{\delta^3}{\sigma\beta_1}\left[q\,[\sum_{n=o}(4n+1)|a_{4n+1}|^2 - \sum_{n=1}(4n-1)|a_{4n-1}|^2] + \frac{q^2}{2}[\sum_{n=o}(4n+1)^2|a_{4n+1}|^2 + \sum_{n=1}(4n-1)^2|a_{4n-1}|^2]\right] \qquad (5.6)$$

Hence, the equilibrium value of q is determined by the minimum of (5.6):

$$q = \frac{\Sigma(4n+1)\ |a_{4n+1}|^2 - \Sigma(4n-1)\ |a_{4n-1}|^2}{\Sigma(4n+1)^2\ |a_{4n+1}|^2 + \Sigma(4n-1)^2\ |a_{4n-1}|^2} \qquad (5.7)$$

the length scale in reciprocal space being $\kappa_o = (\delta/\sigma)$ which is the wavenumber of the modulation in the sinusoidal region. Consistently, eq (5.7) yields q = 1 for a series reduced to the fundamental period (n = 0, in the first sum of (5.7)). Eq. (5.7) also shows that the additional harmonics determine q < 1, i.e. a wavenumber which tends to zero as measure as the importance of the harmonics grows. On the other hand, the value of the coefficients a_{4n+1} and a_{4n-1} can be determined by the recurrence equations resulting from eqs. (5.4), for each choice of the parameters ε and d.

Both the numerical analyses by the Fourier series method and by the finite difference method show that, on approaching T_L, the amplitude ρ of the modulation becomes dependent on the spatial coordinates. Fig. 13 shows the schematic behaviour found. As apparent on this figure, the amplitude ρ is larger in the commensurate regions than within the walls separating these regions. Clearly this decrease of ρ in the walls is favoured by the lock-in term. Indeed, within a soliton, the lock-in term is positive and its contribution will be decreased for smaller values of ρ. On the other hand, numerical calculations [17] show that for certain values of the model's parameters, the amplitude has an "overshoot" above the value taken by ρ in the middle of the C-domains. Finite difference methods also disclose an overshoot in $\theta(X)$ [fig.13]

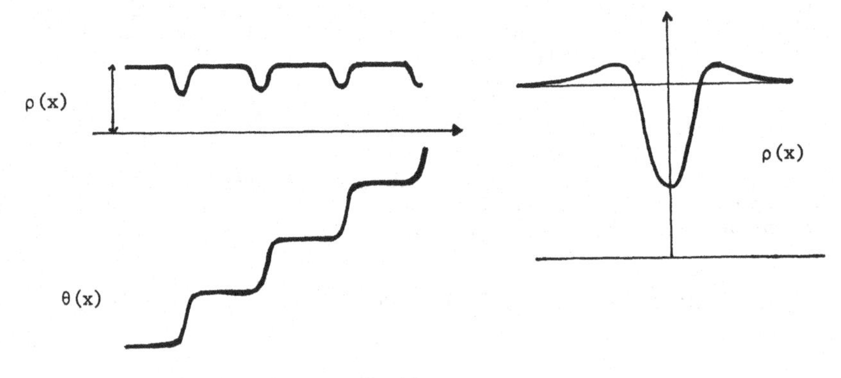

Fig.13

As compared to the PMA, the numerical calculations determine a lock-in transition shifted towards lower temperatures, i.e. a larger stability range for the INC phase. The situation with regard to the thermodynamic order of the transition at T_L is more complex. Finite difference methods [17] provide a phase diagram, as a function of the model parameters, which displays a line of lock-in transitions partly first order (for large lock-in terms) and partly second order (for small lock-in terms). The two portions of lines are separated by a tricritical point (fig.14). On the other hand Fourier series methods always yield a continuous lock-in transition. Possibly, finite difference methods are more precise in the range of parameters for which a discontinuous transition exists.

The occurence of a continuous lock-in transition for a certain range of the model's parameters is in disagreement with the result found in the framework of the PMA with variationally determined amplitude (it agrees with the simpler, but in principle less correct, model with imposed amplitude). As shown in §.4.4.3) the PMA with variational amplitude leads to a discontinuous lock-in transition, for any value of the parameters in the model. This discrepancy with the numerical results denotes the loss of validity of the PMA in the vicinity of T_L, in a temperature range where the spatial dependence of ρ can be pronounced (fig. 13). By contrast the numerical studies yield results close to those determined in the framework of the PMA when the coefficient of the lock-in term is not too large.

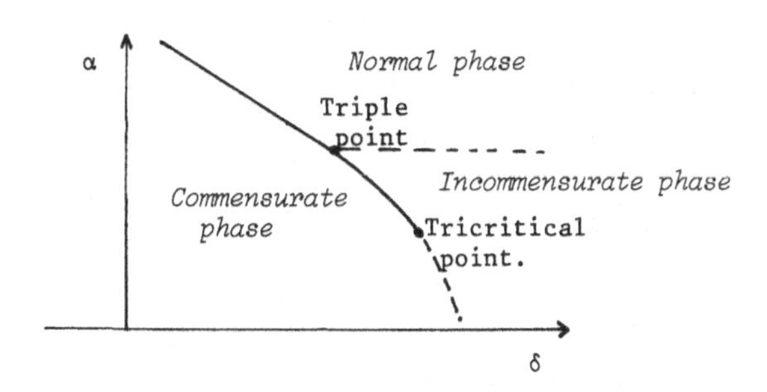

Fig. 14

Schematic phase diagram drawn from ref. 17. and are parameters in eq. (4.1).

5.2.) Phenomenological theory in the absence of a Lifschitz invariant

In this paragraph we examine briefly the theory of INC phases when the OP symmetry (at $k = k_L$) does not allow the existence of a Lifschitz invariant (§.2.3.2). As emphasized in §.3.3.1 the LI is associated to the value of $[d\alpha\,(k)/dk|_{k_L}]$. The absence of a LI denotes that $\alpha(k)$ has a relative extremum for $k = k_L$. Since the existence of the INC phase requires that $\alpha\,(k)$ reaches its absolute minimum for $(k = k_o = k_L + \kappa_o)$, the simplest form of $\alpha(k)$ compatible with those two conditions involves a relative maximum for $k = k_L$ (fig. 4c). The correspondance between the gradient expansion of the free-energy density and the form of $\alpha(k_L + \kappa)$ (§.3.3.1), shows that a maximum at k_L ($\kappa = 0$) implies a negative sign for the coefficient σ of the squared gradient term. It also shows that in order to express the occurence of a minimum at $k_o = (k_L + \kappa_o)$, it is necessary to introduce in the free-energy density additional terms, containing higher-order derivatives of the OP-components. These terms should be of degree two in the OP components and their Fourier transform should be proportional to q^{2n} (with $n > 1$).

Let us examine the case, illustrated by thiourea (§.2.4.3), for which the OP, with wavevector k_L, is one-dimensional. Equation (2.80) provides the required free-energy density. It contains the lowest degree additional term compatible with the preceding conditions : $(\gamma/2)\,[\partial^2\eta/\partial x^2]^2$. Indeed the Fourier transform of the second degree terms in (2.80) provides :

$$\alpha(\kappa) = [\gamma\kappa^4 + \sigma\kappa^2 + \alpha] \qquad (5.8)$$

which, for $\gamma > 0$ and $\sigma < 0$, possesses a maximum at $\kappa = 0$ and a minimum at

$$\kappa_o = (-\sigma/2\gamma)^{1/2} \quad \text{(fig.4.c)} \qquad (5.8)'$$

5.2.1.) Simplified theory |19|

A simplified account of the characteristics of the INC phase is possible on the basis of the free-energy density (2.80). Following the procedure used in §.4, we deduce a variational equation from (2.80) :

$$\gamma \frac{d^4\eta}{dx^4} - \sigma \frac{d^2\eta}{dx^2} + \alpha\eta + \beta\eta^3 = 0 \qquad (5.9)$$

Like the set (5.3)(5.4), the preceding equation can only be solved by numerical methods. However, we can examine its approximate sinusoidal solution. Putting in (5.9) $\eta = \eta_\kappa . \cos(\kappa x)$, and neglecting the higher harmonics generated by $(\beta\eta^3)$, we obtain :

$$(\gamma\kappa^4 + \sigma\kappa^2 + \alpha) + \frac{3\beta}{4} . \eta_\kappa^2 = 0 \qquad \text{for } \kappa \neq 0 \qquad (5.10)$$

and

$$\alpha + \beta\eta_o^2 = 0 \qquad \text{for} \qquad \kappa = 0 \qquad (5.11)$$

Reporting in (2.80) the values of η_κ and η_o deduced from (5.10) and (5.11), we calculate the value of the free-energy :

$$F_\kappa = \int f(x) . dx = - [\gamma\kappa^4 + \sigma\kappa^2 + \alpha]^2/6\beta \qquad \text{for } \kappa \neq 0 \qquad (5.12)$$

and

$$F = - \alpha^2/4\beta \qquad \text{for} \qquad \kappa = 0 \qquad (5.13)$$

Consistently with (5.8)' (5.12) yields the equilibrium value of κ corresponding to the minimum of F, i.e. $\kappa = (-\sigma/2\gamma)^{1/2}$. The equilibrium value of the INC free-energy is

$$F = - (\alpha - \sigma^2/4\,\gamma)^2/6\beta \qquad (5.14)$$

an expression valid for $\alpha < \sigma^2/2\gamma$. The latter condition determines the INC transition temperature $T_I = T_o + \sigma^2/2\alpha_o$. On the other hand, a comparison of the INC and C-equilibrium free-energies (5.14) shows that the INC phase ($\kappa = \kappa_o$) is stable for :

$$T_L = [T_o - \frac{\sigma^2}{4\,\alpha_o\gamma} \cdot \frac{1}{\frac{3}{2} - 1}] < T < [T_o + \sigma^2/4\,\alpha_o\gamma] = T_I \quad (5.15)$$

Below T_L, the C-phase ($\kappa = 0$) is stable. In the above sinusoidal approximation, the lock-in transition at T_L is discontinuous (κ jumps from $\kappa = \kappa_o$ to $\kappa = 0$).

In this model, the different features of the system can be understood qualitatively in the following way : i) below T_I the HS phase is unstable because $\alpha(\kappa_o)$ is negative. ii) between T_I and T_o, while the INC phase with $\kappa = \kappa_o$ is stabilized by fourth degree terms, the C-phase is unstable because α ($\kappa = 0$) > 0. iii) below T_o, both the INC and C-phase correspond to minima of the free-energy. However the value of these minima differ. On the one hand the coefficient of the quadratic term $\alpha(\kappa)$ is more negative for the INC phase,

On the other hand, the spatial averaging of η^4 renormalizes the β coefficient differently for $\kappa \neq 0$ and for $\kappa = 0$. This renormalization determines the stabilization of the C-phase below T_L, when the quartic term becomes sufficiently large. Hence the mechanism of the lock-in transition is, similarly to the one studied in §.3.1.3, related to the different contributions of the fourth degree terms to the INC and C-free energies. However, while in the case formerly studied the renormalization introduced by the spatial averaging only affected the "θ-dependent lock-in term", here it concerns all the contributions to the free-energy density.

5.2.2. Refined theories

a) Numerical solution of the variational equation [20]

A numerical solution of eq. (5.9), in the form of a Fourier series can be worked out by the same method outlined in §.5.1 for the case involving a Lifschitz invariant. It will supply the equilibrium value of κ at each temperature, as well as the form of $\eta(x)$.

Unlike the sinusoidal approximation which yields similar physical results in the absence or presence (§.3.1) of a Lifschitz invariant (constancy of κ , abrupt lock-in transition), the numerical methods yield qualitatively distinct results for the two types of models. The prominent difference is that, in the absence of a Lifschitz invariant, the numerical solution of (5.9) provides results differing little from those obtained in the sinusoidal approximation : i) the lock-in transition is strongly discontinuous [20]; ii) $\kappa(T)$ remains close to κ_o in the whole range of the INC phase [20].iii) harmonics of the basic sinusoidal wave have a small amplitude [20]

This consistency of the sinusoidal description can be illustrated by an explicit determination of the amplitude of the first correction to the sinusoidal approximation. Let us put $\eta(x) = [\eta_\kappa \cos(\kappa x) + \eta_{3\kappa} \cos(3\kappa x)]$ (the reason for the absence of the (2κ) harmonic is the same as for eq.(5.5))

Reporting this form of $\eta(x)$ in eq.(5.9) and expressing this equation for the third harmonic, we obtain :

$$\eta_{3\kappa}\,(81\gamma\kappa^4 + 9\sigma\kappa^2 + \alpha) = -\frac{\beta}{4}\,(\eta_\kappa^3 + 3\cdot\eta_{3\kappa}^3 + 6\cdot\eta_\kappa^2\cdot\eta_{3\kappa}) \qquad (5.16)$$

Replacing in (5.16), η_κ by its value (5.10) in the sinusoidal approximation, and setting $\kappa = \kappa_o$, we obtain :

$$\left|\frac{\eta_{3\kappa}}{\eta_\kappa}\right| \approx \frac{|\alpha_o\,(T_I - T)|}{3|\alpha_o(T-T_I) + 65\cdot\sigma^2/4\cdot\gamma|} \qquad (5.17)$$

This ratio increases on lowering the temperature. Its maximum value is reached for $T = T_L$. Using the value (5.15) of T_L we obtain :

$$\left|\frac{\eta_{3\kappa}}{\eta_\kappa}\right| < 0.03 \qquad (5.18)$$

thus justifying the validity of the sinusoidal approximation even near T_L.

b) **Additional gradient term**

As emphasized in the above discussion, the free-energy density (2.81) cannot account for a pronounced temperature dependence of $\kappa(T)$, while experimentally such a pronounced variation is observed in cases pertaining to the theoretical scheme without a Lifschitz invariant. In order to account for such a feature, it is necessary to introduce additional terms to the free-energy density (2.80). These terms must determine an interaction between the amplitude of the modulation (which has a strong temperature dependence) and the wavenumber $\kappa(T)$. Accordingly, each such term must involve gradients of $\eta(x)$, and be of degree higher than two in the OP components. For the one-dimensional OP considered in this paragraph, it is easy to check that the term of lowest order fulfilling these conditions and compatible with the OP symmetry has the form :

$$\frac{\nu}{2}\cdot\eta^2\cdot\left(\frac{\partial\eta}{\partial x}\right)^2 \qquad (5.19)$$

with $\nu > 0$. This term is a positive contribution to the free-energy whose importance grows on cooling due to its fourth degree in the amplitude of the modulation. Its minimization has the effect of reducing the value of the gradient $|d\eta/dx|$, i.e. of decreasing the wavenumber κ (T). Also, its cancellation in the C-phase favours the stabilization of this phase and determines an upward shift of the lock-in transition.

Let us work out the modifications introduced by the term (5.19) in the results of the sinusoidal approximation. Equations (5.9)(5.10) and (5.12) are respectively replaced by equations (5.20)-(5.22) hereafter :

$$\gamma \frac{d^4\eta}{dx^4} - \sigma \frac{d^2\eta}{dx^2} - \nu\eta \left[\left(\frac{d\eta}{dx}\right)^2 + \eta\left(\frac{d^2\eta}{dx^2}\right) \right] + \alpha\eta + \beta\eta^3 = 0 \qquad (5.20)$$

$$\alpha(\kappa) + \frac{3}{4}\left(\beta + \frac{2}{3}\nu\kappa^2\right)\eta_\kappa^2 = 0 \qquad (5.21)$$

$$F_\kappa = - \frac{[\alpha(\kappa)]^2}{6\left(\beta + \frac{2}{3}\nu\kappa^2\right)} \qquad (5.22)$$

The equilibrium value of κ is determined by $(\partial F_\kappa / \partial\kappa) = 0$. For small values of the ν coefficient, this condition yields a linear temperature dependence for the equilibrium value $\kappa_o'(T)$:

$$\kappa_o'(T) \approx \kappa_o \left[1 - \frac{\nu\alpha_o}{6\beta|\sigma|} (T_I - T) \right] \qquad (5.23)$$

Likewise, a positive shift of the lock-in temperature is determined : we obtain, by comparing (5.22), (5.13) and (5.14) :

$$\Delta T_L = \frac{\nu|\sigma|(T_o - T_L)}{\sqrt{6}\,(\sqrt{6} - 2).\gamma\beta} \qquad (5.24)$$

For larger values of ν, and extending the above treatment by taking $\eta(x)$ in the form of a Fourier series, it is possible to study more precisely the influence of the term (5.19). As illustrated by fig (15), such a study [20] shows that (5.19) improves the progressivity of the lock-in transition : its first order character is attenuated, a steeper variation of κ is obtained in its vicinity, and the contribution of higher harmonics to the shape of $\eta(x)$ is increased. However their relative amplitude remains small, and there is no pronounced structuration into an array of solitons as in the presence of a Lifschitz invariant.

5.2.3) Origin of the negative sign of σ

Th existence of a maximum of $\alpha(k)$ for $k = k_L$ ($\sigma < 0$) requires no explanation in the phenomenological framework considered in this book. However, the occurence of such a maximum very close to a minimum at k_o denotes an anomalous shape of $\alpha(k)$ in the vicinity of k_L for which a justification is necessary. Two types of phenomenological models have been elaborated in this view.

a) Dispersive coupling between irreducible OP [21]

Let us assume that the considered OP belongs to a

"normal" situation, for which $\alpha(k)$ has a minimum at $k = k_L$, and no additional extremum in the vicinity of k_L. Consider, on the other hand another degree of freedom ζ whose symmetry is described by the same one dimensional irreducible representation as $(d\eta/dx)$, i.e. distinct from the symmetry of η. Such a situation can be realized for instance in thiourea. ,as shown in table 4 .

Table 4.

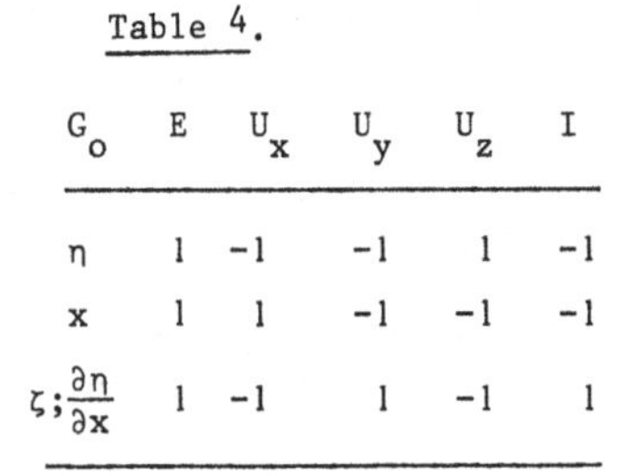

G_o	E	U_x	U_y	U_z	I
η	1	-1	-1	1	-1
x	1	1	-1	-1	-1
$\zeta;\frac{\partial\eta}{\partial x}$	1	-1	1	-1	1

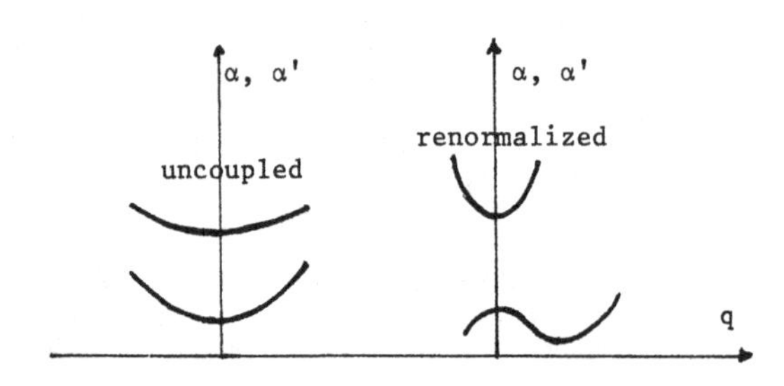

Fig. 15

Due to the identical symmetries of ζ and $\frac{d\eta}{dx}$ (or of η and $\frac{d\zeta}{dx}$) the free-energy density corresponding to both η and ζ will contains terms of the form $\zeta(d\eta/dx)$. The second degree contribution to this free-energy density is :

$$f(x) = \frac{\alpha}{2}\eta^2 + \frac{\sigma'}{2}\left(\frac{d\eta}{dx}\right)^2 + \frac{\alpha''}{2}\zeta^2 + \frac{\sigma''}{2}\left(\frac{d\zeta}{dx}\right)^2 + \mu\left(\zeta\cdot\frac{d\eta}{dx} - \eta\frac{d\rho}{dx}\right) \quad (5.25)$$

(As usual the symmetric combination ($\zeta d\eta/dx + \eta\, d\zeta/dx$) can be dropped out (§.2.3.2)).

The Fourier transform of f(x) determines a density in reciprocal space :

$$f(\kappa) = \frac{(\alpha + \sigma'.\kappa^2)}{2}|\eta_\kappa^2| + \frac{(\alpha'' + \sigma''.\kappa^2)}{2}|\zeta_\kappa^2| + \mu\kappa\eta_\kappa\zeta_\kappa \quad (5.26)$$

We eliminate ζ_κ by minimizing $f(\kappa)$ with respect to ζ_κ. We obtain :

$$f(\kappa) = \frac{1}{2}|\eta_\kappa|^2\left\{\alpha + \kappa^2\left(\sigma' - \frac{4\mu^2}{\alpha'' + \sigma''\kappa^2}\right)\right\} \quad (5.27)$$

Thus, even if $\sigma' > 0$ (i.e. if $\alpha(\kappa)$ is minimum for $\kappa = 0$), the renormalization operated by the elimination of ζ_κ determines and effective coefficient $\alpha'(\kappa)$ of $|\eta_\kappa^2|$ which is negative for $\kappa \to 0$ (i.e. a maximum for $\alpha'(\kappa)$ when $\kappa \to 0$) if the strength μ of the dispersive coupling between η and ζ is large enough. On the other hand, for larger κ, the term $\sigma''\kappa^2$ ($\sigma'' > 0$) has the effect of restoring a positive value to the coefficient of $|\eta_\kappa|^2$. Hence, there will be a minimum in this coefficient for $k = k_o$ close to k_L (fig. 15).

A correct treatment of the preceding model requires taking into account all the invariant terms constructed from $\eta(x)$, $\zeta(x)$, $d\eta/dx$, $d\zeta/dx$, up to 4th degree, and solving the set of two variational equations associated to η and ζ for the complete free-energy density. Such a treatment [21] confirms the qualitative picture given above.

b) **Dispersive coupling between the OP and a strain component** [22]

Another slightly different model can also be considered which differs from the preceding one by the fact that ζ is replaced by a strain component e. The latter quantity is itself the derivative of a displacement. Thus the Fourier transform of a homogeneous strain is proportional to κ. The free-energy density (to second degree) associated to this model, and its Fourier transform, can be written as :

$$f(x) = \frac{\alpha}{2}\eta^2 + \frac{\sigma'}{2}\left(\frac{d\eta}{dx}\right)^2 + \frac{C}{2}e^2 + \mu e\left(\frac{d\eta}{dx}\right) \qquad (5.28)$$

$$f(\kappa) = |\eta_\kappa|^2 \cdot \left[\frac{\alpha + \sigma'\kappa^2}{2}\right] + \frac{C}{2}\kappa^2 . u_\kappa^2 + \mu\kappa^2 . u_\kappa . \eta_\kappa \qquad (5.29)$$

where u_κ is the Fourier transform of the displacement u(x). As in the preceding paragraph the elimination of u_κ leads to :

$$f(\kappa) = |\eta_\kappa|^2 \left[\frac{\alpha}{2} + \frac{\kappa^2}{2}\left(\sigma' - \frac{\mu^2}{C}\right)\right] \qquad (5.30)$$

Similarly to (5.27), if μ is large enough or C small enough (elastic softness), the sign of the coefficient of κ^2 will be changed from positive to negative. Higher order gradients of η, than the ones taken into account in (5.18), have to be considered in order to restore a positive sign for high κ and determine a minimum of the coefficient for $\kappa \neq 0$

5.3) Effect of additional spatially dispersive terms.

a) Alternate mechanism for the INC transition

The stabilization of the INC phase, considered in §.4 and in §.5.2, has its origin, either in the presence of a Lisfchitz invariant, or in the presence of a squared gradient invariant with a negative coefficient. If neither of these features occurs, it is nevertheless possible to account for the onset of an INC phase at T_I by considering additional terms in the free-energy density, which are allowed by the symmetry of the system, and which are spatially dispersive terms of degree higher than two in the OP components. Such an effect is, for instance, produced by a term of the type $(\eta_i^2 \frac{\partial \eta_j}{\partial x})$ of third degree in the OP components. Note that for a one-dimensional OP this term has no physical effect since it is equal to a total derivative (cf.§.2.3.2). We have therefore to consider modulated OP with higher dimensions. As an illustration we investigate a simplified model [23] associated to a 2-component OP, and whose free-energy density is :

$$f(x)=\frac{\alpha}{2}(\eta_1^2+\eta_2^2)+\frac{\beta}{4}(\eta_1^2+\eta_2^2)^2+\gamma(\eta_1^2+\eta_2^2)(\frac{\partial \eta_2}{\partial x})+\frac{\sigma}{2}\left[(\frac{\partial \eta_1}{\partial x})^2+(\frac{\partial \eta_2}{\partial x})^2\right] \quad (5.31)$$

Let us put $\eta_1=\rho[1+\cos\kappa(x)]$ and $\eta_2=-\sin\kappa(x)$, and calculate for this solution the spatial average of $f(x)$. We obtain :

$$F=\int f(x)\cdot dx=(\alpha+\frac{\sigma}{2}\cdot\kappa^2)\rho^2 \quad \gamma\kappa\rho^3+\frac{3}{2}\beta\rho^4 \quad (5.32)$$

The conditions $(\partial F/\partial\kappa)=0$, $(\partial F/\partial\rho)=0$ yield :

$$\kappa_o=\frac{\gamma}{\sigma}\rho_o \quad ; \quad \rho_o^2=\frac{-\alpha}{(3\beta-\frac{\gamma^2}{\sigma})} \quad ; \quad F_o=\frac{-\alpha^2}{6\beta(1-\gamma^2/3\beta\sigma)} \quad (5.33)$$

On the other hand, a homogeneous solution $\eta_1=\rho\cos\theta$, $\eta_2=\rho\sin\theta$, corresponds to an equilibrium free-energy $F_1=-(\alpha^2/4\beta)$. Comparing with the free-energy F_o, we find that the modulated phase will be stable for $(\gamma^2>\beta\sigma)$. This phase will onset for $\alpha<0$ (i.e for $T<T_I$) with zero wavenumber $\kappa_o=\rho_o=0$), and its wavenumber will increase on lowering the temperature. Thus the considered third degree term determines the occurence of an unusual type of INC phase with "reversed" temperature dependence for the wavevector.

b) Thermodynamic order of the lock-in transition [19,24,25]

We have seen in §.5.2.b, that in the sinusoidal limit, a dispersive coupling of the OP to a strain component renormalized the squared gradient term in a sense favouring the onset of an INC phase. Let us show that, in the multisoliton limit (§.4.4.2), i.e. near T_L, this type of coupling can change the thermodynamic order of the lock-in transition from second order (found in the PMA, §.4.4.2.c) to first order.

We consider the free-energy density (4.1) containing a Lifschitz invariant. In this case symmetry will always allow a coupling term of the form $\mu\rho^2\cdot\dot{\theta}\cdot e$, where e is a totally symmetric strain component, and $(\rho^2\cdot\dot{\theta})$ the Lifschitz invariant. Taking into account the elastic energy, the free-energy density is :

$$f \quad = f(\rho,\theta)+\frac{C}{2}\cdot e^2+\mu\rho^2\cdot e\,\frac{d\theta}{dx} \quad (5.34)$$

where $f(\rho,\theta)$ is expression (4.1). Let us assume that e is not modulated, [19], and that ρ and θ are determined in the framework of the PMA with imposed amplitude (§.4.4.2).

Spatially averaging (5.34) we obtain straightforwardly :

$$F = F_{NS} + F_{INT} + \frac{C}{2} e^2 + \mu\rho^2 . e.N.\frac{\pi}{2} \qquad (5.35)$$

where $(F_{NS} + F_{INT})$ is the free-energy of N interacting solitons (eqs. 4.36-4.41). We then eliminate e, by expressing $(\partial F/\partial e) = 0$, and obtain :

$$F = F_{NS} + F_{INT} - (\frac{\mu^2}{8C} \pi^2 \rho^4) N^2 \qquad (5.36)$$

A qualitative argument allows to understand that the additional term of the form $-DN^2$, generated by the coupling to the strain, drives the lock-in transition discontinuous.

Indeed, this term is minimum for $N \to \infty$, and it will therefore favour the persistence of solitons below the lock-in temperature T_L^o in eq.(4.47), in the absence of coupling. If, below T_L^o a continuous lock-in occured at T_L, we could write for $T \to T_L$, $N \to 0$,

$$F_{INC} - F_C \approx AN - DN^2 \qquad (5.37)$$

where $A > 0$ and $D > 0$ (cf.eq.4.36). The minimum of (5.37) with respect to N yields $N \approx (A/2D)$ which increases on cooling in contradiction with the assumption of a continuous lock-in (A is proportional to the single soliton energy which vanishes at T_L^o and becomes positive below).

Numerical investigation of (5.36) confirms the first order-character of T_L. It is worth noting that the above treatment of the strain as a homogeneous quantity is not, strictly speaking, correct. A complete treatment [25] should allow for the variation of e(x) and take into account, as in (eq.5.29) its form as the derivative of a displacement. It can then be shown that the variational equations obtained, lead to a contribution of the form $(-DN^2)$, as found in eq.(5.36). This treatment shows, in addition, that the spatial average of the modulated strain is non-zero, and proportional to N.

5.4.) Higher number of OP components and of modulation directions

In the preceding paragraphs the discussion of the properties of INC systems has been restricted to the cases of two-dimensional OP (for the OP corresponding to k_o) and to a single direction x of modulation.

A certain number of examples exist (cf.§.6) in which the appropriate number of INC OP-components is larger than two. In these examples there are one or several independent modulation directions (cf.§.2.1.5). For instance, we have seen in §.2.1.4 and 2.1.5, that the INC phase in $BaMnF_4$ is associated to a 4-component OP and a single modulation direction. Likewise, the INC phase in 2H-$TaSe_2$ corresponds (§.2.1.4, 2.1.5) to a 6-component OP and a maximum number of two independent modulation directions.

Three main points distinguish the properties of these "complex" cases from those of the standard system considered in §.4 :

i) The onset of the INC phase can generate spontaneous components of macroscopic tensors which break the point symmetry of the HS phase.

ii) The phase diagram can include several INC phases differing by their "effective" point-symmetry as well as by their number of modulation directions.

iii) Spatial solitons can form a complex pattern.

As an illustration of these features, let us first discuss the example of barium sodium niobate.

5.4.1.) Barium sodium niobate

The INC, commensurate, and modulated free-energy expansions pertaining to the INC phase in this material have been derived in §.2.4.2. The INC and C-order-parameters have 4-dimensions and the INC star of $\vec{k}_o$ vectors contains two independent modulation directions.

a) Spontaneous symmetry breaking strain and effective point-symmetry in the INC phase

Consider, first, the irreducible OP corresponding to an INC wavector. Its transformation properties are described by the matrices (2.76). The only translationally invariant polynomials of the OP components are independent of the angular variables θ_i (defined for eq.2.58) and only depend on the two moduli ρ_1 and ρ_2. However these invariants are not all totally symmetric (i.e. rotationally invariant). For instance, there are 2 translationally invariant second-degree terms, $(\rho_1^2 + \rho_2^2)$ and $(\rho_1^2 - \rho_2^2)$. The first is totally symmetric while the second has the same rotational symmetry properties as the strain component $(e_{XX} - e_{YY})$ (X and Y were specified in eq. 2.57). Hence the INC free-energy can contain a term of the form :

$$(e_{XX} - e_{YY})(\rho_1^2 - \rho_2^2) \qquad (5.38)$$

This coupling term is of a type already analyzed in the case of improper ferroelastics (chap III). Its effect is to induce, in the INC phase, a spontaneous value for the shear strain $(e_{XX} - e_{YY})$, which is proportional to $(\rho_1^2 - \rho_2^2)$.

This result is in contrast with the one stated in §.4.5 for two-components OP. In the latter case, the OP could only be coupled to totally symmetric tensorial components, which are already symmetry permitted in the HS-phase. Hence the INC and HS phase could be considered to possess the same "effective" point symmetry which governs the form of macroscopic tensors . In barium sodium niobate, due to the four-dimensionality of the OP, which generates terms as (5.38), the number of non-zero tensor components can be higher in the INC phase, and, accordingly, this phase can possess a lower effective point-symmetry than the HS phase.

b) **Phase diagram**

The INC free-energy (2.78) describes the INC phase in the approximation where no harmonics of the basic periodicity are taken into account.

This free-energy is, similarly to its two-component counterpart (2.6), independent of the angular variables θ_1 and θ_2. Thus, the property pointed out in §.3.1.4 can be extended : the modulation can slide freely. Since θ_1 and θ_2 are respectively associated to wavevectors possessing INC components along the Y and X axes, a free-sliding of the modulation can occur independently along either of these directions.

Let us now examine the stable phases related to eq.(2.78) Since the θ_i angles have arbitrary values, the equilibrium of the system is determined by the minimum of the free-energy (2.78) with respect to the two moduli ρ_1 and ρ_2. This procedure yields two possible low-symmetry phases for $\alpha < 0$ (i.e. for $T < T_I$) : (I) $\rho_1^2 = (-\alpha/\beta_1)$; $\rho_2^2 = 0$ or the equivalent state $\rho_2^2 = -\alpha/\beta_1$; $\rho_1^2 = 0$ (II) $\rho_1^2 = \rho_2^2 = -\alpha/(2\beta_1 + \beta_2)$. The two phases are INC. Phase I corresponds to non-zero values for OP components associated to a single pair of opposite wavevectors, respectively $(\pm \vec{k}_o)$ or $(\pm k_o^{(2)})$ for the two energetically equivalent states. The first state is INC in the Y direction and the second one is INC in the X-directions.

Equation (5.38) shows that in phase I the spontaneous strain $(e_{XX} - e_{YY})$ is non-zero, and that it has opposite values in the two possible states : Phase I is an INC phase modulated in a single direction and it involves "ferroelastic" domains differing by the direction of the modulation as well as by the value of the spontaneous strain.

Phase II corresponds to the freezing-in of all the vectors of the star of $\vec{k}_o$. It is INC along the two directions X and Y. The spontaneous strain (eq.5.38) is zero, and, more generally, no new tensorial components, as compared to the HS-phase, are allowed in the INC phase. Similarly to the case of a two-component OP (§.4.5) the effective point symmetry of the INC phase is the same as that of the HS phase.

c) Commensurate phase

The lock-in term is $(\beta_3/4)\,[\rho_1^4 \cdot \cos(4\theta_1) + \rho_2^4 \cdot \cos(4\theta_2)]$ Minimization of the C-free-energy yields 4 possible phases whose characteristics are summarized on table 4 . Similarly to the situation already encountered for two-component OP (§.3.1.2), each INC phase can be related to two C-phases depending on the sign of β_3. In the case where INC phase I is stable there is no change of the effective point-symmetry between the INC phase and the C-phase in contrast to the situation previously discussed (§.4.5).

d) **Description by a modulated OP with uniform amplitudes (PMA)**

The modulated free-energy density is expressed by eq.(2.79). It contains a Lifschitz invariant depending on the two variables X and Y. As these variables correspond to the directions of incommensurability of $\vec{k}_o$, we can assume, on the basis of the relation established in §.3.3.3, that the Lisfschitz invariant entirely accounts for the occurence of the INC phase, and that no more complex schemes of the types discussed in §.5.2, and 5.3 are of interest.

Let us first show that the results in b) hereabove can also be obtained using the modulated scheme. We put, $\theta_1 = k_1.Y$ and $\theta_2 = k_2.X$ in (2.79), the spatial average of this free energy density is :

$$F = \rho_1^2\,[\frac{\alpha}{2} - \delta k_1 + \frac{(\sigma_1-\sigma_2)}{2}\,k_1^2] + \rho_2^2\,[\frac{\alpha}{2} - \delta k_2 + \frac{(\sigma_1-\sigma_2)}{2}\,k_2^2] + \frac{\beta_1}{4}(\rho_1^4+\rho_2^4) + \frac{\beta_2}{2}\,\rho_1^2.\rho_2^2 \qquad (5.39)$$

Minimizing the coefficients of the second degree terms with respect to k_1 and k_2 yields :

$$k_1 = k_2 = \frac{\delta}{(\sigma_1 - \sigma_2)} \qquad (5.40)$$

Replacing into (5.39) the values of $k_1 = k_2$, we obtain the INC free energy (2.78) and therefore the same phase diagram as derived in b). In addition the wavenumber of the modulation, and the temperature T_I of the INC transition $T_I = T_o + \frac{\delta^2}{2(\sigma_1 - \sigma_2)}$ are expressed as functions of the coefficients of the modulated free-energy.

Let us now examine the stable states of the system beyond the sinusoidal approximation. In this view we write the two pairs of variational equations, respectively associated to the moduli ρ_i and the angles θ_i, in the free-energy density (2.79) :

$$\frac{d}{dX}\,(\frac{\partial f}{\partial \rho_{i,X}}) + \frac{d}{dY}\,(\frac{\partial f}{\partial \rho_{i,Y}}) = \frac{\partial f}{\partial \rho_i} \qquad (5.41)$$

$$\frac{d}{dX}\,(\frac{\partial f}{\partial \theta_{i,X}}) + \frac{d}{dY}\,(\frac{\partial f}{\partial \theta_{i,Y}}) = \frac{\partial f}{\partial \theta_i} \qquad (5.42)$$

We first restrict to the "phase modulation only" approximation, by assuming that the ρ_i moduli are independent of the space coordinates. The explicit form of the θ_i-equations is:

$$(\sigma_1 + \sigma_2)\frac{\partial^2 \theta_1}{\partial X^2} + (\sigma_1 - \sigma_2)\frac{\partial^2 \theta_1}{\partial Y^2} = -\beta_3 \rho_1^2 \sin(4\theta_1) \quad (5.43)$$

$$(\sigma_1 - \sigma_2)\frac{\partial^2 \theta_2}{\partial X^2} + (\sigma_1 + \sigma_2)\frac{\partial^2 \theta_2}{\partial Y^2} = -\beta_3 \rho_2^2 \sin(4\theta_2) \quad (5.44)$$

The form (2.58) of the Lifschitz invariant shows, according to eqs. (3.22)-(3.24), that θ_1 and θ_2 are respectively modulated along Y and X. Putting $\theta_1 = \theta_1(Y)$ and $\theta_2 = \theta_2(X)$ in eqs.(5.43) and (5.44), we find that θ_1 and θ_2 satisfy the equation :

$$(\sigma_1 - \sigma_2)\frac{\partial^2 \theta_j}{\partial X_i^2} = -\beta_3 \rho_j^2 \sin(4\theta_j) \quad (5.45)$$

which is the sine-GORDON equation (4.6) studied in §.4 for the two-component case. The θ_i functions are therefore elliptic amplitudes (4.13) whose arguments μ_i depend of the two moduli ρ_i and of the integration constants of the two equations (5.45). We can easily deduce the free-energy of the system as a function of the ρ_i moduli and of the arguments μ_i, by relying on the similarity of the free-energy densities (2.79) and (4.1) and by using eq.(4.28).We obtain:

$$F = \frac{\alpha}{2}(\rho_1^2 + \rho_2^2) + \frac{(\beta_1 - |\beta_3|)}{4}\cdot(\rho_1^4+\rho_2^4) + \frac{\beta_2}{2}\rho_1^2\rho_2^2 + G(\rho_1,\mu_1) + G(\rho_2,\mu_2) \quad (5.46)$$

with :

$$G(\rho,\mu) = -\frac{\pi}{2}\cdot\delta\sqrt{\frac{|\beta_3|}{\sigma_1-\sigma_2}}\,\frac{\rho^3}{\mu K(\mu^2)} + |\beta_3|\rho^4\,\frac{\mu^2-1+2E(\mu^2)/K(\mu^2)}{2\mu^2} \quad (5.47)$$

The values of μ_1 and μ_2 are determined by $(\partial F/\partial\mu_1) = (\partial F/\partial\mu_2)=0$

The form of eq.(5.47) shows that μ_1 and μ_2 only depend, respectively, of ρ_1 or ρ_2 and that both arguments have the same functional form :

$$\mu_i = g(\rho_i) \quad (5.48)$$

determined by eq.(4.33) as in the case of a two-component OP. Replacing (5.48) into the free-energy (5.46), we obtain :

$$F = F_1(\rho_1) + F_1(\rho_2) + \frac{\beta_2}{2}\cdot\rho_1^2\cdot\rho_2^2 \quad (5.49)$$

which can be shown to define, as in the sinusoidal limit, two possible INC phases depending on the value of β_2 : either $\rho_1 \neq 0$, $\rho_2 = 0$ (equivalently $\rho_2 \neq 0$, $\rho_1 = 0$) or $\rho_1 = \rho_2 \neq 0$

If, for instance $\rho_1 \neq 0$, $\rho_2 = 0$, the θ_2 angle has no physical significance, while θ_1 is determined by eq.(5.45), and the free-energy density (2.79) becomes identical to the form (4.1) relevant to a two-component OP. Thus, all the properties determined in §.4 are valid for the modulated phase considered here : temperature dependent wavenumber, structuration into a multisoliton structure, continuous or discontinuous lock-in transition depending on the procedure used to determine ρ_1 (imposed commensurate value, or variationally determined value).

If $\rho_1 = \rho_2$, eq.(5.48) shows that $\theta_1(Y)$ and $\theta_2(X)$ are identical functions : the INC phase is the superimposition of two modulations directed respectively along Y and X, having the same wavenumber, and the same multisoliton structure (fig. 9) Again the set of eqs (5.43)(5.44) has the same form as the corresponding set of eqs.(4.4)(4.5) discussed in §.4, up to a substitution of coefficients. The lock-in transition will therefore possess the same characteristics as in the standard model (§.4)

e) **Complete variational equations**

If we assume that ρ_1 and ρ_2 depend on X and Y, we must solve a set of 4 variational equations associated to (2.79). As in the case of two-component it is necessary to recur to a numerical method. For eq.(2.79) such a method has not yet been applied. Let us therefore limit our discussion to the specific feature arising from the higher OP dimensionality and the higher number of modulation directions. In this view we express the variational equation relative to ρ_1 :

$$(\sigma_1+\sigma_2)\ddot{\rho}_{1,X^2}+(\sigma_1-\sigma_2)\ddot{\rho}_{1,Y^2} = \rho_1[\alpha +\beta_1\rho_1^2+\beta_2\rho_2^2-2\delta\dot{\theta}_{1,Y}+(\sigma_1+\sigma_2)\dot{\theta}_{1,X}^2 + (\sigma_1-\sigma_2)\dot{\theta}_{1,Y}^2] \qquad (5.50)$$

Unlike eq(5. 45), the preceding equation contains a term, $\beta_2\rho_1\rho_2^2$, which couples the OP components relative to the two modulation directions. Consequently, we cannot express the solutions of (5.50) in the form $\rho_1 = \rho_1(Y)$ and $\rho_2 = \rho_2(X)$ and separate the corresponding equations, unless one of the two moduli is zero. In the latter case, i.e. for a single direction of modulation, the system has the properties derived in (§.5.1) for a two-component OP. If, however, $\rho_1 \neq 0$ and $\rho_2 \neq 0$, the four quantities ρ_1, ρ_2, θ_1, θ_2 will all be functions of both X and Y. Such a complex behaviour can be understood qualitatively. As shown by the numerical calculations in the two component case (fig. 13), the modulation amplitude ρ changes rapidly within a spatial soliton. In the present case such a variation of one modulus will influence the behaviour of the other. In addition, at the crossing between perpendicular solitons there will be coupled variations of the two moduli, due to the term $\beta_2\rho_1^2\rho_2^2$ in (2.79).

5.4.2.) Other systems

Detailed numerical investigations of an INC phase with several modulation directions has been performed for 2H-$TaSe_2$, a system displaying a symmetry scheme similar to that of barium sodium niobate : existence of a Lisfschitz invariant, six-component OP, occurence of two independent modulation directions (§.2.12) in the Y plane, non-linear coupling of the OP components relative to the different directions of modulation.

A simplified free-energy density which includes the preceding features has been investigated by using the Fourier series method explained in §.5.1 for the two-component model.

Some of the results obtained are the analog of those derived for barium sodium niobate : there are two possible INC phases, one is modulated in three directions (two of which are independent) and possesses the same effective point symmetry as the HS phase, while the other is modulated along a single direction and has a lower effective point-symmetry

Other results go beyond the ones obtained in §.5.4.1, since the method used allows the spatial variation of the amplitudes ρ_i. Mainly, it is found that the INC phase consists of a two-dimensional array of C_i-domains separated by boundaries (solitons) in which the moduli ρ_i possess a dip, and where two of the three phase angles undergoe a rapid variation. Also, in contrast with the result obtained by the Fourier series method in the two-component case (§.5.1), the lock-in transition is found to be discontinuous.[26].

In both barium sodium niobate and 2H-$TaSe_2$, the theory relates the existence of the INC modulation and its directions to the presence and the form of a Lifschitz invariant.

However, it is possible to extend the type of theory developped in §.5.2, and account, in the absence of a Lifschitz invariant, for the occurence of an INC phase possessing several directions of modulation and a number of INC components larger than 2.

Thus, the theory developped for quartz generalizes the scheme exposed in §.5.2. It considers a one-dimensional OP, modulated in two independent directions (the actual dimension of the INC order-parameter being 6).[27].

As in §.5.2., the stabilization of the INC phase results from a negative coefficient σ for the squared gradient of the modulated OP and of a positive coefficient γ for a higher-order gradient term (of degree two in the OP).

An important distinctive feature, as compared to §.5.2, is the fact that the coefficient of the squared gradient has to show a pronounced anisotropy in order to account for the directions of the modulation wavevector. The origin of this feature can be related to an extension of the mechanism pointed out in §.5.2.3.b), i.e. in a dispersive coupling between the OP and a strain component.Table 6 hereunder, shows that the set of derivatives ($\partial\eta/\partial x$, $\partial\eta/\partial y$) has the same symmetry properties as the set of strain components ($e_{XX}-e_{YY}$;$-2e_{XY}$). Hence the following terms are an invariant contribution to the free-energy density :

$$\frac{C_1}{2}(e_{XX}+e_{YY})^2+\frac{C_2}{2}[(e_{XX}-e_{YY})^2+4\,e_{XY}^2]+\mu[(e_{XX}-e_{YY})\frac{\partial\eta}{\partial x}-2e_{XY}\frac{\partial\eta}{\partial y}] \quad (5.51)$$

The first two terms replace $(\frac{C}{2}e^2)$ in eq.(5.28), while the last term is the counterpart of the dispersive coupling term $(e\frac{d\eta}{dx})$ in eq.(5.28). Similarly to the procedure used in §.5.2.3.b), we take into account the fact that the strain components e_{ij} are derivatives of displacements u_i, and we calculate the Fourier transform of(5.51). After elimination of the displacement degrees of freedom we obtain a contribution to the second degree terms of the free-energy density in κ-space, which is the extension of eq.(5.30):

$$-\frac{\mu^2}{2}\cdot|\vec{\kappa}|^2\cdot\left[\frac{1}{C_1+2C_2}+\frac{\sin^2 3\,\psi}{2\,C_2}\right]|\eta_{\vec{\kappa}}|^2 \quad (5.52)$$

where ψ is the angle between $\vec{\kappa}$ and $\vec{x}$. It appears that this contribution will be maximum for $|\sin 3\,\psi| = 1$, i.e. for $\psi = (\pi/6) + h(\pi/3)$.

The preceding anisotropic term thus favours the onset of the modulation along definite directions in the (x,y) plane.

When considering several modulation directions, it is possible to encounter an intermediate situation where certain directions of modulation are determined by a Lifschitz invariant and others are determined by a negative (and anisotropic) σ-coefficient. This situation is illustrated by the INC phase in biphenyl[1]where there are two independent modulation directions (x,y). The modulated OP is two-dimensional (and admits a Lifschitz invariant involving the y-variable), while the actual INC order-parameter has 4 components and a wavevector with INC components along x and y. The phenomenological theory then requires taking into account both types of gradient invariants used in eqs (2.74) and (2.80) [28]

6. CONCLUSIONS

A number of theoretical models have been developped in this chapter to account for the essential properties of INC systems which were summarized in §.1.1. The different models examined differ by the fact that the OP is a homogeneous or modulated quantity, by the form of the homogeneous or dispersive terms which are incorporated in the free-energy, by the number of OP components and of modulation directions, and by the nature of the approximation made in order to derive the physical results. We have seen that the simplest of these models (§.3.1-3.2) was able to account for the main properties of INC systems which are, the existence of the lock-in transition, some temperature dependence for the INC wavevector, and the occurence of two regimes (sinusoidal and multisoliton) for the modulation. More elaborate models allow a more reliable derivation of the thermodynamic order of the lock-in transition, of the quantitative temperature dependence of physical quantities (wavevector, susceptibility, and of the detailed spatial variation of the OP components.

The reason for the success of these models does not appear explicitly, since the OP free-energies used are constructed along the same symmetry principles as for ordinary structural transitions, while they allow to account for a much more complex situation. Actually, this success relies on a two-fold implicit assumption : the closeness of the INC wavevector k_o and of the commensurate wavevector k_L, and the association of k_L with a <u>low-degree</u> lock-in term in the free-energy. The physical meaning of this assumption is that the INC system is submitted to a <u>strong</u> underlying potential, commensurate with the crystal's lattice.

Most of the attention in this chapter has been focused on the model of a two dimensional OP, modulated in a single direction, and compatible with the existence of a Lifschitz invariant and of fourth degree-lock-in term. The importance of this model is due to its physical relevance to real systems, and the amount of investigations available for it (§.4 and 5.1).

We have seen that, though this model is one of the simplest considered, the working out of its physical consequences meets technical difficulties related to the resolution of non-linear differential equations, and it is necessary, either to make simplifying assumptions whose validity is uncertain, or to recur to numerical methods. This situation is in contrast to the one of ordinary structural transitions where one can generally derive the main physical results by simple algebraic manipulation of the Landau free-energy. As pointed out in §.5.4, technical difficulties are naturally greater for systems with higher number of OP components or of modulation directions. For the latter models, the analytical derivation of results has often to be restricted to the sinusoidal approximation

For a given set of experimental results it is useful to know which is the least sophisticated model displaying the required properties. In this view we have schematized on fig. 16 the results of the various models with a single direction of modulation and a two-component incommensurate OP as well as the hierarchy of the models. On the other hand, table 5 enumerates the symmetry characteristics of the INC phases in real systems (OP symmetry properties, existence of a Lifschitz invariant, etc...)

Fig.16

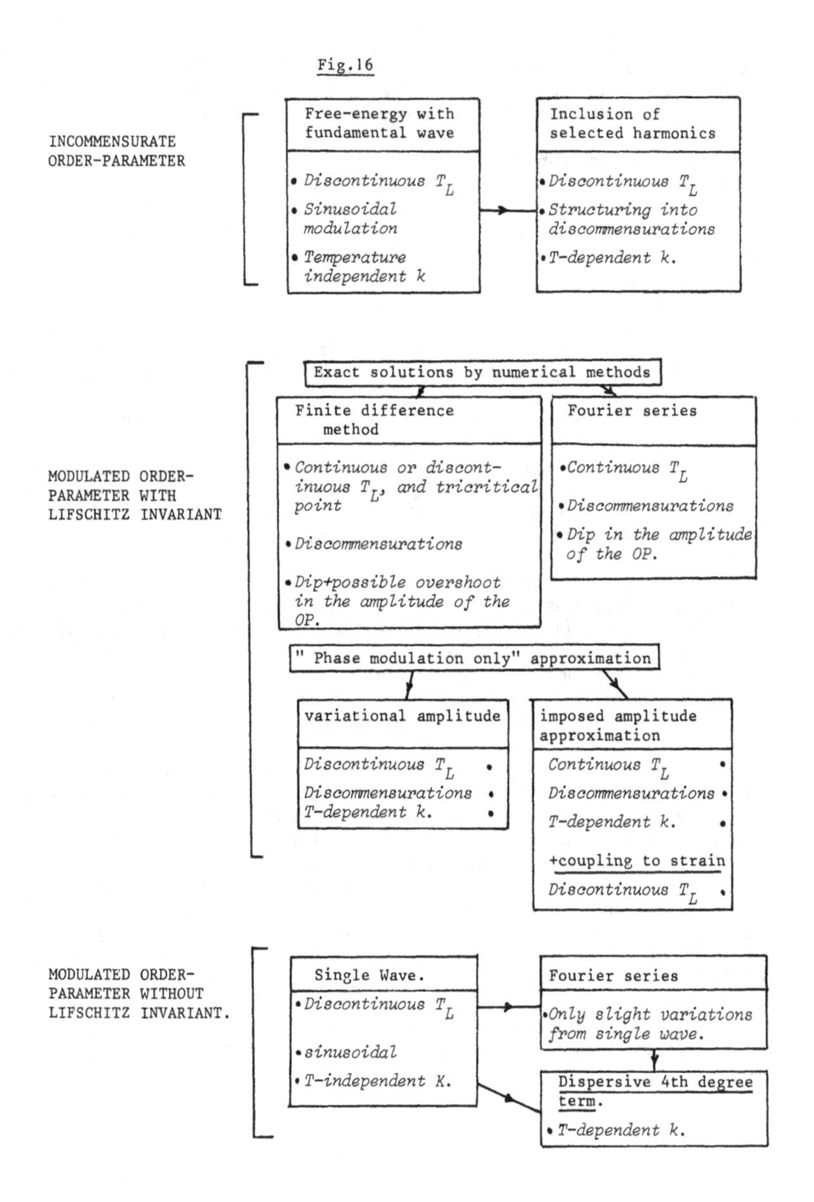
INCOMMENSURATE ORDER-PARAMETER
Free-energy with fundamental wave
• Discontinuous T_L
• Sinusoidal modulation
• Temperature independent k
Inclusion of selected harmonics
• Discontinuous T_L
• Structuring into discommensurations
• T-dependent k.
MODULATED ORDER-PARAMETER WITH LIFSCHITZ INVARIANT
Exact solutions by numerical methods
Finite difference method
• Continuous or discontinuous T_L, and tricritical point
• Discommensurations
• Dip+possible overshoot in the amplitude of the OP.
Fourier series
• Continuous T_L
• Discommensurations
• Dip in the amplitude of the OP.
" Phase modulation only" approximation
variational amplitude
Discontinuous T_L •
Discommensurations •
T-dependent k. •
imposed amplitude approximation
Continuous T_L •
Discommensurations •
T-dependent k. •
+coupling to strain
Discontinuous T_L •
MODULATED ORDER-PARAMETER WITHOUT LIFSCHITZ INVARIANT.
Single Wave.
• Discontinuous T_L
• sinusoidal
• T-independent K.
Fourier series
• Only slight variations from single wave.
Dispersive 4th degree term.
• T-dependent k.

Table 5

Substance	Incom.OP dimension	Modulated OP dimension	degree of lock-in term	Lifshitz invariant	No of modulat. directions
$(NH_4)_2BeF_4$	2	2	4	yes	1
K_2SeO_4	2	2	6	yes	1
TMA-MX_4*	2	2	6-10	yes	1
Na_2CO_3	2	2	12	yes	1
2H-$TaSe_2$	6	6	3	yes	2
$Ba_2NaNb_5O_{15}$	4	4	4	yes	1/2
$BaMnF_4$	4	irrel.	irrel.	irrel.	1
$NaNO_2$	2	1	4	no	1
$SC(NH_2)_2$	2	1	4	no	1
Biphenyl	4	2	4	yes/no	1/2
$ThBr_4$	2	irrel.	irrel.	irrel.	1
Quartz	6	1	3	no	2

* TMA-MX_4 stands for a family of tetramethyl compounds e.g. TMA-$ZnCl_4$.

In spite of the variety of phenomenological approaches used to investigate the properties of INC systems, as exposed in this chapter, the theory has yet to be developped on a certain number of points. Mainly, the phase diagrams for systems with several modulation directions are not reliably known. For the same systems, the configurations of the multisoliton patterns must be further specified. Even in the case of a single modulation direction and of a two-component order parameter, the relevance of higher order gradient invariants (e.g. those considered in §5.3) has to be evaluated. If such terms are important, then more complex phase diagrams than the ones presented are to be expected.

REFERENCES

1 *Incommensurate phases in dielectrics*. Vols.1 and 2. Ed. R.Blinc and A.P. Levanyuk (North Holland, Amsterdam 1986).

2 H. Bestgen .Solid State Commun. 58,197 (1986)

3 L. Michel Preprint. IHES/P/83/21.

4. R.M. Fleming,D.E. Moncton,D.B. Mc Whan, and F.J. Di Salvo Phys. Rev. Letters 45,576 (1980).

5 P.M. De Wolff Acta Cryst. A30,777(1974). T. Janssen,and A. Janner.Physica 126A,163 (1984) and references therein.

6 T.A. Aslanyan ,and A.P. Levanyuk Sov. Phys. Solid State 19,812 (1979).

7 M. Guilluy ,P. Tolédano and J.C. Tolédano Ferroelectrics 36,281 (1981).

8 A. Michelson. Phys. Rev. B18,459 (1978).

9 V. Dvorak and Y. Ishibashi J. Phys. Soc. Jpn 45,775 (1978)

10 D.E. Moncton, J.D. Axe, and F.J. Di Salvo Phys. Rev. B16,801 (1977)

11 I.E. Dzyaloshinskii Sov. Phys. JETP 19, 960 (1964); A.P. Levanyuk,and D.G. Sannikov Sov. Phys. Solid State 18, 245 (1976); W.L. Mc Millan Phys. Rev. B14, 1496 (1976); Y. Ishibashi ,and V. Dvorak J. Phys. Soc. Jpn 44, 32 (1978).

12 J. Przystawa, and J. Lorenc J. Phys. C15, 681 (1982).

13 S. Aubry in *Structures et instabilités* Ed. C.Godrèche (Ed.Physique Paris 1986

14 V. Dvorak,V. Janovec,and Y. Ishibashi J.Phys. Soc. Jpn 52,2053 (1983); J. Schneck,J.C. Tolédano, C. Joffrin, J.Aubrée,B. Joukoff, A. Gabelotaud Phys.Rev. B25,1766 (1982).

15 A. Levstik,P. Prelovsek,C. Filipic,and B. Zeks Phys. Rev. B25,3416 (1982); R. Currat ,and J.D. Axe (unpublished).

16 V.A. Golovko Sov. Physics Solid State 22,1729 (1980)

17 A.E. Jacobs ,and M.B. Walkers Phys. Rev. B21,4132 (1980); P. Prelovsek J.Phys C15,2269 (1982).

18 H. Shiba ,and Y. Ishibashi J. Phys. Soc .Jpn 44,1592 (1978);Y. Ishibashi Ferroelectrics 20,103 (1978).

19 A.D. Bruce,R.A. Cowley,and A.F. Murray J. Phys. C11,3591 (1978)

20 Y. Ishibashi,and H. Shiba J. Phys. Soc. Jpn 45,409 (1978)

21 A.P. Levanyuk,and D.G. Sannikov Sov. Phys. Solid State 18,1122 (1976).

22 T.A. Aslanyan ,and A.P. Levanyuk Solid State Commun. 31,547 (1979).

23 E.B. Loginov, Sov. Phys. Crystallogr. 24,637 (1980).

24 T. Nattermann,and S. Trimper J. Physics C14,1603 (1981).

25 P. Bak ,and J. Timonen J.Phys. C11,4901 (1978)

26 K. Nakanishi ,and H. Shiba J.Phys. Soc Jpn 43,1839 (1977);44,1465 (1978).

27 T.A. Aslanyan,A.P. Levanyuk, M. Vallade,and J. Lajzerowicz J.Phys. C16,6705 (1983).

28 Y. Ishibashi J. Phys. Soc Jpn 50,1255 (1981)

CHAPTER VI

THE LANDAU THEORY OF MAGNETIC PHASE TRANSITIONS

1. INTRODUCTION

In the preceding chapters, in which only structural transitions were discussed, the description of a given phase required use of a mean density of electric charges $\rho(\vec{r})$ remaining invariant under the operations of the crystallographic space-group G of the phase. The scalar function $\rho(\vec{r})$ is associated with the static distribution of electrons which constitutes the "electrical" structure of the phase. Magnetic phases require for their description the additional introduction of a current density $\vec{j}(\vec{r})$ which accounts for the distribution of moving charges.

It is Landau [1] who first noticed that between the electric and magnetic properties of crystals, there exist a fundamental distinction connected with the different behaviours of static charges and currents with respect to time-reversal. Using the time-reversal operator, introduced by Wigner [2] and defined by $R(t) = -t$, yields the equations :

$$R\rho(\vec{r}) = \rho(\vec{r}) \tag{1.1}$$

and

$$R\vec{j}(\vec{r}) = -\vec{j}(\vec{r}) \tag{1.2}$$

Eq.(1.1) expresses the fact that the crystalline structure is independent of the orientation of the electrons on their orbit, i.e. $\rho(\vec{r})$ is always non-zero. In other words, an "electric" structure always exists and one can simply speak of the structure of a given phase. By contrast Eq.(1.2) shows that a phase will display a magnetic structure only if $R(\vec{j}) \neq \vec{j}$, i.e. if $\vec{j} \neq \vec{0}$. Thus, non magnetically ordered phases, namely paramagnetic and diamagnetic phases are characterized by $\vec{j} = \vec{0}$, whereas ferromagnetic or antiferromagnetic phases are not invariant under application of the time-reversal operator and correspond to $\vec{j} \neq \vec{0}$

As was pointed out by Landau [1] the quantity $\int \vec{j}\, dv$ over the whole volume of the crystal should always be equal to zero. Otherwise the corresponding current would create a magnetic field and the crystal would possess some magnetic energy that would increase with an increase in the dimensions of the crystal. Accordingly, it is the mean magnetic moment per unit-cell

$$\vec{m} = \int [\vec{r} \wedge \vec{j}\ ]\, dv \tag{1.3}$$

which allows to distinguish between the two main classes of magnetically ordered structures. When $\vec{m} = \vec{0}$, we deal with an antiferromagnetic ordering, while $\vec{m} \neq \vec{0}$ corresponds to the various types of magnetic structures having a macroscopically non-vanishing magnetic moment, i.e. ferromagnetic, ferrimagnetic, weak ferromagnetic and latent antiferromagnetic structures.

The application of the Landau theory to transitions from paramagnetic to magnetically ordered phases (or between different sorts of magnetically ordered phases), requires as a preliminary to specify the symmetry of the phases and their irreducible degrees of freedom. These questions are discussed in the two following sections of this chapter.

2. MAGNETIC SYMMETRY

The magnetic symmetry of a crystal coincides with the group of symmetry operations which leave invariant the mean density of electric currents $\vec{j}(\vec{r})$. This group can be obtained by combining the rotations, reflections and translations forming the crystallographic group G of the crystal, with the time-reversal operator R. 1651 magnetic groups can be obtained [3-6] which describe paramagnetic, ferromagnetic and antiferromagnetic structures. Let us examine on some selected examples the properties of magnetic groups.

2.1. Magnetic point-groups

At first we show on an example the procedure wich allows to construct all possible magnetic point-groups, denoted M, from a given crystallographic point group G. Figure 1 illustrates the result of such a procedure for G = 4mm (C_{4v}). We can see that three types of magnetic point groups can be distinguished :

1) "Black and white" magnetic groups resulting from the combination of R with half of the symmetry operations of 4mm, the remaining half forming a subgroup H of index 2 of 4mm. Such groups can be written :

$$M = H + (G - H)R \qquad (2.1)$$

2) The "white" group 4mm which also describes a magnetic structure (see §. 2.2)

$$M = G \qquad (2.2)$$

3) The "grey" group labeled 4mm1' which contains the operations of G plus their combination with R

$$M = G + RG \qquad (2.3)$$

Let us note that RE = R is by itself a symmetry operation of the group 4mm1'. Thus, this group describes a paramagnetic structure, i.e. at each magnetic ion, a magnetic moment coincides with its opposite. In the black and white groups, R appears only through combinations with axes, planes or the inversion. These groups, as well as the white groups, correspond either to antiferromagnetic, or to ferromagnetic structures.

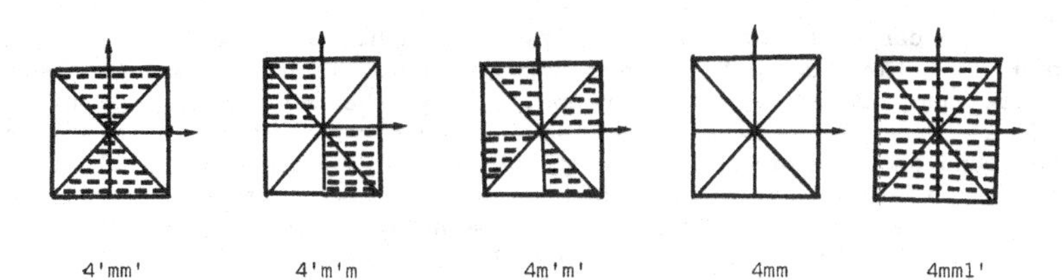

Figure 1 : Black and white, white and grey magnetic point groups M obtained from the crystallographic point group G = 4 mm. The primes denote symmetry operations of the type g.R with $g \in G$. The subgroups H of index 2 associated with 4'mm', 4'm'm and 4m'm' are respectively $m_x m_y 2$, $m_{xy} m_{\bar{x}y} 2$ and 4.

As a second example we show on Figure 2 the stereographic projections of the four black and white, and the white magnetic groups obtained from the crystallographic point group G = 2/m (C_{2h}). From the index 2 subgroups of 2/m, namely 2(C_2), m(C_s) and $\bar{1}$(C_i), one gets the three black and white magnetic groups 2/m', 2'/m and 2'/m'.

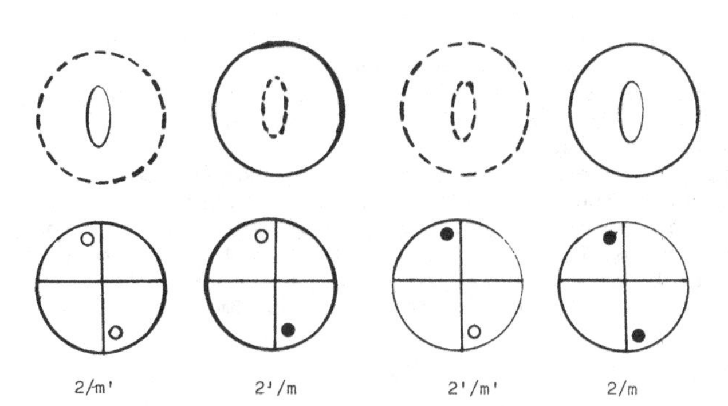

Figure 2 : Stereographic projection of the magnetic points-groups M = 2/m', 2'/m , 2'/m', 2/m, obtained from the crystallographic point group G = 2/m. Symmetry elements (upper figures) are indicated in dashed lines when combined with Time Reversal.The transformation properties of a general point (lower figures) are represented by an empty dot under the effect of R.

It can be noted that the crystallographic point groups G = 1, 3 and 23, which do not possess subgroups of index 2, do not lead to black and white magnetic groups. Along the same line, combination of a three fold axis with the time-reversal operator R may only be found among paramagnetic groups. Extending the procedure illustrated by the two preceding examples to the 32 crystallographic point groups, allows construction of 122 magnetic point groups [3-6] among which 58 black and white, 32 grey and 32 white groups.

2.2. Identification of the type of magnetic ordering associated with a given magnetic group

In order to identify the type of magnetic ordering represented by a given group M (black and white or white) one has to examine the transformation properties of an atomic magnetic moment $\vec{\mu}$ under the effect of the symmetry operations of M. In this respect one must keep in mind that $\vec{\mu}$ is an axial vector, the orientation of which is not reversed by the inversion I or by a plane orthog onal to $\vec{\mu}$ (as it is the case for a polar vector). Conversely $\vec{\mu}$ is reversed by R as well as by a plane parallel to $\vec{\mu}$. Rotation axes have the same effect on axial or polar vectors. Consequently the practical method for determining the type of collinear ordering corresponding to M is to express the transformation properties of the three components μ_x, μ_y and μ_z of $\vec{\mu}$, under the symmetry operations belonging to M. Table 1 illustrates such a method for the ferromagnetic and antiferromagnetic groups deduced, in the preceding section from the crystallographic groups 4mm and 2/m. Figure 3 schematizes the corresponding magnetic structures. One can verify that the groups 4m'm', 2/m and 2'/m' possess a non vanishing magnetic moment $\vec{m}$ per unit cell and thus correspond to a ferromagnetic ordering. By contrast $\vec{m}$ is equal to zero for the groups 4mm, 4'mm', 4'm'm, 2/m' and 2'/m which describe antiferromagnetic structures.

Applying the preceding method to the 90 black and white or white groups, one finds 60 antiferromagnetic point groups an 30 ferromagnetic point groups. The ferromagnetic point groups are the 13 white groups 1, $\bar{1}$, 2, m, 2/m, 4, $\bar{4}$, 4/m, 3, $\bar{3}$, $\bar{6}$, 6 and 6/m, and the 17 black and white groups 2', m', 2'/m', 2'2'2, m'm'2, mm'2', m'mm, 4m'm', 42'2', $\bar{4}$2'm', 4/mm'm', 32, 3m, $\bar{3}$m', 62'2', 6m'm' and $\bar{6}$m'2'.

Table 1 : Transformation properties of the components μ_x, μ_y, μ_z under application of the symmetry operations of the magnetic groups obtained from 4mm and 2/m. For each group, the last column of the table gives the resulting magnetic moment $\vec{m} = \Sigma \vec{\mu}_i$ per unit cell, and the type of ordering.

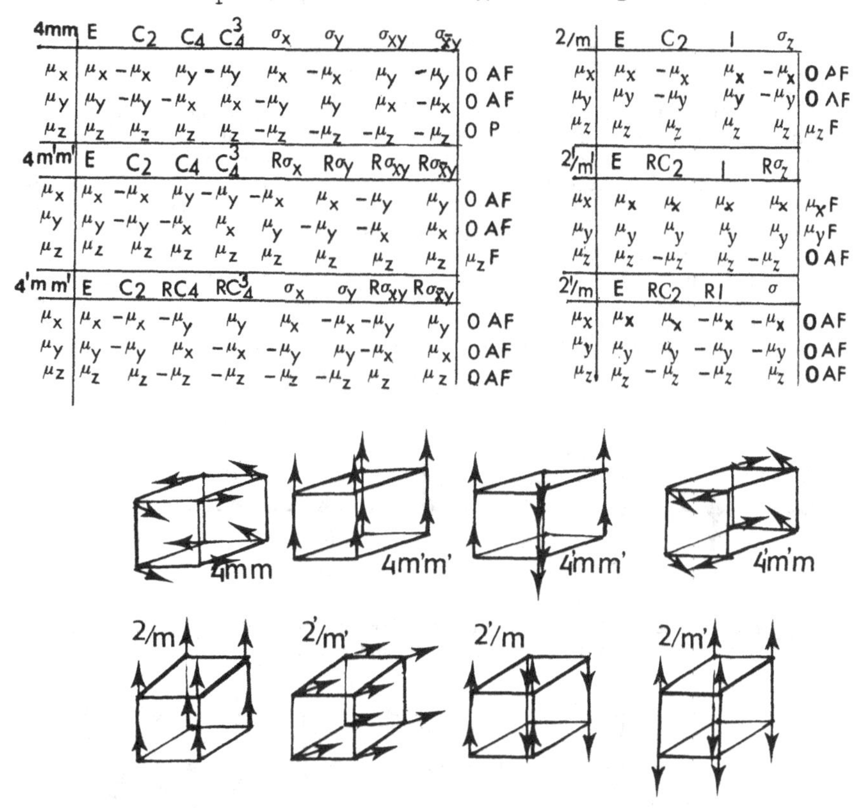

4mm	E	C_2	C_4	C_4^3	σ_x	σ_y	σ_{xy}	$\sigma_{\bar{x}y}$	
μ_x	μ_x	$-\mu_x$	μ_y	$-\mu_y$	μ_x	$-\mu_x$	μ_y	$-\mu_y$	0 AF
μ_y	μ_y	$-\mu_y$	$-\mu_x$	μ_x	$-\mu_y$	μ_y	μ_x	$-\mu_x$	0 AF
μ_z	μ_z	μ_z	μ_z	μ_z	$-\mu_z$	$-\mu_z$	$-\mu_z$	$-\mu_z$	0 P
4m'm'	E	C_2	C_4	C_4^3	$R\sigma_x$	$R\sigma_y$	$R\sigma_{xy}$	$R\sigma_{\bar{x}y}$	
μ_x	μ_x	$-\mu_x$	μ_y	$-\mu_y$	$-\mu_x$	μ_x	$-\mu_y$	μ_y	0 AF
μ_y	μ_y	$-\mu_y$	$-\mu_x$	μ_x	μ_y	$-\mu_y$	$-\mu_x$	μ_x	0 AF
μ_z	μ_z	μ_z	μ_z	μ_z	μ_z	μ_z	μ_z	μ_z	μ_z F
4'mm'	E	C_2	RC_4	RC_4^3	σ_x	σ_y	$R\sigma_{xy}$	$R\sigma_{\bar{x}y}$	
μ_x	μ_x	$-\mu_x$	$-\mu_y$	μ_y	μ_x	$-\mu_x$	$-\mu_y$	μ_y	0 AF
μ_y	μ_y	$-\mu_y$	μ_x	$-\mu_x$	$-\mu_y$	μ_y	$-\mu_x$	μ_x	0 AF
μ_z	μ_z	μ_z	$-\mu_z$	$-\mu_z$	$-\mu_z$	$-\mu_z$	μ_z	μ_z	0 AF

2/m	E	C_2	I	σ_z	
μ_x	μ_x	$-\mu_x$	μ_x	$-\mu_x$	0 AF
μ_y	μ_y	$-\mu_y$	μ_y	$-\mu_y$	0 AF
μ_z	μ_z	μ_z	μ_z	μ_z	μ_z F
2'/m'	E	RC_2	I	$R\sigma_z$	
μ_x	μ_x	μ_x	μ_x	μ_x	μ_x F
μ_y	μ_y	μ_y	μ_y	μ_y	μ_y F
μ_z	μ_z	$-\mu_z$	μ_z	$-\mu_z$	0 AF
2'/m	E	RC_2	RI	σ	
μ_x	μ_x	μ_x	$-\mu_x$	$-\mu_x$	0 AF
μ_y	μ_y	μ_y	$-\mu_y$	$-\mu_y$	0 AF
μ_z	μ_z	$-\mu_z$	$-\mu_z$	μ_z	0 AF

Figure 3 : Type of magnetic ordering associated with the ferromagnetic and antiferromagnetic groups deduced from 4mm and 2/m. A single type of magnetic ion is assumed, which is located at the corners of a primitive unit-cell. The two-fold and four-fold axes are oriented along the z axis.

2.3. Magnetic lattices and magnetic space-groups.

Combination of the time-reversal operator R with a translation $\vec{t}$ is an antitranslation $R\vec{t}$ (or $\vec{t}'$). Two consecutive antitranslations $R\vec{t}$ are equivalent to a translation $2\vec{t}$. Accordingly, in the direction of $\vec{t}$, a magnetic atom with magnetic moment $\vec{\mu}$, will be located between two magnetic atoms with magnetic moments $-\vec{\mu}$ (Figure 4). This property allows to deduce 22 black and white magnetic lattices from the 14 "white" (Bravais) crystallographic lattices [7,8] . On figure 5 we show the 4 tetragonal and 5 monoclinic black and white lattices which can be constructed from the tetragonal (P and I) and monoclinic (P and C) crystallographic lattices with their standard labelling [7,8]. Black and white lattices are always associated with an antiferromagnetic structure, while a white lattice may describe either ferromagnetic or antiferromagnetic structures, depending on the magnetic point group.

All possible combinations of the 36 magnetic lattices (22 black and white + 14 white) with the 122 magnetic point groups yield 1651 magnetic space-groups [3-6] . As for the magnetic point groups they can be classified in four categories : - 230 (type I) magnetic space groups which coincide with the 230 crystallographic space groups.

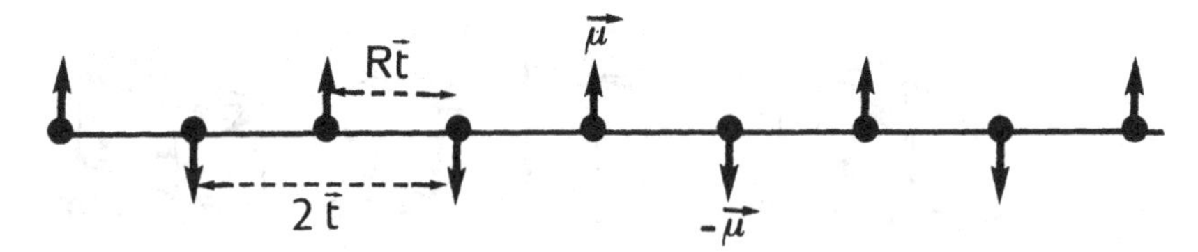

Figure 4 : Linear chain of magnetic atoms ordered antiferromagnetically with an antitranslation $R\vec{t}$ and translation $2\vec{t}$.

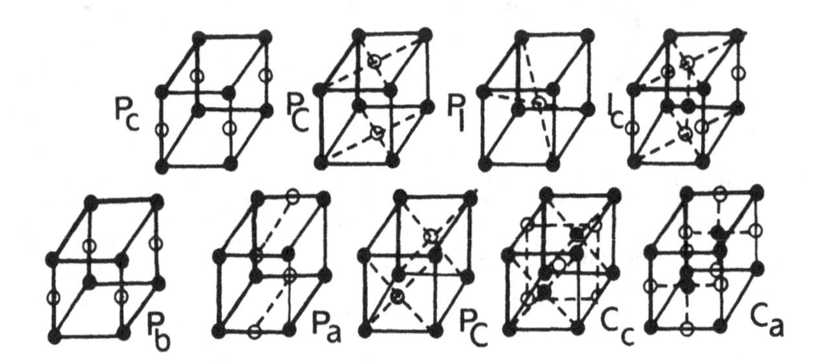

Figure 5 : Black and white magnetic lattices deduced from the P and I tetragonal lattices (above) and the P and C monoclinic lattices (below).

- 230 (type II) grey space groups which can be written $M = G + RG$ where G is a crystallographic space group.
- 674 (type III) black and white magnetic space groups, which are obtained by the combination of the 14 white Bravais lattices with the 58 black and white point groups.
- 517 (type IV) black and white magnetic space groups, which result from the combination of the 22 black and white lattices with the 32 crystallographic point groups.

Type IV groups always correspond to an antiferromagnetic structure, while types I and III groups can be either ferromagnetic or antiferromagnetic groups. Type II groups are paramagnetic. In table 2 we list the four types of magnetic space groups which are respectively deduced from the crystallographic space groups P4mm and P2/m

Table 2 : The four types of magnetic space groups which can be constructed from the space-groups P4mm and P2/m

Type I	Type II	Type III	Type IV
P4mm	P4mm1'	P4'm'm	Pc4mm
		P4'mm'	PC4mm
		P4m'm'	P_I4mm
P2/m	P2/m1'	P2'/m	P_a2/m
		P2/m'	P_b2/m
		P2'/m'	P_C2/m

3. IRREDUCIBLE COREPRESENTATIONS OF THE MAGNETIC GROUPS

The time-reversal operation R is antiunitary. This can be shown by considering the effect of R on a wave function ψ . The behaviour of ψ as a function of time t is determined by the time-dependent Schrödinger equation :

$$\mathcal{H}\psi = ih \frac{\partial\psi}{\partial t} \tag{3.1}$$

where $\mathcal{H}$ is the Hamiltonian of the system. $\mathcal{H}$ is invariant under R because, even though R reverses a magnetic field, the terms in the Hamiltonian only involve products such as $\vec{v} \wedge \vec{B}$ that remain invariant under R. Therefore, the right-hand side of Eq (3.1) must also be invariant under R. This can only be achieved if R not only changes the sign of t but also changes i to -i, i.e. R involves the operation of complex conjugation. Suppose that E_k and ψ_k are the eigenvalues and eigenfunctions of $\mathcal{H}$, that is they are the solutions of the time dependent Schrödinger equation $\mathcal{H}\,\psi = E\,\psi$. The wave function $\psi(t)$ can be expanded in terms of the ψ_k, so that we may write $\psi(t) = \sum_k a_k(t)\,\psi_k$. Therefore, since R involves both the operation of complex conjugation and a change in the sign of t :

$$R\psi(t) = R \sum_k a_k(t)\ \psi_k = \sum_k a_k^* (-t)\ R\psi_k \qquad (3.2)$$

The appearance of the complex conjugate in (3.2) is the mathematical expression of the statement that R is an antilinear operator It is fairly trivial [2] to show that R must be antiunitary than just antilinear.

It follows that all combinations of R with a unitary symmetry operation is an antiunitary operation. Thus, grey and black and white magnetic groups contain both unitary and antiunitary operations. One has then to consider the problem of the representations containing antiunitary elements. This problem was first discussed by Wigner [2] who introduced the concept of the corepresentations of non-unitary groups, i.e. of groups in which half the elements are unitary and the other half are antiunitary. If we write such a group M = G + RG, as for the grey paramagnetic groups, where g_i are the elements of G and $a_i = Rg_i$. If $D(g_i)$ and $D(a_i)$ are the matrices associated with the g_i and a_i respectively, it is not possible to construct matrix representations of M following the usual composition rule for representations of unitary groups : $D(h_i).D(h_j) = D(h_i.h_j)$, but one can form a corepresentation of M using the alternative composition rules :

$$\begin{aligned} D(g_i).\ D(g_j) &= D(g_i.g_j) \\ D(g_i).\ D(a_j) &= D(g_i.a_j) \\ D(a_i).D^*(g_j) &= D(a_i.g_j) \\ D(a_i).D^*(a_j) &= D(a_i.a_j) \end{aligned} \qquad (3.4)$$

Wigner has shown [2] that the irreducible corepresentations (IC's) of a magnetic non-unitary group M, can be easely deduced from the irreducible representations (IR's) of the associated crystallographic subgroup G. Three situations are distinguished by this author depending if the matrices of the considered IR are real or imaginary, and also on the nature of the antiunitary operations pertaining to M. If we only consider here the question of essential interest in the interpretation of magnetically ordered systems, namely the description of transitions from a paramagnetic to a magnetically ordered phase, then we can restrict ourselves to the problem of constructing the IC's of paramagnetic groups. As the time reversal operator R belongs by itself to the grey groups, it is simple to show [9-12] that no degeneracies take place for the energy eigenvalues of the Hamiltonian of the system, so that the representation space has the same dimensionality for the IC and for the corresponding IR. Accordingly a simple rule can be established for constructing the IC's from a given IR, which is summarized in Table 3. As an illustrative example the corepresentations of the group P2/m1' at $\vec{k} = \vec{0}$, are given in Table 4. It can be noted that one has to consider paramagnetic Brillouin zones, in which the wave-vectors are invariant by R. More details about the construction of IC's associated with the various classes of magnetic groups, are given in refs [9-12].

Table 3 The matrices $D(h_i)$ constituting the IC's of the paramagnetic group M are identical to the matrices $\Delta(g_i)$ of the corresponding IR, for $h_i = g_i \in G$, or equal to $\pm \Delta(g_i)$ for $h_i = a_i \in RG$. For transitions to a magnetically ordered phase, only the odd corepresentations Γ^- have to be considered. Note that the $\Delta(g_i)$ are assumed to be real, i.e. they are physically irreducible and written in a suitable basis.

M	$g_i \in G$	$a_i \in RG$
Γ^+	$\Delta(g_i)$	$\Delta(g_i)$
Γ^-	$\Delta(g_i)$	$-\Delta(g_i)$

Table 4 IC's of the group P2/m1' at $\vec{k} = \vec{0}$

P2/m1'	E	C_2	I	σ_z	RE	RC_2	RI	$R\sigma_z$
Γ_1^+	1	1	1	1	1	1	1	1
Γ_2^+	1	1	-1	-1	1	1	-1	-1
Γ_3^+	1	-1	1	-1	1	-1	1	-1
Γ_4^+	1	-1	-1	1	1	-1	-1	1
Γ_1^-	1	1	1	1	-1	-1	-1	-1
Γ_2^-	1	1	-1	-1	-1	-1	1	1
Γ_3^-	1	-1	1	-1	-1	1	-1	1
Γ_4^-	1	-1	-1	1	-1	1	1	-1

4. SPECIFIC FORMULATION OF THE LANDAU THEORY FOR MAGNETIC SYSTEMS AND EXCHANGE SYMMETRIES

4.1. Formulation of the Landau theory.

Let us assume a phase transition from a paramagnetic to a magnetically ordered phase. It is characterized by the onset, below the transition temperature T_c, of a non-zero average density of the magnetic moment $\vec{m}(\vec{r})$*. The thermodynamic potential F of the system is a function of $\vec{m}(\vec{r})$ and the equilibrium value of $\vec{m}(\vec{r})$ is determined by the minimization of F. Assuming a continuous change in $\vec{m}(\vec{r})$ accross the transition point, we deduce that at T_c, $\vec{m}(\vec{r})=\vec{0}$, and near T_c, F can be expanded in power series in $\vec{m}(\vec{r})$.

* It is more likely, in particular for the rare earth, to speak of average density of the magnetic moment rather than of average density of spin.

Let us consider separately the dependence of the components $m^i(\vec{r})$ (i = 1-3) of $\vec{m}(\vec{r})$ on the coordinates and its dependence on the vectorial index i. $m^i(\vec{r})$ can be expanded on a basis $\phi_{n\alpha}(\vec{r})$ which spans the nth IC of the symmetry group M = G + RG describing the paramagnetic phase. The index α enumerates the functions belonging to this IC. Thus, one can write :

$$m^i(\vec{r}) = \sum_{n,\alpha} M^i_{n\alpha}\phi_{n\alpha}(\vec{r}) \tag{4.1}$$

where the coefficients $M^i_{n\alpha}$ in (4.1) transform as the components of an axial (pseudo) vector. As it is usual in the Landau theory [1], we can consider that, except for the transformation over the index i, in a transformation of the coordinates x, y, z, it makes no difference whether the functions $\phi_{n\alpha}$ change and the $M^i_{n\alpha}$ remain unchanged, or the $\phi_{n\alpha}$ are fixed and the $M^i_{n\alpha}$ transform with respect to the index α . In this case, the $M^i_{n\alpha}$ transform as the direct product of the n th IC of the paramagnetic group, and of the axial vectorial representation of this group. Thus, the corepresentation spanning the $M^i_{n\alpha}$ is reducible and we can construct different linear combinations of the $M^i_{n\alpha}$, denoted $S^i_{p\beta}$, $S^i_{q\gamma}$,..... that transform according different IC's p, q,... of the paramagnetic group.

Using (4.1) we can expand F in powers of the $M^i_{n\alpha}$ or of the $S^i_{p\beta}$. F is formed by terms which remain invariant under the symmetry operations of the paramagnetic group M. As the time-reversal operation R belongs to M by itself, we can deduce that no odd terms figure in the expansion of F, since R changes the sign of all of them. In particular no third degree invariant can be constructed i.e. the Landau condition is always fulfilled for the transitions which take place from a paramagnetic phase. Along the same line, let us stress that for such transitions no specific mention must be made concerning the Lifshitz condition (see chapters III and V) : for a given IC, this condition is always fulfilled if the associated IR is active, whereas the condition is violated for inactive IR's. This is a consequence of the facts that the time-reversal operation transforms a Lifshitz invariant in its complex conjugate [13] , and that we only consider physically irreducible corepresentations of the paramagnetic group.

For each IC, there exist only one quadratic invariant and there exist no coupling invariants of the second degree, which transform according to two different IC's. Thus the thermodynamic potential F can be written :

$$F = F_o + \sum_p a_{p\beta} \sum S^2_{p\beta} + \ldots . \tag{4.2}$$

where the index i is omitted from the sum. (4.2) is the standard form of the Landau expansion in terms of the $S_{p\beta}$ which transform according to the different IC's of the paramagnetic group M. However, as will be shown in the examples presented in §.5, it is more convenient to use the $M_{n\alpha}$ despite the fact they transform as a reducible corepresentation. This is a typical feature of the Landau theory of magnetic systems. It allows to separate in the Landau expansion the (isotropic) invariants which express the exchange forces, from the (anisotropic) invariants associated with relativistic spin-spin and spin- lattice forces, and thus to permit an estimate of the order of magnitude of the corresponding coefficients. This question is examined in more details in §.4.2.

As the magnitude of the exchange forces does not change if all the spins are rotated by the same arbitrary angle, the exchange invariants are composed by combinations of the $M^i_{n\alpha}$ which do not change under rotation by an arbitrary angle with respect to the index i and for a given α. It is thus clear that only scalar or mixed products of the types :

$$\vec{M}^i_{n\alpha} \cdot \vec{M}^j_{n\alpha\beta} \quad \text{or} \quad \vec{M}^i_{n\alpha} \cdot (\vec{M}^j_{m\beta} \wedge \vec{M}^k_{p\gamma}) \qquad (4.3)$$

may be used to work out the exchange energy. In particular, for each corepresentation there exist only one exchange invariant of the second order $\sum_{\alpha} M^2_{n\alpha}$. For the construction of the anisotropy energy one can use the standard projector techniques, as described for structural transitions (chapter III).

To the paramagnetic state correspond all $S_{p\beta} = 0$ in (4.2). This can take place only when all $a_p(T) > 0$. Non-zero $S_{p\beta}$ appear if one (an only one) a_p changes sign. By determining the $S_{p\beta}$ from the minimization of F, we can then calculate using (4.1) the equilibrium magnetic moment density, i.e. the magnetic structure of the crystal. The minimization of F must be performed in two steps :

a) minimization of the <u>exchange</u> part of F which determines the <u>angles between the magnetic moments located at the different magnetic atoms.</u>

b) minimization of the <u>anisotropic</u> part of F which provides the <u>orientation of the $\vec{M}_{n\alpha}$ relative to the crystallographic axes.</u>

Our formulation of the Landau theory of magnetic transitions mainly follows the presentation by Dzialoshinskii [14] . A similar approach was proposed by Kovalev [15-17] , in which an additional symmetry requirement was quoted for the selection of the relevant IC's associated with a given magnetic structure. Kovalev's condition consists in writing that the average density of magnetic moments should be non-zero at the magnetic atoms. This statement does not hold for a general formulation of the Landau theory, when no particular atomic structure is considered.

4.2. Exchange symmetry

4.2.1. Exchange and magnetic anisotropy energies

One of the main distinctive feature of the phenomenological approach of magnetic transitions is that one is able to separate in the Landau expansion the respective contributions of exchange and relativistic forces which play an essentially different role in the magnetic ordering below T_c.

a) The relativistic forces, namely the spin-orbit and spin-spin interactions, are connected with the specific structural symmetry of the crystal, i.e. with the space distribution of electrons in the crystalline lattice which determine the orientations of the orbital moments and the localization of the spins. For a crystal possessing a macroscopic magnetization $\vec{M}$, the magnetic anisotropy due to relativistic forces is equivalent to an internal magnetic field of the same order of magnitude than $\vec{M}$ [18].

b) The exchange forces are isotropic as the scalar product $-J\ \vec{S}_i.\vec{S}_j$, where J is the exchange integral and $\vec{S}_i$, $\vec{S}_j$ the total spins of the interacting atoms. They are equivalent to a mean field $\lambda\vec{M}$ with $\lambda \sim 10^3-10^5$.

Accordingly the ratio of the magnetic anisotropy energy to the exchange energy and of the corresponding coefficients in the Landau expansion lies in the range 10^{-3} to 10^{-5}. More precisely we can introduce the ratio :

$$\tau = \frac{E^2 v^2}{T_c c^2} \tag{4.4}$$

wher T_c is the transition temperature, E is of the order of the energy of atomic interactions, v the speed of the electrons in the crystal and c the speed of the light. At temperatures such $T_c - T \simeq \tau$, i.e. near from the transition point, the relativistic terms must be taken into account, whereas for $T_c - T >> \tau$ they can be neglected as the exchange interaction largely dominates. As will be shown in the examples examined in § 5, it is crucial when discussing the Landau expansion to take into account correctly the respective importance of the exchange and magnetic anisotropy energies. An obvious example is the case of ferromagnets, were considering only exchange forces would lead to a random orientation for the magnetization, which actually lies on directions of easy magnetization that are determined by the anisotropy energy. These considerations bring us to reexamine the problem of the symmetry of magnetic structures.

4.2.2. Classification of magnetic structures in terms of exchange symmetry

The classification of magnetic structures presented in §.2 of this chapter and based on the concept of magnetic groups (so-called Heesch-Shubnikov groups) gives a graphic description of the sublattice moments $\vec{\mu}_i$ assuming a localized spin-density distribution. This description accounts correctly for the relativistic interactions, but neglects the fundamental role, stressed hereabove, played by the exchange forces, as the symmetry of these forces is higher than the symmetry of relativistic interactions. Thus, defining the magnetic symmetry of a crystal by its sole magnetic group, means to lose information about the higher symmetry of exchange forces. This shortcoming appears, for example, in the fact that magnetic groups allow only to distinguish between ferromagnetic and antiferromagnetic structures, and do not account for the variety of non compensating antiferromagnets, such as ferrimagnets, weak ferromagnets or latent antiferromagnets [19-21].

Accordingly, in addition to the magnetic group of a

crystal which determines the symmetry of the sublattice arrangement and the magnetic anisotropy part of the Landau expansion, one has to specify its exchange symmetry group, which leaves invariant the exchange contribution to the free-energy expansion. The concept of exchange symmetry was first introduced implicitely by Dzialoshinskii [14] and later more explicitely by Andreev and Marchenko [22,23]. Following the spirit of these authors, we will now examine the classification of magnetic materials with respect to their exchange symmetry.

Let us neglect all relativistic interactions and consider exclusively the exchange symmetry of a magnetic crystal. Since the exchange forces depend only on the relative orientations of the spins (and not on their absolute orientation with respect to the structure of the crystal), beside the time-reversal operator R, which plays the role of inversion in the spin-space, there appears an infinite set of new symmetry operations consisting of all rotations of the spin-space, i.e. rotations of all the spins through the same angle. Under such operations the agregate orientation of the spins with respect to the crystallographic axes becomes arbitrary. In this respect we may consider that the components $M^i_{n\alpha}$ of the average magnetic moment density behave as scalars in all purely spatial transformations, and are components of a vector in the spin space, i.e. they transform as the component of an axial vector, only under the rotation of the spin-space. The exchange symmetry of the crystal is thus determined by the specification of the exchange group G_{ex} which contains all those combinations of purely crystallographic operations, of rotations of the spin-space and of the time-reversal operator with respect to which $\vec{m}(\vec{r})$ is invariant. G_{ex} is a continuous group the specification of which is only of formal interest*. More practically we need to determine the components $M^i_{n\alpha}$ which enter in the products (4.3) for forming the exchange energy. The following group-theoretical considerations simplify the procedure of finding the relevant $M^i_{n\alpha}$ for a given magnetic crystal.

a) As the scalar product $\vec{M}^i_{n\alpha} . \vec{M}^j_{m\beta}$ are spin scalars, they must be invariant under the operations of the paramagnetic symmetry group M. These products transform according to the direct product (denoted X) of the IC's n and m of the group M. By virtue of the antiunitarity of the IC's, we have :

$$\vec{M}^i_{n\alpha} \cdot \vec{M}^j_{m\beta} = C_{in} \ \delta_{ij} \ \delta_{n\bar{m}} \ \delta_{\alpha\beta} \tag{4.5}$$

where C_{in} are some constants, and the δ's are Kronecker products. In (4.5), it is assumed that the complex conjugate corepresentation n and $\bar{n}$ are combined into a single IC, using real magnetic vectors instead of complex conjugate ones. Relation (4.5) shows that different vectors $\vec{M}_{n\alpha}$ are perpendicular to each other and that the maximum number

* It can be noted that if we orient the magnetic moments in a definite manner with respect to the crystallographic axes excluding relativistic interactions, the crystal is then characterized by an exact symmetry corresponding to one of the magnetic space groups defined in §.2

<u>of non vanishing vectors $\vec{M}_{n\alpha}$ corresponds to the dimensionality of the (reducible) corepresentation involved in the transition</u>

b) The magnetic vectors which transform according to non-identity corepresentations make no contribution to the total magnetization and are therefore <u>antiferromagnetic vectors</u>. If there are vectors transforming as the identity corepresentation, then the crystal possesses a non-zero magnetization. By appropriate orthogonal transformation and normalization of the functions $\phi^1_{n\alpha}$ we can assume that one of these vectors coincides with the magnetization and is denoted $\vec{M}$, the remaining vectors composing the identity corepresentation being the antiferromagnetic ones.

c) The constants C_{in} corresponding to the antiferromagnetic vectors may be assumed to be equal to zero or unity, i.e. the non-vanishing antiferromagnetic vectors are unit-vectors. This can be achieved by appropriate normalization of the functions $\phi^1_{n\alpha}$. The non-vanishing antiferromagnetic vectors will be denoted $\vec{L}_i$.

The determination of the $\vec{M}$ and $\vec{L}_i$ vectors for a given magnetic crystal is equivalent to the specification of its exchange symmetry. As will be shown in §.5, it is also equivalent to finding the magnetic sublattices which may be associated with a given exchange magnetic structure.

Introducing the concept of exchange classes [22] which are to exchange space groups what magnetic point groups are to magnetic space-groups, the classification of magnetic structures with respect to their exchange classes can be performed by considering the limited number of IC's (which are one, two or three dimensional) of the 32 paramagnetic (grey) point groups. Such a classification has been completed by Andreev and Marchenko [22] who find four essentially different situations :

a) There exist a single magnetic vector transforming according to the identity representation and it coincides with the magnetization $\vec{M}$. The crystal is a ferromagnet, a collinear ferrimagnet or latent antiferromagnet

b) Only one vector $\vec{L}$ is involved which transforms as a non-identity one dimensional IC. The crystal is a collinear antiferromagnet.

c) One finds the vector $\vec{M}$ and -one antiferromagnetic vector $\vec{L}$ perpendicular to $\vec{M}$, transforming according to a one dimensional, identity or non identity IC or -two vectors $\vec{L}_1$ and $\vec{L}_2$ perpendicular to each other and to M. These vectors transform as the same or different one dimensional IC's, or according to a single two-dimensional IC

The crystal is a <u>non collinear</u> ferrimagnet or a latent antiferromagnet.

d) There are -two mutually perpendicular vectors $\vec{L}_1$ and $\vec{L}_2$ which transform as the same, or different, non identity one dimensional IC's, or according to a single two-dimensional IC

- or three mutually perpendicular vectors $\vec{L}_1$, $\vec{L}_2$, $\vec{L}_3$ transforming according to one, two or three non-identity one dimensional IC's, or according to a non identity one dimensional, <u>and</u> a two dimensional IC, <u>or</u> to a three dimensional IC.

Here, we deal with a non-collinear antiferromagnet.

From the preceding classification have been excluded weak ferromagnets, which require consideration of both exchange and relativistic interactions, and the different types of incommensurate magnetic phases (e.g. helical, cycloïdal, etc...) which can be predicted under additional symmetry considerations (see §.6). Restricting to the above mentioned four categories of exchange magnetic structures, Andreev and Marchenko have found 561 macroscopically distinct magnetic structures differing by their exchange classes.

5. PRACTICAL APPLICATION OF THE LANDAU THEORY TO MAGNETIC TRANSITIONS : SOME EXAMPLES

5.1. Introduction

The procedure that has to be followed for applying the Landau theory to a given magnetic system depends on the available experimental data, and on the type of informations one expects from the theory. The more favourable situation occurs when the symmetries of the paramagnetic and magnetically ordered phases are both known, so the theory is used in order to get more informations about the type of magnetic ordering and interactions which characterize the ordered phase, as well as a more complete picture of the phase diagram, the possible coupling of the magnetic sublattices with secondary macroscopic quantities etc... A less favourable case is when only the paramagnetic symmetry is known (this is the minimal information needed to apply the theory) together with some incomplete data on the low-temperature phases such as the magnetization or the magnetic susceptibility as a function of temperature, the domain structure or the point-group symmetry. Here the Landau theory will provide only the possible symmetries of the ordered phase and a tentative model for the sublattice ordering.

In this section three examples will be successively discussed which illustrate different aspects of the theoretical considerations exposed in §.4.

5.2. Phase transitions from the paramagnetic group $Pca2_1 1'$ at $k = o$

In this first example we assume that the symmetry group and atomic structure of the paramagnetic phase are known as well as the wave-vector $\vec{k}$ of the paramagnetic Brillouin zone associated with the transition. From these data we can deduce the possible symmetry groups and the corresponding magnetic moment configurations of the ordered phase which may arise below a transition. The assumed paramagnetic symmetry is the orthorhombic group $Pca2_1 1'$ which is composed by the operations [24] :

$$g_1 = \{E|000\} \;,\quad g_2 = \{U_z|00\tfrac{c}{2}\} \;,\quad g_3 = \{\sigma_x|\tfrac{a}{2}0\tfrac{c}{2}\},\; g_4 = \{\sigma_y|\tfrac{a}{2}00\}$$

and their combinations with the time reversal operation R. The elementary unit-cell contains four identical atoms in inequivalent positions, labeled 1 to 4 in Figure 6.

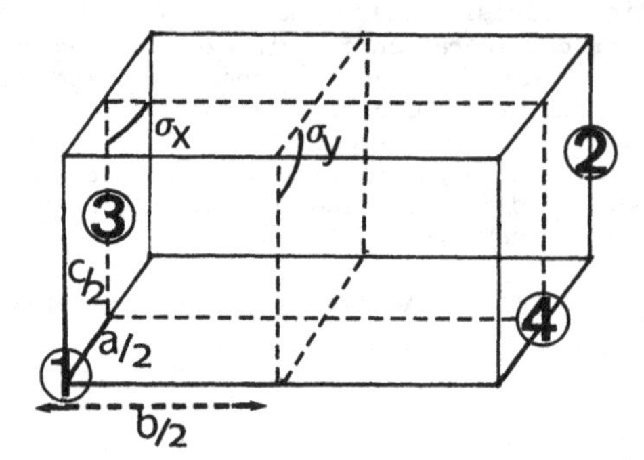

Figure 6 : Schematic unit-cell corresponding to the group $Pca2_1$. Atom 2, 3 and 4 are obtained from atom 1 by application of the operations g_2, g_3 and g_4 respectively.

The wave vector is located at the Γ point of the primitive orthorhombic magnetic Brillouin zone [25] , i.e. the number of atoms in the elementary unit-cell remains unchanged across the transition. The application of the Landau theory can be performed in two steps: a) determination of the magnetic symmetries of the ordered phases which may arise below T_c. This requires only knowledge of the IC's of the paramagnetic group b) specification of the sublattice arrangements characterizing the ordered phases. This second steps necessitates as a preliminary to find the transformation properties of the $M^i_{n\alpha}$ components, in order to construct the Landau expansion in terms of exchange and anisotropy energies.

5.2.1. Symmetries of the magnetically ordered phases

The IC's denoted τ_n (n = 1-4), of the group $Pca2_11'$ at $\vec{k} = \vec{0}$ are given in Table 5. They have been deduced, as indicated in §.3, from the IR's of the crystallographic space group $Pca2_1$, which can be obtained in the available tables [26-28].

Table 5 : The four IC's of the group $Pca2_11'$ at $\vec{k} = \vec{0}$

$Pca2_11'$	g_1	g_2	g_3	g_4	Rg_1	Rg_2	Rg_3	Rg_4	$\vec{t}_1$	$\vec{t}_2$	$\vec{t}_3$
τ_1	1	1	1	1	-1	-1	-1	-1			
τ_2	1	1	-1	-1	-1	-1	1	1	1	1	1
τ_3	1	-1	1	-1	-1	1	-1	1			
τ_4	1	-1	-1	1	-1	1	1	-1			

The four one dimensional IC's τ_i correspond to the same standard one-component order-parameter expansion :

$$F = F_o + \frac{\alpha}{2} \eta^2 + \beta\eta^4 \qquad (5.1)$$

which possesses (see chapter II) only one low-temperature stable phase with $\eta = \pm (-\frac{\alpha}{\beta})^{1/2}$. Thus, one can write for each τ_n the variation of the average density of magnetic moments as :

$$\delta\vec{m}_n = M_n \vec{\phi}_n \qquad (5.2)$$

The symmetry operations of $PCa2_1 1'$ which leave invariant $\delta\vec{m}_n$ for each τ_n are given in Table 6 (column 2). As the translations (primitive and non-primitive) remain unchanged accross the transition, we can easely identify the corresponding magnetic groups which are constructed (see §.2.2) from the space group $Pca2_1$. The labeling of these groups (column 3 of table 6) can be found in the list of magnetic groups by Belov et al [4]. We can verify from the method indicated in §.2.2. that one ferromagnetic group ($Pc'a'2_1$) and three antiferromagnetic groups ($Pca2_1$, $Pca'2_1'$ and $Pc'a2_1'$) may describe the macroscopic symmetry of the phase arising below a transition from the $Pca2_1 1'$ group at $\vec{k} = \vec{0}$. Figure 7 represents the four groups assuming the magnetic moments oriented along the z axis. However, as will be shown in the following section the actual sublattice ordering realizing the preceding symmetries may correspond also to different orientations for the spins, which are determined by considering the relativistic energy.

Table 6 : Group of symmetry operations of the $Pca2_1 1'$ group (column (2)) leaving invariant the average density δm_n for each τ_n (n=1-4) Column (3) : labeling of the groups following the standard notation of Belov et al [4].

τ_1	g_1	g_2	g_3	g_4	$Pca2_1$
τ_2	g_1	g_2	Rg_3	Rg_4	$Pc'a'2_1$
τ_3	g_1	g_3	Rg_2	Rg_4	$Pca'2_1'$
τ_4	g_1	g_4	Rg_2	Rg_3	$Pc'a2_1'$

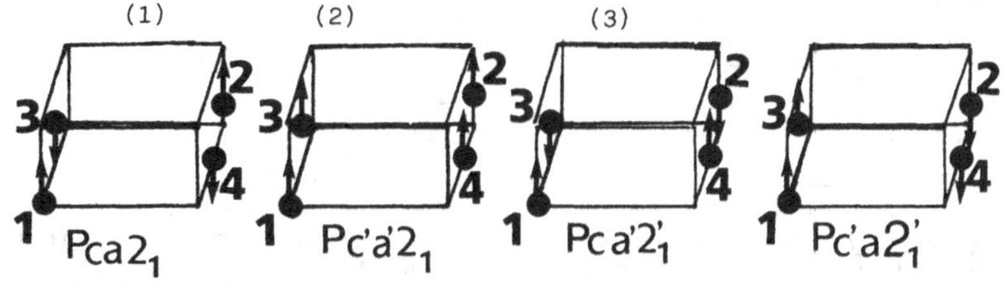

Figure 7 : Unit-cells schematizing the magnetic ordering induced by the IC's τ_1 to τ_4, assuming the two-fold axis along z.

5.2.2. Exchange and relativistic energies : nature of the magnetic order below T_c

Let us now determine the $\vec{M}^i_{n\alpha}$ vectors which define (see Eq. (4.1)) the average density of magnetic moments below T_c. For this purpose, we associate to each of the inequivalent atoms of the $Pca2_1$ unit cell a magnetic moment $\vec{\mu}_i$ (i = 1-4) which represents the total spin of atom i. From the four types of sublattices shown in Figure 7, we can easely see that one can define four vectors denoted $\vec{M}$ and $\vec{L}_i$ (i = 1-3) which represent respectively the total magnetization in the unit-cell and the three possible antiferromagnetic collinear lattices :

$$\begin{aligned} \vec{M} &= \vec{\mu}_1 + \vec{\mu}_2 + \vec{\mu}_3 + \vec{\mu}_4 \\ \vec{L}_1 &= \vec{\mu}_1 - \vec{\mu}_2 + \vec{\mu}_3 - \vec{\mu}_4 \\ \vec{L}_2 &= \vec{\mu}_1 + \vec{\mu}_2 - \vec{\mu}_3 - \vec{\mu}_4 \\ \vec{L}_3 &= \vec{\mu}_1 - \vec{\mu}_2 - \vec{\mu}_3 + \vec{\mu}_4 \end{aligned} \qquad (5.3)$$

Using (5.3) and taking into account the transformation properties of atoms 1-4 under the symmetry operations of the group $Pca2_11'$ one can easely verify, as shown in table 7, that the $\vec{M}$ and $\vec{L}_i$ vectors follow an internal composition law, and thus form a basis for the four IC's of the group $Pca2_11'$ at k = 0.

Table 7 : Transformation properties of the $\vec{M}$ and $\vec{L}_i$ vectors under the operations of the paramagnetic group $Pca2_11'$

	g_1	g_2	g_3	g_4	Rg_1	Rg_2	Rg_3	Rg_4
M	M	M	M	M	-M	-M	-M	-M
L_1	L_1	$-L_1$	L_1	$-L_1$	$-L_1$	L_1	$-L_1$	L_1
L_2	L_2	L_2	$-L_2$	$-L_2$	$-L_2$	$-L_2$	L_2	L_2
L_3	L_3	$-L_3$	$-L_3$	L_3	$-L_3$	L_3	L_3	$-L_3$

Besides as the operations of $Pca2_1$ transform the space coordinates as follows :

$E : \vec{\mu}_i(x,y,z) \rightarrow \vec{\mu}_i(x,y,z)$; $U_z : \vec{\mu}_i(x,y,z) \rightarrow \vec{\mu}_i(-x,-y,z)$

$\sigma_x : \vec{\mu}_i(x,y,z) \rightarrow \vec{\mu}_i(x,-y,-z)$; $\sigma_y : \vec{\mu}_i(x,y,z) \rightarrow \vec{\mu}_i(-x,y,-z)$

one can deduce the components M_α and $L_{i\alpha}$ (α = x,y,z) which are associated with each one of the four IC's of $Pca2_11'$ at $\vec{k} = 0$ (table 8). As stressed in §.4.1, we are able to verify that the $\vec{M}$ and $\vec{L}_i$ vectors transform as <u>reducible</u> corepresentations, i.e. as linear combinations of IC's of $Pca2_11'$.

Table 8 : Components of the $\vec{M}$ and $\vec{L}_i$ vectors which form a basis for the IC's of $Pca2_11'$ at $\vec{k} = \vec{0}$

τ_1	L_{1x}, L_{2y}, L_{3z}
τ_2	L_{1y}, L_{2x}, M_z
τ_3	L_{2z}, L_{3y}, M_x
τ_4	L_{1z}, L_{3x}, M_y

The results of Table 8 allow us to construct the Landau free-energy, separating the exchange and magnetic anisotropy energies. Due to the magnitude of the two contributions we write the exchange terms up to the fourth degree and the relativistic terms up to the second degree :

$$\begin{aligned} F = F_o &+ \sum_i \frac{a_i}{2} L_i^2 + \frac{c}{2} M^2 + \sum_i \frac{b_i}{4} L_i^4 + \frac{d}{4} M^4 \\ &+ \frac{1}{2} \sum_{i,\alpha} (\nu_{i\alpha} L_{i\alpha}^2 + \beta_\alpha M_\alpha^2) \\ &+ \delta_1 L_{1x} L_{2y} + \delta_2 L_{1x} L_{3z} + \delta_3 L_{2y} L_{3z} \\ &+ \delta_4 L_{1y} L_{2x} + \delta_5 L_{2z} L_{3y} + \delta_6 L_{1z} L_{3x} \\ &+ \sigma_1 L_{1y} M_z + \sigma_2 L_{2x} M_z + \sigma_3 L_{2z} M_x \\ &+ \sigma_4 L_{3y} M_x + \sigma_5 L_{1z} M_y + \sigma_6 L_{3x} M_y \end{aligned} \qquad (5.4)$$

The first line in (5.4) represents the exchange energy. It is formed from the scalar products $L_i^2 = \vec{L}_i.\vec{L}_i$ and $M^2 = \vec{M}.\vec{M}$. The five following lines in (5.4) correspond to the relativistic terms. We have separated in the magnetic anisotropy terms the δ_i terms which couple different antiferromagnetic components $L_{i\alpha}$, from the σ_i terms which couple the total magnetization components M_α to the $L_{i\alpha}$. We are now able to discuss in a complete manner the type and nature of magnetic order which may arise below T_c.

a) Let us neglect in first approximation the relativistic terms in (5.4). Minimization of F shows that, when $a_i > 0$ (i =1-3) and c > 0, the equilibrium values are : $|\vec{L}_1| = |\vec{L}_2| = |\vec{L}_3| = |\vec{M}| = 0$ which lead, using (5.3) to : $\vec{\mu}_1 = \vec{\mu}_2 = \vec{\mu}_3 = \vec{\mu}_4 = \vec{0}$. The total spin at each atom is zero. This corresponds to the paramagnetic high-temperature state.

We can now assume that one of the a_i or c becomes zero at T_c, namely a_1, and that a_2, a_3, c remain positive. The equilibrium values obtained for the $\vec{L}_i$ and $\vec{M}$ are :

$$|\vec{L}_2| = |\vec{L}_3| = |\vec{M}| = 0 \qquad (5.5)$$

and $$|\vec{L}_1| = \pm\left(\frac{a_1}{b_1}\right)^{1/2} \tag{5.6}$$

Introducing (5.5) in (5.3) yields :

$$\vec{\mu}_1 = \vec{\mu}_3 = -\vec{\mu}_2 = -\vec{\mu}_4 \tag{5.7}$$

which corresponds to an antiferromagnetic ordering. (5.6) provides the temperature dependence of the $\vec{\mu}_i$. Assuming similarly that a_2, a_3 and c vanish at T_c (with the other quadratic coefficients positive) one gets respectively :

$$\vec{\mu}_1 = \vec{\mu}_2 = -\vec{\mu}_3 = -\vec{\mu}_4 \tag{5.8}$$

$$\vec{\mu}_1 = \vec{\mu}_4 = -\vec{\mu}_3 = -\vec{\mu}_2 \tag{5.9}$$

and $$\vec{\mu}_1 = \vec{\mu}_2 = \vec{\mu}_3 = \vec{\mu}_4 \tag{5.10}$$

(5.8) and (5.9) correspond to exchange antiferromagnetic orders whereas (5.10) to an exchange ferromagnetic structure. Let us note that the equations (5.7) to (5.10) determine only the relative orientation of the $\vec{\mu}_i$ with respect to each other, and not with respect to the crystal axes, which requires consideration of some relativistic terms in (5.4), as will be seen now.

b) Assuming that only $\vec{L}_1$ has non-zero values below T_c, let us determine the possible orientations of the magnetic moments $\vec{\mu}_i$ with respect to the crystallographic axes, i.e. the magnetic groups which may characterize the ordered phase. We consider the reduced form of the Landau expansion :

$$F = F_o + \frac{a_1}{2} L_1^2 + \frac{b_1}{4} L_1^4 + \frac{1}{2} \sum_\alpha \nu_{1\alpha} L_{1\alpha}^2 \tag{5.11}$$

where the exchange and relativistic terms associated with the $\vec{L}_1$ components are taken into account. From the minimization of (5.11) one can see that if $(a_1 + \nu_{1x})$, which has a value close to a_1 as $a_1 \gg \nu_{1x}$, vanishes first at T_c, one will have the magnetic order represented in Figure 8a, with the magnetic moments oriented along the x axis, and corresponding to the magnetic group $Pca2_1(x)$. Analogously, when $(a_1 + \nu_{1y})$ or $(a_1 + \nu_{1z})$ become zero first at the transition, the magnetic order will correspond to the symmetry groups $Pc'a'2_1(y)$ or $Pc'a2_1'(z)$ (Figures 8b and 8c)

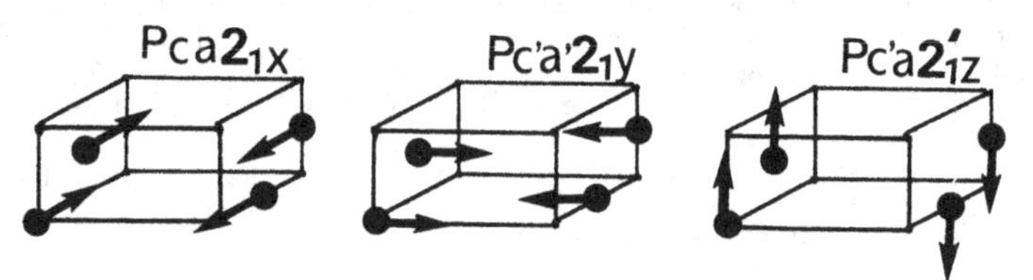

Figure 8 : Possible magnetic orderings corresponding to antiferromagnetic exchange structure defined by $\vec{L}_1 \neq 0$, $\vec{L}_2 = \vec{L}_3 = \vec{M} = 0$.

It can be stressed that the three magnetic structures represented in Figure 8 correspond to the same exchange antiferromagnetic order (5.7). It is the relativistic contribution to (5.11) which determines which one of the components L_{1x}, L_{1y} or L_{1z} become "active" below T_c. Table 9 gives all the possible symmetries corresponding to the exchange structures (5.7) to (5.10)

Table 9 : Magnetic groups (column b) associated with the possible exchange orderings given in column a

(a)	(b)
$\vec{\mu}_1 = \vec{\mu}_3 = -\vec{\mu}_2 = -\vec{\mu}_4$ (AF)	$Pca2_1(x)$, $Pc'a'2_1(y)$, $Pca'2_1'(z)$
$\vec{\mu}_1 = \vec{\mu}_2 = -\vec{\mu}_3 = -\vec{\mu}_4$ (AF)	$Pc'a'2_1(x)$, $Pca2_1(y)$, $Pca'2_1'(z)$
$\vec{\mu}_1 = \vec{\mu}_4 = -\vec{\mu}_3 = -\vec{\mu}_2$ (AF)	$Pc'a2_1'(x)$, $Pca'2_1'(y)$, $Pca2_1(z)$
$\vec{\mu}_1 = \vec{\mu}_2 = \vec{\mu}_3 = \vec{\mu}_4$ (F)	$Pca'2_1'(x)$, $Pc'a2_1'(y)$, $Pc'a'2_1(z)$

c) Let us finally consider a more complete form of (5.4) which takes into account the δ and σ coupling terms. We first assume that a_2 vanishes first at T_c, so we can restrict ourselves to the simplified form of F which contains only invariants associated with the $\vec{L}_2$ components :

$$F = F_o + \frac{a_2}{2} L_2^2 + \frac{b_2}{4} L_2^4 + \frac{1}{2} \sum_\alpha \nu_{2\alpha} L_{2\alpha}^2 + \delta_1 L_{1x}L_{2y} + \delta_3 L_{2y}L_{3z} + \delta_4 L_{1y}L_{2x} + \delta_5 L_{2z}L_{3y} + \sigma_2 L_{2x}M_z + \sigma_3 L_{2z}M_x \tag{5.12}$$

Minimization with respect to L_{2x} gives the equilibrium condition :

$$a_2L_{2x} + b_2L_{2x}^3 + \nu_{2x}L_{2x} + \delta_4L_{1y} + \sigma_2M_z = 0 \tag{5.13}$$

Near below T_c the exchange b_2 and relativistic ν_{2x} terms can be neglected with respect to the exchange a_2 term. Thus equation (5.13) reveals the existence of two possible types of ordering of <u>relativistic origin</u>.

i) an <u>induced</u> magnetization :

$$M_z \simeq -\frac{a_2}{\sigma_2} L_{2x} \tag{5.14}$$

corresponding to a <u>weak ferromagnetic</u> order [20] along the z axis. As shown in Figure 9a), it can be graphically schematized as the result of a <u>canting</u> of the antiferromagnetic sublattice moments $\vec{\mu}_i$ with respect to the axis, in the (x,z) plane. The canting angle is : $\alpha \sim tg\alpha \simeq \frac{M_z}{L_{2x}} = -\frac{a_2}{\sigma_2}$

and can be predicted to be of the order of 10^{-2} - 10^{-5}. It should be emphasized that, whereas the weak-ferromagnetic structure has the magnetic symmetry Pc'a'2_1(z), the exchange antiferromagnetic basic structure is described by the magnetic group Pc'a'2_1(x).

ii) an <u>induced antiferromagnetic order</u> along the y axis, expressed by the equation :

$$L_{1y} \simeq -\frac{a_2}{\delta_4} L_{2x} \qquad (5.15)$$

which results from a canting of the exchange sublattice moments in the (x,y) plane (Figure 9.b). This "weak" antiferromagnetic order, corresponding to the symmetry group Pc'a'2_1(y), is however unlikely to be detected as it should be hindered by the exchange antiferromagnetic basic sublattice directed along the x axis, which is of several orders of magnitude larger.

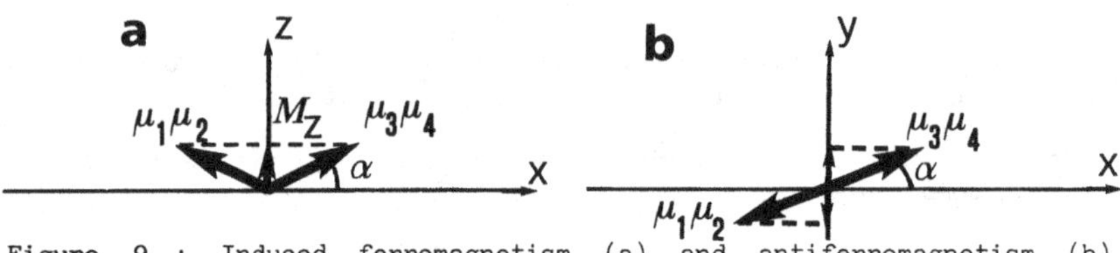

Figure 9 : Induced ferromagnetism (a) and antiferromagnetism (b) of relativistic origin.

Minimization with respect to L_{2y} and L_{2z} provides the equations:

$$a_2L_{2y} + b_2L_{2y}^3 + \nu_{2y}L_{2y} + \delta_1L_{1x} + \delta_3L_{3z} = 0 \qquad (5.16)$$

$$a_2L_{2z} + b_2L_{2z}^3 + \nu_{2z}L_{2z} + \delta_5L_{3y} + \sigma_3M_x = 0 \qquad (5.17)$$

from which can be predicted a variety of induced relativistic phenomena. For example, from Eq.(5.17), one can derive a weak ferromagnetic component M_x induced by a canting of the exchange antiferromagnetic sublattice L_{2z}. Similar deductions can be made when considering the relativistic couplings with the $\vec{L}_1$, $\vec{L}_3$ or $\vec{M}$ components.

The preceding discussion brings us to the following summarizing conclusions :

1. Although a minimization of the Landau potential with respect to the exchange terms, lead to, the basic underlying magnetic structure below T_c (i.e. the relative orientation of the $\vec{\mu}_i$ with respect to each other), it is the relativistic quadratic invariant which determines the absolute orientation of the magnetic moments with respect to the crystallographic axes, and thus the magnetic group describing the ordered phases. Besides, relativistic coupling terms can be responsible of induced orderings with magnetic sublattice moments which are several orders of magnitude smaller than the exchange sublattice moments.

2. In comparison to the phenomenological approach to structural transitions (see chapter III), an essential distinctive feature of the Landau theory of magnetic systems, is that the nature of the interactions are known, and that the two main forces (exchange and relativistic) can be formally separated in the thermodynamic potential. In this respect, the exchange variables, namely the L_i and $\vec{M}$ vectors, play the role of the primary order-parameters, whereas the relativistic variables (i.e. the $L_{i\alpha}$ and M_α components) are treated as secondary parameters, which influence the transition through relativistic couplings. The main difference with the approach to structural systems is here, that it is both the primary and secondary parameters which are responsible of the symmetry breaking taking place at T_c, while it is only the primary order-parameter which determines the symmetry of the low-temperature phases in structural systems.

Transitions from the $Pca2_11'$ paramagnetic phase to exchange antiferromagnetic, or weak ferromagnetic phases have been found in four members of the boracite family, namely Co-Br, Co-I, Ni-Cl and Ni-Br boracites [29-31] . However, due to the complexity of their structure, the actual sublattice arrangement characterizing the ordered phases of these substances, still remain to be elucidated. In the following example, we will show how the knowledge of the ordered magnetic structure allows a more accurate discussion and a more extensive elaboration of the phenomenological model.

5.3. Phase transitions in α - Fe_2O_3

α- Fe_2O_3 possesses in its paramagnetic phase the structural symmetry $R\bar{3}c$ (D_{3d}^6) with four Fe^{+3} ions located along the diagonal of the rhombohedron (Figure 10a). Neutron diffraction studies have established [32-34] that there exist two low-temperature magnetically ordered phases, denoted I and II, which are stable below 250 K and in the range 250 K-950 K respectively. These two phases display an antiferromagnetic sublattice ordering (Figure 10b) with the spins pointing along the crystal axis (state I) and in one of the vertical planes of symmetry, at a small angle to the basal plane (III) (state II). In state II, α -Fe_2O_3 exhibits a small spontaneous magnetization of about 2.10^{-2} from the nominal value, i.e. the value of the saturation magnetization with all the magnetic moments parallel. These magnetic moments disappear at the transition from state II to state I.

In this section we successively determine the IC's which are associated with the symmetries observed in state I and II of α-Fe_2O_3 (§.5.3.1), the sublattice arrangement and temperature dependence of the magnetic moments (§.5.3.2) the phase diagram (§.5.3.3) and the behaviour of α-Fe_2O_3 under application of a magnetic field (§.5.3.4). The three last subsections were discussed by Dzialoshinskii in his classical paper on the theory of weak ferromagnetism[20].

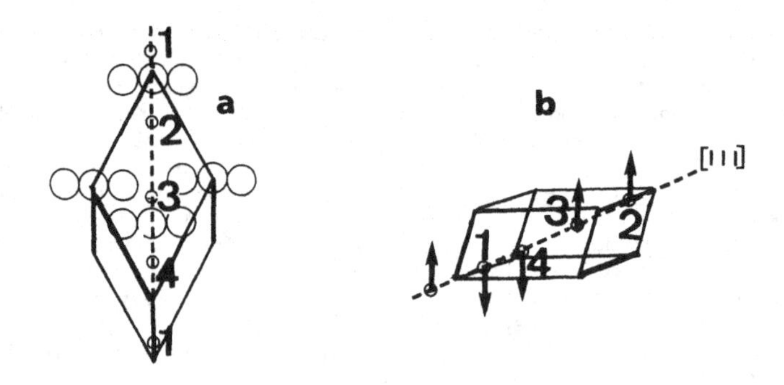

Figure 10 (a) Elementary unit-cell of α-Fe_2O_3 : the large circles denote oxygen ions, the small circles Fe^{+3}. (b) Antiferromagnetic structure of α-Fe_2O_3 .

5.3.1. Symmetry of the ordered phases

The experimental data [32-34] show that the magnetic unit-cells of the low temperature phases have the same volume than the crystallographic cells. Thus the transitions from the paramagnetic phase to the I and II phases take place at the center of the rhombohedral Brillouin zone. At $\vec{k} = \vec{0}$, the paramagnetic group $R\bar{3}c1'$ possesses six odd IC's, labelled τ_1 to τ_6 ,four of which are one-dimensional ($\tau_1 \rightarrow \tau_4$) and two bi-dimensional (τ_5, τ_6). The corresponding matrices are given in Table 10.

Table 10 : Matrices of the six IC's of the $R\bar{3}c$ group at $\vec{k} = 0$. The matrices of the Rg_i elements are obtained from the matrices of the g_i elements by the rule indicated in Table 3.

	E	C_3	C_3^2	σ_x	σ_{xy}	σ_y	I	S_6	S_6^5	U_x	U_{xy}	U_y
τ_1	1	1	1	1	1	1	1	1	1	1	1	1
τ_2	1	1	1	-1	-1	-1	1	1	1	-1	-1	-1
τ_3	1	1	1	1	1	1	-1	-1	-1	-1	-1	-1
τ_4	1	1	1	-1	-1	-1	-1	-1	-1	1	1	1
τ_5	$\begin{bmatrix}1 & 0\\0 & 1\end{bmatrix}$	$\frac{1}{2}\begin{bmatrix}-1 & -\sqrt{3}\\\sqrt{3} & -1\end{bmatrix}$	$\frac{1}{2}\begin{bmatrix}-1 & \sqrt{3}\\-\sqrt{3} & -1\end{bmatrix}$	$\frac{1}{2}\begin{bmatrix}-1 & \sqrt{3}\\\sqrt{3} & 1\end{bmatrix}$	$\begin{bmatrix}1 & 0\\0 & -1\end{bmatrix}$	$\frac{1}{2}\begin{bmatrix}-1 & -\sqrt{3}\\-\sqrt{3} & 1\end{bmatrix}$	$\begin{bmatrix}1 & 0\\0 & 1\end{bmatrix}$	$\frac{1}{2}\begin{bmatrix}-1 & -\sqrt{3}\\\sqrt{3} & -1\end{bmatrix}$	$\frac{1}{2}\begin{bmatrix}-1 & \sqrt{3}\\-\sqrt{3} & -1\end{bmatrix}$	$\frac{1}{2}\begin{bmatrix}-1 & \sqrt{3}\\\sqrt{3} & 1\end{bmatrix}$	$\begin{bmatrix}1 & 0\\0 & -1\end{bmatrix}$	$\frac{1}{2}\begin{bmatrix}-1 & -\sqrt{3}\\-\sqrt{3} & 1\end{bmatrix}$
τ_6	$\begin{bmatrix}1 & 0\\0 & 1\end{bmatrix}$	$\frac{1}{2}\begin{bmatrix}-1 & -\sqrt{3}\\\sqrt{3} & -1\end{bmatrix}$	$\frac{1}{2}\begin{bmatrix}-1 & \sqrt{3}\\-\sqrt{3} & -1\end{bmatrix}$	$\frac{1}{2}\begin{bmatrix}1 & -\sqrt{3}\\-\sqrt{3} & -1\end{bmatrix}$	$\begin{bmatrix}-1 & 0\\0 & 1\end{bmatrix}$	$\frac{1}{2}\begin{bmatrix}1 & \sqrt{3}\\\sqrt{3} & -1\end{bmatrix}$	$\begin{bmatrix}-1 & 0\\0 & -1\end{bmatrix}$	$\frac{1}{2}\begin{bmatrix}1 & \sqrt{3}\\-\sqrt{3} & 1\end{bmatrix}$	$\frac{1}{2}\begin{bmatrix}1 & -\sqrt{3}\\\sqrt{3} & 1\end{bmatrix}$	$\frac{1}{2}\begin{bmatrix}-1 & \sqrt{3}\\\sqrt{3} & +1\end{bmatrix}$	$\begin{bmatrix}1 & 0\\0 & -1\end{bmatrix}$	$\frac{1}{2}\begin{bmatrix}-1 & -\sqrt{3}\\-\sqrt{3} & 1\end{bmatrix}$

The one-dimensional IC's induce a single low-temperature stable phase, whereas the two-dimensional IC's, which are associated with the image C_{6v} (see chapter II) can give rise to two possible stable phases corresponding to the equilibrium values of the order-parameter components (η_1,η_2) : $(\eta_1 \neq 0,\ \eta_2 = 0)$ and $(\eta_1 = 0,\ \eta_2 \neq 0)$. Following the procedure described in chapters II and III, we can thus determine the magnetic groups, subgroups of $R\bar{3}c1'$, which leave invariant the probability densities : $\delta\vec{m}_{1-4} = M_{1-4}\ \vec{\phi}_{1-4}$ (for τ_1 to τ_4) and $\delta\vec{m}_{5-6} = M^{(1)}_{5-6}\ \vec{\phi}^{(1)}_{5-6} + M^{(2)}_{5-6}\ \vec{\phi}^{(2)}_{5-6}$. These groups are given in Table 11.

Table 11 : Magnetic groups describing the symmetry of the ordered phases induced by the IC's of the R3c1' group. In line (2) the corresponding type of ordering is indicated (AF or F)

(1)	τ_1	τ_2	τ_3	τ_4	τ_5	τ_6
(2)	$R\bar{3}c$(AF)	$R\bar{3}c'$(F)	$R\bar{3}'c$(AF)	$R\bar{3}'C'$(AF)	{C2/c(F), C2'/c'(F)}	{C2/c'(AF), C2'/c(AF)}

In α-Fe_2O_3 the two states I and II observed below 950 K correspond respectively to the magnetic groups $R\bar{3}c$ and C2/c [34] . From Table 11 we can deduce that they are connected with the IC's τ_1 and τ_5 respectively. Let us note however, that the ferromagnetic symmetry C2/c does not impose any limitation on the magnitude of the magnetic moments so that it may describe a pure ferromagnetic structure of exchange origin. In order to understand the weak value exhibited by the magnetic moment, we need to perform a more complete analysis of the Landau potential in terms of exchange and relativistic energies.

5.3.2. Exchange antiferromagnetism and weak ferromagnetism in α-Fe_2O_3

As for the example treated in §.5.2, the structure of α-Fe_2O_3 possesses four inequivalent atoms in the unit-cell (figure 10a) A similar discussion than in §.5.2.2. brings us to introduce four vectors $\vec{M}$ and $\vec{L}_i$ (i = 1-3) defined by Eq.(5.3), which express all possible configurations for the magnetic moments $\vec{\mu}_i$ (i = 1-4). It can easely be found here that the $\vec{M}$ and $\vec{L}_i$ components have the transformation properties given in Table 12.

Table 12 : Components of the $\vec{M}$ and $\vec{L}_i$ vectors transforming as the six IC's of the R3c1' group.

τ_1	τ_2	τ_3	τ_4	τ_5	τ_6
L_{3z}	M_z	L_{2z}	L_{1z}	$\{(M_x,M_y),\ (-L_{3y},L_{3x})\}$	$\{(L_{1x},L_{1y}),\ (L_{2y},-L_{2x})\}$

we are now able to write the general form of the thermodynamic potential F, separating the exchange interactions (first line) from the relativistic ones (second and third lines) :

$$F = F_o + \sum_i \frac{a_i}{2} L_i^2 + \frac{c}{2} M^2 + \sum_i \frac{b_i}{4} L_i^4 + \frac{d}{4} M^4$$
$$+ \frac{1}{2} \sum_{i,\alpha} (\nu_{i\alpha} L_{i\alpha}^2 + \beta_\alpha M_\alpha^2) \tag{5.18}$$
$$+ \delta_1 (L_{3x} M_y - L_{3y} M_x) + \delta_2 (L_{1x} L_{2y} - L_{2x} L_{1y})$$

In the paramagnetic phase, every a_i and c are positive. Assuming a purely exchange symmetry, the transition would occur when one of the a_i or c becomes zero. Actually we know from the magnetic structure of α-Fe_2O_3 [34] (figure 10b) that the antiferromagnetic order realized in this material corresponds to :

$$\vec{\mu}_1 = \vec{\mu}_4 = - \vec{\mu}_3 = - \vec{\mu}_2 \tag{5.19}$$

namely to the vector $\vec{L}_3$. We can thus restrict ourselves to consider in (5.18) only those terms in which the $\vec{L}_3$ components are involved :

$$F = F_o + \frac{a_3}{2} L_3^2 + \frac{b_3}{4} L_3^4 + \frac{c}{2} M^2$$
$$+ \frac{1}{2}[\nu_{3z} L_{3z}^2 + \nu_{3xy}(L_{3x}^2 + L_{3y}^2) + \beta_z M_z^2 + \beta_{xy}(M_x^2 + M_y^2)]$$
$$+ \delta_1 (L_{3x} M_y - L_{3y} M_x) \tag{5.20}$$

Because of the coupling of L_{3x} and L_{3y} to M_y and M_x respectively we have also retained in (5.20) the square invariants of $\vec{M}$ and its components. The minimum of F with respect to the $L_{3\alpha}$ and M_α are determined by the equations :

$$\begin{cases} L_{3z} (a_3 + b_3 L_{3z}^2 + \nu_{3z}) = 0 \\ L_{3x} (a_3 + b_3 L_{3x}^2 + \nu_{3xy}) + \delta_1 M_y = 0 \\ L_{3y} (a_3 + b_3 L_{3y}^2 + \nu_{3xy}) - \delta_1 M_x = 0 \end{cases} \tag{5.21}$$

and

$$\begin{cases} (c + \beta_z) M_z = 0 \\ (c + \beta_{xy}) M_x - \delta_1 L_{3y} = 0 \\ (c + \beta_{xy}) M_y + \delta_1 L_{3x} = 0 \end{cases} \tag{5.22}$$

The system has two solutions :

I. $M_x = M_y = M_z = 0$, $L_{3x} = L_{3y} = 0$

and

$$L_{3z} = \pm \left[- \frac{(a_3 + \nu_{3z})}{b_3}\right]^{1/2} \tag{5.23}$$

It corresponds to the state I, where the spins of the ions are directed along the crystal axis (figure 10b). In this purely exchange antiferromagnetic state, only the L_{3z} component is active. L_{3z} spans the one-dimensional IC τ_1 which induces the R3c symmetry actually observed below 250 K.

II. $M_z = 0, \quad L_{3z} = 0$

$$M_x = \frac{\delta_1}{c + \beta_{xy}} L_{3y}, \qquad M_y = \frac{-\delta_1}{c + \beta_{xy}} L_{3x} \tag{5.24}$$

$$\text{with } L_{3x} = L_{3y} = \pm\left[-\left(\frac{a_3 + \nu_{3xy} - \delta_1^2/c + \beta_{xy}}{b_3}\right)\right]^{1/2} \tag{5.25}$$

which corresponds to state II of $\alpha\text{-}Fe_2O_3$. Here the spins lie in the (III) plane and it appears a spontaneous magnetization moment perpendicular to the vector $\vec{L}_3$ of magnitude

$$|M_x| \simeq |M_y| \simeq \left|\frac{\delta_1}{c}\right| \left(-\frac{a_3}{b_3}\right)^{1/2} \tag{5.26}$$

(in (5.24) and (5.25) we can neglect the relativistic coefficients ν_{3xy} and β_{xy} with respect to the exchange coefficients a_3 and c respectively. As the $\frac{\delta_1}{c}$ ratio is small ($10^{-2} - 10^{-5}$) it follows that $\frac{M_{x,y}}{L_{y,x}} \simeq 10^{-2}\text{-}10^{-5}$ in agreement with the experimental value of 2.10^{-4} [32,33] . Thus, in state II the magnetic moments $(\vec{\mu}_1,\vec{\mu}_4)$ and $(\vec{\mu}_2,\vec{\mu}_3)$ no longer compensate each other, but are turned with respect to each other through a small angle of order 10^{-4} (figure 11).

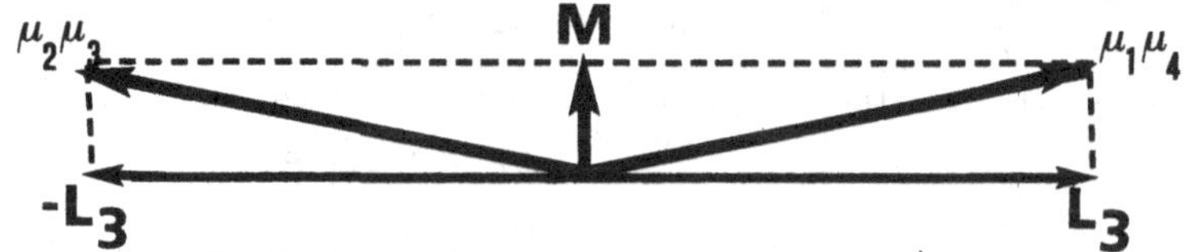

Figure 11 : Projection of the magnetic moments $\vec{\mu}_i$ (i = 1-4) of the ions in $\alpha\text{-}Fe_2O_3$ on the (III) plane.

It must be emphasized that the direction of the vector $\vec{L}_3$ is not determined by the terms in the expansion (5.20). Actually one has to expand F up to the sixth degree in L_{3x} and L_{3y} in order to find a symmetry breaking invariant. The fourth and sixth degrees invariants are :

$$\mu(L_{3x}^2 + L_{3y}^2)^2, \quad \gamma_1(L_x^2 + L_y^2)^3 \quad , \quad \gamma_2(L_x^6 + L_y^6) \tag{5.27}$$

Depending on the sign of the anisotropic invariant γ_2, $\vec{L}_3$ will lie either along one of the two-fold axes ($\gamma_2 > 0$) or in one of the symmetry planes ($\gamma_2 < 0$). It is easy to verify that these two possible stable states have respectively the symmetries C2'/c' and C2/c induced by τ_5. This latter symmetry is the one actually observed in state II of α-Fe_2O_3.

5.3.3. Phase diagram of α-Fe_2O_3

a) Let us first determine the range of stability of phases I and II as a function of the coefficient ν_{3z}. Introducing the equilibrium values (5.23) in (5.20), we get for phase I :

$$F_I = F_o - \frac{(a_3 + \nu_{3z})^2}{4b_3} \qquad (5.28)$$

In phase II we can assume the relativistic coefficients ν_{3xy} and β_{xy} negligible with respect to the exchange coefficients a_3 and c respectively. Substituting conditions (5.24) and (5.25) in (5.20) gives :

$$F_{II} = F_o - \frac{(a_3 - \delta_1^2/c)^2}{4b_3} \qquad (5.29)$$

Since ν_{3z} and δ_1 are of the same order of magnitude we can approximate (5.28) and (5.29) by :

$$F_I \simeq F_o - \frac{a_3^2}{4b_3} - \frac{a_3\,\nu_{3z}}{2b_3} \qquad (5.30)$$

and

$$F_{II} \simeq F_o - \frac{a_3^2}{4b_3} \qquad (5.31)$$

Consequently either phase I or II will take place depending on the sign of ν_{3z}. As $a_3 < 0$ in the two phases, we verify that $\nu_{3z} > 0$ will correspond to state II ($F_{II} < F_I$) whereas $\nu_{3z} < 0$ in state I ($F_I < F_{II}$).

If T_c is the transition temperature from the paramagnetic phase to phase II, we can write $a_3 = a_o(T - T_c)$ ($a_o > 0$). Close from T_c in phase II we thus have from (5.24), and (5.25) :

$$|L_{3x}| = |L_{3y}| \simeq \left[\frac{a_o(T_c - T)}{b_3}\right]^{1/2} \qquad (5.32)$$

$$|M_x| = |M_y| \simeq \frac{|\delta_1|}{c}\left[\frac{a_o(T_c - T)}{b_3}\right]^{1/2} \qquad (5.33)$$

b) In order to complete the phase diagram of α-Fe_2O_3, it is necessary to evaluate the preceding components at temperatures far from T_c. At such temperatures the L_α components are no longer small and expansion (5.20) becomes incorrect. However it can be replaced by an expansion in which the unit-vector $\vec{n}$ in the direction of $\vec{L}_3$ can be used as a variable. By contrast, an expansion in powers of $\vec{M}$ is always possible due to the smallness of this quantity. As in 5.3.2. we must distinguish the purely exchange invariants, the relativistic invariants and the coupling invariants. In the selection of the invariants and their powers, we must keep in mind that the ratio of the coupling-invariant energy to the exchange energy is proportional to $\frac{v^2}{c^2}$ (see §.4.2.1) whereas the ratio of the relativistic and exchange energies is proportional to $\frac{v^4}{c^4}$. Taking into account the magnitude of the various sorts of invariants, we can write :

$$\begin{aligned} F = F_o &+ \frac{\nu'_z}{2} M_z^2 + \gamma'_1 (n_x^6 + n_y^6) + \delta'_1 (n_x^3 - n_y^3) n_z \\ &+ \frac{\beta_z}{2} M_z^2 + \delta'_2 (n_x^3 + n_y^3) M_z + \delta'_3 (n_x M_y - n_y M_x) \\ &+ \frac{c}{2} M^2 + \frac{e}{2} (\vec{n}.\vec{M})^2 \end{aligned} \quad (5.34)$$

where only have been retained the invariants which a posteriori appear to be relevant for the discussion of the equilibrium values of the n_α ($\alpha \equiv x,y,z$) components of the unit vector $\vec{n}$, and of the components of $\vec{M}$. Thus only a second degree term has been written for n_z and one sixth degree invariant for (n_x, n_y). δ'_1, δ'_2 and δ'_3 express couplings between the n_α and with the M_α components. The two first lines in (5.34) contain relativistic invariants, while the third line is constituted by exchange terms. We will use the thermodynamic potential (5.34) for the determination of the magnetic structure of $\alpha - Fe_2O_3$ at temperatures far from T_c. Introducing polar corrdinates for the unit vector $\vec{n}$, (5.34) takes the form :

$$\begin{aligned} F = F_o &+ \frac{\nu'_z}{2} \cos^2\theta + \gamma'_1 \cos 6\phi \sin^6\theta + \delta'_1 \cos\theta \sin^3\theta \sin^3\phi \\ &+ \frac{\beta_z}{2} M_z^2 + \delta'_2 M_z \sin^3\theta \cos^3\phi + \delta'_3 \sin\theta(M_y \cos\phi - M_x \sin\phi) \\ &+ \frac{c}{2} M^2 + \frac{e}{2} [M_z \cos\theta + \sin\theta(M_x \cos\phi + M_y \sin\phi)]^2 \end{aligned} \quad (5.35)$$

Minimization of (5.35) with respect to the $\vec{M}$ components yields the equilibrium values :

$$M_x = \frac{\delta'_3}{c} \sin\theta \sin\phi, \quad M_y = -\frac{\delta'_3}{c} \sin\theta \cos\phi, \quad M_z = -\frac{\delta'_2}{c} \sin^3\theta \cos^3\phi \quad (5.36)$$

Besides, minimization of (5.35) with respect to θ and ϕ shows that three absolute minima are possible :

I. $\theta = 0$

From (5.36) one finds $\vec{M} = \vec{0}$. This is the antiferromagnetic state I observed in α-Fe_2O_3 below 250 K, with all the spins directed along the [III] axis

II. $\theta = \frac{\pi}{2} - \frac{\delta_1'}{\nu_z'}$, $\phi = \frac{\pi}{2}$. It corresponds to $M_x = \frac{\delta_3'}{c}$, $M_y = M_z = 0$.

This is the state II realized in α-Fe_2O_3 with the antiferromagnetic part of the spins lying in one of the symmetry planes at a small angle (of the order $\frac{\delta_1}{\nu_z'} \sim \frac{v^2}{c^2}$) to the (III) plane. The spontaneous magnetic moment, with $|\vec{M}| = \frac{|\delta_3'|}{c} \sim \frac{v^2}{c^2}$, is directed along a two fold axis, perpendicular to the antiferromagnetic sublattice.

III. $\theta = \frac{\pi}{2}$, $\phi = 0$, $M_y = -\frac{\delta_3'}{c}$, $M_z = -\frac{\delta_2'}{c}$, $M_x = 0$. In this state, which is not observed in α-Fe_2O_3, the antiferromagnetic sublattice is directed along one of the twofold axes with a spontaneous magnetic moment lying in a symmetry plane perpendicular to the two fold axis, at a small angle to the (III) plane.

At a given temperature and pressure, the one of the three preceding states which is realized, corresponds to the smallest value of F. Introducing the equilibrium conditions I, II, III in (5.35) gives :

$$F_I = \frac{\nu_z'}{2} , \quad F_{II} = -\frac{\delta_3'^2}{2c} - \frac{\delta_1'^2}{4\nu_z'} - \gamma_1' , \quad F_{III} = -\frac{\delta_3'^2}{2c} + \gamma_1' \qquad (5.37)$$

As $|\nu_z'| >> \frac{\delta_3'^2}{2c}$, $\frac{\delta_1'^2}{4|\nu_z'|}$ and γ_1', we can conclude that for $\nu_z' > 0$, state II will be more stable than state I and III if $\gamma_1' > -\frac{\delta_1'^2}{8\nu_z'}$

For the inverse inequality state III should be the more stable (a situation which does not take place in α-Fe_2O_3). For $\nu_z' < 0$, it is state I which is the more stable.

c) We can finally discuss the behaviour of α-Fe_2O_3 near the transition from phase II to phase I where ν_z' is close from zero. It can be noted that as these phases are induced by different IC's of the R$\bar{3}$c1' group,the transition is necessarily of the first order [see chapter IV]. For its description we can consider the following reduced potential :

$$F = F_o + \frac{\nu_z'}{2} n_z^2 + \frac{\mu_z}{4} n_z^4 + \delta_1' (n_x^3 - n_y^3) n_z + \frac{c}{2} M^2 + \frac{e}{2} (\vec{n}.\vec{M})^2 + \delta_3' (n_x M_y - n_y M_x) \qquad (5.38)$$

discarding from (5.34) all the terms which after discussion appear as unnecessary. Minimizing (5.38) with respect to $\vec{M}$ and replacing their equilibrium values in F lead to the expression of F in polar coordinates for the $\vec{n}$ variable :

$$F = F_o - \frac{\delta_3'^2}{2c} + \frac{1}{2} (\nu_z' + \frac{\delta_3'^2}{c}) \cos^2\theta + \frac{\mu_z}{4} \cos^4\theta + \delta_1' \cos\theta \sin^3\theta \sin 3\phi \qquad (5.39)$$

Again, one can minimize (5.39) with respect to θ and ϕ . We find that in phase I, $\cos\theta = \pm 1$; and in state II, $\sin\phi = \pm 1$. Thus in phase I and II we have respectively :

$$F_I = -\frac{\delta_3'^2}{2c} + \frac{1}{2}\left(\nu_z' + \frac{\delta_3'^2}{c}\right) + \frac{\mu_z}{4} \qquad (5.40)$$

and

$$F_{II} = F_I - \frac{1}{2}\left(\nu_z' + \frac{\delta_3'^2}{c}\right)\cos^2\theta - \frac{\mu_z}{4}(1+\cos^2\theta)\sin^2\theta$$
$$- |\delta_1'| \cos\theta \sin^3\theta \qquad (5.41)$$

where θ is determined in phase II by the equilibrium condition :

$$\left(\nu_z' + \frac{\delta_3'^2}{c}\right)\cos\theta + \mu_z \cos^3\theta - |\delta_1'| \sin^3\theta + 3|\delta_1'|\cos^2\theta \sin\theta = 0 \quad (5.42)$$

At the first order transition between phases I and II, we have $F_I = F_{II}$ and $\frac{\partial F_{II}}{\partial\theta} = 0$ which provide the equilibrium value of θ in phase II :

$$tg\,\theta_o = -\frac{\mu_z}{4|\delta_1'|} + \left[\left(\frac{\mu_z}{4|\delta_1'|}\right)^2 + 1\right]^{1/2} \qquad (5.43)$$

From (5.36) we find the equilibrium magnetization in phase II as a function of θ_o :

$$|\vec{M}_o| = \frac{|\delta_3'|}{c}\sin\theta_o \qquad (5.44)$$

On the other hand the jump in entropy at the I-II transition is :

$$\Delta S = -\frac{\partial}{\partial T}(F_I - F_{II}) = -\frac{1}{2}\left[\frac{\partial(\nu_z' + \delta_3'^2/c)}{\partial T}\right]_{T_N} \sin^2\theta_o \qquad (5.45)$$

where $T_N \simeq 250$ K is the transition temperature. From (5.44) and (5.45) we deduce the transition latent heat :

$$\Delta Q = -\frac{1}{2}T_N\left[\frac{\partial(\nu_z' + \delta_3'^2/c)}{\partial T}\right]_{T_N}\left(\frac{M_o C}{|\delta_3'|}\right)^2 \qquad (5.46)$$

We can conclude this section by giving a summarizing picture of the phase diagram in α-Fe_2O_3 : at $T_c \simeq 950$ K a second order transition takes place from the paramagnetic phase to a monoclinic phase where the magnetic moments located at the iron ions have the approximative antiferromagnetic order : $\vec{\mu}_1 = \vec{\mu}_4 = -\vec{\mu}_2 = -\vec{\mu}_3$, and are directed along the intersection line of the (III) plane and of one of the symmetry planes. More accurately, the $\vec{\mu}_i$ are rotated toward one another about the crystal axis [III] by a small angle, of order 10^{-4}. This canting angle give rise to a weak ferromagnetic magnetization $\vec{M}$ amounting 4/100 of the nominal moment and directed along a two-fold axis. Upon cooling the angle between the $\vec{\mu}_i$ and the [III] axis decreases. At $T_N = 250$ K, it reaches a critical value θ_o and the spontaneous magnetization falls to a value $|\vec{M}_o|$. A first-order transition takes place at which the spins change discontinuously to the rhombohedral [III] direction, and the spontaneous magnetization vanishes. For $T < T_N$, the crystal has a rhombohedral purely antiferromagnetic order, with magnetic moments oriented along the crystal axis.

5.3.4. Behaviour of α-Fe_2O_3 in a magnetic field : magnetic susceptibilities, induced magnetization and hysteresis

To complete the description of the phenomenological properties of α-Fe_2O_3 we will discuss the behaviour of this material under application of a magnetic field $\vec{H}$. For this purpose we use the thermodynamic potential :

$$G = F - \vec{M}\,\vec{H} \qquad (5.47)$$

where the expression (5.35) of F in polar coordinates is assumed.

a) Let us first consider the case of very weak fields, whose intensity is small in comparison with the anisotropy energy. The latter is expressed in (5.35) through the invariants δ_1' and γ_1'. Assuming $H^2 \ll \nu_z' c$ and a small value for $\cos\theta$, of the order of $\frac{\delta_1'}{\nu_z'}$, the anisotropy energy has the order of magnitude of $\gamma_1' + \frac{\delta_1'^2}{\nu_z'}$. As the energy associated to the applied field H can be taken as $|\vec{H}|\frac{|\delta_3'|}{c}$, where $|\vec{M}| = \frac{|\delta_3'|}{c}$ is the spontaneous magnetization, a criterion expressing the weakness of the field is :

$$|\vec{H}|\ \frac{|\delta_3'|}{c} \ll \gamma_1' + \frac{\delta_1'^2}{\nu_z'} \qquad (5.48)$$

which we will write as :

$$|\vec{H}| \ll |\vec{H}_o| = \frac{6c}{|\delta_3'|}\left(\gamma_1' + \frac{\delta_1'^2}{\nu_z'}\right) \qquad (5.49)$$

Let us assume that the system is in state II. In this state the spontaneous magnetic moment is directed along the x axis and the coordinates are (see 5.3.3) : $\theta = \frac{\pi}{2} - \frac{\delta_1'}{\nu_z'}$, $\phi = \frac{\pi}{2}$. Introducing the variables : $M_x' = M_x - \frac{|\delta_3'|}{c}$, $u = \frac{\pi}{2} - \phi$, $v = \frac{\pi}{2} - \frac{\delta_1'}{\nu_z'} - \theta$, $M_y' = M_y + \frac{\delta_3' u}{c}$, $\gamma'' = \gamma_1' + \frac{\delta_1'}{4\nu_z'}$, we can, dropping non essential terms, expand G up to second degree :

$$G = F_o + \frac{\nu_z'}{2} v^2 + 18\,\gamma'' u^2 + \frac{\delta_1'\delta_3'}{\nu_z'} v M_x' + \frac{c}{2} M_x'^2 + \frac{c + e}{2} M_y'^2$$
$$+ \frac{c + \beta_z}{2} M_z^2 + \frac{e\delta_1'}{\nu_z'} M_z M_y' - 3\delta_2' u M_z - H_x M_x' - H_z M_z - H_y M_y' + \frac{\delta_3'}{c} H_y u \qquad (5.50)$$

The minimization of G with respect to v, u, M_x', M_y' and M_z give the equilibrium conditions :

$$\nu_z' v + \frac{\delta_1'\delta_3'}{\nu_z'} M_x' = 0 \qquad c M_x' + \frac{\delta_1'\delta_3'}{\nu_z'} v - H_x = 0$$

$$36\,\gamma''u + \frac{\delta_3'}{c} H_y - 3\,\delta_2' M_z = 0, \quad (c + e)M_y' + \frac{e\,\delta_1'}{\nu_z'} M_z - H_y = 0$$

$$(c + \beta_z)M_z + \frac{e\delta_1'}{\nu_z'} M_y' - 3\,\delta_2' u - H_z = 0 \qquad (5.51)$$

which allow to work out expressions for u, v, M_x, M_y and M_z :

$$u = -\frac{\delta_3'}{36\,\gamma''c} H_y + \frac{\delta_2'}{12\,\gamma''c} H_z, \quad v = -\frac{\delta_1'\delta_3'}{\nu_z'^2 e} H_x \qquad (5.52)$$

$$M_x = \frac{|\delta_3'|}{c} + \frac{H_x}{c}, \quad M_y = \frac{\delta_3'^2 H_y}{36\,\gamma''c^2} - \frac{\delta_3'\delta_2' H_z}{12\,\gamma'' c^2}$$

$$M_z = \frac{H_z}{c + \beta_z} - \frac{\delta_3'\,\delta_2'\,H_y}{12\,\gamma''\,c^2} \qquad (5.53)$$

From (5.53) we can deduce the components of the magnetic susceptibility tensor :

$$\chi_{xx} = \frac{1}{c}, \quad \chi_{zz} = \frac{1}{c + \beta_z}, \quad \chi_{yy} = \frac{\delta_3'^2}{36\,\gamma''c^2}, \quad \chi_{yz} = -\frac{\delta_3'\delta_2'}{12\,\gamma''\,c^2} \qquad (5.54)$$

and $\chi_{xy} = \chi_{xz} = 0$

(5.54) shows that χ_{xx}, χ_{zz} and χ_{yz} are equal, in order of magnitude to the susceptibility $1/c$ in the paramagnetic phase, whereas χ_{yy} has a much greater value of order v^2/c^2. The vanishing of χ_{xy} and χ_{xz} can be predicted on the basis of the monoclinic symmetry 2/m of phase II. According to (5.54) one can write (5.50) as :

$$G = F_o - |\vec{M}|\,H_x - \frac{1}{2}\chi_{xx} H_x^2 - \frac{1}{2}\chi_{yy}H_y^2 - \frac{1}{2}\chi_{zz}H_z^2 - \chi_{yz}H_yH_z \qquad (5.55)$$

b) Let us now assume large values $|\vec{H}| >> |\vec{H}_o|$ for the magnetic field. In this case the anisotropy energy can be neglected in the thermodynamic potential. Furthermore, since $H^2 << \nu_z'\,c$, the deviation of θ from $\frac{\pi}{2}$ can be neglected. Thus G takes the simplified form :

$$G = F_o - \frac{c}{2} M^2 + \beta_z M_z^2 + \delta_3'(M_y \cos\Phi - M_x \sin\Phi) + \frac{e}{2}(M_x \cos\Phi + M_y \sin\Phi)^2 - \vec{M}.\vec{H} \qquad (5.56)$$

Minimization with respect to M_x, M_y and M_z give their dependence as a function of the magnetic field :

$$M_x = \frac{\delta_3'}{c}\sin\Phi + \frac{H_x}{c} - \frac{e}{c + e}(H_x \cos\Phi + H_y \sin\Phi)\cos\Phi$$

$$M_y = -\frac{\delta_3'}{c}\cos\Phi + \frac{H_y}{c} - \frac{e}{c + e}(H_x \cos\Phi + H_y \sin\Phi)\sin\Phi \qquad (5.57)$$

$$M_z = \frac{H_z}{c + \beta_z}$$

Introducing in the xy plane polar coordinates for H_x and H_y ($H_x = h \cos\psi$, $H_y = h \sin\psi$) and eliminating the $\vec{M}$ components in G, we get :

$$G = F_o - \frac{\delta_3'}{c} h \sin(\psi - \phi) + \frac{e\,h^2}{2c(c+e)} \cos^2(\phi - \psi) \qquad (5.58)$$

where terms independent of ϕ have been dropped. The components of $\vec{M}$ parallel and perpendicular to $\vec{h}$ are :

$$M_{//} = -\frac{\delta_3'}{c} \sin(\psi - \phi) + \frac{h}{c} - \frac{eh}{c(c+e)} \cos^2(\psi - \phi)$$
$$M_{\perp} = \frac{\delta_3'}{c} \cos(\psi - \phi) - \frac{eh}{c(c+e)} \cos(\psi - \phi)\sin(\psi - \phi) \qquad (5.59)$$

The equilibrium values of ϕ are determined by the equation $\frac{\partial G}{\partial \phi} = 0$ which has two solutions :

$$\text{I} : \cos(\psi - \phi) = 0 \qquad \text{and II} : \sin(\psi - \phi) = \frac{\delta_3'(c+e)}{eh} \qquad (5.60)$$

to which correspond the equilibrium values :

$$G_1 = -\frac{|\delta_3'|h}{c} \qquad \text{and} \qquad G_2 = \frac{eh^2}{2c(c+e)} + \frac{\delta_3'^2(c+e)}{2\,ce} \qquad (5.61)$$

from which we see that the sign of $G_2 - G_1$ coincides with the sign of e. Thus, for large fields, the crystal can be in either of the two states (5.60) depending on the sign of e.

<u>In state I</u> :

$$e > 0,\ \phi = \psi + \frac{\pi}{2},\ M_{//} = \frac{|\delta_3'|}{c} + \frac{h}{c},\ M_{\perp} = 0,\ M_z = \frac{H_z}{c + \beta_z} \qquad (5.62)$$

The spontaneous magnetization lies in the xy plane and is directed along the field. The antiferromagnetic sublattice moments are directed perpendicular to it so that the magnetic susceptibility is practically isotropic, in contrast to the case of ordinary antiferromagnets. This is the situation found in α-Fe_2O_3 [32] and represented schematically in Figure 12 a)

<u>In state II</u> :

$$e < 0,\ M_{//} = \frac{h}{c+e},\ M_{\perp} = 0,\ M_z = \frac{H_z}{c + \beta_z} \qquad (5.63)$$

The susceptibility is sharply anisotropic in the direction of the crystal axis, and the spontaneous magnetization is absent, i.e. the crystal behaves as an ordinary antiferromagnet (Fig. 12b)

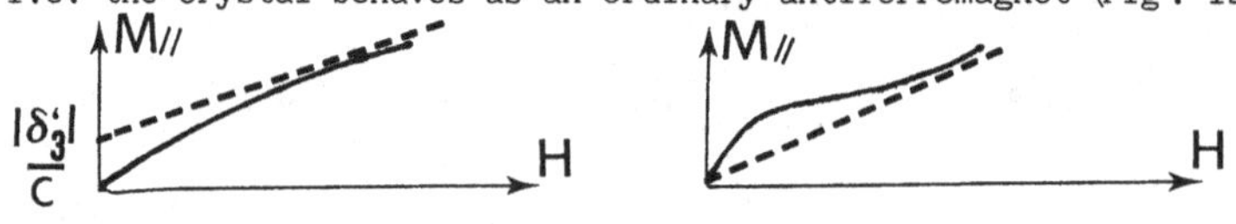

<u>Figure 12</u> : Magnetization curve in state I (a) and state II (b).

c) Finally we can consider moderate values of the magnetic field with $|\vec{H}| \sim |\vec{H}_o|$. Following a similar procedure, it can be shown that the form (5.58) for the thermodynamic potential must be supplemented will a term of the form $\gamma'' \cos 6\phi$ representing the anisotropy energy. Expressions (5.59) for the components of $\vec{M}$ remains valid. Besides the term in h^2 in (5.58) is small in comparison with the term in h, since $|\vec{H}_o| \approx c \frac{\gamma''}{\delta'_3}$ and $\frac{\delta'_3}{c} |\vec{H}_o|$ are of the order of $\frac{v^6}{c^6}$, whereas $\frac{H_o^2}{c} \sim \frac{v^8}{c^8}$. Thus, we can write :

$$G = F_o + \frac{\delta'_3}{c} h \sin(\psi - \phi) + \gamma'' \cos 6\phi \qquad (5.64)$$

and the minimization equation with respect to ϕ :

$$\delta'_3 h \cos(\psi - \phi) + 6 \gamma'' c \sin 6\phi = 0 \qquad (5.65)$$

Besides the equilibrium solutions Eqs.(5.6.4) and (5.6.5) have non equilibrium solutions resulting in an hysteresis curve. This curve can be obtained by using, in addition to Eq.(5.65) the limit stability condition :

$$\frac{\partial^2 G}{\partial \phi^2} = \delta'_3 h \sin(\psi - \phi) + 36 \gamma''_c \cos 6\phi = 0 \qquad (5.66)$$

As a conclusion for the case of α-Fe_2O_3, let us emphasize that this example has illustrated the necessity, in the application of the Landau theory to magnetic systems, to take into account carefully the respective magnitude of the invariants allowed by symmetry, and forming the thermodynamic potential. It has also shown that a comprehensive model can be performed when the magnetic symmetry of all the phases are known with accuracy. In the next example we will examine what sorts of informations can be worked out from the Landau approach, when the preceding data are incomplete.

5.4. Latent antiferromagnetism in nickel-iodine boracite

5.4.1. Identification of the order-parameter symmetry

$Ni_3B_7O_{13}I$ (Ni-I) undergoes at T_c = 61.5 K a continuous transition to a magnetically ordered state with a small spontaneous magnetization of 0.9 G [35-42]. The transition has the remarkable property of being simultaneously ferroelectric and ferroelastic [43-46]. While the symmetry of the paramagnetic phase is well established ($F\bar{4}3c1'$), neutron diffraction experiments [41,42] could not solve the magnetic structure of the low temperature phase. However, its point-group symmetry has been elucidated from magnetic [35], dielectric [35], magnetoelectric [36], piezoelectric [37] and birefringence [35] measurements. In particular the domain pattern observed below T_c shows the existence of twelve ferroelectric domains and twenty-four ferromagnetic domains [31,35] which are consistent with the structural $\bar{4}3m1' \rightarrow m1'$ and magnetic $\bar{4}3m1' \rightarrow m'$ point-group modifications.

As not direct indications exist on the breaking of translational symmetry at T_c that would allow to find out the wave-vector corresponding to the transition, one has to examine systematically the symmetry changes induced by all the IC's associated with the F$\bar{4}$3c1' group, in order to find which one may give rise to a phase possessing a monoclinic symmetry with point-group m'. However, as the continuous character of the 61.5 K transition is firmly established [35-37] as well as the commensurate ordering below T_c, one can restrict to consider the sole IC's which fulfill the Lifshitz condition. In this respect, it has been shown [47,48] that only the IC's of the F$\bar{4}$3c1' group pertaining to two high symmetry points of the face-centered cubic Brillouin-zone are active. These are five IC's (one-two-and three dimensional) at the Γ point ($\vec{k} = \vec{0}$) and five IC's (three-and-six-dimensional) at the X point ($\vec{k} = (0, 0, 2\frac{\pi}{a})$) of the Brillouin-zone surface. The images, thermodynamic potentials, as well as the results of the minimization of these potentials, corresponding to the ten above mentioned active IC's are given in table 13a. Using these data we can find the symmetries induced by each of the IC's following the standard procedure described in Chapters II and III of this book. The conclusions of the symmetry analysis is shown in Table 13 b.

<u>Table 13a</u> : Thermodynamic potentials corresponding to the one (a) two (b), three (c) and six-dimensional (F) IC's at the Γ and X points (first column). Second column : equilibrium values of the order-parameters for the low-temperature stable states.

Thermodynamic potential	Equilibrium values
$a = \frac{\alpha}{2}\eta^2 + \frac{\beta}{4}\eta^4$	$\eta \neq 0$
$b = \frac{\alpha}{2}\rho^2 + \frac{\beta}{4}\rho^4 + \frac{\gamma_1}{6}\rho^6 + \frac{\gamma_2}{6}\rho^6 \cos 6\psi \; (\eta_1 = \rho\cos\psi, \; \eta_2 = \rho\sin\psi)$	I $\psi = 0, \rho \neq 0$ II $\psi = \frac{\pi}{2}, \rho \neq 0$
$c = \frac{\alpha}{2}\sum_{i=1,3}\eta_i^2 + \frac{\beta_1}{4}\sum_{i=1,3}\eta_i^4 + \frac{\beta_2}{2}\sum_{i<j=1,3}\eta_i^2 \cdot \eta_j^2$	I $\eta_1 \neq 0, \eta_2 = \eta_3 = 0$ II $\eta_1 = \eta_2 = \eta_3 \neq 0$
$F = \frac{\alpha}{2}\sum_{i=1,3}\rho_i^2 + \frac{\beta_1}{4}\sum_{i=1,3}\rho_i^4 + \frac{\beta_2}{4}\sum_{i=1,3}\rho_i^4\cos 4\psi_i + \frac{\beta_3}{4}(\rho_1^2\rho_2^2 + \rho_1^2\rho_3^2 + \rho_2^2\rho_3^2)$ $+ \frac{\beta_4}{2}(\rho_1^2\rho_2^2 \sin 2\psi_1 \sin 2\psi_2 + \rho_1^2\rho_3^2 \sin 2\psi_1 \sin 2\psi_3 + \rho_2^2\rho_3^2 \sin 2\psi_2 \sin 2\psi_3)$ $+ \frac{\beta_5}{2}[\rho_1^2\rho_2^2(\sin 2\psi_1 - \sin 2\psi_2) + \rho_2^2\rho_3^2(\sin 2\psi_2 - \sin 2\psi_3) + \rho_1^2\rho_3^2(\sin 2\psi_3 - \sin 2\psi_1)]$ $[\eta_{2P+1} = \rho_{P+1}\cos\psi_{P+1} \; (P = 0,1,2) \; ; \; \eta_{2P} = \rho_P \sin\psi_P \; (P = 1,2,3)]$	I $\rho_1^2 = -\alpha/\beta_1+\beta_2, \rho_2 = \rho_3 = 0, \psi_1 = 0$ II $\rho_1^2 = -\alpha/\beta_1-\beta_2, \rho_2 = \rho_3 = 0, \psi_1 = \frac{\pi}{4}$ III $\rho_1^2 = \rho_2^2 = -\alpha/\Delta_1, \rho_3 = 0, \psi_1 = -\psi_2 = \frac{\pi}{4}$ IV $\rho_1^2 = \rho_2^2 = \rho_3^2 = -\alpha/\Delta_2, \psi_1 = \psi_2 = \psi_3 = \frac{\pi}{4}$ V, VI $\rho_1^2 = \rho_2^2 = \rho_3^2 = -\alpha/\Delta_3$, V: $\psi_1 = \psi_2 = \psi_3 = \frac{\pi}{2}$; VI: $\psi_1 = \psi_2 = \psi_3 = 0$ VII $\rho_1^2 = -\alpha/\Delta_7, \rho_2^2 = \rho_3^2 = -\alpha/\Delta_8, \psi_1 = 0, \psi_2 = \frac{\pi}{4}, \psi_3 = \frac{3\pi}{4}$ VIII $\rho_1^2 = -\alpha/\Delta_5, \rho_2^2 = \rho_3^2 = -\alpha/\Delta_6, \psi_1 = 0, \psi_2 = 3\frac{\pi}{4} \; \psi_3 = \frac{\pi}{4}$

Table 13b: Magnetic symmetry groups induced by the IC's of the $F\bar{4}3c1'$ group at the Γ and X points of the face centered cubic Brillouin zone (k_{11} and k_{10} in the Kovalev notation [26]). In the third column the groups are indicated together with the type of magnetic ordering (AF or F). Fourth column : primitive translations of the corresponding elementary unit-cell as a function of the translations of the cubic F elementary unit-cell.

Group	Point	Ordering	Magnetic groups	IC	Primitive translations	
$F\bar{4}3c1'$	$\Gamma(k_{11})$	AF	$F\bar{4}3c(\tau_1)$, $F\bar{4}'3c'(\tau_2)$		t_1, t_2, t_3	(V)
		AF	(I $I\bar{4}c2$, II $I\bar{4}'c'2$) (τ_3)			
		AF	(I $I\bar{4}'c2'$, II $R3c$) (τ_4)			
		F	(I $I\bar{4}c'2'$, II $R3c'$) (τ_5)			
	$X(k_{10})$	AF	I $P_I\bar{4}c2(\tau_2,\tau_5)$, $P_I\bar{4}b2(\tau_3,\tau_4)$		$t_2-t_1, t_3, t_1+t_2-t_3$ Rt_1, Rt_2	(2V)
		AF	II $P\bar{4}3n(\tau_2,\tau_5)$, $P\bar{4}'3n'(\tau_3,\tau_4)$		$t_1\pm t_2\mp t_3$ $t_2+t_3-t_1$	(4V)
		AF	I P_Ic2_1a, II C_A222_1	(τ_1)	$t_2-t_1, t_3, t_1+t_2-t_3$ Rt_1, Rt_2	(2V)
		AF	III $P\bar{4}2_1c$, IV $P2_13$, V $R3c$	(τ_1)	$t_1\pm t_2\mp t_3$ $t_2+t_3-t_1$	(4V)
		F	VI $R3c'$ VII Cc' VIII Cc	(τ_1)		

From the results contained in Table 13b, we can deduce that the paramagnetic to ferromagnetic $F\bar{4}3c1' \rightarrow m'$ symmetry change observed in Ni-I boracite can be unequivocally related to the six-dimensional IC, labelled τ_1, at the X point of the face-centered Brillouin zone. Actually as it can be verified in Table 13b, the other IC's of the $F\bar{4}3c1'$ group lead either to antiferromagnetic structures or to tetragonal and rhombohedral ferromagnetic groups. Furthermore, we are able to specify : i) the magnetic space group describing the ordered phase, namely the group Cc' which consists in the identity and the glide plane $(R\sigma_{xz} | 0 0 \frac{a}{2})$ ii) the lattice modification, which corresponds to a fourfold multiplication of the cubic primitive cell as shown in Figure 13.

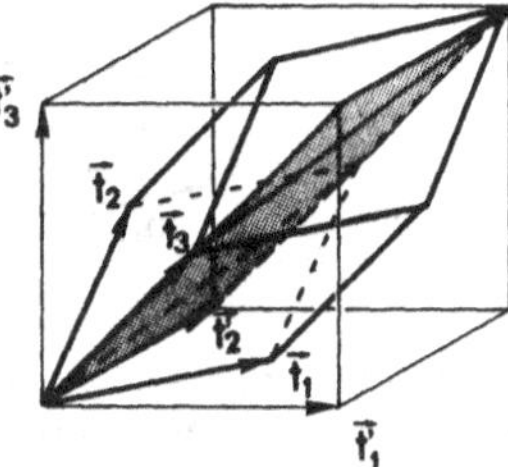

Figure 13 : Lattice modification at the $F\bar{4}3c1' \rightarrow Cc'$ transition assumed for Ni-I boracite. The monoclinic cell contains four cubic primitive cells. The magnetic and crystallographic monoclinic cells have the same primitive translations : $\vec{t}'_1=\vec{t}_1-\vec{t}_2+\vec{t}_3$, $\vec{t}'_2=\vec{t}_1+\vec{t}_2-\vec{t}_3$, $\vec{t}'_3=-\vec{t}_1+\vec{t}_2+\vec{t}_3$

5.4.2. Tentative model for the ordered phase of Ni-I boracite : latent antiferromagnetism

In the paramagnetic phase, the cubic primitive cell of Ni-I boracite contains two formula units. Thus the conventional cell $F\bar{4}3c1'$ contains eight formula units with 24 magnetic ions N_i^{2+}. As shown on figure 13, the monoclinic Cc' unit-cell has the same volume as the conventional F-cubic cell. Let us assume, without loss of generality, that the lowering of symmetry taking place at T_c = 61.5 K is entirely connected with the displacement of the nickel ions. The 24 metallic ions will thus be distributed among 12 independent monoclinic sublattices. The positions of the ions forming each sublattice are given in table 14.

Table 14 : coordinates of the 24 nickel ions in the monoclinic $Cc(C_s^4)$ cell of Ni-I boracite. The coordinates are given with respect to the cubic axes x,y,z

1.	1/4, 1/4, 0	9.	1/2, 1/4, 1/4	17.	1/4, 0 , 3/4
2.	3/4, 3/4, 0	10.	1/2, 1/4, 3/4	18.	1/4, 0 , 1/4
3.	1/4, 3/4, 0	11.	1/2, 3/4, 1/4	19.	3/4, 0 , 3/4
4.	3/4, 1/4, 0	12.	1/2, 3/4, 3/4	20.	3/4, 0 , 1/4
5.	0 , 1/4, 1/4	13.	1/4, 1/4, 1/2	21.	1/4, 1/2, 3/4
6.	0 , 1/4, 3/4	14.	3/4, 3/4, 1/2	22.	1/4, 1/2, 1/4
7.	0 , 3/4, 1/4	15.	3/4, 1/4, 1/2	23.	3/4, 1/2, 3/4
8.	0 , 3/4, 3/4	16.	1/4, 3/4, 1/2	24.	3/4, 1/2, 1/4

In figures 14a) and 14b) the magnetic ions are represented in

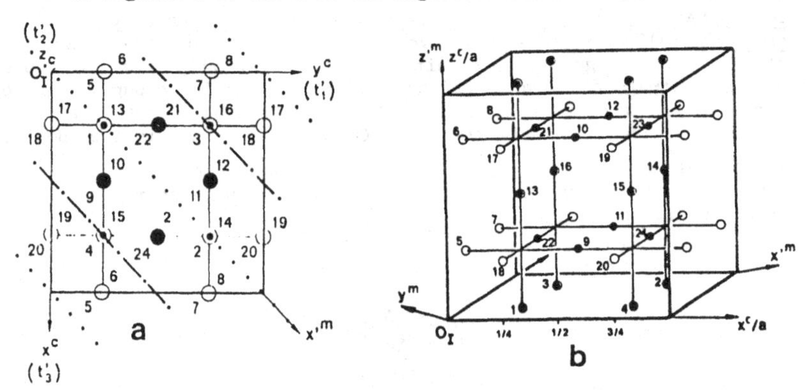

Figure 14 Position of the 24 nickel ions in the monoclinic Cc phase of Ni-I boracite, neglecting the monoclinic deformation with respect to the cubic phase : (a) projection on the xy cubic plane (b) positions of the metals in the volume of the monoclinic cell. In the two figures the cubic (x^c, y^c, z^c) and monoclinic (x'^m, y'^m, z'^m) axes have the same origin (O_I) as in Ref. 49

projection on the pseudo cubic plane (00I) and within the monoclinic cell respectively. Having regard to their structural environment, the 12 independent nickel ions form three groups, denoted (1,2,3,4), (5,6,7,8) and (9,10,11,12) in figures 14, which are located inside mixed oxygen halogen octahedra, the axes of which are parallel or at 45 % to the monoclinic plane. The three groups of atoms lie respectively in planes perpendicular to the cubic z direction, and to the x direction (x = 0 and x = 1/2).

The magnetic structure of the crystal in the ordered phase will be determined if the magnetic moments of the ions are given. Let us symbolize these moments by $\vec{\mu}_1, \vec{\mu}_2, \ldots, \vec{\mu}_{12}$ and denote $\vec{\mu}_{13}, \vec{\mu}_{14}, \ldots, \vec{\mu}_{24}$ the moments obtained by reflection in the monoclinic plane $R\sigma_{\overline{xz}}$. As in §.5.2. and §.5.3 let us introduce auxiliary magnetic vectors expressing the possible sublattice arrangements. Following the method detailed in §.5.2, we find twelve such vectors labelled $\vec{M}_j^\alpha$ and $\vec{L}_j^\alpha$ (j,α = 1,2,3), which are defined by the equations :

$$\vec{L}_1^1 = \vec{\mu}_1 + \vec{\mu}_{13} + \vec{\mu}_2 + \vec{\mu}_{14} - \vec{\mu}_3 - \vec{\mu}_{15} - \vec{\mu}_4 - \vec{\mu}_{16}$$
$$\vec{L}_2^1 = \vec{\mu}_1 + \vec{\mu}_{13} - \vec{\mu}_2 - \vec{\mu}_{14} + \vec{\mu}_3 + \vec{\mu}_{15} - \vec{\mu}_4 - \vec{\mu}_{16}$$
$$\vec{L}_3^1 = \vec{\mu}_1 + \vec{\mu}_{13} - \vec{\mu}_2 - \vec{\mu}_{14} - \vec{\mu}_3 - \vec{\mu}_{15} + \vec{\mu}_4 + \vec{\mu}_{16} \qquad (5.67)$$
$$\vec{L}_j^2 = \vec{L}_j^1\,(\vec{\mu}_{i+4}),\ \vec{L}_j^3 = \vec{L}_j^1\,(\vec{\mu}_{i+8}) \quad (j = 1,2,3,\ i = 1\text{-}4 \text{ and } 12\text{-}16)$$
$$\vec{M}^1 = \vec{\mu}_1 + \vec{\mu}_{13} + \vec{\mu}_2 + \vec{\mu}_{14} + \vec{\mu}_3 + \vec{\mu}_{15} + \vec{\mu}_4 + \vec{\mu}_{16}$$
$$\vec{M}^2 = \vec{M}^1\,(\mu_{i+4}),\ \vec{M}^3 = \vec{M}^1\,(\mu_{i+8}) \quad (i = 1\text{-}4 \text{ and } 12\text{-}16)$$

The vector $\vec{M} = \vec{M}^1 + \vec{M}^2 + \vec{M}^3$ represents the total magnetization moment of the monoclinic unit-cell below T_c. One can verify that it transforms as the three-dimensional vector corepresentation of the $F\bar{4}3c1'$ group at the Γ point, labelled τ_5 in Table 13. On the other hand, the reducible corepresentation spanning the antiferromagnetic vectors $\vec{L}_i^\alpha$ decomposes into two IC's of the $F\bar{4}3c1'$ group at the X point of the corresponding Brillouin zone. More precisely, (L_{1x}^α, L_{2y}^α, L_{3z}^α) transform as the three dimensional IC denoted τ_3 in Table 13 b, whereas the six-dimensional IC τ_1 describes the transformation properties of the components (L_{1y}^α, L_{1z}^α, L_{2x}^{α}, L_{2z}^α, L_{3x}^α, L_{3y}^α). It can be verified that each of the sets of $\vec{L}_j^\alpha$ projections are distributed over the arms of the star $\vec{k}_{10}^*$, namely $\vec{k}_{10}^1 = (0, 0, \frac{2\pi}{a})$, $\vec{k}_{10}^2 = (0, \frac{2\pi}{a}, 0)$, $\vec{k}_{10}^3 = (\frac{2\pi}{a}, 0, 0)$.

As τ_1 was found to induce the phase transition in Ni-I boracite (§.5.4.1), we can work out the linear relationships between the L_i^α components and the abstract order-parameter components η_i (i=1-6) which spans the IC τ_1 (X). Identifying the basis of τ_1 in terms of the η_i and the L_{ju}^α (u = x,y,z) yields :

$$\eta_1 = L_{3x}^\alpha + L_{3y}^\alpha,\ \eta_2 = L_{3x}^\alpha - L_{3y}^\alpha,\ \eta_3 = L_{1z}^\alpha - L_{1y}^\alpha \qquad (5.68)$$
$$\eta_4 = L_{1y}^\alpha + L_{1z}^\alpha,\ \eta_5 = L_{2z}^\alpha - L_{2x}^\alpha,\ \eta_6 = -L_{2x}^\alpha - L_{2z}^\alpha$$

where $\alpha = 1,2,3$. Using (5.68) we can express the expansion F given in Table 13a in terms of the $\vec{L}^{\alpha}$, $\vec{M}^{\alpha}$, M_u^{α}, L_{ju}^{α} separating its exchange and relativistic parts. We find :

$$F = \sum_{\alpha=1-3} (F_{ex}^{\alpha} + F_r^{\alpha}) \tag{5.69}$$

with (omitting the superscript α) :

$$\begin{aligned} F_{ex} = {} & \frac{a}{2} \sum_j L_j^2 + \frac{\beta_1}{4} [\sum_j L_j^2]^2 + \frac{\beta_2}{4} \sum_j L_j^4 + \frac{\beta_3}{4} \sum_{i \neq j} (\vec{L}_i . \vec{L}_j)^2 \\ & + \frac{c}{2} M^2 + D[(\vec{M}.\vec{L}_1).\vec{L}_2.\vec{L}_3 + (\vec{M}.\vec{L}_2).\vec{L}_1.\vec{L}_3 + (\vec{M}.\vec{L}_3).\vec{L}_1.\vec{L}_2] \\ & + \frac{E_1}{2} M^2 [\sum_j L_j^2] + \frac{E_2}{2} \sum_j (\vec{M}\vec{L}_j)^2 \end{aligned} \tag{5.70}$$

and

$$\begin{aligned} F_r = {} & \frac{1}{2} \alpha (L_{1y}^2 + L_{1z}^2 + L_{2x}^2 + L_{2z}^2 + L_{3x}^2 + L_{3y}^2) \\ & + \frac{\beta_1}{4} (L_{1y}^4 + L_{1z}^4 + L_{2x}^4 + L_{2z}^4 + L_{3x}^4 + L_{3y}^4) \\ & + \frac{\beta_2}{2} (L_{1y}^2 L_{1z}^2 + L_{2x}^2 L_{2z}^2 + L_{3x}^2 L_{3y}^2) \\ & + \frac{\beta_3}{2} [(L_{1y}^2 + L_{1z}^2)(L_{2x}^2 + L_{2z}^2 + L_{3x}^2 + L_{3y}^2) + (L_{2x}^2 + L_{2z}^2)(L_{3x}^2 + L_{3y}^2)] \\ & + \frac{\beta_4}{2} [(L_{3x}^2 - L_{3y}^2)(L_{1z}^2 - L_{1y}^2 + L_{2z}^2 - L_{2x}^2) + (L_{1z}^2 - L_{1y}^2)(L_{2x}^2 - L_{2z}^2)] \\ & + \frac{\beta_5}{2} [(L_{3x}^2 - L_{3y}^2 - L_{1z}^2 + L_{1y}^2)(L_{1z}^2 + L_{1y}^2 - L_{2x}^2 - L_{2z}^2) + (L_{2z}^2 - L_{2x}^2)(L_{3x}^2 + L_{3y}^2 - L_{1y}^2 - L_{1z}^2)] \\ & + \delta_1 [M_x (L_{1z} L_{2z} L_{3x} + L_{1y} L_{2x} L_{3y}) + (L_{3x} + L_{3y})(L_{1y} L_{2x} M_y - L_{1y} L_{2z} M_z)] \\ & + \delta_2 [M_x (L_{1y} L_{2x} L_{3y} - L_{1z} L_{2z} L_{3x}) + M_y L_{1y} L_{2x} (L_{3x} + L_{3y}) + M_z (L_{1y} L_{2z} L_{3x} - L_{1z} L_{2x} L_{3y}) \end{aligned} \tag{5.71}$$

We will summarize the main results obtained from the minimization of (5.70) and (5.71) in connection with the interpretation of the transition in Ni-I boracite :

a) At $T_{c\alpha} = 61.5$ K, where $a = a_o(T-T_c)$ vanishes, some of the components L_i^{α} become non zero. They vary as $|L_i^{\alpha}| \sim (T_c-T)^{1/2}$. Since F_{ex}^{α} contains an invariant linear in M^{α}, <u>a non zero spontaneous magnetization will arize simultaneously with the appearance of the L_i^{α}, that will be of exchange origin.</u> The expression for $\vec{M}^{\alpha}$ is :

$$\vec{M}^{\alpha} = - \frac{D}{C} [\vec{L}_1^{\alpha} (\vec{L}_2^{\alpha}.\vec{L}_3^{\alpha}) + \vec{L}_2^{\alpha}(\vec{L}_1^{\alpha}.\vec{L}_3^{\alpha}) + \vec{L}_3^{\alpha}(\vec{L}_1^{\alpha}.\vec{L}_2^{\alpha})] \tag{5.72}$$

which is proportional to $(T_c-T)^{3/2}$ (Figure 15). The absolute minimum of F_{ex} associated with this non zero magnetization corresponds to :

$$\vec{L}_1^\alpha = \pm \vec{L}_2^\alpha = \pm \vec{L}_3^\alpha \tag{5.73}$$

Introducing (5.73) in (5.67) leads to :

$$\vec{\mu}_i + \vec{\mu}_{i+12} = \frac{1}{4} (\vec{M}^\alpha \pm 3\vec{L}_1^\alpha)$$
and
$$\vec{\mu}_j + \vec{\mu}_{j+12} = \frac{1}{4} (\vec{M}^\alpha \pm \vec{L}_1^\alpha) \quad (j \neq i) \tag{5.74}$$

where the number i is determined by the signs in Eq.(5.73) (e.g. for $L_1^\alpha = L_2^\alpha = L_3^\alpha$: $i = 1$, $j = 2,3,4$)

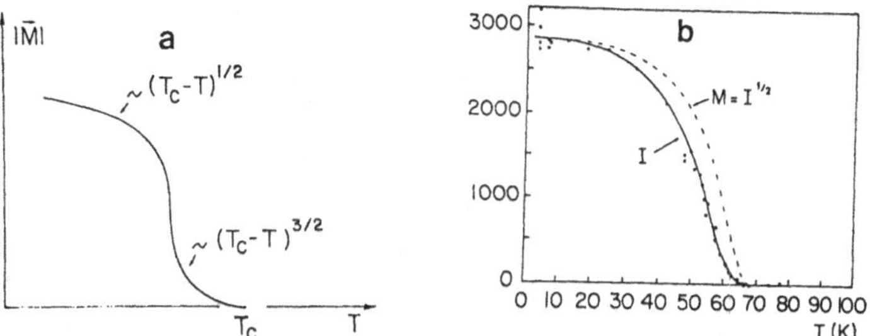

Figure 15 : Temperature dependence of the magnetization in Ni-I boracite (a) Theoretical curve ; (b) experimental curve deduced from the intensity I of the magnetic (III) neutron reflection [40].

From Eq.(5.74) it can be seen that the average magnetic moments of the ions can be divided into two groups, the absolute magnitude of the moments differing from one group to the other. Such a property is usual for ferrimagnets. However, here the moments are associated with one identical type of magnetic ions, which is found in equivalent crystallographic position in the paramagnetic phase. This is in contrast with the situation found in standard ferrimagnets such as ferrites or garnets [50,51]. Dzialoshinskii and Man'ko [21] suggested to denominate this new type of uncompensated antiferromagnetism, latent antiferromagnetism. As noted by these authors, no confusion should be made with ferrimagnetism because of the peculiar temperature variation of the magnetization (with a $\frac{3}{2}$ critical exponent) in the vicinity of T_c. From the experimental data on Ni-I [40], it appears that the interval in which the $(T_c-T)^{3/2}$ law holds is very narrow, and may escape detection. Another distinctive feature of latent antiferromagnetic materials, is that despite its exchange origin, the magnetization must be expected to assume weak values at any temperature below T_c. This is connected with the improper character of the transition, i.e. with the fact that $\vec{M}$ results from a coupling to the third power of the antiferromagnetic sublattices.

In Ni-I boracite, the magnetization at 4.2 K is found to be about 0.9 G [38,39] , which represents 1 % of the nominal value.

In the ferrimagnets Fe_3O_4 [52] and $Y_3Fe_4O_{12}$ [53], the numerical values found for $\vec{M}$ are respectively, about 5.10^3 and 2.10^3 G. No other experimental example of latent antiferromagnetic material displaying a spontaneous magnetization is known, that would allow to check if the weak value found for $|\vec{M}|$ in Ni-I boracite corresponds to some standard order of magnitude.

b) The orientation of the vectors $\vec{L}_i$ below T_c is determined by the relativistic invariants contained in (5.71). As shown in table 13a, the monoclinic phase corresponds to the equilibrium values of the order parameter : $\eta_1 \neq 0$, $\eta_2 = 0$, $\eta_3 = \eta_4 = -\eta_5 = \eta_6 \neq 0$, $\eta_1 \neq \eta_i$ (i = 3-6). In terms of the L^α_{iu}, one attains from (5.68):

$$\begin{aligned} & L^\alpha_{1y} = 0, \quad L^\alpha_{2x} = 0, \quad L^\alpha_{3x} = L^\alpha_{3y}, \quad L^\alpha_{1z} = -L^\alpha_{2z} \\ & L^\alpha_{3x} \neq (L^\alpha_{1x}, L^\alpha_{2y}, L^\alpha_{1z}, L^\alpha_{2x}, L^\alpha_{3z}) \\ & M^\alpha_x \neq M^\alpha_z \neq 0, \quad M^\alpha_y = 0 \end{aligned} \tag{5.75}$$

Substituting (5.75) in (5.67) yields the following relationships among the sublattices corresponding to α= 1.

$$\begin{aligned} \mu_{1x}+\mu_{13x} &= \tfrac{1}{4}[M^1_x+L^1_{1x}+L^1_{3x}] , & \mu_{2x}+\mu_{14x} &= \tfrac{1}{4}[M^1_x+L^1_{1x}-L^1_{3x}] \\ \mu_{1y}+\mu_{13y} &= \tfrac{1}{4}[L^1_{2y}+L^1_{3x}] , & \mu_{2y}+\mu_{14y} &= \tfrac{1}{4}[-L^1_{2y}-L^1_{3x}] \\ \mu_{1z}+\mu_{13z} &= \tfrac{1}{4}[M^1_z+L^1_{3z}] , & \mu_{2z}+\mu_{14z} &= \tfrac{1}{4}[M^1_z+2L^1_{1z}-L^1_{3z}] \\ \mu_{3x}+\mu_{15x} &= \tfrac{1}{4}[M^1_x-L^1_{1x}-L^1_{3x}] , & \mu_{4x}+\mu_{16x} &= \tfrac{1}{4}[M_x-L^1_{1x}+L^1_{3x}] \\ \mu_{3y}+\mu_{15y} &= \tfrac{1}{4}[L^1_{2y}-L^1_{3x}] , & \mu_{4y}+\mu_{16y} &= \tfrac{1}{4}[-L^1_{2y}+L^1_{3x}] \\ \mu_{3z}+\mu_{15z} &= \tfrac{1}{4}[M^1_z-2L^1_{1z}-L^1_{3z}] , & \mu_{4z}+\mu_{16z} &= \tfrac{1}{4}[M^1_z+L^1_{3z}] \end{aligned} \tag{5.76}$$

and similar relationships for the average moments $\vec{\mu}_{i+4}$ and $\vec{\mu}_{i+8}$ (i = 1-4, 12-16) with α = 2 and α = 3 respectively. Eq.(5.76) provide some information about the possible magnetic structure of the monoclinic phase in Ni-I boracite. Thus, the non-compensation of the antiferromagnetic sublattice appears on the x and z projections, whereas the average spin ordering compensates along the y direction. Furthermore, projections of unequal magnitude (i.e. $M^\alpha_z + L^\alpha_{3z}$, and $M^\alpha_z - 2L^\alpha_{1z} - L^\alpha_{3z}$) are found along the z direction, as well as <u>between</u> the x, y and z directions. Such results contradict the three-sublattice arrangement proposed by Von Wartburg from neutron diffraction data [41].

Although this last example is an illustration of the complexity of the Landau approach when a large number of magnetic sublattices exist in the system, it also shows that the approach can be used to elaborate a tentative model for the magnetic structure, to be checked experimentally.

6. APPLICABILITY OF THE LANDAU THEORY TO MAGNETIC SYSTEMS : ORDER-PARAMETER SYMMETRIES, FIRST-ORDER TRANSITIONS AND INCOMMENSURATE MAGNETIC PHASES

6.1. Order-parameter symmetries for second and first order transitions in magnetic systems

As illustrated in the preceding section, the Landau theory is a powerful tool for a phenomenological description of magnetic systems. However, since the pioneer works of Dzialoshinskii [19-21] and Kovalev [15-17] there has been a relatively small number of models devoted to magnetic systems, based on this theory (see for example refs [54-58]. This is due on the one hand to the popularity of the statistical mechanics approach to such systems [59,60], which is fruitful for systems with a small number of degrees of freedom, and on the other hand to the fact that for systems with a large number of degrees of freedom, the magnetic structure of the ordered phase is difficult to solve experimentally.

Among the few studies which have investigated the properties of magnetic transitions in the framework of the Landau theory, the more comprehensive work is due to Clin et al [58], who have established systematic tables providing all the possible symmetry modifications and thermodynamic potentials associated with the IC's fulfilling the Lifshitz condition, and belonging to the 230 paramagnetic groups. Comparing the results obtained by these authors with the available experimental data [61] allows a general view on the capability of the Landau theory to interpretate correctly the behaviour of magnetic transitions. This comparison leads to the following conclusions.

a) All the well established second-order transitions from a paramagnetic phase to a ferromagnetic or antiferromagnetic phase, can be unambiguously connected to a single active IC of the paramagnetic group.

This property can be verified in a large number of materials belonging to all classes of magnetic transitions. In Tables 15 to 18 we have listed some representative examples of ferromagnets, ferrimagnets, antiferromagnets and weak ferromagnets respectively undergoing a second order transition accompanied by a symmetry modification related to a single IC of the high-temperature phase.

Table 15 : Examples of materials undergoing a continuous paramagnetic to ferromagnetic transition. For the materials listed in column (a), columns (b), (c) and (d) give respectively the symmetries of the high and low temperature phases and the transition temperature in Kelvins. Columns (e), (f) and (g) provide the IC, the order-parameter dimensionality and the image of the IC which allows construction of the thermodynamic potential. The labelling of the IC's follows the notation of the associated irreducible representation in Zak's et al [27]. In tables 15 and 16 all the IC's belong to the Brillouin-zone center.

(a)	(b)	(c)	(d)	(e)	(f)	(g)
$CoFe_2Se_4$	$C2/m1'$	$C2'/m'$	125	τ_2		
DyNi	$Pnma1'$	$Pnm'a'$	62	τ_2	1	C_i
TbGa	$Cmcm1'$	$Cm'c'm$	158	τ_4		
MnAl	$P4/nmm1'$	$P4/nm'm'$	518	τ_2		
$NiMnO_3$	$R\bar{3}1'$	$P\bar{1}$	435	$\tau_2+\tau_3$	2	C_6
Nd CO_3	$R\bar{3}m1'$	$R\bar{3}m'$	395	τ_2		
$CrBr_3$	$R\bar{3}c1'$	$R\bar{3}c'$	35	τ_2		
$NdCO_5$	$P6/mmm1'$	$P6/mm'm'$	913	τ_2	1	C_i
Mn_5GeO_3	$P6_3/mcm1'$	$P6_3/mc'm'$	304	τ_2		
α- Co	$P6_3/mmc1'$	$P6_3/mm'c'$	1383	τ_2		
U_3P_4	$I\bar{4}3d1'$	$R3c'$	144	τ_5		
Ni	$Fm3'm$	$R\bar{3}m'$	633	τ_5	3	O_h
$NdAl_2$	$Fd3'm$	$I4_1/am'd'$	76	τ_5		
α- Fe	$Im3'm$	$I4/mm'm'$	1042	τ_5		

Table 16 : Examples of materials undergoing a continuous paramagnetic to ferrimagnetic transition. Columns have the same meaning as in Table 15

(a)	(b)	(c)	(d)	(e)	(f)	(g)
Fe_2VSe_4	$C2/m1'$	$C2'/m'$	155	τ_2	1	C_i
HoAl	$Pbcm1'$	$Pb'cm'$	26	τ_2		
$CuCr_2O_4$	$I\bar{4}2d1'$	$Fd'd2'$	133	τ_5	2	C_{4v}
Mn_2Sb	$P4/nmm1'$	$P4/nm'm'$	550	τ_2	1	C_i
Cr_5S_6	$P\bar{3}c1'$	$C2/c'$	305	τ_6	2	C_{6v}
$ErCo_3$	$R\bar{3}m1'$	$R\bar{3}m'$	401	τ_2		
$HoCo_5$	$P6/mmm1'$	$P6/mm'm'$	1000	τ_2	1	C_i
Ni_3Mn	$Pm3'm$	$P4/mm'm'$	610	τ_5		
MnV_2O_4	$Fd3'm$	$I4_1/am'd'$	52	τ_5	3	O_h
$Dy_3Fe_5O_{12}$	$Ia3'd$	$R\bar{3}c'$	551	τ_5		

Table 17 : Examples of second order paramagnetic to antiferromagnetic transitions. Columns have the same meaning as in Table 15.

(a)	(b)	(c)	(d)	(e)	(f)	(g)
DyOOH	$P2_1/m1'$	$P2_1/m'$	7.2	$\tau_3(\Gamma)$		
$LiCuCl_3 2H_2O$	$P2_1/c1'$	$P2_1/c$	6.7	$\tau_4(\Gamma)$		
$HoCoO_3$	Pnma1'	Pn'm'a'	2.4	$\tau_5(\Gamma)$		
$NiCrO_4$	Cmcm1'	Cmcm	23	$\tau_1(\Gamma)$		
$C_2(Zn)_2$	Imma1'	Im'ma	7.5	$\tau_6(\Gamma)$	1	C_i
$CuFeS_2$	$I\bar{4}2d1'$	$I\bar{4}2d$	815	$\tau_1(\Gamma)$		
CoF_2	$P4_2/mnm1'$	$P4_2/mnm'$	37	$\tau_3(\Gamma)$		
$DyPO_4$	$I4_1/amd1'$	$I4_1/a'm'd$	3.4	$\tau_8(\Gamma)$		
$\alpha-Fe_2O_3$	$R\bar{3}c1'$	$R\bar{3}c$	956	$\tau_1(\Gamma)$		
$HoMn_3$	$P6_3cm1'$	$P6_3cm$	76	$\tau_1(\Gamma)$		
FeS	$P\bar{6}2c1'$	$P\bar{6}'2'c$	599	$\tau_5(\Gamma)$		
NiS	$P6_3/mmc1'$	$P6_3/m'm'c$	720	$\tau_4(\Gamma)$		
Er_2O_3	Ia3'	Ia3	4	$\tau_1(\Gamma)$		
Mn_3GaN	Pm3'm	$R\bar{3}m$	298	$\tau_4(\Gamma)$	3	O_H
$NiCl_2 6(H_2O)$	C2/m1'	C_c2/c	5.34	$\tau_2(Y)$		
$NiWO_4$	P2/c1'	P_a2/c	67	$\tau_1(A)$		
$RbFeF_4$	$Pca2_1 1'$	P_bna2_1	-	$\tau_3(Y)$	1	C_i
$CaMn_2O_4$	Pbcm1'	P_abcm	225	$\tau_1(X)$		
$CoUO_4$	Imma1'	C_c2/m	12	$\tau_1(T)$	2	C_{4v}
UAs_2	P4/mmm1'	P_c4/nca	283	$\tau_2(Z)$	1	C_i
Cs_2MnCl_4	I4/mmm1'	C_Amca	52	$\tau_3(X)$	2	C_{4v}
$CoTiO_3$	$R\bar{3}1'$	$Ps\bar{1}$	37	$\tau_2+\tau_3(Z)$	2	C_6
Yb_2O_2S	$P\bar{3}m1'$	C_c2/m	3	$\tau_3(A)$	2	C_{6v}
$DyGa_2$	P6/mmm1'	C_cmcm	15	$\tau_5(A)$	2	C_{6v}
CuMnSb	$F\bar{4}3'm$	R_I3c	55	$\tau_2(L)$	4	109.01
$KFeF_3$	Pm3'm	$R_I\bar{3}c$	112	$\tau_5(R)$	3	O_h
$NdIn_3$	Pm3'm	P_C4/mbm	7	$\tau_2(M)$	3	O_h
$PbCrO_3$	Pm3'm	I_c4/mcm	240	$\tau_3(R)$	2	C_{6v}
NdSe	Fm3'm	$R_I\bar{3}c$	14	$\tau_2(L)$	4	109.01

Table 18 : Examples of second order paramagnetic to weak-ferromagnetic transitions. Columns have the same meaning as in Table 15

(a)	(b)	(c)	(d)	(e)	(f)	(g)
$CO_3B_7O_{13}Br$	$Pca2_11'$	$Pc'a2_1'$	16	$\tau_4(\Gamma)$	1	C_i
$Ni_3B_7O_{13}Cl$	$Pca2_11'$	$Pca2_1$	25	$\tau_1(\Gamma)$	1	
NiF_2	$P4_2/mnm$	Pnn'm'	73	$\tau_5(\Gamma)$	2	C_{4v}
$Co_3B_7O_{13}Cl$	R3c1'	Cc	9-11	$\tau_3(\Gamma)$	2	C_{6v}
α-Fe_2O_3	$R\bar{3}c1'$	C2/c	950	$\tau_5(\Gamma)$	2	

From the data contained in Tables 15-18, it can be noticed that for all the experimentally observed transitions to ferromagnetic, ferrimagnetic or weak-ferromagnetic phases, the symmetry change can be connected with an IC of the corresponding Brillouin-zone center (the Γ point). However there exist a number of paramagnetic-ferromagnetic symmetry modifications which are predicted theoretically for IC's of high-symmetry points of the rhombohedral, hexagonal and cubic Brillouin-zone surfaces (see ref. 58). This "improper" ferromagnetism, of which $Ni_3B_7O_{13}I$ is presently the only confirmed example (see §.5-4), was also invoked for the potential case of UO_2 [21]

As shown in ref. 58 the order-parameter symmetries corresponding to the active IC's of the 230 paramagnetic groups do not differ in an essential manner from the symmetries found for structural transitions (see chapter III). Due to the fact that a number of three-and six-dimensional irreducible representations discarded by the Landau condition are active IC's, one finds two additional six-dimensional images, and one additional six-simensional thermodynamic potential (i.e. at the M point of the cubic $P4_33'2$ paramagnetic group). They are discussed in ref. 62.

b) Most of the first-order transitions observed in magnetic systems can also be related to a single active IC.

Although continuous transition are found the more commonly in real magnetic systems (in contrast to the situation encountered among structural transitions, as mentioned in chapter IV), there exist a number of magnetic first order transitions, reported experimentally. Table 19 provides some examples of first-order transitions from a paramagnetic phase, induced by an active IC of the corresponding paramagnetic group. Various theoretical models have been proposed to explain the first-order character of the transition in the materials listed in Table 19, among which can be mentioned an exchange magnetostrictive mechanism (i.e. for MnAs) [63, 64] and different sorts of specific interactions between the magnetic system and the crystal lattice [65-67]. For systems having a number of order-parameter components $n > 4$ (such as UO_2, ErSb or MnO) a more general explanation was suggested in the framework of the Renormalization group approach [68,69] , namely that these systems should not have

Table 19 : Examples of first-order paramagnetic to ferro or antiferromagnetic transitions, induced by active IC's columns have the same meaning as in Table 15.

(a)	(b)	(c)	(d)	(e)	(f)	(g)
MnAs	$P6_3/mmc1'$	$P6_3/mm'c'$	313	$\tau_4(M)$	3	O_h
V_2O_3	$R\bar{3}c1'$	$C2'/c$	161	$\tau_6(\Gamma)$	2	C_{6v}
FeRh	$Pm3m$	P_I4_132	350	$\tau_3(X)$	6	L_7
MnO	$Fm3'm$	C_c2/m	125	$\tau_6(L)$	8	M_2
UO_2	$Fm3'm$	$Pa3$	30.8	$\tau_5(X)$	6	L_7
NiO	$Fm3'm$	C_c2/m	523	$\tau_6(L)$	8	M_2
ErSb	$Fm3'm$	$P_s\bar{1}$	3.5	$\tau_3(L)$	8	M_2
$LiO.SFe_{2.5}O_8$	$Fd3'm$	$P4_332$	931	$\tau_3(X)$	6	L_4
NiS	$P6_3/mmc1'$	$P6_3'/m'm'c$	263	$\tau_4(\Gamma)$	1	C_i

a stable fixed point of the Renormalization group recursion relations. In Ref. 62, this conjecture is analysed in more details, and it is shown that it is far from being confirmed by the whole set of experimental data.

As for structural transitions (see chapter IV), phenomenological models of first-order transitions induced by active IC's, necessitate to take into account higher degree invariants in the thermodynamic potential. However, as shown in the example of α-Fe_2O_3 (§.5.3), the selection of such invariants depends on their exchange and relativistic nature and on the region of the phase diagram, i.e. close or far from the transition point, that one has to describe.

IC's which do not fulfil the Lifshitz condition have not been considered in ref. 58. As for structural transitions (see chapter IV) they can be shown to be connected with second or first order transitions to incommensurate magnetic structures, or to first order transitions towards commensurate phases. Before giving some illustrative experimental examples of such situations, let us examine in more details the specifity of the Landau theory of incommensurate magnetic phases.

6.2. Specific features of the Landau theory of incommensurate magnetic systems

The Landau theory of incommensurate systems was formulated in the pioneer series of paper by Dzialoshinskii (14). Although it does not differ formally in its main features, from the approach to structural incommensurate systems examined in chapter V, it has a number of particularities that must be underlined.

6.2.1. Prediction for incommensurate structures in magnetic systems

Let us first examine the specific arguments which lead to predict the existence of incommensurate magnetic states for the inactive IC's violating the Lifshitz condition (for a detailed derivation of this condition see chapters III and V). We can start from Eq.(4.1) which expresses the components $\overset{i}{m}(\vec{r})$ (i = x,y,z) of the average density of magnetic moment as a function of the basis functions $\phi_{n\alpha}(\vec{r})$ spanning the IC's belonging to the paramagnetic group. In (4.1) the wave-vector $\vec{k}_1$ associated with the considered IC's was assumed to be fixed, and located at a high symmetry point of the Brillouin-zone. It determined the breaking of translational symmetry in the low-temperature ordered phase, the unit-cell of which is an integer multiple n of the paramagnetic unit-cell. In ref. 58, it was shown that for active IC's, n can take the values n = 1, 2, 3, 4, 6, 8 and 32, as it is the case for structural transitions.

Let us now assume that there exist situations in which the wave vector $\vec{k}$ varies with temperature, and takes values which remain close to $\vec{k}_1$, where $\vec{k}_1$ corresponds to one of the precedingly mentioned high-symmetry points. Then, for each value of $\vec{k} = \vec{k}_1 + \delta\vec{k}$, one will have a set of basis functions $\phi^{\vec{k}}_{n\alpha}(\vec{r})$which can be written under the form :

$$\phi_{\vec{k}}(\vec{r}) = u_{\vec{k}}(r)\, e^{i\vec{k}.\vec{r}} \qquad (6.1)$$

where the indices n and α are omitted. $u_{\vec{k}}(\vec{r})$ is a periodic function of the lattice. Along the same line, the coefficient a_P of the quadratic invariant in the thermodynamic potential (4.2) becomes not only a function of temperature but also a function of $\vec{k}$. For the coefficient a_{Po} which vanishes at the transition, one can assume in the close vicinity of T_c, a <u>linear</u> dependence as a function of $\vec{k} - \vec{k}_1$:

$$a_{Po}(\vec{k},T) = a_o(\vec{k}_1,T) + \delta(\vec{k} - \vec{k}_1) \qquad (6.2)$$

<u>If the δ coefficient is anomalously small</u>, then, in addition to the linear term in (6.2), it is necessary to take into account higher degree terms. Restricting ourselves to the second degree, we can write :

$$a_{Po}(\vec{k},T) = a_o(\vec{k}_1,T) + \delta(\vec{k} - \vec{k}_1) + \frac{\sigma}{2}(\vec{k} - \vec{k}_1)^2 \qquad (6.3)$$

where $\frac{\delta}{\sigma} << \frac{1}{d}$, with d of the order of the interatomic spacing. Expansion (6.3), with $\sigma > 0$, allows a minimum for $a_{Po}(\vec{k},T)$ at the value :

$$|\vec{k}| = |\vec{k}_1| - \frac{\delta}{\sigma} \qquad (6.4)$$

Accordingly, the presence of terms linear in $(\vec{k} - \vec{k}_1)$ with small coefficients in the k-expansion of a_{Po}, leads to a phase transition that proceeds not to the (commensurate) state described by an IC associated with $\vec{k}_1$, but to a state with wave vector $|\vec{k}_1| - \frac{\delta}{\sigma}$ that will correspond to a period many times greater than the period of the lattice as determined by $\vec{k}_1$, i.e. there will be a superposition of two periods (with wave vectors k and $k_1 - \frac{\delta}{\sigma}$) incommensurate with respect to each other.

In magnetic systems there exist two distinct situations, both arising for IC's which do not fulfil the Lifshitz condition, in which the δ coefficient can be found to be small with respect to σ :

a) there exist relativistic invariants of the type

$$S^i_{Po,k_1} \cdot \frac{\partial S^j_{Po,k_1'}}{\partial x} - S^j_{Po,k_1'} \cdot \frac{\partial S^i_{Po,k_1}}{\partial x} \qquad (6.5)$$

where S^i_{Po,k_1} and $S^j_{Po,k_1'}$ are two different components of the order-parameter which transform as the IC denoted P_o, associated with the wave-vector k_1.k_1' belongs to the star of the wave vector k_1 (it can identify to k_1 if the little IC is multidimensional) and x is the space coordinate corresponding in real space to the direction determined by $\vec{k}_1$ in the reciprocal space.

As shown in chapter V, invariants of the type (6.5) express in the representation space the existence of linear terms in the k-dependence of a_{Po}. Thus the δ coefficient in (6.3) is connected with forces of the spin lattice and spin-spin type, and is consequently always small compared to the coefficient σ of the square term in (6.3), which corresponds in the order-parameter space, to invariants of the type

$$\left(\frac{\partial \vec{S}_{Pok_1}}{\partial x}\right)^2 \qquad (6.6)$$

which are obviously of exchange origin. Accordingly the existence of incommensurate phases of this type can be predicted on the sole basis of symmetry , i.e. for the IC's associated with k_1 vectors, allowing invariants of the type (6.5).

b) There exist exchange invariants of the form

$$\vec{S}_{Po,k_1} \cdot \frac{\partial \vec{S}_{Po,k_1'}}{\partial x} - \vec{S}_{Po,k_1'} \cdot \frac{\partial \vec{S}_{Po,k_1}}{\partial x} \qquad (6.7)$$

transforming as the IC corresponding to the coefficient $a_{Po}(k_1)$. Here there is no general reason for which the coefficient should be smaller than σ , and no transition should take place. However, δ may be small accidentally, i.e. for some particular reason connected with the substance under consideration, for which there would exist a sharp anisotropy of the exchange interaction leading to a minimum for $a_{Po}(k,T)$ for values of k where symmetry does not require it. Such cases would be recognized experimentally by the fact that the wave vector should exhibit a drastic change with temperature, by contrast to case a) where k should vary only weakly.

Symmetry conditions allowing existence of Lifshitz invariants of the form (6.5) or (6.7) are given in ref. 14. Thus relativistic invariants (6.5) can be found for the IC's associated with k-vector the invariance point group of which is 222,$\bar{4}$2m, $\bar{6}$, 32, $\bar{6}$m2, $\bar{4}$, 422, 622, T, Td and O, while invariants (6.7) exist when the point-group of the wave-vector is 1, m, 2, 3, 4, 6, mm2, 3m, 4mm and 6mm. With some exceptions (i.e. there exist active IC's for which the point-group of the wave-vector is $\bar{4}$2m, $\bar{6}$m2 and $\bar{3}$m [58]) the preceding conditions are <u>sufficient conditions</u> to find a Lifshitz invariant. However these conditions <u>are not necessary ones</u>, as invariants of the type (6.5) or (6.7) can also be found for IC's associated with wave-vectors at the surface of the Brillouin-zone which possess the inversion in their group of the wave-vector (see chapter III).

6.2.2. Examples of incommensurate magnetic phases

Let us first examine two concrete examples of the situations a) and b) indicated in 6.2.1. As an illustration of the situation a) we can consider the case of $MnAu_2$ which undergoes at T_N = 370 K a transition to an antiferromagnetic helimagnetic state [70]. In this incommensurate state, all Mn atoms in the same layer have their moments mutually parallel and normal to the c axis. These moments make and angle of about 51° from one plane to another [71]. In its paramagnetic phase $MnAu_2$ has the symmetry I4/mmm1' and the helical ordering corresponds to a wave-vector located at the A point $[\vec{k}_{12} = (\frac{\pi}{a}, \frac{\pi}{a}, \frac{\pi}{c})]$ of the body centered tetragonal Brillouin zone. The group of the wave vector is $\bar{4}$2m, and following §.6.2.1. one should find a relativistic Lifshitz invariant of the type (6.5)

To the wave vector $\vec{k}_{12}$ corresponds the function $e^{i\frac{\pi(x+y)}{a} + \frac{i\pi z}{c}}$, in place of which we can also take

$$\phi^+ \sim \cos\frac{\pi x}{a} \quad \cos\frac{\pi y}{a} \quad \cos\frac{\pi z}{c} \tag{6.8}$$

which has the same translational properties. ϕ^+ remains invariant under all symmetry operations of I4/mmm1' that leave in place the Mn ions located on the corners of the unit-cell, and transforms to the function :

$$\phi^- \sim \sin\frac{\pi x}{a} \quad \sin\frac{\pi y}{a} \quad \sin\frac{\pi z}{c} \tag{6.9}$$

under the operations of the paramagnetic group that carry the corners of the unit-cell to its center. Accordingly ϕ^+ and ϕ^- constitute a basis for the IC's of I4/mmm1', to which we can associate the magnetic vectors :

$$\vec{L}^{+} = \vec{L}_o \cos\frac{\pi x}{a} \cos\frac{\pi y}{a} \cos\frac{\pi z}{c} \qquad (6.10)$$

and

$$\vec{L}^{-} = \vec{L}_o \sin\frac{\pi x}{a} \sin\frac{\pi y}{a} \sin\frac{\pi z}{c}$$

The corepresentation realized by the vectors $\vec{L}^{+}$ and $\vec{L}^{-}$ is reducible and decomposes into three of the five IC's of the paramagnetic group at the A point, namely τ_1, τ_3 and τ_5 in the notation of the tables of Zak et al [27], which are respectively of dimension two (for τ_1 and τ_3) and four (for τ_5). The components of $\vec{L}^{+}$ and $\vec{L}^{-}$ which span the preceding IC's are respectively : (L_z^+, L_z^-); $(L_x^+ + L_y^+, L_x^- - L_y^-)$ and $(L_x^+ - L_y^+, L_x^- + L_y^-)$. Thus we can write the thermodynamic potential separating the exchange and relativistic parts, as :

$$\begin{aligned} F = F_o &+ \frac{1}{2} a \left[(\vec{L}^{+})^2 + (\vec{L}^{-})^2\right] + \frac{1}{2}\sigma \left[\left(\frac{\partial \vec{L}^{+}}{\partial z}\right)^2 + \left(\frac{\partial \vec{L}^{-}}{\partial z}\right)^2\right] \\ &+ \frac{1}{2}\nu_1 \left[(L_x^+ + L_y^+)^2 + (L_x^- - L_y^-)^2\right] \\ &+ \frac{1}{2}\nu_2 \left[(L_x^+ + L_y^+)^2 + (L_x^- + L_y^-)^2\right] \\ &+ \delta\left(L_x^+ \frac{\partial L_y^-}{\partial z} + L_y^+ \frac{\partial L_x^-}{\partial z} - L_y^- \frac{\partial L_x^+}{\partial z} - L_x^- \frac{\partial L_y^+}{\partial z}\right) \end{aligned} \qquad (6.11)$$

where the relativistic δ term is much smaller than the exchange σ term. Let us note that, for the helical structure to take place, one has to assume that in the vicinity of T_N the δ term should be dominant compared to the anisotropic relativistic terms ν_1 and ν_2.

As an example of incommensurate phases corresponding to an exchange Lifshitz invariant we can consider the case of VF_2 which undergoes at T_N = 7 K a spiral structure propagating along the c axis, with magnetic moments perpendicular to c, the turn angle being about 96 ± 0.5° [72]. The paramagnetic group of VF_2 is P4/mnm1' and the wave-vector is located at the Z point $[\vec{k}_{19} = (0, 0, \frac{\pi}{c})]$. Although the group of the wave-vector is 4/mmm and possesses an inversion center, the Lifshitz condition is not fulfilled. It is easy to see that to the vector $\vec{k}_{19}$ corresponds a corepresentation given by two vectors transforming as :

$$\vec{L}^{+} = \vec{L}_o \cos\frac{\pi z}{c} \quad \text{and} \quad \vec{L}^{-} = \vec{L}_o \sin\frac{\pi z}{c} \qquad (6.12)$$

which is reducible. The thermodynamic potential can be written under the form :

$$\begin{aligned} F = F_o &+ \frac{1}{2} a \left[(\vec{L}^{+})^2 + (\vec{L}^{-})^2\right] + \frac{1}{2}\sigma\left[\left(\frac{\partial \vec{L}^{+}}{\partial z}\right)^2 + \left(\frac{\partial \vec{L}^{-}}{\partial z}\right)^2\right] \\ &+ \frac{1}{2}\nu_1 \left[(L_x^+ + L_y^+)^2 + (L_x^- - L_y^-)^2\right] \end{aligned}$$

$$+ \frac{1}{2} \nu_2 [(L_x^+ - L_y^+)^2 + (L_x^- + L_y^-)^2]$$
$$+ \delta (\vec{L}^+ \frac{\partial \vec{L}^-}{\partial z} - \vec{L}^- \frac{\partial \vec{L}^+}{\partial z}) \qquad (6.13)$$

For the incommensurate structure to take place, we have to assume that the coefficient δ is small with respect to σ. As for the example of $MnAu_2$, we will not go further in the discussion of (6.13), as it leads to lengthy calculations of the type shown, in chapter V. As for magnetic transitions between strictly periodic (commensurate) phases, the main difference with the approach to structural incommensurate systems, is that, while the breaking of symmetry is induced by a single IC, one has still to take into account in the thermodynamic potential terms pertaining to the different IC's under which the components of the magnetic vectors (6.10) or (6.12) transform Besides a careful distinction must be made in the discussion between exchange and relativistic terms.

Examples of experimentally observed incommensurate magnetic phases are listed in Table 20. As shown in column (f) they correspond to a large variety of modulated structures : helicoïdal waves, with the magnetic moments turning in a plane perpendicular to the propagation vector as in MnP, $TbAu_2$ or Lu_2Fe_{17} ; sinusoidal structures where the magnitude of the magnetic moments vary sinusoïdally, transversally or longitudinally to the wave-vector as in Cr, Nd or Tb_3Co ; different sorts of spirals as in Mn_3B_4, Mn_3O_4 or Ho. The underlying magnetic ordering is in most case anti-ferromagnetic, but basic ferromagnetic arrangement of the magnetic moments is found in MnP, Ho or $TbMn_2$.

It can be stressed that in the large majority of the cases given in table 20, one finds an exchange Lifshitz invariant of the type (6.7), and according to the prediction of Dzialoshinskii [14] the wave vector exhibits drastic variations as a function of temperature, or composition. We can also note that, although the transitions from the paramagnetic phase are found the more commonly to be continuous, some first-order transitions to the incommensurate state have been evidenced, as in $BiFeO_3$ FeI_2, Nd, Ho, β-MnS, MnS_2, Cr and Eu. Finally, in contrast to the situation observed for structural systems (see chapter V), a large fraction of the incommensurate structures remain stable down to low temperatures, i.e. do not become unstable with respect to lock-in commensurate phases, a fact that can be related to the respective magnitude of the Lifshitz invariant and anisotropic terms involved in the thermodynamic potential.

6.2.3. Dielectric and metallic magnetic systems : non-localized spin density

In §.6.2.1, the existence of stable incommensurate states was based on the implicite assumption that the coefficient $a\rho_o$ of the quadratic order-parameter invariant, and the thermodynamic potential F itself, are analytic functions of the wave-

<u>Table 20</u> : Examples of incommensurate magnetic systems. For each material listed in column (a), are given the paramagnetic group (b), the transition temperature in Kelvin (c), and type of basic ordering (d), the wave vector or wave vector direction (e) and the modulation characteristics (f).

(a)	(b)	(c)	(d)	(e)	(f)
$ErMn_2O_5$	Pbam1'	1.5	AF	$\frac{1}{2}$,0,0.247	helical
CrAs	Pnma1'	280	AF	0,0,0.354	double spiral
Tb_3Co	Pnma1'	82	F	0.208,0,0	sinusoïdal
$MnSO_4$	Cmcm1'	15	AF	// x	cycloïdal spiral
MnP	Pbnm1'	50	F	// x	helical
Mn_3B_4	Immm1'	226	AF	// y	Screw spiral
$TbZn_2$	Imma1'	55	AF	// z	linear transverse spin wave
VF_2	$P4_2$/mnm1'	7	AF	// z // x	spiral
$DyAg_2$	I4/mmm1'	15	AF	// x	linear transverse spin wave
HoC_2	I4/mmm1'	26	AF	// x	elliptic helical
$TbAu_2$	I4/mmm1'	55	AF	// z	helical
Mn_3O_4	$I4_1$/amd1'	42	AF	// y	spiral
MnI_2	P$\bar{3}$m1'	9.3	AF	// [307]	helical
$BiFeO_3$	R3c1'	595	F	0.0.0.372	cycloïdal spiral
$BaSc_xFe_{12-x}$	$P6_3$/mmc1'	713	F+AF	// z	cone spiral
Nd	$P6_3$/mmc1'	19	AF	0.12b*	sinusoïdal
Ho	$P6_3$/mmc1'	133 23	AF F	// z // y	helical spiral
Dy	$P6_3$/mmc1'	179	AF	// z	screw spiral
Ho_xEr_{1-x}	$P6_3$/mmc1'	35	F	// z	cone spiral
$RbNiCl_3$	$P6_3$/mmc1'	11	AF	// x	screw spiral
Lu_2Fe_{17}	$P6_3$/mmc1'	235	F	// z	helical
Au_2MnAl	Pa31'	65	AF	// x	spiral
MnS_2	Pa31'	48	AF	0, $\frac{1}{2}$, 0	
β.MnS	F$\bar{4}$3'm	150	AF	0, $\frac{1}{2}$, 0	
CeSb	Fm3m	18	AF	// x	cosinusoïdal
NpP	Fm3m	130	AF	// x	sinusoïdal
$Ag_{0.5}In_{0.5}Cr_2S_4$	Fd3m	17	AF	// z	helical
$F_2Cr_2O_4$	Fd3m	35	F	//[110]	spiral
$TbMn_2$	Fd3m	40	F	//[110]	spiral
α-Cr	Im3m	311.5 120	AF	// x	sinusoïdal transverse longitudinal
Eu	Im3m	91	AF	// x	helical

vector $\vec{k}$ associated with the magnetic structure, i.e that a_{po} and F can be expanded in power series of $\vec{k}$ in the vicinity of any point of the paramagnetic Brillouin zone. This allows to investigate the symmetry properties of F as a function of $\vec{k}$, with respect to the operations pertaining to the paramagnetic group. As shown by Dzialoshinskii [14], such an assumption is perfectly valid for dielectric magnetic systems, where the magnetic structure is completely determined by the values of the magnetic moments $\vec{\mu}_i$ of the magnetic ions, at each lattice site i. The corresponding density of magnetic moments (here we can speak equivalently of spin-density) takes the form :

$$\vec{m}(\vec{r}) = \sum_i \vec{\mu}_i \, \delta(\vec{r} - \vec{r}_i) \qquad (6.14)$$

where the δ function accounts for the sharp maximum at $\vec{r} = \vec{r}_i$. Indeed, a magnetic non-conducting crystal can be represented as a system of localized "point" spins firmly affixed to the lattice sites, and a more refined model taking into account a form factor for each ion, would not change in an essential manner the results obtained when starting with an average spin-density under the simplified form (6.14). It is this form which strongly narrows the choice of the possible IC's which are liable to induce incommensurate structures, i.e. of the wave-vectors which minimize F(k) and a_{po} (k) in dielectrics.

Eq.(6.14), or equivalently the assumption of an analytic k-dependence of F, are not anymore justified in metals and alloys. In this case, F(k) and a_{po}(k) always have singularities, occuring as a result of the interaction of the ions spins with the conduction electrons lying near the electronic Fermi surfaces, which result in minima in the k-dependence of these functions. The position of the corresponding singular points are not only determined by the symmetry properties of the lattice but also by the extremal diameters of the Fermi-surfaces (or by multiples of the latter). In other words, due to the fact that the electrons are able to move freely through the lattice, the average density of magnetic moments can have extrema not only at lattice points, but also in interstices. Accordingly, there exist no limitation _a priori_, on the IC's belonging to the expansion of $\vec{m}(\vec{r})$, other than those connected with the behaviour of $\vec{m}$ under time reversal. Hence, the possible types of incommensurate magnetic structures in metals should be much more diverse than in insulators. In particular, the wave-vector may take arbitrary asymmetric positions, a circumstance confirmed for a number of the examples contained in table 20.

We will not review here the specific considerations developed by Dzialoshinskii [14] for the incommensurate structures in conductors, as these considerations have essentially a microscopic content. Let us however give the general conclusions reached by this author, regarding the applicability of the Landau theory to such systems :

i) when $\vec{k}$ is near from singular points, one cannot construct a purely phenomenological approach of metallic incommensurate phases, as one must take into account the energy spectrum of the conduction electrons, the shape of the corresponding Fermi surfaces, the amplitudes of the scattering of the electrons by the magnetic ions etc...

ii) When $\vec{k}$ is far from singular points, the Landau theory applies to metallic systems as well as to insulators, i.e. the thermodynamic potential can be assumed to be a functional of the mean magnetic moment density, providing the specific symmetry properties of $\vec{m}(\vec{r})$ are taken into account. For this purpose, $\vec{m}(\vec{r})$ can be written as the sum :

$$\vec{m}(\vec{r}) = \vec{m}_l(\vec{r}) + \vec{m}_{nl}(\vec{r}) \tag{6.15}$$

of a localized spin-density $\vec{m}_l$ of the ions, which transform according to one of the IC's allowed for the system of "point" spins, and a non localized part $\vec{m}_{nl}$ associated with the conduction electrons. This latter has the transformation properties of IC's which are forbidden for the "point" spins, but correspond to the same wave vector $\vec{k}$ than the IC's associated with $\vec{m}_l$. Assuming that the symmetry of $\vec{m}_l$ does not allow construction of a Lifshitz invariant of the types (6.5) or (6.7) on may still construct, when $\vec{m}_l$ and $\vec{m}_{nl}$ have specific symmetry properties (in particular, different parities with respect to inversion), invariants of the type :

$$S^i_{lk} \frac{S^j_{nlk}}{x} - S^j_{nlk} \frac{S^i_{lk}}{x} \tag{6.16}$$

that will stabilize incommensurate states providing the corresponding coefficient is small with respect to the invariant of the type (6.6). The influence of non localized spin densities on the behaviour of magnetic systems is discussed in more details in refs 14, and 55.

7. COUPLING OF THE MAGNETIC ORDER-PARAMETER TO NON-MAGNETIC PHYSICAL QUANTITIES : SPONTANEOUS MAGNETOSTRUCTURAL EFFECTS

7.1. Introduction

In the preceding sections, the magnetic ordering arising below the transition from a paramagnetic phase, was assumed to take place without any distortion of the lattice. Accurate studies of the modification of the lattice constants across the transitions in magnetic systems reveal in fact the existence of a distortion which is called spontaneous magnetostriction [73-75] . For most magnetic materials, the relative magnitude of this distortion of the lattice constants is of about 10^{-5}-10^{-4} so that the paramagnetic structural symmetry can be considered, within a first approximation, to remain unchanged. In some classes of magnetic materials however, such as the rare earth-Fe_2 compounds, the spontaneous magnetization induces through the magnetostrictive effect strain components of the order of 10^{-2},

[76] i.e. of the order of magnitude of spontaneous strains in ferro-elastic transitions (see chapter III). In this case, the structural modification cannot be neglected and must be taken into account in a phenomenological model of the transition.

In the framework of the Landau theory, the spontaneous magnetostrictive effect can be expressed by introducing coupling terms in the thermodynamic potential, which relate linear combinations of the spontaneous strain components to the square of the magnetic vectors components L_i^α or(and) M_α. In §.7.2., we introduce the spontaneous magnetostrictive effect through a theoretical example, and discuss of its compatibility with the breaking of symmetry occuring at a paramagnetic to ferro or antiferromagnetic transition .

A number of other induced effects have been evidenced experimentally in which non-magnetic macroscopic tensors (e.g. polarization, strain, etc...) take place, as the result of a coupling to the ferromagnetic or antiferromagnetic order-parameter. In §.7.3. and §.7.4 respectively, we examine two of these effects, namely the spontaneous piezomagnetic and spontaneous magnetoelectric effects. Finally we discuss briefly the problem of the mutual correlation of structural and magnetic order-parameters in materials undergoing magnetic and structural transitions (§.7.5)

7.2. Spontaneous magnetostriction

Let us consider again the example discussed in §.5.2, assuming an orientation of the magnetic moments along the z axis. As shown in Table 8, the IC's $\tau_1-\tau_4$ of the $Pca2_11'$ paramagnetic group transform respectively as the components L_{3z}, M_z, L_{2z} and L_{1z} of the magnetic vectors defined by Eq. (5.3). Accordingly the thermodynamic potential (5.4)reduces to :

$$F_M = F_o + \sum_i \frac{a_i}{2} L_i^2 + \frac{c}{2} M^2 + \sum_i b_i L_i^4 + \frac{d}{4} M^4 + \frac{\nu_{1z}}{2} L_{1z}^2 + \frac{\nu_{2z}}{2} L_{2z}^2 + \frac{\nu_{3z}}{2} L_{3z}^2 + \frac{\beta_z}{2} M_z^2 \quad (7.1)$$

In order to construct the magnetoelastic energy which accounts for the interaction between the magnetic anisotropy and the spontaneous strains arizing at the transition, we have to consider the coupling between the spontaneous strain components $e_{\alpha\beta}$ (α,β = x, y, z) and the L_{iz} and M_z components. It is easy to find [77] that the even IC's τ_1 to τ_4 are spanned by the $e_{\alpha\beta}$ components as follows :

$$\tau_1 : e_{xx}, e_{yy}, e_{zz} \; ; \; \tau_2 : e_{yz} \; ; \; \tau_3 : e_{xy} \; ; \; \tau_4 : e_{xz} \quad (7.2)$$

(the odd IC's τ_1 to τ_4 do not transform as strain components as these components do not change sign under the time reversal operator). Accordingly the contribution to the thermodynamic potential of the magnetoelastic interaction can be written:

$$F_{ME}=\delta_1 L_{1z}^2 e_{xz}+\delta_2 L_{2z}^2 e_{xy} + (\delta_3 e_{xx}+\delta_4 e_{yy}+\delta_5 e_{zz}) L_{3z}^2+\delta_6 M_z^2 e_{yz} \quad (7.3)$$

to which must be added the elastic energy of the paramagnetic crystal :

$$F_E = \frac{1}{2}(C_{11}e_{xx}^2 + C_{22}e_y^2 + C_{33}e_{zz}^2) + C_{12}e_{xy}^2 + C_{13}e_{xz}^2 + C_{23}e_{yz}^2 \quad (7.4)$$

The minimization of the thermodynamic potential $F = F_M + F_{ME} + F_E$ with respect to strains and magnetic vector components lead to three different situations regarding magnetoelastic effects :

1/ the coefficient c vanishes first at the transition the a_i coefficients remaining positive : an exchange ferromagnetic order corresponding to the symmetry Pc'a'2_1 takes place (see §.5.2) with the spontaneous magnetization oriented along the z axis. Due to the δ_6 coupling term in (7.3) a spontaneous strain arises simultaneously with the equilibrium value :

$$e_{yz} = -\frac{\delta_6}{C_{23}} M_z^2 \quad (7.6)$$

Eq.(7.6) expresses the spontaneous magnetostrictive effect at the paramagnetic-ferromagnetic transition under consideration : the onset of a spontaneous magnetization induces a spontaneous strain component. Using the vocabulary of structural transitions (see chapter III) we deal with an improper ferroelastic transition where the primary order-parameter is the spontaneous magnetization component M_z and the secondary order-parameter the shear strain e_{yz}. It can be noted that the structural symmetry of the paramagnetic phase (Pca2_1) is lowered to P2_1. As mentioned in §.7.1 the magnitude of the effect may be so small that it may be undetectable within accuracy of X-ray measurements. From (7.6) one can see that this magnitude depends on the value of the coupling coefficient δ_6 which generally speaking can be both of exchange and relativistic origin as at the transition we have both a change in the magnitude and in the direction of the magnetic moments.

Let us stress that the spontaneous magnetostrictive effect represented by Eq.(7.6) is a particular case of the magnetostriction phenomenon which is understood more generally as the quadratic deformation produced in magnetic bodies on magnetization, and represented by the general relationship :

$$E_{\alpha\beta} = A_{\alpha\beta\gamma\delta} M_\gamma M_\delta \quad (7.7)$$

where $A_{\alpha\beta\gamma\delta}$ is the fourth order tensor of magnetostrictive constants[78, 79]. Initially [80,81] the word magnetostriction designated exclusively the deformation of ferromagnetic bodies under the effect of a magnetic field. Although its detection in antiferromagnets is more difficult it must be taken into account for the interpretation of the structural modifications occuring in a number of magnetically ordered phase with a basic antiferromagnetic lattice [82]. This brings us back to the discussion of the thermodynamic potential (7.1)

2/ If a_1 vanishes first, then, as shown in §.5.2.2., the paramagnetic phase is destabilized with respect to an antiferromagnetic phase corresponding to the sublattice (5.7) and the magnetic group $Pc'a2_1'$. From the minimization of (7.3) and (7.4), one finds that the onset of the antiferromagnetic lattice with magnetic vector L_{1z}, induces a spontaneous shear strain :

$$e_{xz} = -\frac{\delta_1}{C_{13}} L_{1z}^2 \tag{7.8}$$

which lowers the structural symmetry of the paramagnetic phase to Pb(y). Similarly the vanishing of a_2 would lead to the onset, simultaneously with the L_{2z} antiferromagnetic order, of a spontaneous shear strain :

$$e_{xy} = -\frac{\delta_2}{C_{12}} L_{2z}^2 \tag{7.9}$$

associated with the structural modification $Pca2_1 \rightarrow Pb(x)$. The magnetostrictive effect associated with an antiferromagnetic ordering has been measured in a number of antiferromagnets ferrimagnets or weak-ferromagnets [83].

3/ The common feature of the spontaneous magnetostrictive effect given by Eqs. (7.6), (7.8) and (7.9) is that <u>the onset of a spontaneous strain component breaks the structural symmetry of the paramagnetic phase</u>. This is not always true. An illustrative example of cases where <u>the structural symmetry is preserved</u>, can be given when the a_3 coefficient in (7.1) vanishes at the transition, while the coefficients a_1, a_2 and c remain positive. Here, the onset of the L_{3z} antiferromagnetic ordering which corresponds to the magnetic groups $Pca2_1$, is accompanied by a magnetostrictive effect expressed by :

$$e_{xx} = -\frac{\delta_3}{C_{11}} L_{zz}^2, \quad e_{yy} = -\frac{\delta_4}{C_{22}} L_{3z}^2, \quad e_{zz} = -\frac{\delta_5}{C_{33}} L_{3z}^2 \tag{7.10}$$

which does not break the orthorhombic structural symmetry of the paramagnetic phase, i.e. e_{xx}, e_{yy}, and e_{zz} are not symmetry breaking components of the strain tensor. Thus one cannot speak of <u>spontaneous</u> magnetostriction but of <u>volumetric magnetostriction</u>[84], as the lattice constants undergo a change, proportional in magnitude to the square of the antiferromagnetic vector L_{3z}, which preserves the orthorhombic symmetry. In other words, spontaneous magnetostriction occurs only when the IC associated with the magnetic ordering is constructed from an irreducible representation which differs from the identity representation τ_1.

The symmetry of magnetostrictive properties of magnetic crystals have been discussed by a number of authors [79,85] , as well as the microscopic (quantum) theory of magnetostriction [86,87]. In table 21 we give numerical values obtained for the magnetostriction of some relevant ferromagnets and ferrimagnets.

Table 21 : Magnetostriction λ of some magnetic materials at room temperature (column (b)) and at OK (column (d)) from [88].

(a) materials	(b) $10^6 \lambda$	(c) materials	(d) $10^6 \lambda$
Ni	- 33	$SmFe_2$	- 2100
Co	- 62	$TbFe_2$	2460
Fe	- 9	$DyFe_2$	1260
$NiFe_2O_4$	- 26	$HoFe_2$	200
Fe_3O_4	40	$ErFe_2$	- 300
YCO_3	0.4	$TmFe_2$	- 210
Pr_2CO_{17}	336		
YFe_2	1.7		
$SmFe_2$	- 1560		
$TbFe_2$	1753		
$DyFe_2$	433		
$TbFe_3$	693		
Ho_2Fe_{17}	- 106		

7.3. Spontaneous piezomagnetism

The quadratic character of magnetostriction allows to distinguish this effect from the linear piezomagnetic effect, which corresponds to the onset of magnetization under the effect of applied stress. In terms of conjugate variables spontaneous piezomagnetism can thus be defined as the onset of spontaneous strains-components as the result of ferro or antiferromagnetic order (i.e. in the absence of applied magnetic field), the strain components being proportional to the M or L_i components [89,90]:

$$e_{\alpha\beta} = B_{\alpha\beta\gamma} M_{\gamma} \qquad (7.11)$$

The phenomenon can be predicted from symmetry considerations at ferro and antiferromagnetic transitions, for which the symmetry of the paramagnetic phase allows construction of a mixed invariant, which couple linearly the strain and magnetic vector components. In terms of structural transitions, it is a pseudoproper ferroelastic transition [91] where the primary order parameter is the ferro or antiferromagnetic ordering, the spontaneous strain being a secondary parameter.

After its prediction by Dzialoshinskii [92], the piezomagnetic effect was evidenced for the first time by Borovik-Romanov, in CoF_2 and MnF_2 [93], an then found in a number of other magnetic materials undergoing ferro or antiferromagnetic transitions [94] . The magnetic classes allowing this effect, as well as the form of the piezomagnetic tensor $B_{\alpha\beta\gamma}$ can be found in ref [95].

As an illustrative example, let us consider the paramagnetic-weak ferromagnetic transition which takes place at 950 K in α-Fe_2O_3. In §.5.3, it was shown that the corresponding symmetry change $R\bar{3}c1' \rightarrow C2/c$ was related to the two dimensional IC τ_5 of the $R\bar{3}c1'$ group, at the center of the rhombohedral Brillouin-zone. From the matrices of τ_5, given in table 10, it is easy to show that the following invariants can be constructed :

$$F_{ME} = \delta_1[(e_{xx}-e_{yy})M_x - 2e_{xy}M_y] + \delta_2[e_{xz}M_y - e_{yz}M_x] \tag{7.12}$$

on the other hand, the elastic energy for the rhombohedral group $\bar{3}m$ is [95] :

$$\begin{aligned} F_E = & \frac{1}{2}\lambda_{zzzz}e_{zz}^2 + 2\lambda_{\xi\eta\xi\eta}(e_{xx}+e_{yy})^2 \\ & +\lambda_{\xi\xi\eta\eta}[(e_{xx}-e_{yy})^2 + 4e_{xy}^2)] + 2\lambda_{\xi\eta zz}(e_{xx}+e_{yy})e_{zz} \\ & + 4\lambda_{\xi z\eta z}(e_{xz}^2+e_{yz}^2) + 4\lambda_{\xi\xi\xi z}[(e_{xx}-e_{yy})e_{xz} - 2e_{xy}e_{yz}] \end{aligned} \tag{7.13}$$

where $\xi = x + iy$ and $\eta = x - iy$. From the minimization of the thermodynamic potential $F = F_M + F_{ME} + F_E$, where F_M is given by (5.18), one finds in state II of α- Fe_2O_3, which corresponds to the equilibrium values (5.24) for M_x and M_y, the following expressions for the spontaneous strain components :

$$\begin{aligned} e_{xx}-e_{yy} &= -\frac{\delta_2}{\lambda_1}M_y + 2\frac{\delta_1\lambda_{\xi z\eta\xi}}{\lambda_1\lambda_{\xi\xi\xi z}}M_x \\ e_{xz} &= -\frac{\delta_1}{4\lambda_{\xi\xi\xi z}}M_x - \frac{\lambda_{\xi\xi\eta\eta}}{4\lambda_{\xi\xi\xi z}}(e_{xx}-e_{yy}) \\ e_{xy} &= \frac{\delta_1\lambda_{\xi z\eta\xi}}{\lambda_2\lambda_{\xi\xi\xi z}}M_y + \frac{\delta_2}{\lambda_2}M_x \\ e_{yz} &= \frac{\lambda_{\xi\xi\eta\eta}}{\lambda_{\xi\xi\xi z}}e_{xy} - \frac{\delta_1}{8\lambda_{\xi\xi\xi z}}M_y \end{aligned} \tag{7.14}$$

where $\lambda_1 = 4\lambda_{\xi\xi\xi z} - 2\frac{\lambda_{\xi z\eta\xi}}{\lambda_{\xi\xi\xi z}}\lambda_{\xi\xi\eta\eta}$, and $\lambda_2 = 8(\frac{\lambda_{\xi\xi\eta\eta}\lambda_{\xi z\eta z}}{\lambda_{\xi\xi\xi z}} - \lambda_{\xi\xi\xi z})$

(7.14) expresses the spontaneous piezomagnetic effect at the 950 K transition in α-Fe_2O_3, i.e the onset of spontaneous strains as the result of the weak-ferromagnetic order. It must be stressed that, although the strain components are given as a function of the weak-magnetization components M_x and M_y, the primary order-parameter is actually the antiferromagnetic ordering represented by the antiferromagnetic vectors L_{3x} and L_{3y}, given by (5.25) and corresponding to the sublattice (5.19). Following Dzialoshinskii [92], the piezomagnetic effect can be also expressed in term of the conjugated variables. Thus, (7.12) is replaced by :

$$F_{ME} = -\delta_1[(\sigma_{xx}-\sigma_{yy})H_x - 2\sigma_{xy}H_y] - \delta_2(\sigma_{xz}H_y-\sigma_{yz}H_x) \qquad (7.15)$$

where the $\sigma_{\alpha\beta}$ are stress-tensor components, and H_x, H_y components of the magnetic field conjugated to M_x and M_y. Accordingly one finds the following equilibrium equations :

$$M_x = \delta_1(\sigma_{xx}-\sigma_{yy}) - \delta_2\sigma_{yz} \;, \quad M_y = -2\delta_1\sigma_{xy} + \delta_2\sigma_{xz} \qquad (7.16)$$

which express the onset of magnetization components under application of stresses.

7.4. Spontaneous magnetoelectricty

A crystal may acquire a magnetic moment when an electric field is applied, and conversely a crystal may develope an electric dipole moment when placed in a magnetic field. This is the magnetoelectric effect which was predicted from general symmetry considerations by Landau and Lifshitz [18] and more specifically in Cr_2O_3 by Dzialoshinskii [97] . When the relationship between the moment and the fields are linear, the constitutive equations for such effects can be written :

$$P_i = K_{ij}\,E_j + \alpha_{ij}\,H_j$$

and

$$M_i = \chi_{ji}H_j + \alpha_{ji}\,E_j \qquad (7.17)$$

where α_{ij} are the components of the magnetoelectric tensor [98] . The linear magnetoelectric effect in Cr_2O_3 was verified for the first time by Astrov [99], then found in a large number of ferro and antiferromagnetic materials (see for example ref. 100).

In a more narrow sense the spontaneous magnetoelectric effect can be defined as the onset of an induced spontaneous polarization at a ferro or antiferromagnetic transition. This effect may occur only if the structure of the low temperature phase allows existence for a symmetry breaking component of the polarization. The magnetic point-group modifications which permit the spontaneous magnetoelectric effect have been listed by Kovalev for ferromagnetic [101] and antiferromagnetic [102] transitions. As shown by this author, the effect can be accounted phenomenologically by the existence in the thermodynamic potential of invariants which couple the spontaneous components of the polarization to the square of the ferro or antiferromagnetic vectors. One can thus deduce that the polarization is always an induced effect, i.e. cannot be the primary order-parameter, as coupling terms linear in the magnetic parameters, and of higher degree in the polarization components, are forbidden by symmetry (as least for transitions from a paramagnetic phase).

There exist only a few experimentally confirmed examples of materials in which the spontaneous magnetoelectric effect was found. The more studied is Nickel-iodine boracite that was discussed in §.5.4. In this material a spontaneous polarization takes place at 61.5 K, simultaneously with the latent antiferro magnetic order [35-37]. The straight coupling between electric and magnetic properties was evidenced by the switching of electric and magnetic domains under conjugated fields [36] and by a whole set of measurements which clearly show a spontaneous magnetoelectric effect[35,36,103] Taking into account the cubic to monoclinic point group change $\bar{4}3m1' \to m$, one finds the following coupling between the polarization P_α and weak magnetization components M_α [31]:

$$F_{ME} = \alpha_o(P_x M_y M_z + P_y M_x M_z + P_z M_x M_y) \tag{7.18}$$

The minimization of the free energy $F + F_{ME} + F_E$ where F is given by (5.69) to (5.71), and F_E has the form [31]:

$$\begin{aligned} F_E = & \delta_1 (P_x \rho_1^2 \cos 2\psi_1 + P_y \rho_2^2 \cos 2\psi_2) \\ & + \delta_2 (P_x \rho_3^2 \cos 2\psi_3 + P_z \rho_2^2 \cos 2\psi_2) \\ & + \delta_3 (P_y \rho_3^2 \cos 2\psi_3 + P_z \rho_1^2 \cos 2\psi_1) \\ & + \delta_4 (P_x \rho_2^2 \cos 2\psi_2 + P_y \rho_1^2 \cos 2\psi_1 + P_z \rho_3^2 \cos 2\psi_3) \\ & + \frac{1}{2\chi_o^E}(P_x^2 + P_y^2 + P_z^2) - \vec{E}.\vec{P} \end{aligned} \tag{7.19}$$

yields the equation of state :

$$\frac{P_z}{\chi_o^E} - E_z \simeq -\alpha_o (M_s^{\perp})^2 \sin\theta\cos\theta \tag{7.20}$$

where $M_s^{\perp}$ is the component of the spontaneous magnetization located in the (xy) plane, θ is the angle between $M_s^{\perp}$ and the x axis (figure 16a). From (7.20) one can see that a change in sign of the polarization under application of a suitable electric field ($E_z \to -E_z, P_z \to -P_z$) should result in a 90° rotation of $M_s^{\perp}$ in the xy plane ($\theta \to \theta + \frac{\pi}{2}$). This effect was observed by Ascher et al [36] by application of a ∿ 5 kv/cm field at 56 K. The inverse effect, i.e. the reversal of P_z when an external magnetic field is applied perpendicular to the magnetization, evidenced by the same authors near T_c [36, 44], is expressed by the equation :

$$[(\chi_o^M)^{-1} M_s - H_y] \sin\theta \simeq -\alpha_o M_s^{\perp} P_z \cos\theta \tag{7.21}$$

obtained by the minimization of $F + F_{ME} + F_M$ with respect to M_y, where F_M contains the terms which express the coupling of the magnetization to the six-component order-parameter in (5.70) and (5.71). Eq.(7.21) is illustrated by Figure 16b, which shows that when turning the magnetization by 90° from the initial position $\theta = \frac{\pi}{4}$, under application of the corresponding magnetic field ($\theta \to \theta + \frac{\pi}{2}$), one has to reverse the sign of P_z for Eq.(7.21) to remain unchanged.

From Eq.(7.21) one can deduce the magnetoelectric coefficient

$$\alpha_{zz} = \left[\frac{\Delta P_z}{H_y}\right]_{\substack{E=0 \\ H_x=H_z=0}} \simeq \frac{\chi_o^E \alpha_o}{\left(\frac{M}{\chi_o}\right)^{-2}} (\delta_1 + \delta_2)\rho_1\rho_2\rho_3 \qquad (7.22)$$

Replacing the ρ_i by their equilibrium values given in Table 13a, one obtains $\alpha_{zy} = 0$ for $T > T_c$ and $\alpha_{zy} \simeq (T_c - T)^{3/2}$ for $T < T_c$, which shows that the temperature dependence of the magnetoelectric coefficients follow the same variation law as the induced magnetization (5.72).

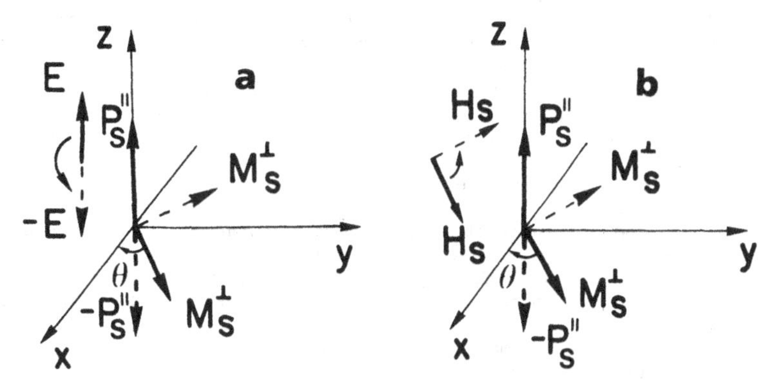

Figure 16 : Magnetoelectric properties of Ni-I boracite : a) rotation of the spontaneous magnetization component lying in the xy plane ($\vec{M}_s^{\perp}$) when the polarization parallel to $\vec{z}$ changes sign under suitably applied electric field $\vec{E}$. b) Reversal of the polarization $P_s^{//}$ when $\vec{M}_s^{\perp}$ is turned in the xy plane under applied magnetic field $\vec{H}_s$.

7.5. Coupling between structural and magnetic transitions

In the preceding subsections, a number of magnetostructural effects have been described, in which the magnetic ordering appears as primarily responsible of the structural modifications arising at the transition. Besides such induced structural effects, there exist in a number of magnetic materials, a close connection between structural and magnetic transitions which, although they occur at different temperatures, influence themselves mutually. For example in the large class of magnetic materials such as $KCuF_3$, $LaMnO_3$, Mn_3O_4, $CuFe_2O_4$ or CrF_2, in which the Jahn-Teller effect plays an important role [104], this effect determines not only the structural properties of these systems but also their magnetic properties[105]. In turn the exchange interaction may substantially affect the lattice modifications. Another example is constituted by the so-called ferroelectromagnets [106] , where the ferroelectric and ferroelastic properties appear as strongly related to the ferro or antiferromagnetic properties, as in $BaMnF_4$, $BiFeO_3$, Cr_2BeO_4 or β-$Tb_2(MoO_4)_3$.

In the framework of the Landau theory, the description of a phase diagram in which structural and magnetic transitions occur at different temperatures, requires to introduce two distinct (primary) order-parameters, and to take into account their mutual correlation through coupling terms, as for sequences of structural transitions associated with more than one order-parameter (see chapter IV). As an example of system in which structural and magnetic transitions are obviously coupled, let us briefly discuss the case of manganese arsenide.

At atmospheric pressure MnAs undergoes a succession of two phase transitions [107,108] : a structural second-order transition at T_1 = 394 K from the high-temperature $P6_3/mmc$ phase to the orthorhombic Pnma phase, followed by a strongly first-order transition at T_2 = 313 K to a ferromagnetic phase of magnetic symmetry $P6_3/m\hat{m}'c'$. The succession of phases has been explained by an exchange magnetostrictive mechanism [65] which assumes an indirect exchange interaction [109]. Following Galkina et al [108,110], one can use for a phenomenological approach, two different order-parameters : a three component order-parameter ϕ which describes the ferroelastic $P6_3/mmc \rightarrow P6_3/mm'c'$ indirect ferromagnetic transition which has the symmetry of the $IC\tau_2$, at the center of the hexagonal Brillouin zone. It can be shown (see for example ref. 58) that only one of the three components of ϕ is non-vanishing below T_1. Besides ϕ and M can form a coupling invariant of the form $\phi^2 M^2$. Thus the effective thermodynamic potential describing the two consecutive transitions reduces to :

$$F = F_o + \frac{1}{2}\alpha\phi^2 + \frac{1}{4}\beta\phi^4 + \frac{1}{2}aM^2 + \frac{1}{4}bM^4 + \frac{1}{6}cM^6 + \frac{\delta}{2}\phi^2 M^2 \qquad (7.23)$$

where sixth degree terms in M are taken onto account, due to the first-order character of the 313 K transition (i.e. $b < 0$). Using the results of the discussion performed in chapter IV for a potential of the form (7.23) we can infer that there exist four possible stable states in the corresponding phase diagram :

I : $M = 0$, $\phi = 0$; II : $M = 0$, $\phi \neq 0$; III : $M \neq 0$, $\phi = 0$,
IV : $M \neq 0$, $\phi \neq 0$.

State I corresponds to the paramagnetic-paraelastic phase of MnAs, whereas phases II and III can be identified as the ferroelastic and ferromagnetic phases found in this material. A more detailed discussion of the pressure-temperature diagram in MnAs can be found in ref. [110].

REFERENCES

1. L.D. Landau, Zh.Eksper. Teor. Fiz. 7, 19 (1937) ; 7, 627 (1937). Translated in "collected Papers of L.D. Landau", Edited by D.Ter Haar, p. 193, Pergamon Press, London (1965W)
2. E.P. Wigner, "Group theory and its application to the quantum mechanics of atomic spectra", Academic Pres, New-York (1959)
3. A.M. Zamorzaev, Sov. Phys. Crystallogr. 2, 10 (1957)
4. N.V. Belov, N.N.Neronova, and T.S.Smirnova,Sov.Phys.Crystallogr.2,311(1957)
5. W.Opechowski and R. Guccione in Magnetism (vol IIA), Edited by G.T. Rado and H.Suhl, p.105, Academic Press, New-York (1965)
6. V.A. Koptsik, "Shubnikov groups. Handbook on the symmetry and physical properties of crystal structures", University Press,Moscow (1966)
7. N.V. Belov, N.N.Neronova and T.S. Smirnova, Trudy Inst. Kristall. 11, 33 (1955)
8. A.V. Shubnikov, and N.V. Belov "Coloured symmetry", Pergamon Press, Oxford (1964)
9. C.J. Bradley and A.P.Cracknell, "The mathematical theory of symmetry in solids" Clarendon Press, Oxford (1972)
10. J.O. Dimmock and R.G.Wheeler, Phys.Rev.127, 391 (1962)
11. J.O. Dimmock, J.Math. Phys. 4, 1307 (1963)
12. C.J. Bradley and B.L.Davies, Rev.mod.Phys.40, 359 (1968)
13. Y.Takagi, Phys. Rev. B17, 2965 (1978)
14. I.E.Dzialoshinskii, Sov.Phys.JETP 19, 960 (1964) ; 20, 223 (1965) ; 20, 665 (1965)
15. O.V.Kovalev, Sov. Phys. Solid state, 5, 2309 (1964) and 5, 2315 (1964)
16. O.V.Kovalev, Fiz. Metal. Metalloved, 17, 490 (1964)
17. O.V.Kovalev, Sov.Phys.Solid State, 7, 77 (1965)
18. L.D. Landau and E.M. Lifshitz, "Electrodynamics of continuous media" Pergamon Press, Oxford (1960)
19. L.Neel, Annls Phys. 3, 137 (1948)
20. I.E. Dzialoshinskii, Sov. Phys. JETP 5, 1259 (1957)
21. I.E. Dzialoshinskii, and V.I. Man'ko, Sov.Phys.JETP 19, 915 (1964)
22. A.F. Andreev, and V.I. Marchenko, Sov. Phys. JETP 43, 794 (1976)
23. A.F. Andreev, and V.I. Marchenko, Sov. Phys. Usp. 23, 21 (1980)
24. International Tables for X-ray Crystallography (Kynoch, Birmingham, 1952)
25. A.P. Cracknell, "Magnetism in Crystalline materials", Pergamon Press, Oxford (1975)
26. O.V.Kovalev, "Irreducible representations of the space groups", Gordon and Breach, New-York (1965)
27. J.Zak, A. Casher, H. Glück, and Y.Gur, "The irreducible representations of space groups", Benjamin, New-York (1969)
28. S.C. Miller and W.F. Love "Tables of irreducible representations of space groups and corepresentations of magnetic space groups".Pruett, Boulder, Colorado (1967)
29. J.P. Rivera and H.Schmid, Ferroelectrics 55, 295 (1984)
30. M.E. Mendoza-Alvarez, H.Schmid and J.P.Rivera, Ferroelectrics 55, 227 (1984)
31. P.Tolédano,H.Schmid,M.Clin and J.P.Rivera,Phys.Rev.B32,6006 (1985)

32. C.G.Shull, W.A.Straser and E.O.Wollan, Phys.Rev.83, 333 (1951)
33. B.N.Brockhause, J.Chem. Phys. 21, 961 (1953)
34. R.Nathaus, S.J.Pickart, H.A.Alperin, and P.J.Brown,Phys.Rev.136A, 1641 (1964)
35. J.P.Rivera and H.Schmid, Ferroelectrics 36, 447 (1981)
36. E.Ascher,H.Rieder,H.Schmid and H.Stössel,J.Appl.Phys.37,1404 (1966)
37. J.P.Rivera and H.Schmid, Ferroelectrics 54, 103 (1984)
38. I.S.Zheludev,T.M.Perekalina,E.M.Smirnovskaya,S.S.Fonton and Yu.N. Yarmukhamedov, JETP Letters 20, 129 (1974)
39. G. Quézel and H.Schmid, Solid State Commun. 6, 447 (1968)
40. P. Fischer, Würenlingen (unpublished)
41. W.Von Wartburg, Phys. Status Solidi A 21, 557 (1974)
42. W. Schäefer and G.Will, Phys. Status Solidi A28, 211 (1975)
43. L.N. Baturov and B.I. Al'Shin, Sov.Phys.Crystallogr. 25, 448 (1981)
44. H.Schmid, Rost.Krist. 7, 32 (1969)
45. W.Rehwald, J.Phys. C11, L157 (1978)
46. J.P. Rivera and H. Schmid, Ferroelectrics, 42, 35 (1982)
47. M.Clin, Thesis, University of Paris VII (unpublished, 1983)
48. J.C.Tolédano and P.Tolédano, Phys. Rev. B21, 1139 (1980)
49. T.Ito,N.Morimoto and R.Sadanaga, Acta Crystallogr. 4,310 (1951)
50. A.Herpin,"Theorie du Magnetisme",Presses Universitaires de France, Paris, (1968)
51. K.J. Standley, "Oxyde magnetic materials", Clarendon Press (Oxford, 1972)
52. E.W. Gorter, Philips Res. Rep. 9, 403 (1954)
53. E.F.Bertaut, F.Forret, A.Herpin and P.Meriel, Rep. CEA 1,597 (1956)
54. J.Solyom, J.de Physique 32, C1 471 (1971)
55. Yu.M.Gufan and I.E.Dzialoshinskii, Sov.phys.JETP 25, 395 (1967)
56. A.P. Cracknell,J.Lorenc, and J.A.Przystawa, J.Phys.C9,1731 (1976)
57. V.E.Naish and V.N.Syromiatnikov, Sov.Phys.Crystallogr.21, 627 (1976)
58. M.Clin, Thesis (University of Paris VII, 1983), unpublished ; P.Toledano, M.Clin and M.Hedoux, Ferroelectrics 35, 239 (1981) ; M.Clin, M.Hedoux and P.Tolédano, Ferroelectrics 53, 265 (1984), and to be published.
59. G.T.Rado, and H.Suhl, "Magnetism" vols 1-4, Academic Press,New-York (1962-1966)
60. D.C. Mattis "The theory of magnetism. An introduction to the study of cooperative phenomena", Harper and Row, New-York (1965)
61. A.Olès, F.Kajzar, M.Kucab and W.Sikora, "Magnetic structures determined by neutron diffraction", Panstwowe wydawnictwo Naukowe, Warshaw (1976)
62. E.Meimarakis and P.Toledano, Jap.Jour of Appl.Phys. 24, 737 (1985)
63. N.P.Grazhdankina, Sov.Phys. Usp 11, 727 (1969)
64. L.Pal, Acta Phys.Acad. Scient.Hungarical, 27, 47 (1969)
65. C.P.Bean, and D.S.Rodbell, Phys.Rev. 126, 104 (1962)
66. E.L.Nagaev, and A.A. Kovalenko, JETP Letters 29, 492 (1979)
67. J.B.Goodenough and J.A. Kafalas, Phys.Rev.157, 389 (1967)
68. P.Bak, S.Krinsky, and D.Mukamel, Phys.Rev.Letters 36, 52 (1976)
69. S.A.Brazovskii,S.A.Dzialoshinskii, and B.G.Kukharenko, Sov.Phys. JETP 43, 1178 (1976)
70. A.Herpin, P.Meriel, Compt.Rend, 250, 1450 (1960)
71. A.Herpin, P.Meriel, J.Phys.Rad., 22, 337 (1961)

72. H.Y.Lau, J.W.Stout, W.C.Koehler and H.R.Child, J.Appl.Phys. 40, 1136 (1969)
73. N.Akulov and E.Koudorsky, Z.Phys. 85, 661 (1933)
74. R.Becker and M. Kornetski, Z.Phys. 88, 634 (1934)
75. S.V. Vonsovskii,"Magnetism", Halsted, New-York (1975)
76. A.E.Clark,H.S.Belson and R.E.Strakna,J.Appl.Phys. 44, 2913 (1973)
77. V.Janovec, V.Dvorak, and J.Petzelt, Czech.J.Phys.B25, 1362 (1975)
78. L.A.Shuvalov and B.A.Tavger,Sov.Phys.Crystallogr. 3, 765 (1959)
79. R.R.Birss, "Symmetry and magnetism", North-Holland, Amsterdam (1966)
80. N.S.Akulov, Z.Physik 52, 389 (1928)
81. R.Becker and W.D. Doring "Ferromagnetismus" (Springer,Berlin),1939
82. M.Mita and T.Miyaji, Jour.of Magn. and Magn.mater.31-34,852 (1983)
83. K.P. Belov, and A.M.Kadomtseva, Sov.Phys. Usp. 14, 154 (1971)
84. M.Kornetzki, Z.Physik 97, 662 (1935)
85. J.F.Nye. Physical properties of crystals, Clarendon Press, Oxford (1957)
86. A.A.Gusev, Sov.Phys. JETP 2, 126 (1956)
87. S.V. Vonsovskii, J.Exper. Theoret.Phys.USSR 10, 762 (1940)
88. A.E. Clark, in "Ferromagnetic materials" Vol. 1, p.531. Edited by E.P. Wohlfarth, North-Holland, Amsterdam (1980)
89. W.Voight, Lehrbuch der Kristall physik, Leipzig (1928)
90. B.A.Tavger and V.M.Zaitsev, Sov. Phys. JETP 3, 430 (1956)
91. P.Tolédano, M.M.Fejer and B.A.Auld, Phys.Rev. B27, 5717 (1983)
92. I.E.Dzialoshinskii, Sov.Phys.JETP 6, 621 (1958)
93. A.S.Borovik-Romanov, Sov.Phys.JETP 9, 1391 (1959)
94. A.S.Borovik-Romanov, Sov.Phys.JETP 11, 786 (1960)
95. A.I.Mitsek and V.G.Shavrov, Sov.Phys.solid state 6, 167 (1964)
96. L.D.Landau and E.M.Lifshitz "Theory of elasticity" Pergamon, London (1959)
97. I.E.Dzialoshinskii, Sov. Phys. JETP 10, 628 (1960)
98. V.G.Shavrov, Sov. Phys. JETP 21, 948 (1965)
99. D.N.Astrov, Sov. Phys. JETP 11, 708 (1960)
100. "Magnetoelectric interaction phenomena in crystals", edited by A.J. Freeman and H.Schmid, Gordon and Breach, London (1975)
101. O.V.Kovalev, Sov.Phys.Solid state 14, 826 (1972)
102. O.V.Kovalev, Sov.Phys.Crystallogr. 18, 137 (1973)
103. L.N.Baturov, and B.I. Al'shin, Sov.Phys.Solid state 21, 1 (1979)
104. G.A.Gehring and K.A.Gehring, Rep.Progr.Phys. 38, 291 (1975)
105. K.I.Kugel and D.I. Khomskii, Sov. Phys. Usp. 25, 231 (1982)
106. G.A.Smolenskii and I.E.Chupis, Sov.Phys.Usp. 25, 475 (1982)
107. I.M.Vitebskii,V.I.Kamenev and D.A. Yablonskii, Sov.Phys.Solid State 23, 121 (1981)
108. E.G.Galkina, E.A.Zavadskii,V.I.Kamenev and D.A.Yablonskii . Sov.Phys.Solid state 25, 1017 (1983)
109. C.Kittel, J.Appl.Phys. 39, 637 (1968)
110. E.G. Galkina, E.A.Zavadskii,V.I.Kamenev and D.A.Yablonskii , Sov. Phys. Solid state 25, 43 (1983).

CHAPTER VII

THE LANDAU THEORY OF LIQUID CRYSTALS

1. INTRODUCTION

The great variety of mesomorphic phases and of their properties, evidenced in the recent years, require a generalization of the specific concepts which have been introduced initially for their interpretation. In this respect, the early statistical and phenomenological models proposed for the description of particular transitions in liquid crystals (LC) have to be inserted in a more general presentation. This is indeed the case for the Landau theory of LC. A first indication for applying the theory to LC phases can be found in the original paper of Landau [1], where a short paragraph is devoted to the form of the probability density that defines a nematic state. Subsequently the Landau approach of phase transitions was used intensively [2-6] as a phenomenological justification of almost all the new facts brought by experiment in the field of LC, such as the multilayer ordering identified in smectic phases, the various types of incommensurate modulations, ferroelectricity, and the specific orderings (i.e. tilting, bond-orientational and positional orders) which appear in the more structured smectic states.

It is undoubtedly in the papers of Indenbom, Pikin and Loginov [7-10] that one can find the first attempt to apply systematically the group-theoretical and thermodynamic concepts forming the Landau theory, to LC systems. In contrast to the intuitive phenomenological approaches to such systems, used previously, these authors start from the full space-symmetry of the phases and of their irreducible representations (IR's) to propose a comprehensive picture of the phase diagrams and macroscopic properties of LC. In this chapter we will mainly systematize the results contained in refs. [7-10] . As for magnetic transitions we must precise at first, the specific symmetries of LC systems and the corresponding irreducible degrees of freedom.

2. SYMMETRY GROUPS OF LIQUID CRYSTALS

2.1. An introductory example

Let us start by the experimental example of TBPA (i.e. terephtal-bis-pentyaniline), which has been detailed in a recent review by Prost [11]. In this substance the following sequence of phases is observed :

isotropic liquid - Nematic - Smectic A-Smectic C-Smectic F-Smectic G-Crystal

Above 232° C the sample is totally isotropic, perfectly black between crossed polarizers as for an ordinary liquid. The centers and orientations of the molecules are randomly distributed. The corresponding symmetry group is :

$$S = R^3 \wedge O(3) \tag{2.1}$$

which represents the semi-direct product of the group R^3 of continuous translations in three dimensional space, by the full orthogonal group, i.e. the invariance group of the sphere which contains the three-dimensional rotational group SO(3) and its multiplication by the inversion I.

Below 232° C, a nematic phase takes place, characterized by a bright threaded texture, which flows almost as well as water. However, this liquid phase possesses an orientational order (figure 1a) revealed by a strong birefringence : the elongated molecules glide freely over each other, but stay parallel on the average. Rotational symmetry has been broken, whereas continuous translational symmetry has been kept in all directions. The microscope reveals a $D_{\infty h}$ symmetry i.e. the invariance group of the cylinder (∞.m : m in the Shubnikov notation [12]) with C_∞ axes perpendicular to the glass slides in which the sample is inserted. Here the space group of the phase is :

$$S = R^3 \wedge D_{\infty h} \tag{2.2}$$

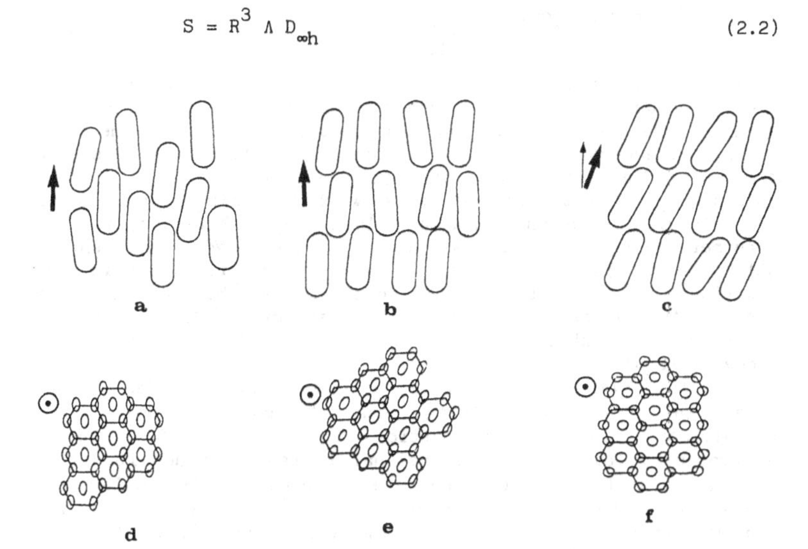

Figure 1 Schematic representation of the molecular ordering in the nematic (a) and smectic phases A (b), C(c),(F,G)(d)I(e)and B_h(f)

On further cooling the sample texture changes again abruptly at 213.5° C. The threads disappear and give way to a so-called focal conic texture, due to the existence of layers of constant thickness which can freely glide over each other. The microscope indicates that the molecules have kept the direction perpendicular to the glass slides (Figure 1.b). The fact that the layers glide freely over each other denotes the absence of long-range translational order in their plane. Such a phase, called smectic A, is thus typified, with respect to the nematic phase, by a sole breaking of the translational symmetry along the perpendicular to the glass slides. The space-group of the smectic A phase is thus :

$$S = (R^2 \times Z) \wedge D_{\infty h} \qquad (2.3)$$

where Z symbolizes the group of discrete translations along the z axis and x the direct product with the group R^2 of two-dimensional continuous translations.

Below 182.5° C the texture of TBPA changes again, the observation of biaxiality revealing that the optical axis is now tilted with respect to the layers (figure 1c). The macroscopic symmetry is lowered to C_{2h} (with the two-fold axis within the smectic layers), while the same discrete one-dimensional order of the layers remains. The symmetry group of the phasecalled the smectic C phase becomes :

$$S = (R^2 \times Z) \wedge C_{2h} \qquad (2.4)$$

When the temperature is decreased below 153.5° C, a faint front sweeps through the sample denoting another phase transition. Although the texture of the new phase, called the smectic F phase, resembles that of the smectic C phase and possesses apparently the same monoclinic symmetry, X-ray data (with X-ray beam parallel to the molecules) show that within a given layer, the molecules are distributed on an hexagonal lattice (figure 1d). More refined X-ray data reveal that this long-range bond-orientational ordering is translationally short-range, i.e. the in-plane positional order extends only over a few hundred Angtröms. Despite the differences between the structures of the smectic C and smectic F phases, the latter possesses the same macroscopic symmetry (2.4).

Below 144° C, the in-plane translational order becomes long-range. The corresponding phase (smectic G) possesses a two-dimensionally ordered lattice within the layers and thus can be considered as a three dimensional system, although it differs from a crystal by a considerable disorder in the orientation of the molecules along their long axes (the layer distributions are not sharp), as well as by some displacement disorder (only the lower-order Bragg reflections are observed). Besides the point-group symmetry of the phase remains C_{2h}, as for the smectic C or F phases. As we now have a quasi long range discrete translational order within the layers (with a correlation length of about 10^3-10^4 Å) the space symmetry of the smectic G phase is :

$$S = Z^3 \wedge C_{2h} \tag{2.5}$$

The same group of symmetry (2.5) characterizes the smectic I phase which is observed in TBDA [13], although this phase differs from the smectic G phase by the orientation of the molecules which are tilted towards the vertices of the hexagons instead of pointing towards their side, as shown in figures 1.d and 1.e.

The preceding example shows that in order to recognize the structure of the various LC phases, one cannot always restricts to their macroscopic symmetry. This is true in particular when the local order is not preserved globally, as in the hexatic smectic F phase (where there exist only a local positional order), or in the smectic B_h phase (schematized in fig. 1.f). Actually, the following data are necessary to identify completely the structure of a given phase : 1) its macroscopic space group:

$$S = T \wedge G \tag{2.6}$$

where T is the group of (continuous or discrete) translations, which indicates the long-range positional order in three dimensions, and G the (continuous or finite) point-group which leaves invariant the molecular sub-units and their packing.

2) the symmetry of the individual subunits, their in-plane arrangement (bond-orientational and positional orders) and orientation with respect to the optical axis (tilting order) <u>when these data are not accounted by the space group S</u>, i.e. when the global (macroscopic) symmetry does not reflect the local symmetry.

The determination of the preceding data constitute the first indispensable step for a phenomenological theory of transitions between mesomorphic phases. In the following subsection the main LC phases are classified with respect to their macroscopic and local symmetries. Before we discuss the possible symmetries for the S groups.

2.2. The macroscopic space groups of liquid crystals

As we deal with elongated molecular subunits, and if we except the isotropic phase for which G = O(3) or SO(3), G is necessarily a continuous or discrete subgroup of $D_{\infty h}$. The classification of the G groups indicated in figure 2 shows that there exist five continuous point group symmetries, and seven classes of discrete (finite) point groups. The schematization of the corresponding subunits, illustrated in figure 3 allows to distinguish between non-chiral subunits ($D_{\infty h}$, $C_{\infty v}$, $C_{\infty h}$, D_{nh}, D_{nd}, C_{nh}, C_{nv}, S_{2n}) and chiral ones (D_∞, C_∞, D_n, C_n). The term <u>chiral</u> (from the greek word for "hand") being used to designate structures in which a center of inversion or a miror plane are absent, i.e. in which right and left handed subunits are not equivalent. Let us stress, that in figure 3 it is assumed that the G group describes the molecular subunits, while it may be associated with the in-plane lattice for phases possessing a bond orientational or positional order (e.g. D_{6h} for the smectic B_h phase).

The translation group T can be composed of discrete (Z) or continuous (R) translations. Thus we will have : $T = R^3$ for the isotropic and nematic phases, $T = R^2 \times Z$ for one dimensional smectics, $T = Z^3$ for three-dimensional smectics and $T = R \times Z^2$ for discotic phases.

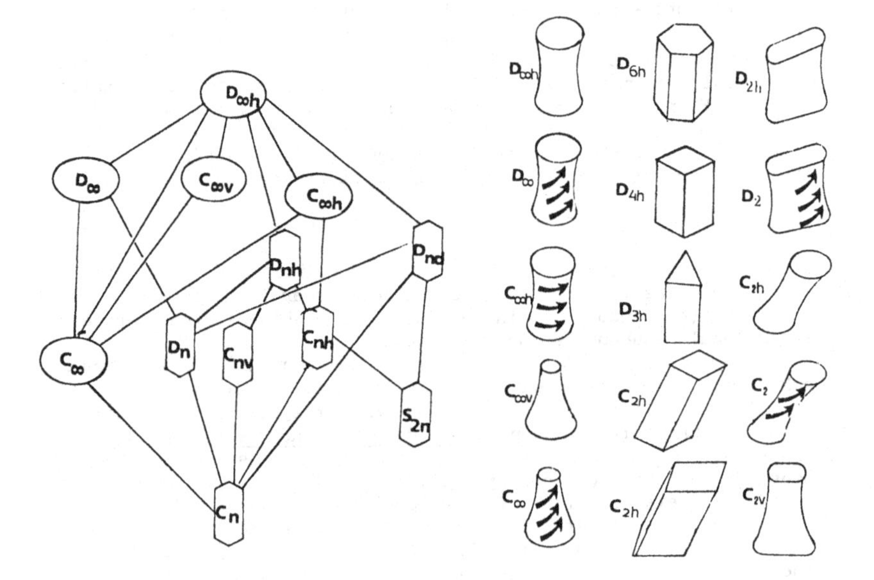

Figure 2 : Classification of the continuous and discrete macroscopic point-groups G, of LC phases

Figure 3 : Schematization of some subunits forms corresponding to continuous or discrete G groups

In table 1 the structural characteristics of the main LC phases are given, namely the translation T and G groups, and their eventual bond-orientational, positional and tilting orders. Cholesterics and incommensurate smectics do not figure in this table, as on a phenomenological basis they can be obtained as distortions from nematic and commensurate smectics respectively. It must be emphasized that this table indicates only the phases, the symmetry of which is well established at the present time. From figure 2 it can be inferred that the possibilities for experimental discovery of new LC structures are still far from being exhausted.

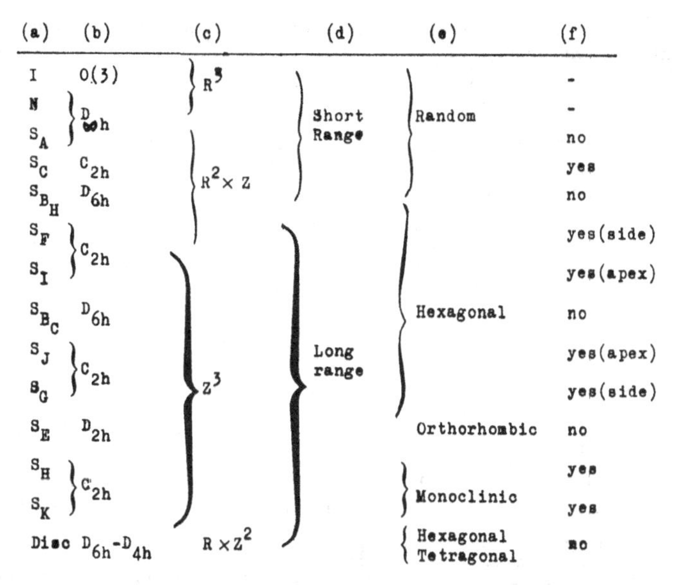

(a)	(b)	(c)	(d)	(e)	(f)
I	O(3)	R^3			-
N	$D_{\infty h}$		Short Range	Random	-
S_A	$D_{\infty h}$				no
S_C	C_{2h}	$R^2 \times Z$			yes
S_{B_H}	D_{6h}				no
S_F	C_{2h}				yes(side)
S_I	C_{2h}				yes(apex)
S_{B_C}	D_{6h}			Hexagonal	no
S_J	C_{2h}		Long range		yes(apex)
S_G	C_{2h}	Z^3			yes(side)
S_E	D_{2h}			Orthorhombic	no
S_H	C_{2h}			Monoclinic	yes
S_K	C_{2h}				yes
Disc	D_{6h}-D_{4h}	$R \times Z^2$		Hexagonal Tetragonal	no

Table 1 : Structural characterization of the main LC phases. (a) type of phase (b) macroscopic point-group G (c) group T of translations (d) bond-orientational ordering (e) type of lattice (f) tilting order

3. THE IRREDUCIBLE REPRESENTATIONS OF LIQUID CRYSTAL PHASES

Once that the space-symmetry S of the high-temperature parent phase has been determined for a LC system, the second step for applying the Landau theory consists in the construction of the irreducible representations (IR's) associated with S. There exist no available tables providing explicitely the IR's of the S groups. However their construction is easy from the IR's of the point-groups G for which a number of character tables can be found [14,15] . In this section we recall the method for constructing the IR's of the continuous and discrete G groups (§.3.1), and S groups (§.3.2)

3.1. Irreducible representations of the G point groups

a) Let us start from the simplest group C_n. All elements of this group represent rotations about the z axis. The n-th power of any element of the group C_n is equal to the unit element. Hence it contains only rotations through the angles :

$$0, \frac{2\pi}{n}, \frac{4\pi}{n}, \ldots\ldots, \frac{2\pi}{n}(n-1) \tag{3.1}$$

The number of these rotations is equal to n, and the elements of the group are :

$$C_1,\ C_n,\ C_n^2, \ldots\ldots,\ C_n^{n-1} \qquad (3.2)$$

As the C_n group is Abelian [16] , all its IR's are one dimensional and their number is equal to the order of the group;i.e. to n. Let τ be one the IR's of this group. Since :

$$\tau^n (C_n) = \tau(C_n^n) = \tau(C_1) = 1 \qquad (3.3)$$

$\tau(C_n)$ takes on one of the values :

$$e^{\frac{2\pi ir}{n}} \quad (r = 0, 1, \ldots., n-1) \qquad (3.4)$$

Thus we obtain all n IR's of the group C_n by :

$$\tau_r (C_n^m) = e^{\frac{2\pi irm}{n}} \qquad (r,m = 0, 1, \ldots, n-1) \qquad (3.5)$$

If n is odd we will thus have the identity IR (denoted A_1) for the values $r = m = 0$, and an even number $n - 1$ of complex IR's. These can be grouped by pair, i.e. the IR $e^{\frac{2\pi irm}{n}}$ and its complex conjugate $e^{\frac{2\pi i(n-r)m}{n}} = e^{-\frac{2\pi irm}{n}}$, in such a manner that we will have an odd number $\frac{n-1}{2}$ of real two dimensional representations, irreducible on R, denoted Er_1, the matrix of which are :

$$\begin{pmatrix} e^{\frac{2\pi irm}{n}} & 0 \\ 0 & e^{-\frac{2\pi irm}{n}} \end{pmatrix} \qquad (3.6)$$

If n is even, we will have the identity IR A_1, a real one-dimensional IR (denoted A_2) associated with the number ± 1 (+ 1 when m is even and -1 when m is odd), and $\frac{n-2}{2}$ two dimensional representations Er_1 irreducible on R. The character tables for the C_n group are given in Table 2 :

(n odd)

C_n	C_n^m
A_1	1
Er_1	$2\cos\frac{2\,mr\pi}{n}$

(n even)

	C_n^m
A_1	1
A_2	ε
Er_1	$2\cos\frac{2\,mr\pi}{n}$

Table 2 : Character tables for the C_n group. $\varepsilon = +1$ for m even and $\varepsilon = -1$ for m odd.

Let us now consider the cyclic group S_{2n}, which consists of the powers of the rotation-inversion $S_{2n} = C_{2n} \cdot \sigma_z$. It has 2n elements $C_1, S_{2n}, S_{2n}^2, \ldots, S^{2n-1}$, the even powers of S_{2n} forming a subgroup which coincides with the group C_n. As S_{2n} is isomorphic to the group C_{2n}, its representations can be found from the formula :

$$\tau_r(S_{2n}^m) = e^{\frac{2\pi i r m}{2n}} \quad (r, m = 0, 1, \ldots, 2n-1) \qquad (3.7)$$

If n is odd, each of the IR of C_n will correspond to two IR's for S_{2n}, whereas for n even, S_{2n} and C_{2n} will have the same number of IR's than C_n.

The group D_n consists of all rotations which transform a regular n-sided prism into itself. It has one n-th order C_n axis and n second-order axes perpendicular to it. These axes are denoted $u_1, u_2, \ldots, u_n$, the angle between two adjacent axes being $\frac{\pi}{n}$. As the rotations C_n^m and C_n^{n-m} are mutually conjugate, and the axes $u_1, u_3, u_5 \ldots$ or $u_2, u_4, u_6 \ldots$ transform into eachother under the rotations C_n^{m} (m = 1, 2, 3,...), it follows [16] that the number of IR's for the D_n group is : $\frac{n}{2} + 3$ for n even, and $\frac{n+3}{2}$ for n odd. One can show [16] that among these IR's one has 1) for n odd : two one-dimensional IR's and $\frac{n-1}{2}$ two dimensional ones ; 2) for n even, four one-dimensional IR's and $\frac{n}{2} - 1$ two dimensional ones. The one-dimensional IR's of D_n (n even or odd) are obtained from the one-dimensional IR's of C_n^n (with same parity for n) by the direct product with the two one-dimensional IR's of the C_2 group, whereas the two-dimensional IR's of D_n, correspond to the same two-dimensional matrices than for the C_n group for the C_n^m elements, whereas the axis u_2^m, obtained by rotating the u_x axis by C_n^m, is represented by the matrix :

$$\begin{bmatrix} 0 & e^{\frac{2 i m r \pi}{n}} \\ e^{-\frac{2 i m r \pi}{n}} & 0 \end{bmatrix}$$

The character tables for the D_n group are given in Table 3

(n odd)

D_n	C_n^m	U_2^m
A_1	1	1
A_2	1	- 1
Er_1	$2\cos\frac{2mr\pi}{n}$	0

(n even)

D_n	C_n^m	U_2^m
A_1	1	1
A_2	ε	ε
A_3	1	- 1
A_4	ε	$-\varepsilon$
Er_1	$2\cos\frac{2mr\pi}{n}$	0

Table 3 : Character tables for the D_n groups. $\varepsilon = 1$ if m is even and $\varepsilon = -1$ if m is odd.

The groups C_{nv}, containing one n-th order axis C_n and n vertical planes $\sigma_1, \sigma_2, \ldots, \sigma_n$ passing through the C_n axis, are isomorphic to the groups D_n, and therefore, have the same IR's. The group C_{nh} is equal to the direct product of C_n and σ_z. Thus its IR's are obtained by multiplying the IR's of C_n by the two one-dimensional IR's of $C_s = \{C_1, \sigma_z\}$. Similarly the IR's of the groups D_{nh} and $D_{(2n+1)d}$ are respectively obtained by multiplying the IR's of D_n and D_{2n+1} by the IR's of the groups C_s and $C_i = \{E, I\}$. For D_{2nd}, by similar arguments than for D_n, one finds that it has 2n-1 two-dimensional IR's with the generative matrices :

$$Er_1(S_{4n}^m) \begin{bmatrix} e^{\frac{i\pi rm}{2n}} & 0 \\ 0 & e^{-\frac{i\pi rm}{2n}} \end{bmatrix} \quad Er_1(\sigma_1) \begin{bmatrix} 0 & 1 \\ 1 & 0 \end{bmatrix} \tag{3.8}$$

and four one-dimensional IR's generated by :

$$\begin{aligned} A_{1,3}(S_{4n}^m) &= 1 \quad , \quad A_{2,4}(S_{4n}^m) = -1 \\ A_{1,3}(\sigma_1) &= \pm 1 \quad , \quad A_{2,4}(\sigma_1) = \pm 1 \end{aligned} \tag{3.9}$$

b) The continuous groups contained in figure 2, correspond to the limiting value $n \to \infty$. Let us note C_ϕ the rotation of an angle ϕ around the z axis ($\phi \in [0, 2\pi]$), $U_{2\phi}$ the two fold axis perpendicular to z, and making an angle ϕ with the x axis, σ_ϕ the vertical plane perpendicular to $U_{2\phi}$, and $S_\phi = C_\phi . \sigma_z$. One can obtain straightforwardly the IR's of the five continuous groups C_∞, D_∞, $C_{\infty h}$, $C_{\infty v}$ and $D_{\infty h}$ the character tables of which are given in table 4, by generalizing the corresponding results obtained for discrete groups. The matrices associated with the IR's Er_1 of the D_∞ group are thus

$$\begin{bmatrix} e^{ir\phi} & 0 \\ 0 & e^{-ir\phi} \end{bmatrix} \text{ for } C_\phi \text{ , and } \begin{bmatrix} 0 & e^{2ir\phi} \\ e^{-2ir\phi} & 0 \end{bmatrix} \text{ for } U_{2\phi} \tag{3.10}$$

and the matrices associated with the two dimensional IR's of the groups $C_{\infty v}$, $C_{\infty h}$ and $D_{\infty h}$ can be worked out by considering that 1/ $C_{\infty v}$ is isomorphic to D_∞, 2/ $C_{\infty h}$ is the direct product of C_∞, and C_s, and 3/ $D_{\infty h}$ can be obtained by multiplying D_∞ by C_s or C_i.

C_∞	C_ϕ
A_1	1
Er_1	$2 \cos r\phi$

$C_{\infty v}$ / D_∞	C_ϕ / C_ϕ	σ_ϕ / $U_{2\phi}$
A_1	1	1
A_2	1	-1
Er_1	$2 \cos r\phi$	0

$D_{\infty h}$	C_ϕ	σ_ϕ	S_ϕ	$U_{2\phi}$
A_1	1	1	1	1
A_2	1	-1	1	-1
A_3	1	1	-1	-1
A_4	1	-1	-1	1
Er_1	$2\cos r\phi$	0	$-2\cos r\phi$	0
Er_2	$2\cos r\phi$	0	$-2\cos r\phi$	0

$C_{\infty h}$	C_ϕ	S_ϕ
A_1	1	1
A_2	1	-1
Er_1	$2\cos r\phi$	$2\cos r\phi$
Er_2	$2\cos r\phi$	$-2\cos r\phi$

Table 4 :Table of characters for the IR's of the continuous groups G.

3.2. Irreducible representations of the space groups S

Discarding from this section the IR's of the isotropic liquid which are considered in §.9, we can restrict ourselves to describe the method allowing to construct the IR's associated with the S groups of the nematic and one-dimensional (1-d) smectic phases, as these phases appear to be the relevant parent phases in LC phase diagrams.

a) As indicated in Table 1, the space-group of the nematic phase can be written :

$$S = R^3 \wedge G \tag{3.11}$$

where G is one of the point groups shown in Figure 2. Three classes of IR's can be constructed, associated with a nematic phase, depending if the lower symmetry phase appearing below the nematic phase possesses 1) no discrete translational order (i.e. the nematic-nematic or nematic-cholesteric transitions) ; a discrete translational order 2) in one direction (i.e. the nematic-1-d smectic transition) 3) in two directions (the nematic-discotic transition) or 4) in three directions (the nematic-3d smectics).

1/ In the simplest case where no breaking of the continuous translational order is involved, one has to consider only the IR's of the point-group G associated with the parent nematic phase. As will be shown in §.4 and 5, the one dimensional IR's denoted A_i in Table 4, will lead to nematic phases with continuous rotational symmetry, whereas the two-dimensional IR's labelled Er_i (i = 1,2) will give rise either to nematic phases with discrete rotational order, of the biaxial type [17] , or to cholesteric phases, when a Lifshitz invariant can be constructed (e.g. for the IR's of the chiral groups C_∞ and D_∞).

2/ When the continuous translational order of the parent nematic phase is broken in only one direction, one has to consider as describing the symmetry of the parent phase the group :

$$S = R \wedge G \tag{3.12}$$

where R symbolizes the direction of space (e.g. the z axis) along which the translational order becomes discrete at low-temperature. This is the situation encountered at the nematic-1-d smectic transition. Here one has to construct the IR's associated with a fictitious one-dimensional Brillouin-zone of length $\frac{2\pi}{d}$ where d is the interlayer distance in the smectic phase. Posing $k = \frac{2\pi}{\alpha d}$ with $\alpha > 1$, one gets the IR's of the groups (3.12) by multiplying the IR's of the point-group G by the factor :

$$e^{ikd} \tag{3.13}$$

Accordingly, from each of the one-dimensional IR's A_i of the group G, one constructs a two dimensional IR denoted E_{ki}, while the two dimensional IR's E_{ri} of the group G lead to two or to four-dimensional IR's labelled E_{rki} or G_{rk}. Table 5 gives the character tables for the groups of the type (3.12) when G is a continuous group. The elements belonging to the S groups are denoted by analogy with the elements of the cristallographic space-groups [18] namely:

$$\{A(\phi)|d\} \tag{3.14}$$

where $A(\phi)$ can be a rotation axis $C(\phi)$ along z, a second order axis perpendicular to z ($U_2(\phi)$), a rotation-reflection $S(\phi)$ or a reflection $\sigma(\phi)$ in a vertical plane containing $C(\phi)$. d denotes the translation of length d along z.

Table 5 : Character tables for the nematic groups $S = R \wedge G$ where G is a continuous group, assuming a one-dimensional Brillouin-zone with discrete length $\frac{2\pi}{d}$.

$R \wedge C_\infty$	$\{C(\phi)\|d\}$
E_{k1}	$2\cos kd$
E_{rk1}	$2\cos(r\phi + kd)$
E_{rk2}	$2\cos(r\phi - kd)$

$R \wedge D_\infty$ / $R \wedge C_{\infty h}$	$\{C(\phi)\|d\}$ / $\{C(\phi)\|d\}$	$\{U_2(\phi)\|d\}$ / $\{S(\phi)\|d\}$
E_{k1}	$2\cos kd$	0
G_{rk}	$4\cos r\phi \cos kd$	0

$R \wedge C_{\infty v}$	$\{C(\phi)\|d\}$	$\{\sigma(\phi)\|d\}$
E_{k1}	$2\cos kd$	$2\cos kd$
E_{k2}	$2\cos kd$	$-2\cos kd$
G_{rk}	$4\cos r\phi \cos kd$	0

$R \wedge D_{\infty h}$	$\{C(\phi)\|d\}$	$\{\sigma(\phi)\|d\}$	$\{S(\phi)\|d\}$	$\{U_2(\phi)\|d\}$
E_{k1}	$2\cos kd$	$2\cos kd$	0	0
E_{k2}	$2\cos kd$	$-2\cos kd$	0	0
G_{rk}	$4\cos r\phi \cos kd$	0	0	0

A multiplication by the factor (3.13) gives similar results for the S groups with discrete G groups. Table 6 gives the character tables for the groups $R \wedge C_n$ and $R \wedge D_n$ with n even and n odd. Let us stress that the concept of a fictitious Brillouin-zone is justified physically by the observation of pretransitional effects in the vicinity of the nematic-1-d-smectic transition, when approaching the transition in the nematic phase [19,20], which clearly show the tendency of the molecular subunits towards a layer ordering with discrete interlayer distance d.

Table 6 : Character tables for the nematic groups $S = R \wedge G$, where G is one of the discrete groups C_n or D_n. $\varepsilon = 1$ if m even and $\varepsilon = -1$ if m odd

$R \wedge C_n$	$\{C_n^m \vert d\}$
n odd	
E_{k1}	$2 \cos kd$
E_{rk1}	$2 \cos (\frac{2mr\pi}{n} + kd)$
E_{rk2}	$2 \cos (\frac{2mr\pi}{n} - kd)$
n even	
E_{k1}	$2 \cos kd$
E_{k2}	$2 \varepsilon \cos kd$
E_{rk1}	$2 \cos (\frac{2mr\pi}{n} + kd)$
E_{rk2}	$2 \cos (\frac{2mr\pi}{n} - kd)$

$R \wedge D_n$	$\{C_n^m \vert d\}$	$\{U_2^m \vert d\}$
n odd		
E_{k1}	$2 \cos kd$	0
G_{rk}	$4 \cos \frac{2mr\pi}{n} \cos kd$	0
n even		
E_{k1}	$2 \cos kd$	0
E_{k2}	$2 \varepsilon \cos kd$	0
G_{rk}	$4 \cos \frac{2mr\pi}{n} \cos kd$	0

3/ Let us now consider the case, realized at the nematic-discotic transition [21-23] where a two-dimensional discrete order takes place. The group of the nematic phase is :

$$S = R^2 \wedge G \tag{3.15}$$

where R^2 symbolizes the two-dimensional continuous order which is broken at the transition, and can be assumed to be in the (x,y) plane, while the translational order along the z axis remains continuous. As in the preceding case one has to consider fictitious two dimensional Brillouin-zones, in which the transition wave vector k is selected, and construct the IR's of the considered group associated with the chosen k-vector. We will not enter into the details of the construction of these IR's. Let us only mention that it consists in multiplying the IR's of the point group G associated with S by a factor of the form :

$$e^{i\vec{k}(d_1\vec{u} + d_2\vec{v})} \tag{3.16}$$

where $\vec{u}$ and $\vec{v}$ are the unit-vectors along the x and y axes, d_1 and d_2 are the lattice constants corresponding to the elementary two-dimensional unit-cell, and k the wave-vector which is located at a high-symmetry point on the boundary of one of the two-dimensional Brillouin-zones.

The list of wave-vectors which have to be considered in the monoclinic, orthorhombic, tetragonal and hexagonal two-dimensional Brillouin-zones can be found in Refs. 24,25 . Let us stress that if one assumes a strictly periodic two-dimensional lattice, the in-plane ordering will necessary correspond to one of the seventeen space-groups in two dimensions [26].

4/ Although there exist no experimental examples of a direct transition from the nematic to a three-dimensionally ordered smectic (see §.5), this case can be treated theoretically along the same line than the two preceding ones. Starting from the S groups :

$$S = R^3 \wedge G \tag{3.17}$$

one has to consider the k-vectors located on the surface of one of the fourteen three-dimensional crystallographic Brillouin-zones. The factors (3.13) and (3.16) become in three dimensions :

$$e^{i\vec{k}.\vec{t}} \tag{3.18}$$

where $\vec{t}$ is a combination of the basic lattice vectors describing one of the fourteen Bravais lattices, and $\vec{k}$ a wave vector which can be found in the tables of IR's of the 230 space-groups [27-29] However it is still unclear from the experimental data on three-dimensional smectics [30,31], if their structure can be merely assimilated to a crystallographic space-group, and reciprocally, if there should not exist some specific rules selecting the crystallographic groups which are liable to describe a three-dimensional smectic phase.

b) The construction of the IR's associated with one-dimensional smectic phases (i.e. smectic A or C phases) follow the same scheme than for nematic phases. If the transition takes place between one-dimensional smectics (i.e. no continuous translational order is broken at the transition), one has to construct the IR's of the group :

$$S = Z \wedge G \tag{3.19}$$

where Z is the discrete translational order perpendicular to the smectic layers. If d is the distance between the layers in the parent smectic phase, the corresponding Brillouin zone is the segment of length $\frac{2\pi}{d}$. Accordingly three classes of IR's can be constructed depending if the transition wave-vector $\vec{k}$ associated with the transition is located at the center (k = 0), inside ($-\frac{\pi}{d} < k < \frac{\pi}{d}$) or on the surface ($k = \pm\frac{\pi}{d}$) of the Brillouin-zone. Multiplying the IR's of the G group by factors of the form (3.13), one can thus easely deduce the matrices associated with the IR's of the S groups. The character tables for $S = Z \wedge D_{\infty h}$ and $S = Z \wedge C_{nv}$ (n even) are given as illustrative examples in Table 7.

Table 7 : Character tables for the $Z \wedge D_{\infty h}$ and $Z \wedge C_{nv}$ (n even) 1-d smectic groups. $\varepsilon = +1$ for m even, and $\varepsilon = -1$ for m odd.

$Z\wedge D_{\infty h}$	$\{C(\phi)\|d\}$	$\{\sigma(\phi)\|d\}$	$\{S(\phi)\|d\}$	$\{U_2(\phi)\|d\}$
$k=0$				
A_1	1	1	1	1
A_2	1	-1	1	-1
A_3	1	1	-1	-1
A_4	1	-1	-1	1
E_{r1}	$2\cos r\phi$	0	$2\cos r\phi$	0
E_{r2}	$2\cos r\phi$	0	$-2\cos r\phi$	0
$k=\pm\frac{\pi}{d}$				
A_5	-1	-1	-1	-1
A_6	-1	1	-1	1
A_7	-1	-1	1	1
A_8	-1	1	1	-1
E_{r3}	$-2\cos r\phi$	0	$-2\cos r\phi$	0
E_{r4}	$-2\cos r\phi$	0	$2\cos r\phi$	0
$-\frac{\pi}{d}<k<\frac{\pi}{d}$				
E_{k1}	$2\cos kd$	$2\cos kd$	0	0
E_{k2}	$2\cos kd$	$-2\cos kd$	0	0
G_{rk}	$4\cos r\phi\cos kd$	0	0	0

$Z\wedge C_{nv}$	$\{C_n^m\|d\}$	$\{\sigma^m\|d\}$
$k=0$		
A_1	1	1
A_2	ε	ε
A_3	1	-1
A_4	ε	$-\varepsilon$
E_{r1}	$2\cos\frac{2mr\pi}{n}$	0
$k=\frac{\pi}{d}$		
A_5	-1	-1
A_6	$-\varepsilon$	$-\varepsilon$
A_7	-1	1
A_8	$-\varepsilon$	ε
E_{r2}	$-2\cos\frac{2mr\pi}{n}$	0
$-\frac{\pi}{d}<k<\frac{\pi}{d}$		
E_{k1}	$2\cos kd$	$2\cos kd$
E_{k2}	$2\varepsilon\cos kd$	$2\varepsilon\cos kd$
E_{k3}	$2\cos kd$	$-2\cos kd$
E_{k4}	$2\varepsilon\cos kd$	$-2\varepsilon\cos kd$
G_{rk}	$4\cos\frac{2mr\pi}{n}\cos kd$	0

Finally, we consider the case where the transition occurs from a 1-d to a 3-d smectic phase, i.e. when the continuous two-dimensional in plane smectic order is broken. The smectic-group of the parent phase is :

$$S = (R^2 \times Z)\wedge G \tag{3.20}$$

The construction of the IR's associated with a group of the type (3.20) follows a procedure which is analogous to the one already discussed for the nematic-3d-smectic transition, i.e. one has to consider a two or three-dimensional fictitious Brillouin zone (depending if the discrete order along z is broken at the transition), and multiply the IR's of the corresponding group G by factors of the type (3.18), where $\vec{k}$ belongs to the two or three-dimensional Brillouin zone.

To conclude this section, it should be emphasized that in the construction of the IR's associated with the various parent phases considered hereabove, only the translational (positional) and rotational symmetries have been taken into account explicitely.

In the following sections we show how to describe on a phenomenological basis, the tilting and bond-orientational orders (which are additional degrees of freedom) taking place in smectic phases.

4. PROBABILITY DENSITIES, ORDER-PARAMETERS AND THERMODYNAMIC POTENTIALS FOR LIQUID CRYSTAL TRANSITIONS

In this section the three quantities which play an essential part in the Landau theory, namely the probability density, the order-parameter, and the Landau free-energy, are discussed in terms adapted to LC transitions (§.4.1. to 4.3). Two examples are then given as an illustration of the practical procedure that one may follow to apply the theory (§.4.4).

4.1. Probability densities

As shown in chapters II and III, the symmetry of crystals can be described via a single probability density $\rho(x,y,z)$ of the particles forming the crystal (electrons, atoms, molecules), which expresses the average statistical distribution of the centers of gravity of the particles. In LC systems, which are formed by elongated molecules, one needs (as for magnetic systems) an additional distribution function, namely a vector density of probability $\vec{u}(x,y,z)$, in order to account for the mean orientation of the molecular subunits. Accordingly the nematic phase will correspond to constant value and direction for $\rho(x,y,z)$ and $\vec{u}(x,y,z)$ respectively, whereas smectic A phases will be characterized by : $\rho(x,y)$ = cte, $\vec{u}(\vec{r}) = \vec{u}_o // z$, $\rho(z)$ being a periodic function with a periodicity corresponding to the lattice spacing.

More generally, the probability density adapted to the description of all mesomorphic phases should be liable to account for the various types of orders which appear in the more structured LC phases, namely the positional (layered and in-plane), bond orientational and tilting orders. To this end, Landau [1] introduced a pair correlation function $\rho_{12}(\vec{r}_{12})$ where $\vec{r}_{12}$ is the distance between molecules 1 and 2, and $\rho_{12}\, dv_2$ is the probability of finding molecule 2 in the volume dv_2 when molecule 1 has a given position*. As pointed out by Goshen and Mukamel [32] the preceding correlation function possesses even parity and is not adapted for non centrosymmetric (chiral) molecules. Subsequently, Indenbom and Pikin [8] proposed to describe chiral molecules by the pair correlation function $\rho_{12}(\vec{r}_{12}, \vec{l}_1, \vec{l}_2)$ which depends on the distance $\vec{r}_{12}$ between the centres of gravity of the molecules and on the orientations $\vec{l}_1$ and $\vec{l}_2$ of the long axes of the molecules 1 and 2. As suggested by these authors ρ_{12} can be formed by pseudo scalar products $(\vec{l}_1 \wedge \vec{l}_2).\vec{r}_{12}$ for polar molecules, or by the product $[(\vec{l}_1 \wedge \vec{l}_2).\vec{r}_{12}]\,(\vec{l}_1.\vec{l}_2)$ for non-

*As noted by Indenbom and Pikin [8], the use of a single density function $\rho(x,y,z)$ in ordinary crystals, implicitely assumes that the crystal axes are fixed, which is equivalent to take many particle correlations into account.

polar molecules. It must be finally emphasized that a still more complicated description is needed for LC phases with discrete symmetries for the molecular subunits (e.g. biaxial nematics [17] or pyramidal mesophases [33])as the short axes of the molecules have to be also taken into account.

As indicated in chapter II, the variation of the probability density $\delta\rho$ associated with a phase transition can be considered as a vector in the representation space, and the components of vector in the basis of this space, are the values of the order parameter which minimize the thermodynamic potential. The more general basis functions among LC systems are the one which transform as the degrees of freedom of the isotropic phase, i.e. the functions [34]:

$$\phi_k^m (\vec{r},\theta,\phi) = e^{i\vec{k}\vec{r}} \cdot Y_m^1 (\theta,\phi) \tag{4.1}$$

where $\vec{k}$ is the wave-vector of the reciprocal space corresponding to the breaking of translational symmetry at the transition, $\vec{r}$ is the position vector at the centre of gravity of the molecules, and $Y_m^j(\theta , \phi)$ are the spherical harmonics which depend on the spherical angles θ and ϕ determining the molecular orientation. Along this line, expressions for the probability density modifications at the various LC transitions were given by Rosciszewski [35]. The formulation of the degrees of freedom corresponding to the functions (3.21) for each LC phase, in terms of IR's, allows a practical and unified method for working out the symmetry modification which take place at LC transitions and the corresponding order-parameter symmetries and thermodynamic potentials. We will focus on these questions in the following paragraphs.

4.2. Primary and secondary order-parameters

Let us characterize the symmetry of the order-parameters associated with the transitions in LC systems. We will start by the isotropic-nematic transition, which is considered in more details in §.9. It is traditionally assumed [4] that the corresponding primary order-parameter has the symmetry of a second-rank traceless tensor [9], the components of which can be written :

$$Q_{ik}(\vec{r}) = l_i l_k - \frac{1}{3} (l_i l_j)\delta_{ik} = Q(r) \; [u_i(\vec{r})u_k(\vec{r}) - \frac{1}{3} \delta_{ik}] \tag{4.2}$$

It represents local averages of bilinear combinations formed by projections of the unit vector $\vec{l}$ of the long-molecular axis, or by analogous combinations of the director components u_i which determine the local orientation of the molecules. The quantity $Q(\vec{r})$ defines the fraction of molecular axes pointing along $\vec{u}$ at a given point. Thus, in the isotropic phase $Q(\vec{r}) = 0$, and in the nematic phase $Q(\vec{r}) \neq 0$. Let us note that Eq.(4.2) applies only for molecular subunits with symmetry $D_{\infty h}$ and D_∞, i.e. in which the two directions $\vec{u}$ and $-\vec{u}$ are equivalent.

In cholesterics phases which appear below the isotropic or nematic phases the bending and twisting of the molecules are taken into account (see §.6) through Lifshitz invariants of the form :

$$\vec{u}\ \mathrm{curl}\ \vec{u}\ , \quad \text{or} \quad u_x \frac{\partial u_y}{\partial z} - u_y \frac{\partial u_x}{\partial z} \qquad (4.3)$$

and thus one may expect cholesteric phases to take place only when the order-parameter symmetry allows invariants of the type (4.3) i.e. for the molecular symmetries C_∞ and D_∞.

The onset of a smectic A phase below on isotropic or nematic phase, corresponds to the formation of a density wave [31]:

$$\phi = \phi_o \cos (kz + \omega) \qquad (4.4)$$

where $k = \frac{2\pi}{d}$, ϕ_o and ω being respectively the amplitude and phase of the wave. ϕ is a secondary order-parameter which couples to the primary order-parameter (4.2) (see §.4.3). By contrast in transitions from the smectic A phase to more structured smectic modifications, ϕ can be assumed to be the primary order-parameter. Thus, in order to describe the transition from the smectic A to a tilted smectic (e.g. the smectic C phase), the tilting will be introduced as a secondary order-parameter, denoted θ , where θ has the symmetry of the product [8]:

$$\theta \sim \widehat{(\vec{u}.\vec{k})}\ [\vec{u} \wedge \vec{k}\,] \qquad (4.5)$$

and couples to the primary order-parameter ϕ . Analogously the bond-orientational order, which can be introduced via the angle ψ between the short axes of the molecular subunits, can be accounted on a phenomenological basis through a coupling to the primary order-parameter ϕ (see §.7). This is also the case for various induced phenomena such as ferroelectricity or flexoelectricity (see §.8).

Generally speaking the symmetry of the transition order-parameter is determined by the IR inducing the transition. According to §.3 we can thus deduce that for nematic-to-nematic or cholesteric transitions, as well as for nematic to 1-d smectics and for transitions between 1-d smectic phases, the dimensionality of the primary order parameters may be only 1, 2 or 4. By contrast this dimensionality can be infinite for transitions from the isotropic phase (see §.9), and may amount 48 when three dimensional Brillouin zones are taken into account. In the following sections such numbers are illustrated by some concrete examples.

4.3. Thermodynamic potentials

Once the IR associated with a given LC transition has been constructed, one can establish the form of the thermodynamic potential F, which provides by minimization, the equilibrium values of the order-parameter components for each stable state. The more general form for F is :

$$F(T,P,\eta_i,X_j\frac{\partial\eta_i}{\partial u_j},\frac{\partial X_j}{\partial u_k}) = F_o(T,P) + F_1(\eta_i)+F_2(X_j) + F_3(\eta_i,X_j) + F_4(\eta_i,X_j,\frac{\partial\eta_i}{\partial u_j},\frac{\partial X_j}{\partial u_k}) \quad (4.6)$$

where $F_1(\eta_i)$ contains homogeneous polynomials of the primary order-parameter components η_i invariant by the S group associated with the parent phase. $F_2(X_j)$ contains invariants of the secondary order-parameters (polarization, tilting, bond-orientational angle, etc...), and $F_3(\eta_i,X_j)$ invariants describing the coupling between the primary and eventual secondary parameters. F_4 must be included when gradient terms are required for the description of low-temperature inhomogeneous states, i.e. for cholesteric phases of incommensurate smectics. As for incommensurate phases in crystals (see chapter V), when the order-parameter is an inhomogeneous function of the space coordinates, (4.6) has the meaning of a free-energy density, and the thermodynamic potential takes the form :

$$\delta = \int_V F\, dV \quad (4.7)$$

where the integral is calculated over the volume V of the system. The phenomenological approach to cholesteric and incommensurate smectic phases is discussed in §.6. General results concerning the thermodynamic potentials associated with nematic and 1-d smectic parent phases are given in ref. 42. In the following paragraph we show the practical method for determining the F_1 contribution to the thermodynamic potential (4.6)

4.4. Practical procedure for the determination of the symmetry changes which take place at Liquid Crystal transitions.

As for structural transitions (see chapter III) the procedure which allows to work out the symmetries of the low-temperature phases appearing below a LC phase of given symmetry S, involves three steps : 1) selection of the IR associated with the group S which corresponds to a definite wave-vector $\vec{k}$ of the Brillouin-zone and construction of its matrices.

2) Construction of the primary order-parameter contribution $F_1(\eta_i)$ to the thermodynamic potential, and determination of the possible low-temperature stable states, corresponding to the equilibrium values η_i^o which minimize $F_1(\eta_i)$.

3) Determination of the subgroups S' of S which leave invariant the probability density change :

$$\delta\rho = \sum_i \eta_i^o \phi_i$$

in the order parameter space, spanned by the basis vectors ϕ_i.

One further step may be the determination of the equilibrium values for the secondary parameters X_i. It requires construction of the $F_2(X_i)$ and $F_3(\eta_i,X_i)$ contribution to the free-energy. We will now make clear such a procedure by considering two examples:

1) The transitions from the 1-d smectic group $S = Z \wedge C_{\infty v}$, and
2) the transitions from the nematic group $S = R \wedge D_{\infty h}$.

4.4.1. Transitions from the smectic group $S = Z \wedge C_{\infty v}$

Let us consider a smectic phase of symmetry $S = Z \wedge C_{\infty v}$ and determine the transitions that may take place from this phase to other 1-d smectic phases, i.e. without any breaking of the continuous in-plane symmetry R^2. Accordingly, only the rotational symmetry or the discrete translational symmetry along the z axis may be broken. The parent smectic phase is schematized in Figure 4. It corresponds to a non-centrosymmetric chiral smectic A phase, for which the two states with directors $\vec{u}$ and $-\vec{u}$ are inequivalent. The characters of the IR's of S are listed in Table 8.

Table 8 : IR's of the smectic $Z \wedge C_{\infty v}$ group.

$Z \wedge C_{\infty v}$	$\{c(\phi)\,\vert d\}$	$\{\sigma(\phi)\vert d\}$
k=0 A_1	1	1
A_2	1	-1
Er_1	$2 \cos r\phi$	0
$k=\pm\frac{\pi}{d}$ A_3	-1	-1
A_4	-1	1
Er_2	$-2 \cos r\phi$	0
$-\frac{\pi}{d}<k<\frac{\pi}{d}$		
E_{k1}	2 cos kd	2 cos kd
E_{k2}	2 cos kd	-2 cos kd
G_{rk}	$4 \cos r\phi \cos kd$	0

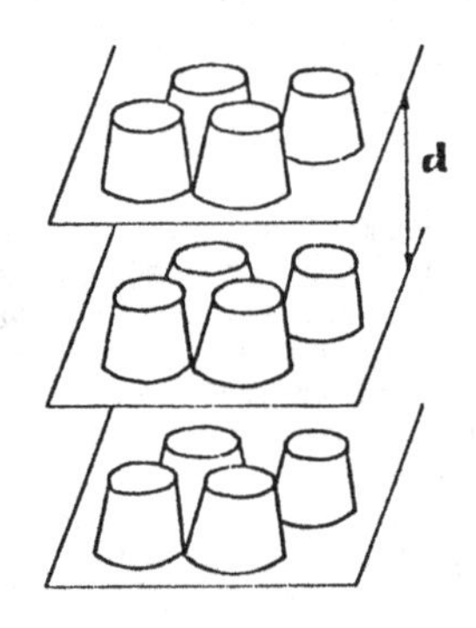

Figure 4 : Schematization of the smectic A phase with symmetry $Z \wedge C_{\infty v}$.

Three main situations can be distinguished for the symmetry changes induced by the IR's listed in Table 8, depending on the values of the wave-vector k

a) At the centre of the one dimensional Brillouin-zone (k = 0) the IR's induce no modification of the layer periodicity d. The identity representation A_1 induces also no modification of the rotational symmetry. The corresponding isomorphous transition (see chapter IV) may be interpreted as a reentrant smectic A-smectic A transition.

A_2 leads to the chiral smectic phase of symmetry $S' = Z \wedge C_{\infty}$, with a loss of the planes $\sigma(\phi)$ perpendicular to the layers. This results are obtained directly by considering for A_1 and A_2, the elements of the $S = Z \wedge C_{\infty v}$ group which leave invariant the one dimensional probability density variation :

$$\delta\rho = \eta\phi \qquad (4.8)$$

and the corresponding standard one-dimensional order-parameter expansion.

For the two-dimensional IR, denoted E_{r1}, the elements of S are represented by the matrices :

$$\{C(\phi)|d\} \to \begin{bmatrix} e^{ir\phi} & 0 \\ 0 & e^{-ir\phi} \end{bmatrix} \quad \text{and} \quad \{\sigma(\phi)|d\} \to \begin{bmatrix} 0 & e^{2ir\phi} \\ e^{-2ir\phi} & 0 \end{bmatrix} \qquad (4.9)$$

using the projection technique detailed in chapter II, we find the second and fourth degree invariants of the two-component order-parameter (η_1,η_2) :

$$\eta_1\eta_2 \quad \text{and} \quad \eta_1^2\eta_2^2 \qquad (4.10)$$

Such invariants refer to the basis (ϕ_1,ϕ_2) from which the complex matrices (4.9) are expressed. In the basis :

$$\phi_1' = \phi_1 + i\phi_2 \quad , \quad \phi_2' = \phi_1 - i\phi_2 \qquad (4.11)$$

one gets the usual invariants :

$$\eta_1^2 + \eta_2^2 \quad \text{and} \quad (\eta_1^2 + \eta_2^2)^2 \qquad (4.12)$$

for the order-parameter expansion $F_1(\eta_1,\eta_2)$. It can be noted that no anisotropic invariant figures in $F_1(\eta_1,\eta_2)$, as the result of the isotropy of the $C_{\infty v}$ group in the (x,y) plane at k = 0. Minimisation of $F_1(\eta_1,\eta_2)$, shows that there exist only one stable state ($\eta_1=\eta_2\neq 0$) with non-zero values for the order-parameter components (η_1,η_2). The corresponding symmetry group S', subgroup of S, is obtained by selecting the matrices (4.9) which leave invariant, for r fixed, the probability density :

$$\delta\rho = \eta_1\phi_1' + \eta_2\phi_2' \qquad (4.13)$$

one easely finds that only the matrices with $\phi = \frac{2\pi}{r}$ may leave invariant $\delta\rho$. Thus E_{11}, E_{21}, E_{31},...., E_{r1} will respectively induce the groups $S' = Z \wedge G'$, with $G' = C_s$, C_{2v}, C_{3v},..., C_{rv}.

b) Transitions corresponding to the IR's on the surface of the Brillouin-zone ($k = \pm\frac{\pi}{d}$) undergo a doubling of the layer periodicity. In terms of the matrices associated with the IR's, it appears in the fact that the parent phase elementary translation d, corresponds to a multiplication factor $e^{ikd} = -1$, so that the d translation is "lost" at the transition and replaced by the elementary translation 2d, associated with the factor $e^{2ikd}=+1$ From the numbers representing the one-dimensional IR'S A_3 and A_4, one can conclude that the symmetry of each layer remains

$C_{\infty v}$ for A_3, while for A_4, it is lowered to C_∞, <u>with the appearance of glide planes</u> $\{\sigma(\phi)^4|d\}$. Such glide planes belong also to the discrete low-symmetry group obtained from the two-dimensional IR E_{r2}. Using the Shubnikov notation [12], which is more adapted to such results, one can denote the group induced by E_{r2} :

$$S' = m.(2r)_r(\tilde{2}d) \tag{4.14}$$

where $(2r)_r$ indicates an axis of order r which is also a screw axis of order 2r, and $(\tilde{2}d)$ a primitive translation (2d) and a non primitive translation d, which combines with the miror plane m (i.e. a glide plane). Bilayer smectics have been observed in a large number of LC systems [36-38].

The matrices associated with E_{r2}, which allow to work out the symmetry (4.14) can be written :

$$\{C(\phi)|Pd\}\rightarrow\begin{bmatrix}(-1)^P e^{ir\phi} & \\ 0 & (-1)^P e^{-ir\phi}\end{bmatrix} \text{ and } \{\sigma(\phi)|Pd\}\rightarrow\begin{bmatrix}0 & (-1)^P e^{2ir\phi}\\ (-1)^P e^{-2ir\phi} & 0\end{bmatrix} \tag{4.15}$$

one can easely verify that the same invariants (4.12) than for E_{r1} are constructed, and thus the same order-parameter expansion $F_1(\eta_1,\eta_2)$ which has a single minimum for $T < T_c$ corresponding to $\eta_1 = \eta_2$. Among the matrices (4.15), the ones which leave the probability density (4.13) invariant are those with :

1) $P = 1$, $\phi = \frac{\pi}{r}$, i.e. a screw axis of order 2r $\{C_{2r}|d\}$, and a glide plane $\{\sigma_{\pi r}|d\}$ 2) $P = 2$, $\phi = \frac{2\pi}{r}$, i.e. an axis of order r $\{C_r|2d\}$, and a mirror plane $\{\sigma_{2\pi r}|2d\}$.

It can be pointed out, that for all the preceding IR's (except A_1) at $k = 0$ or $k = \pm\frac{\pi}{c}$ no cubic invariant of the order-parameter can be constructed. In other words, the Landau condition is fulfilled and the corresponding transitions can be continuous. One can also check that the Lifshitz condition is also satisfied, so that one should not expect any inhomogeneity (modulation) in the amalgamation of the layers, or among the distribution of the subunits within the layers, for the low-temperature smectic phases. By contrast, phase transitions associated with the two and four-dimensional representations E_{k1}, E_{k2} and G_{rk} lead generally to an incommensurate ordering as it will be shown now.

c) The matrices associated with the IR's E_{k1} and E_{k2}, which are constructed from A_1 and A_2 for k lying inside the Brillouin-zone, are :

$$\begin{bmatrix}e^{ikPd} & 0\\ 0 & e^{-ikPd}\end{bmatrix} \text{ for } \{C(\phi)|Pd\} \text{, and } \varepsilon\begin{bmatrix}e^{ikPd} & 0\\ 0 & e^{-ikPd}\end{bmatrix} \text{ for} \{\sigma(\phi)|Pd\} \tag{4.16}$$

with $\varepsilon = +1$ for e_{k1} and $\varepsilon = -1$ for E_{k2}. Two cases must be distinguished:

1) $k = \pm\frac{2\pi}{nd}$, n integer > 2. From (4.16) one can find the order-parameter

expansion :

$$F_1(\eta_1,\eta_2)=\frac{\alpha}{2}(\eta_1^2+\eta_2^2)+\frac{\beta}{4}(\eta_1^2+\eta_2^2)^2 + \delta(\eta_1 \frac{\partial \eta_2}{\partial z} - \eta_2 \frac{\partial \eta_1}{\partial z}) + F_{AN}(\eta_1,\eta_2) \quad (4.17)$$

where F_{AN} contains anisotropic invariants of the order-parameter (η_1,η_2). More precisely :

$$F_{AN} = \gamma_1 \eta_1^{2n} + \gamma_2 \eta_2^{2n} \quad (4.18)$$

for E_{k1}, and for E_{k2} if n is even, while if n is odd for E_{k2}, F_{AN} has the form :

$$F_{AN} = \gamma_1\eta_1^n + \gamma_2\eta_2^n \quad (4.19)$$

Using the results obtained for two-component order-parameter expansions in Chapters II and V, one can draw the following conclusions for the minimization of (4.17): if the transition is continuous or close to the second order, it will generally lead to an incommensurate structure. Let us note that here the modulation necessarily occurs in the stacking of the layers (i.e. the interlayer distance is modulated incommensurately). For large values of the order-parameter components at lower temperatures, a lock-in phase will eventually take place. From (4.16) it is easy to see that the commensurate symmetry of this lock-in phase is in Shubnikov notation :

$$S' = \infty \,.\, m\ (nd) \quad (4.20)$$

i.e. the $C_{\infty v}$ parent smectic phase symmetry is preserved but a multilayer ordering takes place, the periodicity along the z axis becomes nd.

If the transition is first-order, it will generally lead directly (see chapter IV) to the preceding low-temperature lock-in commensurate phase.

2) $k = \pm \frac{2\pi}{xd}$, with x irrational number > 2. The F_{AN} term vanishes in (4.17). One can predict a transition to an incommensurate phase characterized by a modulation of the interlayer distance, the stability of which will depend on the temperature variation of the wave-vector and of its eventual locking at a rational fraction of $\pm \frac{2\pi}{d}$.

d) The matrices associated with the four-dimensional IR's denoted G_{rk}, are :

$$\begin{bmatrix} e^{ir\phi}e^{ikdp} & & & 0 \\ & e^{-ir\phi}e^{-ikdp} & & \\ & & e^{ir\phi}e^{-ikdp} & \\ 0 & & & e^{-ir\phi}e^{ikdp} \end{bmatrix} \text{for}\{C(\phi)|pd\},\text{and} \begin{bmatrix} 0 & & & e^{ir\phi}e^{ikdp} \\ & & e^{-ir\phi}e^{-ikdp} & \\ & e^{ir\phi}e^{-ikdp} & & \\ e^{-ir\phi}e^{ikdp} & & & 0 \end{bmatrix} \text{for}\{\sigma(\phi)|pd\} \quad (4.21)$$

The corresponding order-parameter expansion can be written in polar coordinates ($\eta_1 = \rho_1 \cos\theta_1, \eta_2 = \rho_1 \sin\theta_1$, $\eta_3 = \rho_2 \cos\theta_2$, $\eta_4 = \rho_2 \sin\theta_2$)[1] :

$$F_1(\rho_i,\theta_i) = \frac{\alpha}{2}(\rho_1^2+\rho_2^2) + \frac{\beta_1}{4}(\rho_1^4+\rho_2^4) + \frac{\beta_2}{2}\rho_1^2\rho_2^2 + \delta(\rho_1^2\frac{\partial\theta_1}{\partial z} - \rho_2^2\frac{\partial\theta_2}{\partial z}) + F_{AN} \quad (4.22)$$

where F_{AN} is formed by anisotropic invariants. For n even in $k = \frac{2\pi}{nd}$ (n > 2), one has :

$$F_{AN} = \gamma_1\rho_1^{n/2}\rho_2^{n/2}\cos\frac{n}{2}(\theta_1-\theta_2) + \gamma_2\rho_1^{n/2}\rho_2^{n/2}\sin\frac{n}{2}(\theta_1-\theta_2) \quad (4.23)$$

whereas for n odd :

$$F_{AN} = \gamma_1\rho_1^n\rho_2^n\cos n(\theta_1-\theta_2) + \gamma_2\rho_1^n\rho_2^n\sin n(\theta_1-\theta_2) \quad (4.24)$$

Thus, for r fixed, G_{rk} will induce an incommensurate phase which may be characterized both by an incommensurate modulation of the layer spacing, and by a layer to layer incommensurability of the rotational ordering. For large values of the order-parameter components (i.e. if the transition is first-order, or at lower temperatures) a lock-in phase may appear for $k = 2\pi/nd$ (n > 2). A minimization of F_1 given by (4.22) shows that a commensurate phase possesses two possible symmetries, corresponding respectively to the equilibrium values of the order-parameter components : I $\eta_1=\eta_2\neq 0$, $\eta_3=\eta_4=0$ and II $\eta_1=\eta_2=\eta_3=\eta_4\neq 0$. These symmetries are, for fixed r, obtained for $\phi = \frac{2\pi}{r}$, and P = n. They will correspond to a n-layer ordering of smectic planes, each plane having the discrete symmetry group : C_r (for stable phase I) or C_{rv} (for stable phase II). In table 9, the symmetries associated with the various

Table 9 : Symmetries of the low-temperature smectic phases induced by the IR's of the Z ∧ $C_{\infty v}$ group (∞ .m(d)).

k = 0		k=±π/d		k = 2π/nd (n > 2)	
A_1 :	∞.m(d)	A_3 :	∞.m(2d)	E_{k1} :	∞.m(nd)
A_2 :	∞(d)	A_4 :	∞($\tilde{2}$d)	E_{k2} :	r(nd)
E_{r1} :	r.m(d)	Er_2 :	m.$(2r)_r(\tilde{2}d)$	G_{rk} :	r .m(nd)

IR's of the Z ∧ $C_{\infty v}$ group are summarized using the Shubnikov notation.

4.4.2. Transitions from the nematic group $S = R \wedge D_{\infty h}$

Let us discuss the transitions that may take place from the nematic phase with space-group $S = R \wedge D_{\infty h}$ assuming that the continuous translational symmetry is either unbroken, or broken only along the direction of the molecular axis, i.e. the transition leads to a nematic, cholesteric or 1-d smectic phase. In this respect, one has to consider (see §.3) a one dimensionalfictitious Brillouin-zone and consider two classes of IR's : 1) at $k = 0$: the continuous translational symmetry is preserved, and only the rotational symmetry may be broken at the transition. 2) at $k \neq 0$: a discrete translational order takes place corresponding to an interlayer distance $d = \frac{2\pi}{k}$. The character of the IR's corresponding to $k = 0$, and $k \neq 0$, are given in Tables 4 and 5 respectively. Following a procedure, similar to the one developed in the preceding example, we will now determine the symmetry modifications induced by such IR's.

a) As shown in Table 4 there exist four one-dimensional IR's, denoted A_1 to A_4, and two-dimensional IR's, labelled E_{r1} and E_{r2}, associated with the group S at $k = 0$. As no breaking of the continuous translational symmetries is involved, we have to determine only the subgroups G' of $G = D_{\infty h}$ induced by each IR. For the identity IR, G' = G, so A_1 may be used for the interpretation of a reentrant nematic-nematic transition [39]. From the characters of A_2 and A_3, it is clear that the subgroups G' connected with these IR's are $C_{\infty h}$ and $C_{\infty v}$ respectively, i.e. we deal with transitions between nematic phases with a lowering of the molecular packing symmetry. In contrast, A_4 can be used for the interpretation of a nematic-cholesteric transition, due to the existence in its one-component order-parameter expansion of an invariant of the form :

$$\eta u_z \left(\frac{\partial u_y}{\partial x} - \frac{\partial u_x}{\partial y}\right) \sim \eta \vec{u} \cdot \text{curl } \vec{u} \qquad (4.25)$$

Such an invariant can be worked out by considering that the director components u_z and (u_x, u_y) transform respectively as the IR's A_3 and E_{11} (and the order-parameter η, as A_4). It can be shown that terms of the type (4.25) play the role of a Lifshitz invariant and produce inhomogeneities in the spatial dependence of the order-parameter. Accordingly A_4, will be connected with the onset of a nematic phase displaying a modulation along z, i.e. a cholesteric phase. Let us note that when neglecting the invariant (4.25), A_4 would induce a chiral nematic phase of symmetry D_∞.

The matrices associated with E_{r1} and E_{r2} are similar to the ones given in (4.9) and (4.15), which allow construction of the same invariants (4.12). Thus, E_{r1} will induce a unique nematic stable phase of discrete symmetry D_{rh} (for r fixed), while E_{r2} leads for each value of r to a nematic phase of symmetry D_{2rd}, which possesses rotation-reflections of order 2r. Such discrete nematic phases may correspond to biaxial nematics.

b) The two IR's E_{k1} and E_{k2} corresponding to $\vec{k} \neq \vec{0}$ are associated with the following matrices :

$$\pm \begin{bmatrix} e^{ikd} & 0 \\ 0 & e^{-ikd} \end{bmatrix} \quad \text{for } \{C(\phi)|d\} \text{ and } \{\sigma(\phi)|d\}$$

and

$$\pm \begin{bmatrix} 0 & e^{ikd} \\ e^{-ikd} & 0 \end{bmatrix} \quad \text{for } \{S(\phi)|d\} \text{ and } \{U_{2\phi}|d\} \qquad (4.26)$$

where $d = \frac{2\pi}{|\vec{k}|}$. For constructing the matrices (4.26) one must keep in mind that the star of the wave-vector $\vec{k}$ has two branches $\vec{k}$ and $-\vec{k}$ and that the group of the wave-vector $\vec{k}$ is $G_{\vec{k}} \equiv C_{\infty v}$. It follows from (4.26) that E_{k1} induces no breaking of the rotational symmetry, but gives rise to the onset of a discrete translational order along z, with interlayer distance d. We have thus transitions from a nematic to a smectic A phase, with rotational symmetry $D_{\infty h}$. By contrast, the smectic A phase induced by E_{k2} has the rotational symmetry $C_{\infty h}$, and a glide plane containing the non-primitive translation d. It can be noted that the order-parameter expansion corresponding to E_{k1} and E_{k2} has the form (4.17) with $F_{AN} = 0$, i.e. it contains a Lifshitz invariant. Accordingly, the transition may possibly lead to a modulated smectic A phase.

The four-dimensional IR denoted G_{rk} in Table 5 correspond to the matrices (4.21) for $\{C(\phi)|pd\}$ and $\{\sigma(\phi)|pd\}$, and to the matrices generated by :

$$\begin{bmatrix} 0 & 0 & 1 & 0 \\ 0 & 0 & 0 & 1 \\ 1 & 0 & 0 & 0 \\ 0 & 1 & 0 & 0 \end{bmatrix} \qquad (4.27)$$

The corresponding order-parameter expansion has the form (4.22) with $F_{AN} = 0$. Depending on the values of the coefficients β_1 and β_2, it leads to five possible low-temperature phases with discrete symmetries : $\{C_r|d\}$, $\{D_r|d\}$, $\{C_{rh}|d\}$ $\{C_{rv}|d\}$ and $\{D_{rh}|d\}$ which correspond to smectic phases with interlayer distance d, and a bond-orientational order within the smectic planes (see §.7). As a Lifshitz invariant figures in the expansion (4.22), one may have a modulation of the smectic layers (see §.6). The results of the symmetry analysis performed for the IR's of the nematic group $S = R \wedge D_{\infty h}$ are summarized in Table 10

Table 10 : Symmetries induced by the IR's of the nematic group $S = R \wedge D_{\infty h}$. N, C and S denote a nematic, cholesteric and smectic phase respectively. The Shubnikov notation is used [8,12]

k = 0	A_1 : m.∞ : m(N)		k ≠ 0	E_{k1} : m.∞ : m(d)	(S)
	A_2 : ∞ : m(N)			E_{k2} : ∞ : m($\tilde{d}$)	(S)
	A_3 : ∞.m (N)			G_{rk}: r(d); r : 2(d); r : m(d); r.m(d); m.r : m(d)	(S)
	A_4 : ∞ : 2(N or C)				
	E_{r1} : m.r : m(N)				
	E_{r2} : m.($2\tilde{r}$) : 2(N)				

5. APPLICABILITY OF THE LANDAU THEORY TO PHASE-TRANSITIONS IN LIQUID CRYSTALS

The procedure followed in the preceding section can be applied to all nematic and smectic groups (continuous or discrete). This work has been performed recently [40-45] and is partially reproduced in Tables 11 and 12 for the groups possessing a continuous rotational symmetry. It allows, by comparison with the experimental data, to draw a number of conclusions regarding the applicability of the Landau theory.

Table 11 : Transitions from the nematic groups $R \wedge C_\infty$, $R \wedge D_\infty$, $R \wedge C_{\infty v}$ and $R \wedge C_{\infty h}$, to nematic, cholesteric or 1-d smectic phases. The notations have the same meaning as in Table 10.

	$R \wedge C_\infty$	
k = 0	A_1	∞ (N or C)
	E_{r1}	r (N or C)
k ≠ 0	E_k	∞(d) (S)
	E_{rk1}, E_{rk2}	r(d) (S)

	$R \wedge D_\infty$	
k = 0	A_1	∞ : 2 (N or C)
	A_2	∞ (N)
	E_{r1}	r : 2 (N or C)
	E_k	∞ : 2 (d) (S)
	G_{rk}	r(d); r : 2(d) (S)

	$R \wedge C_{\infty v}$	
k = 0	A_1	∞.m(N)
	A_2	∞(N or C)
	Er_1	r.m (N)
k ≠ 0	E_{k1}	∞.m(d) (S)
	E_{k2}	∞ ($\tilde{d}$) (S)
	G_{rk}	r.m(d); r(d) (S)

	$R \wedge C_{\infty h}$	
k = 0	A_1	∞ : m (N)
	A_2	∞ (N or C)
	E_{r1}	r : m (N)
	E_{r2}	r (N)
k ≠ 0	E_k	∞ : m(d) (S)
	G_{rk}	r : m(d); r(d) (S)

Table 12 : Transitions between 1-d-smectic groups, for the parent smectic groups $Z \wedge C_\infty$, $Z \wedge D_\infty$, $Z \wedge C_{\infty h}$. The notations are the same as in §.4 $(n_o d)$ means : primitive translation of length $n_o d$. $(n_o r)_r$ means:an axis of order r which is also a screw axis of order $n_o r$. The results for $Z \wedge D_{\infty h}$ are given in Ref.[9].

$Z \wedge C_\infty$	
$k = 0$	
A_1	$\infty(d)$
E_{r1}	$n(d)$
$k = \pm\frac{\pi}{d}$	
A_2	$\infty(2d)$
E_{r2}	$(2n)_n(2d)$
$k = \frac{2\pi}{n_o d}$	
E_{k1}	$\infty(n_o d)$
E_{rk1}	$(n_o n)_n(n_o d)$
E_{rk2}	$(n_o n)_n(n_o d)$

$Z \wedge D_\infty$	
$k = 0$	
A_1	$\infty: 2(d)$
A_2	$\infty(d)$
Er_1	$r : 2(d)$
$k = \frac{\pi}{d}$	
A_3	$\infty : 2(2d)$
A_4	$\infty : 2(2d)$
E_{r2}	$(2n)_n:2(2d)$
$k = \frac{2\pi}{n_o d}$	
E_k	$\infty:2(n_o d)$ or $\infty(n_o d)$
G_{rk}	$(n_o n)_n(n_o d)$ $(n_o n)_n:2(n_o d)$

$Z \wedge C_{\infty h}$	
$k = 0$	
A_1	$\infty:m(d)$
A_2	$\infty(d)$
E_{r1}	$n:m(d)$
E_{r2}	$n(d)$
$k = \pm\frac{\pi}{d}$	
A_3 A_4	$\infty:m(2d)$
E_{r3} E_{r4}	$(2n)_n:m(2d)$
$k = \frac{2\pi}{n_o d}$	
E_k	$\infty:m(n_o d)$ or $\infty(n_o d)$
G_{rk}	$(n_o n)_n(n_o d)$ $(n_o n)_n:m(n_o d)$

Let us analyse the results given in the preceding tables at the light of the main experimental phenomena involving nematic and smectic phases which have been evidenced in the recent years. A number of questions are left for a more detailed discussion in §.6 to §. 9 .

5.1. The nematic to smectic A transition

The nematic-smectic A (N-A) transition has been the subject of intense theoretical and experimental efforts since Kobayashi [46] and Mc Millan [47] predicted, on the basis of mean field calculations, that the transition should be first order unless $T_{AN}/T_{NI} < 0.87$, where T_{AN} and T_{NI} are respectively the N-A and Isotropic-nematic transition temperatures. Thus for a special value of the molecular length corresponding to $T_{AN}/T_{NI} = 0.87$ a tricritical point was predicted [48]. The interest for the transition was still enhanced by the subsequent work of Gennes [2] who stressed analogies in the phenomenological description of the N-A transition and of the superfluid and superconducting transitions : in the Landau-type model of de Gennes the two-dimensional order-parameter of the smectic phase is a complex phase function $\phi(\vec{r})\ e^{2i\pi z/d}$, similar to the phase funciton in superfluids, which couples to the fluctuations in the director of the nematic phase, in the same way that the electromagnetic vector potential couples to the superconducting order-parameter. Thus the order of the transition depends on the strength of the coupling.

The existence of two possible regimes (2nd and first-order) and of a tricritical point received a number of experimental confirmations (e.g. in binary mixtures of 80.3 and 20.3 or 9 CB- 10 CB) [49-53]. Theory [54-56] and experiments [53,57] showed also that the critical behaviour at T_{NA} was more complex than predicted by de Gennes model (i.e. a behaviour governed by a two-component Heisenberg fixed point) : the transition is never under the influence of a single fixed point, and there exist an anisotropic critical behaviour characterized by two different critical exponents associated with different correlation lengths.

In the framework of the unified phenomenological approach presented in this chapter, one can see from tables 10 and 11 that the transition is necessarily connected with the two dimensional IR's denoted E_{k1} and E_{k2} for the nematic groups $R \wedge D_{\infty h}$ and $R \wedge C_{\infty v}$, and E_k for the other nematic groups. E_{k1} and E_{k2} can be distinguished by the fact that for E_{k2}, the smectic A phase possesses a glide plane ($\hat{d}$). Before developing the corresponding thermodynamic model let us emphasize that the concept of a fictitious Brillouin-zone, assumed for the construction of the IR's associated with the nematic phase, is here fully justified experimentally by the observation, in the vicinity of the N-A transition, of significant fluctuations leading to the formation of small smectic clusters, so called cybotactic groups [58], that were found, for example, in the mixtures 5 CB-6050 [59] and also by the evidence of smectic A layers in the bulk nematic phase, with a penetration depth equal to the longitudinal smectic A correlation length [60]

As the primary order-parameter for the N-A transition is the orientational ordering, i.e. the second rank tensor Q_{ij} given by (4.2), one has to include the isotropic phase in the phase diagram containing the N-A transition line, as non-zero equilibrium values of Q_{ij} appear at the isotropic-nematic (I-N) transition. Accordingly the main features of the N-A transition, namely the possibility for the transition to be second or first order, the existence of a tricritical point separating the two regimes etc..., will be described by the two order-parameter potential :

$$F(Q_{ij}, \phi) = F_1 (Q_{ij}) + F_2(\phi) + F_3 (Q_{ij}, \phi) \tag{5.1}$$

where ϕ is the secondary two-component order-parameter (4.4) associated with the layer ordering in the smectic A phase, which transforms as E_{k1} or E_{k2}, and identifies to the mass density wave introduced by de Gennes [2].

In §.9, we show that $F_1(Q_{ij})$ can take the simplified form:

$$F_1(Q_{ij}) = \frac{\alpha}{2} Q^2 + \frac{B}{3} Q^3 + \frac{\beta}{4} Q^4 + \ldots \tag{5.2}$$

where $Q(\vec{r})$, defined by Eq (4.2), accounts for the orientational order in the nematic and smectic A phases, i.e. it increases continuously as the temperature is lowered from the I-N transition.

The transformation properties of E_{k1} or E_{k2} provide the following form for $F_2(\phi)$:

$$F_2(\phi) = \frac{a}{2}\phi_o^2 + \frac{b}{4}\phi_o^4 \qquad (5.3)$$

where ϕ_o is the amplitude of the order-parameter ($\phi = \phi_o e^{i\theta(z)}$). Let us note that E_{k1} and E_{k2} allow existence for a Lifshitz invariant of the form $\phi_o^2 \frac{\partial\theta}{\partial z}$ which is neglected in (5.3), otherwise it would describe an incommensurate modulation of the smectic layers (see §.6). In other words the assumption is made that an homogeneous smectic A phase takes place.

We will write the mixed term $F_3(Q_{ij}, \phi)$ as :

$$F_3(Q_{ij}, \phi) = -Q\,\phi_o^2\left(\frac{\delta_1}{2} + \frac{\delta_2}{4}\phi_o^2\right) \qquad (5.4)$$

The two invariants $\delta_1 Q\phi^2$ and $\delta_2 Q\phi^4$ are allowed by the respective symmetries of the IR's inducing the I-N, and N-A transitions. We have restricted ourselves to select the lower degree invariants with respect to the orientational parameter Q, as this parameter is supposed to have large values at the N-A transition. Besides, the coupling terms (5.4) allow to show the existence on the N-A transition line, of a tricritical point separating a first-order, and a second-order transition regimes. δ_1 and δ_2 are assumed to be positive, which means that additional orientational order induced by the appearance of the smectic layers, corresponds to an increase of the molecular interactions, i.e. it decreases the energy of the system.

In the expansion (5.1) only one coefficient is assumed to vary with temperature, namely $\alpha = \alpha_o(T-T_c)$, whereas the other coefficients remain constant. Without loss of generality we will take $B < 0$, $\beta > 0$, $a > 0$, $b > 0$. Minimizing F_1 gives the equilibrium value of Q below the I-N transition :

$$Q^e = \frac{-B + (B^2 - 4\,\alpha\beta)^{1/2}}{2\beta} \qquad (5.5)$$

Non-zero values of Q^e appear at the first-order I-N transition, which takes place at the temperature (see chapter IV)

$$T_{I-N} = T_c + \frac{2\,B^2}{9\,\alpha_o\beta} \qquad (5.6)$$

with a corresponding discontinuity of Q^e equal to :

$$\Delta Q^e = -\frac{B}{3\beta} \qquad (5.7)$$

In addition to the preceding stable state with ($Q^e \neq 0$, $\phi_o^e = 0$) minimization of (5.1) provides two other possible stable phases corresponding to ($Q^e = 0$, $\phi_o^e \neq 0$) and ($Q^e \neq 0$, $\phi_o^e \neq 0$). The state ($Q^e = 0$, $\phi_o^e \neq 0$) has no physical meaning in the present case, as it would be associated with a layer ordering without orientational order. ($Q^e \neq 0$, $\phi_o^e \neq 0$) is actually the equilibrium conditions realized in the smectic A state. The equilibrium values for ϕ_o, are deduced for the absolute minima conditions :

$$a - \delta_1 Q^e + \phi_o^2 \quad (b - \delta_2 Q^e) = 0 \qquad (5.8)$$

$$a - \delta_1 Q^e + 3\,\phi_o^2 \,(b - \delta_2 Q^e) \; > \; 0 \qquad (5.9)$$

which show that the T_{NA} transition line defined by the vanishing of the renormalized square term coefficient :

$$a - \delta_1 \, Q^e \, (T_{NA}) = 0 \qquad (5.10)$$

is a second order transition line if the condition :

$$\frac{b}{\delta_2} \; > \; Q^e \, (T_{NA}) \qquad (5.11)$$

is fulfilled. When the inequality (5.11) is reversed, the transition is first-order, and one has to include a sixth-order invariant in (5.3) to find the equilibrium value of ϕ_o. Thus, the coordinates of the tricritical point separating the two portions of the T_{NA} line is defined by (5.10) and by the equality :

$$\frac{b}{\delta_2} = Q^e \, (T_{NA}{}^{tric}) \qquad (5.12)$$

When the discontinuity in the orientational parameter Q^e is sufficiently large, the line of the I-N transition defined by $\alpha(T_{I-N}) = \frac{2B^2}{9\beta}$ comes together with the N-A line, the two transition lines intersecting when :

$$3 \;\; \beta a + B \, \delta_1 = 0 \qquad (5.13)$$

With further increase of the Q^e parameter the smectic A phase is formed directly from the isotropic phase as a result of a first-order transition. Following ref. 47, we have represented in Figure 5, the schematic temperature T -molecular length l phase-diagram, as the coefficients in the potential (5.1) depend mainly on l.

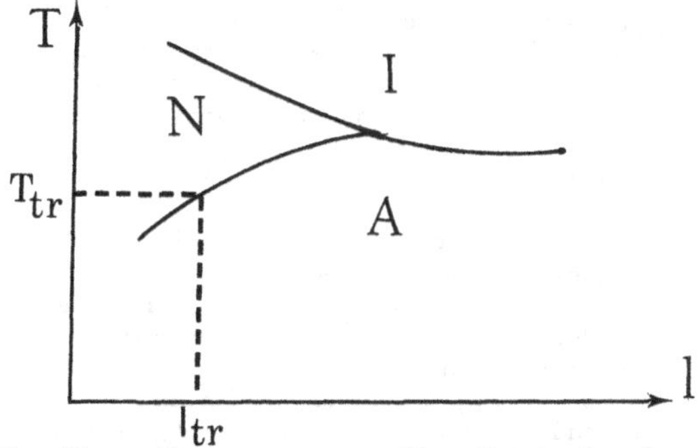

Figure 5 : Phase diagram representing the isotropic, nematic and smectic A phases.

5.2. The nematic-smectic A-smectic C phase diagram

Smectic A and smectic C L C can be described as orientationally ordered fluids with one-dimensional mass density waves either along or at an angle to the unique orientational axis. In appropriate binary LC mixtures it is possible to have lines of second order N-A and A-C transitions crossing over to a line of first-order N-C transition. The point at which the three lines meet has been labelled the NAC multicritical point [61] . First predicted theoretically by Chen and Lubensky [61] and by Chu and Mc Millan [62] this point was realized experimentally by Johnson et al [63] and by Sigaud et al [64] in binary mixtures of $\overline{7}$S5-$\overline{8}$S5 and 70 NE-80 CB respectively.

Different versions of the Landau theory were proposed for the NAC phase diagram. In one version [61,65] there exist only one order-parameter which is the density wave modulation, with a temperature dependent wave-vector. The NAC point is here a Lifshitz point of the Hornreich et al type (see chapter IV). In a second approach [62,66,67] two order-parameters are involved: the density wave modulation and the tilt angle. The model predicts the existence of a tricritical point on the A-C line near the NAC point. A comparison of the various theories with experiment show that none of the existing models can adequatly account for the behaviour over the entire phase diagram. In particular, there is a lack of qualitative agreement for each model with the data in important areas, such as the observed X-ray line shapes, the phase diagram topology, the evolution of thermal anomalies, or the correlation lengths along the nematic-smectic A phase boundaries [68-73] . The renormalization group analysis [74,75] , although predicting a more complex situation than the Landau theory, does not bring more clarification on the desagreement between theory and experiment.

In the framework of the approach presented in this chapter, the NAC phase diagram requires two coupled order-parameters :
1) a primary order-parameter associated with the layer ordering taking place from the nematic phase, in the smectic A or C phases. As in §.5.1 it corresponds to the density function $\phi = \phi_o e^{i\vec{k}\vec{r} + \omega}$ and its complex conjugate, and transforms as the IR denoted E_{k1} or E_{k2} in Tables 10 and 11. 2) a secondary order-parameter which characterizes the tilting order in the C phase, and transforms as the product (see §.7) :

$$\vec{u}.\vec{k} \, . \, [\vec{u} \wedge \vec{k}]\phi \qquad (5.14)$$

which defines the tilting angle θ between the director $\vec{u}$ and the wave vector $\vec{k}$. (5.14) transforms as the product of the IR denoted E_{11} in Table 12 and ref. 8, and of E_{k1} or E_{k2}. In the three phases the degree of orientational order Q is supposed to remain constant (i.e. a more refined model considering an increase of Q when the temperature is lowered implies to include the isotropic phase in the diagram).

Assuming the tilt angle θ to be small, and a constant wave vector $\vec{k} \simeq k_o$, one can approximate (5.14) by $\theta k_o^2 \phi_o \simeq \theta\phi_o$ so that the thermodynamic potential can be written as the sum :

$$F(\phi_o,\theta) = F_1(\phi_o) + F_2(\theta\phi_o) + F_3\,(\phi_o,\theta\phi_o) \qquad (5.15)$$

F_1 describes the nematic-smectic A transition and identifies to (5.3). F_2 corresponds to the smectic A-smectic C transition and can be written as :

$$F_2\,(\theta\phi_o) = \frac{a'}{2}\,\theta^2\phi_o^2 + \frac{\theta^4}{4}\,(b'\phi_o^2 + c'\,\phi_o^4) \qquad (5.16)$$

$F_3\,(\phi_o,\theta\phi_o)$ expresses the coupling between the ordering parameters in the smectic A and smectic C phases. It transforms as the product $E_{ki} \times E_{11}$. Restricting to the lower degree term, one can write :

$$F_3\,(\phi_o,\theta\phi_o^e) = -\,\delta\theta^2\,\phi_o^4 \qquad (5.17)$$

All the coefficients, except a (in (5.3)) and a' (in (5.16)) are assumed to be positive constants. The positiveness of the δ coefficient means that a uniform tilt of the molecules in the C phase corresponds, on the average to an increase of the attraction between the molecules in the layers, and correspondingly, to an increase of the amplitude ϕ_o of the density wave. The positiveness of the fourth degree coefficient b and b' significate that the nematic-smectic A, and smectic A-smectic C transitions can take place as second-order transitions. The temperature dependence of a and a' are :

$$a(T) = a_o(T-T_o) \quad \text{and} \quad a' = a_o(T-T_o'), \quad \text{with } a_o > 0, a_o' > 0 \qquad (5.18)$$

where T_o and T_o' are functions of the concentration x in the mixtures. For a certain value $x = x_o$, it is assumed that :

$$T_o\,(x_o) = T_o'\,(x_o) = T_c \qquad (5.19)$$

while for $x < x_o$, the inequality $T_o > T_o'$ is realized.

Minimization of (5.3), (5.16) and (5.17) with respect to ϕ_o and θ provide the equilibrium states and the transition temperatures Thus, for $x < x_o$ the N-A and A-C transition temperatures are :

$$T_{NA} = T_o \qquad (5.20)$$

and

$$T_{AC} = \frac{\delta a_o\,T_{NA} + a_o'\,b\,T_o'}{\delta a_o + a_o'\,b} \qquad (5.21)$$

one can verify that for $x = x_o$, one has $T_{NA}\,(x_o) = T_c$.

For $x > x_o$, the N-A transition does not occur since the onset of the density wave ϕ_o is always accompanied by a spontaneous non-zero tilt θ of the director $\vec{u}$. Minimizing (5.16) with respect to θ, one gets the equilibrium value for θ :

$$\theta^e = \pm\,(-\,\frac{a'}{b'}\,)^{1/2} \qquad (5.22)$$

reintroducing (5.22) in (5.15) renormalizes the coefficient of the square term ϕ_o^2. At the N-C transition one has the relationship :

$$a'^2(T_{NC}) - 4a\,(T_{NC})b' = 0 \qquad (5.23)$$

and the equilibrium value for ϕ_o in the C phase :

$$\phi_o^e = \pm \left[\frac{b'\,(a'^2 - 4\,ab')}{4\,(a'^2\,c' + \delta a'b' + bb'^2)}\right]^{1/2} \qquad (5.24)$$

Thus the N-C transition occurs at a second order transition, the amplitude ϕ_o^e arizing continuously at T_{NC}, the density wave making a finite angle $|\theta^e|$ with the director $\vec{u}$. The tilt angle θ^e given by (5.22) increases with increasing concentration x, and $\theta^e(x_o) = 0$. The preceding results, which are summarized in Figure 6, hold only if the conditions :

$$a'^2 > 4\,ab' \qquad (5.25)$$

$$b^* = b + \frac{a'\,(a'c' + \delta b')}{b'^2} > 0 \qquad (5.26)$$

are fulfilled. If the inequality (5.26) is reversed (i.e. if the fourth degree ϕ_o^4 renormalized coefficient b* in the thermodynamic potential becomes negative at a given concentration), the nematic-smectic C and smectic A-smectic C transitions become first-order. One has then to include in the expansion (5.3) a sixth degree term $\frac{c}{6}\phi_o^6$ with $c > 0$, for a phenomenological description of the NAC phase diagram. If b* changes sign for $x_c > x_o$, then a tricritical point takes place on the line T_{NC} (x) defined by Eq.(5.23). This point is determined both by Eq (5.23) and by $b^*(x_c) = 0$. A different situation arises if b* becomes negative when $x_c < x_o$. In this case a tricritical point arises on the line of second order A-C transition defined by Eq.(5.21). Accordingly for $x > x_c$ the A-C transition occurs as a first order transition 8. The various possibilities depictated in figures 6 and 7 cover the [63,64,68-73] obtained experimentally.

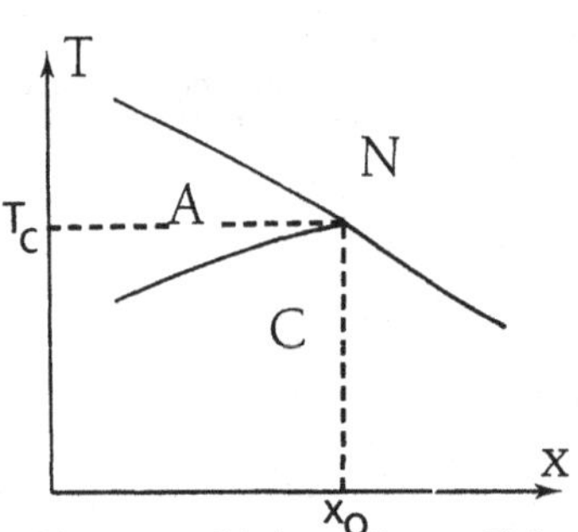

Figure 6 : NAC phase diagram T(x): T_{NA}, T_{AC} and T_{NC} are lines of second order transitions.

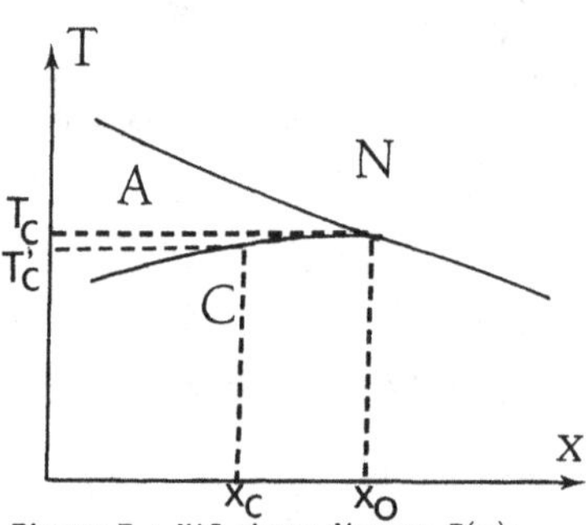

Figure 7 : NAC phase diagram T(x): the T_{AC} line becomes a first-order transition line for $x_c < x < x_o$. The T_{NC} line is first order for $x > x_o$

5.3. Reentrant nematic phases

The partial sequence of transitions Nematic (N)-Smectic A (SmA)-Nematic (N_{re}) was observed for the first time by Cladis [39] in the mixture HBAB-CBOOA, then in pure compounds by Cladis et al [76], and Liebert and Daniels [77]. Various sequences of transitions where nematic phases alternate with bilayer, partially bilayer, monolayer SmA or SmC phases were reported in other mixtures or pure compounds such as CBHHA-CBNA, 8OCB-6OCB, 9OBCAB, DB_9ONO_2 or CPMCB homologues [78-85] . A one-component order-parameter model, assuming an optimum density for the smectic phases, and a sixth-degree invariant for the thermodynamic potential, was proposed by Pershan and Prost [86]. Empirical thermodynamic [87] or lattice [88] models were also discussed.

If one assumes that the N and N_{re} phases have identical structures (a fact that seems to be supported by the similirity of the pretransitional phenomena observed when approaching the intermediate SmA phase either from the high or from the low temperature side), the theoretical results contained in tables 10 and 11 suggest two possible interpretations for the N-SmA-N_{re} succession of phases :

1) a single order-parameter model which involves the two-dimensional IR denoted E_{k1} in Tables 10 and 11. For $\vec{k} \neq \vec{0}$, E_{k1} induces a layered smectic phase having the same point group symmetry than the parent nematic phase. Assuming that the k-vector varies with temperature (or pressure or composition) one will get isomorphous nematic phases (i.e. reentrant nematic phases) when $\vec{k}$ drops to zero, while for $\vec{k} \neq \vec{0}$ one may obtain bilayered ($k = \frac{\pi}{d}$, monolayered ($k = \frac{2\pi}{d}$) or partially bilayered ($\frac{\pi}{d} < k < \frac{2\pi}{d}$) smectic A phases. The transition free-energy can be written as (4.17). For $\vec{k} \neq \vec{0}$, the δ and F_{AN} terms in (4.17) vanish by symmetry, while only the δ term vanishes for values of k corresponding to the mono-or bilayer smectic A phase. In this latter case F_{AN} is equal to (4.18) or (4.19). For a partially bilayered smectic phase $F_{AN} = 0$ and the δ term is non-zero.

2) A coupled order-parameter model in which two IR's are involved, namely E_{k1} and the identity IR denoted A_1 in tables 10 and 11. Depending on the degree n of the one-component order-parameter potential associated with A_1, one gets one (n = 4), or two (n = 8) reentrant nematic phases (see chapter IV). E_{k1} with $\vec{k} \neq \vec{0}$ determines the onset of the intermediate smectic modifications.

It should be noted that for the IR's associated with the nematic groups $R \wedge C_\infty$ and $R \wedge D_\infty$, as an invariant of the type (4.25) is allowed by symmetry, one may interpretate sequences of cholesteric reentrant phases which have been reported by Billard [89], Tinh et al [90] and Vaz et al [91].

5.4 The uniaxial to biaxial nematic transition

Predicted on a theoretical basis by Freiser [92] , and later by Alben [93] , the uniaxial to biaxial nematic transition was evidenced for the first time by Yu and Saupe [94] in the lyotropic LC, potassium laurate 1-decanol D_2O mixture, and recently in a thermotropic compound [95]. The biaxiality can be attributed to the existence of a short axis for the molecular subunits, i.e. the nematic phase has a discrete rotational symmetry. Thus, one cas straightforwardly deduce from tables 10 and 11, that the IR's inducing the uniaxial to biaxial nematic transition belongs to the classes denoted E_{r1} or E_{r2}. Therefore the corresponding order-parameter has two-dimensions, and the thermodynamic potential associated with the transition (that may be second order, as predicted by Freiser [92] , is of the form (4.17) with $F_{AN} = 0$, and $\delta = 0$ except for the chiral nematics. In this latter case, which has not yet been observed, the transition from the uniaxial nematic phase would lead to a biaxial cholesteric ordering that was discussed theoretically by Schröder [96].

5.5. Transitions between smectic phases

The symmetry modifications listed in Tables 9 and 12 (see also refs 8, 42-45 and 97) allow to interpretate the main sequences of 1-d, 2d or 3-d smectic phases.

a. Transitions between smectic A phases

They are induced by the one-dimensional IR's denoted A_i at the center or at the boundary of the one-dimensional Brillouin-zone, or by the two-dimensional IR's labelled E_k with k lying inside the Brillouin-zone. Considering the example of the parent smectic A phase with symmetry $Z \wedge C_{\infty h}$, one can see in Table 12 that A_1 corresponds to an isomorphous first-order Smectic A-Smectic A transition whereas A_2 leads to a smectic A phase with chiral symmetry $Z \wedge C_{\infty}$. For $k = \pm \frac{\pi}{d}$, A_3 and A_4 are associated with a monolayer-bilayer transition of the type SA_1-SA_2 first reported by Hardouin et al [98], and later found in a large number of LC mixtures [99,100]. For $k = \frac{2\pi}{n_o d}$ the IR labelled E_k induces a Smectic A phase possessing either a n_o -layer ordering (with $n_o > 3$) or a partially bilayer ordering ($1 < n_o < 2$) of the type reported by Levelut et al [101] and Leadbetter et al [102], which may possess (for irrational values of n_o) a long range modulation of the layer ordering [103]. The onset of a bilayer ordering below an intermediate partially bilayed phase, can be described phenomenologically as a lock-in mechanism taking place below an incommensurate structure. In the thermodynamic potential (4.17), such a mechanism results (see chapter V) from the competition of the δ and F_{AN} terms, the F_{AN} term arising by symmetry for rational values of n_o.

b. The Smectic A-smectic C transition

Although, one should include the A-C transition in the more general N-A-C phase diagram discussed in §.5.2, one can formally describe the A-C transition, as the result of the IR's E_{ri} associated with the A phase at k = 0. As shown in Table 7 for the group $Z \wedge D_{\infty h}$, E_{11} induces the symmetry $Z \wedge C_{2h}$ with the two-fold axis within the smectic layers, i.e. a smectic C type phase. In §.5.2, the corresponding tilting of the molecules is expressed as a secondary order-parameter which couples to the primary mechanism, which is the stacking of the layers. In §.8, we discuss in more details the transition to chiral smectic C* ferroelectric phases, which take place from the chiral smectic A phases $Z \wedge C_{\infty}$ and $Z \wedge D_{\infty}$.

c. Transitions from the smectic A to 2-d and 3-d ordered smectic modifications

The various smectic modifications which are listed in Table 1 can be obtained in the framework of the Landau theory, as the result of the IR's E_{ri} (k = 0) for monolayer ordering and E_{ri}, G_{rk} (k ≠ 0) for a multilayer ordering. The discrete groups induced by the preceding IR's can be interpreted as the acquisition by the molecular subunits of a well defined ordering of their short axes, i.e. an azimuthal symmetry. The correlation function $\rho_{12}(r_{12})$ which is determined from X-ray data, acquires the corresponding asymmetry at short or long distances. The resulting tilting and bond orientational orders are described as secondary order-parameters. In figure 8 we show some smectic configurations and the corresponding IR's of the parent smectic A phase. In §.7, the example of the hexatic ordering is discussed in more details. As already mentioned in §.3.2, a true long-range positional order within the smectic layers, necessitates to take into account wave-vectors pertaining to three dimensional Brillouin-zones.

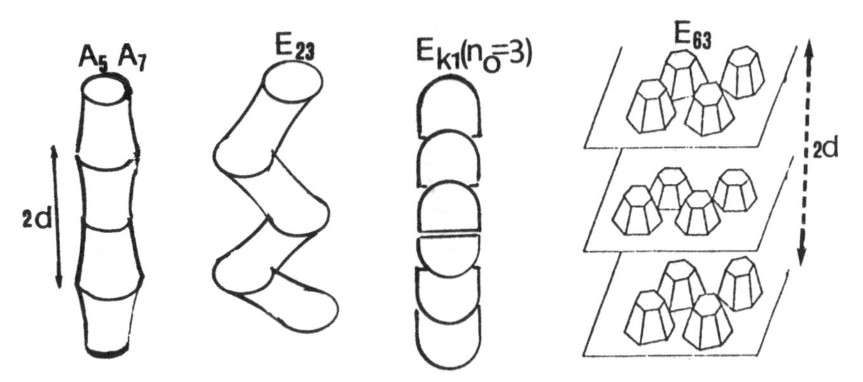

Figure 8 : Schematization of the subunit configurations induced by some IR's of the $Z \wedge D_{\infty h}$ smectic A phase at k ≠ 0.

6. MODULATED LIQUID CRYSTAL PHASES

Let us briefly analyse the conditions under which a modulated structure can be predicted in LC systems, leaving out the specific case of the ferroelectric smectic C* modulation, which is discussed in §.8.

a. Cholesteric phases

The cholesteric state is characterized by an helicoidal twisting of the director $\vec{u}(\vec{r})$ over distances large compared with the molecular length l. In this situation, which is schematized in Figure 9, $\vec{u}(\vec{r})$ is a slowly varying function of the coordinates and one has $|k|\ l \ll 1$, where $\vec{k}$ is the modulation wave-vector. By considering the ratio between the period λ of the helicoïd and the molecular length, one can speak of an incommensurate modulation, although cholesteric structures are not usually classified as incommensurate systems.

Cholesteric phases are found to take place either below the isotropic liquid, or as the result of a transition from the nematic phase. In both cases [104-107] their existence can be accounted phenomenologically by including in the thermodynamic potential density a Lifshitz invariant of the form (4.3). When the parent phase is the isotropic state the invariant appears in the potential through a coupling term of the form :

$$Q^2\ (\vec{k} + \vec{u}\ \mathrm{rot}\ \vec{u})^2 \qquad (6.1)$$

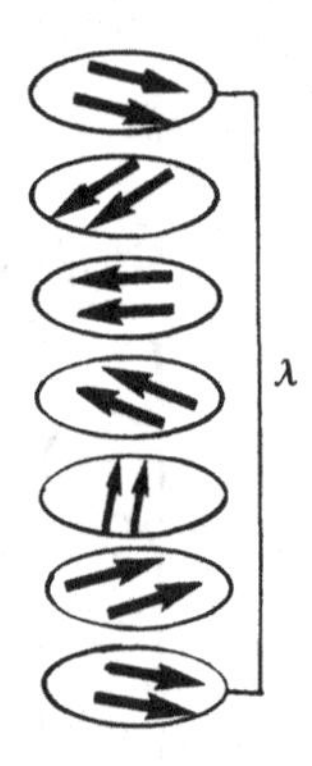

Figure 9 : Helical cholesteric structure : λ is the helical pitch. The arrows show the average orientation of the molecules

which provides after minimization, the equilibrium values :

$$Q^e \neq 0 \quad \text{and} \quad u_z = 0, \quad u_x = \cos kz, \; u_y = \sin kz \qquad (6.2)$$

expressing the simultaneous onset of the orientational order, reminiscent of a nematic state, and the twisting of the director. When the parent phase is considered as the nematic phase, Q has already non-zero values, and the Lifshitz term (4.3) appears by itself as an invariant in the thermodynamic potential density. From tables 10 and 11 one can see that cholesteric phases are always characterized by the chiral point groups C_∞ or D_∞ . Thus a cholesteric phase can be predicted to appear from symmetry considerations : i) when the parent phase correspond to the chiral nematic groups $R \wedge C_\infty$ and $R \wedge D_\infty$, for the one dimensional IR's denoted A_i (i = 1,2) and the two-dimensional IR, E_{r1}.

ii) when the parent phase is non-chiral, for the IR's inducing chiral symmetries (e.g. A_4 for $R \wedge D_{\infty h}$, or A_2 for $R \wedge C_{\infty v}$ and $R \wedge C_{\infty h}$).

Let us note that for the nematic-cholesteric transition, it can easely be shown that the pitch λ of the helix in the cholesteric state is proportional to the inverse of the order-parameter, i.e. it diverges as : $(T_c-T)^{1/2}$ where T_c is the transition temperature.

b. Modulated smectic phases

There exist a large variety of situations in which a modulated smectic phase may take place. From symmetry considerations two classes of incommensurate structures can be predicted.

i) At k = 0 or $k = \pm \frac{\pi}{d}$, for the two-dimensional IR's labelled E_{r1} and E_{r2}, associated with the chiral groups $Z \wedge C_\infty$ and $Z \wedge D_\infty$ Here, the modulation takes place in the orientational properties of the molecules, e.g. it corresponds to a twisting of the long or short axes of the molecules (Figures 10a and 10b), while the distance between the layers remain unchanged. This situation is analyzed in more details in §.8 as it is characteristic of the ferroelectric smectic C* phase.

ii) For all smectic groups for the two- and-four dimensional IR's labelled E_{ki}, Er_{ki} and Gr_{ki} associated with wave-vectors $\vec{k}$ located inside the one-dimensional smectic Brillouin-zone. Here the modulation may be associated not only with the orientational properties of the molecules but also to the stacking of the layers (Figure 10c).

The phase diagrams corresponding to the IR's E_{ki}, Er_{ki} and Gr_{ki} depend on the value of r, and of $n_o = \frac{2\pi}{|\vec{k}|d}$. They have been discussed by Indenbom and Loginov [10] , and are completely analogous to the standard diagrams discussed in Chapter V for incommensurate structural systems.

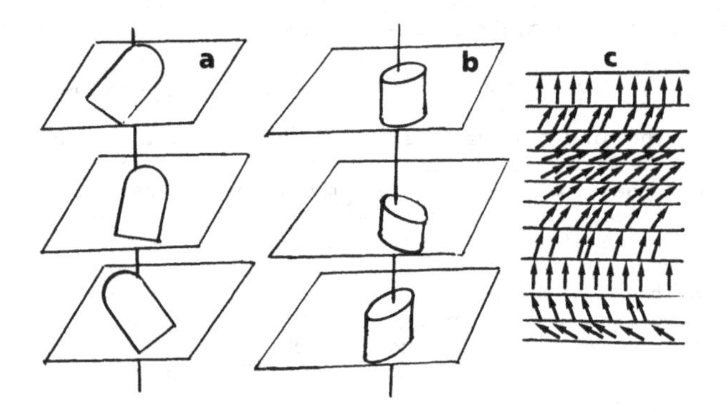

Figure 10 : Different configurations for incommensurate smectic phases. a) Twisting of the long axis. b) Twisting of the short axis. c) Incommensurate ordering in the analgamation of the layers.

A more specific type of modulated structures has been found in smectic A phases. It is obtained by an uniform stretching along the crystal axis z, and consists in periodically curved smectic layers, with preservation of the local symmetry of the phase (Figure 11). Such a situation was discussed experimentally by Clark et al [108,109]and Ribotta et al [110], and is connected with the stability of smectic A type phases under weak external perturbations, e.g. mechanical as discussed in refs. 108-110. As shown by Pikin and Indenbom [8] the contribution to the thermodynamic potential which expresses the bending of the layers under stress can be written as the coupling term :

$$F_3(Q, {}_o, e_{uv}) = \frac{\delta_1}{2} Q^2 \left(\frac{\partial e_{xz}}{\partial x} + \frac{\partial e_{yz}}{\partial y}\right)^2 + \frac{\delta_2}{2} \phi_o^2 \left(2 e_{zz} - e_{xz}^2 - e_{yz}^2\right) \tag{6.3}$$

where Q and ϕ_o are the order-parameters associated with the orientational and layer ordering respectively, and e_{uv} (u,v = x,y,z) the components of the induced strain. A term of the type (6.3) allows to show that a small uniform stretching of the smectic A phase results in an modulated smectic phase as depicted in Figure 11. As noted in ref. 8 this type of induced modulation is typical for systems where the density function is periodic in one dimension. Although they should be unstable against thermal fluctuations [111] , the elastic interactions between layers of given thickness make them stable.

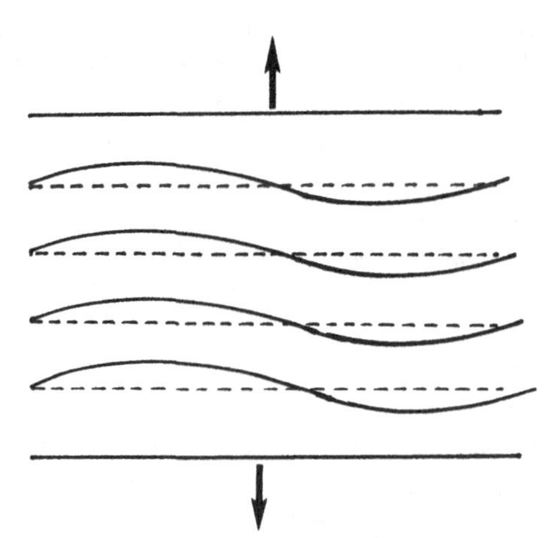

Figure 11 : Bending of the layers of a smectic A phase under uniform stretching along the z axis.

7. TILTING ORDER AND BOND-ORIENTATIONAL ORDER IN SMECTIC PHASES

From Table 1 (see also Figure 1), one can verify that most of the smectic modifications possess a long-range bond-orientational order within the smectic layers. Only for the smectic phases labelled F and B_H this order is short range, while for the A and C phases there is no definite azimuthal symmetry, i.e. the ordering of the subunit short axes is random. On the other hand, a long-range tilting order is reported for the phases denoted C, F, I, J, H and K, whereas for the A, B_H, B_c and E the long axes of the molecular subunits are, in the average, perpendicular to the layers.

In §.5.2, it has been shown that the tilting order in the C phase could be expressed phenomenologically as a secondary mechanism resulting from a coupling to the primary order-parameter which consists in the layer ordering. A similar approach can be defined in order to work out the bond-orientational ordering. As an example let us consider the Smectic A-Smectic B_c transition which corresponds to a $D_{\infty h} \rightarrow D_{6h}$ lowering in the point group symmetry of the phases. Assuming a monolayer ordering for the two phases (i.e. the transition takes place at k = 0), the preceding symmetry change can be connected with the IR denoted E_{61}, of the group $Z \wedge D_{\infty h}$ (see for example refs 8 and 42). The basis functions for the corresponding two-component order-parameter are :

$$\eta_1 = \eta_o\, e^{6i\phi} \qquad \text{and} \qquad \eta_2 = \eta_o e^{-6i\phi} \tag{7.1}$$

where ϕ is the rotation angle around the z axis, and η_o measures the average orientational order in the B_c phase, i.e. it has the meaning of an average angle between the short axes of the subunits. Accordingly the thermodynamic potential associated with the A-B_c transition can be written :

$$F(\phi_o^e, \eta_o) = F_1(\phi_o^e) + F_2\,(\phi_o^e \eta_o) \tag{7.2}$$

where ϕ_o^e is the equilibrium value of the ordering parameter for the nematic-smectic A transition, i.e. it transforms as the IR E_{k1} (or E_{k2}) of the nematic group $R \wedge D_{\infty h}$ with $k = \frac{2\pi}{d}$.

$F_2(\phi_o^e \eta_o)$ transforms as the product $E_{k1} \times E_{61}$ and can be written :

$$F_2(\phi_o^e \eta_o) = \frac{a'}{2} \eta_o^2 \phi_o^{e^2} + \frac{c'}{4} \eta_o^4 \phi_o^{e^4} - \frac{\delta}{2} \eta_o^2 \phi_o^{e^4} \qquad (7.3)$$

which is analogous to the expressions (5.16) and (5.17). Accordingly, minimization of (7.3) provides the equilibrium value for the azimuthal parameter in the B_c phase :

$$\eta_o^{e^2} = \frac{\delta \phi_o^{e^2} - a'}{c' \phi_o^{e^2}} \qquad (7.4)$$

with $a' = a_o (T - T_{AB})$, where T_{AB} is the A-B_c transition temperature, and $\phi^{e2} = -\frac{a}{b}$ as can be deduced from (5.3). The δ coefficient vanishes by symmetry in the A phase, in such a manner that (7.4) expresses a first-order transition from the A to the B_c phase, with a jump in the order-parameter equal to $\Delta\eta_o^{e2} = \frac{\delta}{c'}$ at T_{AB}^c.

Smectic phases possessing both a tilting order and a bond orientational order can be described analogously, either by considering the smectic C as the parent phase, or by introducing two secondary parameters ϕ and η_o, the spontaneous values of which appear simultaneously as the result of a coupling to the smectic A order-parameter ϕ_o. When a true long-range positional ordering exists in the low-temperature smectic phase, one has to consider, as mentioned in §.3.2, a three-dimensional Brillouin-zone for obtaining the in-plane lattice vectors.

8. FERROELECTRIC, FLEXOELECTRIC AND PIEZOELECTRIC EFFECTS

It is usually admitted that ferroelectricity cannot take place in the isotropic liquid phase of LC formed by polar molecules, because, due to their high molecular weight, the dipole interaction between the molecules is weaker than the Van der Waals interaction. Along the same line, although one can predict on a symmetry basis (see §.9) the existence of transitions from the isotropic LC phase to polar cholesteric phases (of symmetry C_∞) or polar nematic phases (of symmetry $C_{\infty v}$), such types of ferroelectric orderings are unlikely, and have not yet been observed experimentally. By contrast the stability of ferroelectric chiral smectic phases was predicted by Meyer et al [112] and received immediate experimental confirmations [112-116]. Subsequently the field of smectic ferroelectrics developed considerably (see for example ref. 117), mainly after the demonstration by Clark and Lagerwall [118] of a fast bistable electroopic switch, based on the ferroelectric properties of the chiral smectic C phase of macroscopic symmetry C_2 (denoted C*). In this phase the very weak value measured for the spontaneous polarization (one of the higher values obtained for the polarization e.g. in DOBAIMBC is of the order of 10^{-8} C/cm^2) clearly establishes that this quantity is not the primary mechanism of the smectic A* -

Smectic C* (or Smectic C - Smectic C*) transition.

Following Indenbom et al [7] , let us show that the onset of a spontaneous polarization $\vec{P}_s$ corresponds to a pseudo-proper ferroelectric transition to the smectic C* state, i.e. $\vec{P}_s$ appears as the result of a linear coupling with the primary order-parameter assumed to be the spontaneous tilt of the director $\vec{u}$ in space. For this purpose we will consider a transition in which the parent phase is a Smectic A* phase, of symmetry D_∞, and discuss the symmetry properties of the IR, denoted E_{11} (at $k \cong 0$) in table 4. The basis for this two dimensional IR is equivalently the polarization components (P_x, P_y) or the products $(u_z u_y, -u_z u_x)$. In other words, under the symmetry operations $C(\phi)$ and $U_2(\phi^z)$ pertaining to D_∞, P_x transforms as $u_z u_y$ and P_y as $-u_z u_x$. Posing :

$$\eta_1 = u_z u_x \quad , \quad \eta_2 = u_z u_y \tag{8.1}$$

one can write the thermodynamic potential F :

$$\begin{aligned} F = & \frac{\alpha}{2} (\eta_1^2 + \eta_2^2) + \frac{\beta}{4} (\eta_1^2 + \eta_2^2)^2 + \delta(\eta_1 \frac{\partial \eta_2}{\partial z} - \eta_2 \frac{\partial \eta_1}{\partial z}) \\ & + \frac{\sigma}{2} [(\frac{\partial \eta_1}{\partial z})^2 + (\frac{\partial \eta_2}{\partial z})^2] \\ & - \mu(P_x \eta_2 - P_y \eta_1) + \frac{1}{2\chi_o} (P_x^2 + P_y^2) \\ & - \nu(P_x \frac{\partial \eta_1}{\partial z} + P_y \frac{\partial \eta_2}{\partial z}) \end{aligned} \tag{8.2}$$

The first two lines in (8.2) express the order-parameter expansion which has the transformation properties of E_{11}. The δ -Lifschitz - invariant produces an helicoïdal twisting of the axes of the molecules. The macroscopic point-group symmetry of the phase is C_2. The third line in (8.2) expresses the piezo-electric linear coupling between the spontaneous polarization and the spontaneous tilt of the molecules in the layers. A partial minimization of (8.2) gives :

$$P_x = -\sigma\chi_o \eta_2 \quad , \quad P_y = \sigma\chi_o \eta_1 \tag{8.3}$$

which accounts for this piezoeffect. The constant $\sigma\chi_o$ should be very small due to the absence of any substantial dipole moment in the subunits. The helicoidal distribution of the spontaneous polarization can be viewed directly by the invariant :

$$P_x \frac{\partial P_y}{\partial z} - P_y \frac{\partial P_z}{\partial z} \tag{8.4}$$

which has the same symmetry properties as the δ Lifshitz invariant, and can be deduced for (8.3). The ν term in line 4 of (8.2) expresses the flexoelectric effect predicted by Meyer [119] , discussed by a number of authors [120-122] and observed experimentally [123,124].

Assuming the degree of order Q of the long axes of the molecules to be constant, and the tilt angle θ of the director to be small, one can take as local equilibrium values for the incommensurate order-parameter (η_1,η_2) (see chapter V) :

$$\eta_1 \simeq \theta \cos kz \quad , \quad \eta_2 \simeq \theta \sin kz \tag{8.5}$$

where k is the wave-vector of the modulation.

Introducing (8.5) in the equations which minimize (8.2) with respect to P_x and P_y, one gets :

$$P_x = (\mu-\nu k)\theta \sin kz \quad \text{and} \quad P_y = (\nu k-\mu)\theta \cos kz \tag{8.6}$$

which leads to the renormalized form for the thermodynamic potential in the ferroelectric phase as a function of θ and k :

$$F = [\frac{\alpha}{2} - \frac{1}{2}\chi_o \mu^2 - \frac{(\delta+\chi_o\mu\nu)^2}{2(\sigma-\chi_o\nu^2)}]\theta^2 - \frac{1}{2}(\sigma-\chi_o\nu^2)\theta^2(k-k_c)^2 \tag{8.7}$$

where

$$k_c = \frac{\delta + \chi_o\mu\nu}{\chi_o\nu^2-\sigma} \tag{8.8}$$

is the value of k which minimizes the k-expansion of F in the ferroelectric phase.

In (8.7) one can see that the last term becomes negative for large values of ν. Accordingly (8.7) and (8.8) are valid only if $\chi_o\nu^2 < \sigma$. If this condition is not fulfilled one has to consider higher degree terms in the potential F. Assuming small values for the ν constant, one can deduce from (8.7) and (8.8) the transition temperature to C* phase :

$$T_c = T_o + \frac{\chi_o\,\mu^2}{2\,a_o} + \frac{1}{2\,a_o}(\sigma - \chi_o\nu^2)\,k_c \tag{8.9}$$

At T_c there appears a finite tilt angle θ which gives rise to an induced polarization, the amplitude of which is :

$$|\vec{P}_s| \sim \chi_o|\mu - \nu k|\theta \tag{8.10}$$

The direction of $\vec{P}_s$ is perpendicular to the z axis and rotates within the smectic layers with a pitch $\frac{2\pi}{k}$. The effective dipole moment per molecule is proportional to the wave vector k and amounts to a small fraction of the magnitude of the constant dipole moment characterizing the individual molecule.

In summary, in the preceding model, developed by Indenbom et al [7], a modulated ferroelectric phase results from the existence of two Lifshitz invariants : one with coefficient δ which corresponds to the order-parameter wave, and another one, with coefficient ν which originates in a piezoelectric (flexoelectric) effect. If one of this invariant dominates the other one may have either a modulated structure of piezoelectric origin (if $\nu \gg \delta$) or connected with the intrinsic property of the order parameter (i.e. when

$\delta >> \nu$). When the contribution of the two invariants cancel, a mascroscopically uniform ferroelectric smectic state may stabilize.

Additional refinements of the model are discussed in ref. 7, namely, the change of the pitch of the helical structure under the effect of thermal fluctuations, the possible existence of a piezoeffet in the smectic A* parent phase under the effect of a uniform external electric field, and the behaviour of the modulated ferroelectric phase under application of an electric field.They will not be discussed here. Let us finally emphasize that another model can be proposed, based of the four component IR's G_{rk}, in which the ferroelectric effect results from an improper ferroelectric transition (and not a pseudoper one).

9. TRANSITIONS FROM THE ISOTROPIC PHASE

The development of uniaxial nematic, cholesteric and smectic phases from the isotropic phase also admits a phenomenological approach of the type considered in this chapter. This approach was briefly discussed in the original paper of Landau [1] and in more details in refs. 105-107, 125 and 126. Let us summarize its main features.

a) The point symmetry of an isotropic liquid formed by centrosymmetrical molecules, is described by the full orthogonal group O(3) ($\infty/\infty m$ in the Shubnikov notation).The space symmetry of such a liquid phase is expressed by (2.1).

The point symmetry of an isotropic liquid with chiral molecules is described by the rotation group SO(3), and the corresponding space group is :

$$S = R^3 \wedge SO(3) \tag{9.1}$$

Let us first recall, following the presentation of ref. 125, the method which allows to construct the IR's of the point groups O(3) and SO(3). The group SO(3) of all rotations about a fixed point in a three-dimensional space (i.e. the invariance group of the sphere) is locally isomorphic to the group SU(2) of all 2 x 2 unitary matrices with a determinant equal to unity. SO(3) and SU(2) have the same Lie algebra [127] constructed on the three Hermitian infinitesimal operators J_1, J_2 and J_3 which satisfy the Lie-Cartan relations :

$$[J_\alpha, J_\beta] = i \sum_\gamma \varepsilon_{\alpha\beta\gamma} J_\gamma \tag{9.2}$$

where $\varepsilon_{\alpha\beta\gamma}$ is the completely antisymmetric pseudo-tensor of rank three. SO(3) and SU(2) are of rank one and consequently there exists only one nonlinear function of the infinitesimal operators which commutes with all of them. This invariant is the Casimir operator $J_1^2 + J_2^2 + J_3^2$, which can be used to label the IR's. These IR's denoted D^j are labelled by an index j such that 2j is integer. The dimension of D^j is 2 j + 1. Only D^j IR's with 2j even are to be considered. Once the IR's of SO(3) are known, the IR's of O(3) can be easely worked out as O(3) is the direct product of SO(3) and C_i. Thus to each IR D^j of SO(3) correspond two IR's D^j_+ and D^j_- of O(3), which both have the dimension 2j + 1. In D^j_+ proper and improper rotations are represented by the same matrix, while in D^j_- proper and improper rotations are represented by matrices which differ in sign. In the IR D^j of SO(3), the character $\chi^j(\phi)$ of a rotation through an angle ϕ is :

$$\chi^j(\phi) = \frac{\sin (2j + 1) \phi/2}{\sin \phi/2} \qquad (9.3)$$

one cas thus straightforwardly deduce the characters of the IR's $D^j_\pm$ of O(3) by multiplication by the two IR's of the group C_i.

The construction of the 2j + 1 dimensional matrices associated with an IR of SO(3) with given j, requires to find the transformation properties of the 2j + 1 basis vectors spanning the representation space of SO(3). Examples of such a construction are given in refs. 127-131. The theoretical basis for constructing the IR's of the space-groups (9.1) and (2.1) can also be found in the preceding references. However, for LC systems we do not need to construct the more general IR's of the groups $R^3 \wedge$ SO(3) and $R^3 \wedge$ O(3) : such IR's would necessitate to consider a spherical Brillouin zone of radius k; and the corresponding stars of the wave-vectors would possess an infinite number of branches in all directions of the reciprocal space [129,131]. For transitions from the isotropic to nematic, cholesteric and smectic phases, we can assume that the direction of the transition wave vector is fixed a posteriori by the orientation of the director $\vec{u}$, a simplification which is physically justified by the pretransitional effects observed in the vicinity of the transition from the liquid state.

Denoting $\vec{k}_z$ the direction of the wave-vector, the rotational part of the group of the wave-vector will contain only rotations around $\vec{k}_z$ (i.e. it is C_∞). As a consequence, the basis functions which span the IR's of the group $S = R^3 \wedge$ SO(3) identify to (4.1). As in the spherical harmonics $Y^l_m(\theta, \phi)$, one obtains the same IR for fixed m and different l, the representations of S will be denoted D^{km}. As shown in refs. 128-131, all the representations D^{km} are irreducible, except D^{oo} which can be decomposed into the sum of the IR's D^l of the group R^3. When passing to the IR's of the group $S = R^3 \wedge$ O(3), one must take into account that, as the invariance group of $\vec{k}_z$ becomes $C_{\infty v}$ the factors $e^{im\phi}$ and $e^{-im\phi}$ belong to the same IR. Thus the basis functions of S can be written :

$$\phi^{\pm m}_k = e^{i\vec{k}_z.\vec{r}} \, Y^l_{\pm m}(\theta, \phi) \qquad (9.4)$$

b) Let us now, indicate following Ref. 125 some general results concerning the Landau theory of transitions from the isotropic state to LC phases. From the form of the characters (9.3) of the IR's of SO(3) and of the characters $\chi^j_\pm$ of O(3) one can immediately deduce : 1) that <u>for any even</u> j D^j_+ induces a symmetry change leading to the group $D_{\infty h}$ and all its subgroups, while D^j_- induces the group D_∞ and all its subgroups. 2) <u>for any odd</u> j, D^j_+ induces the group $C_{\infty h}$

and all its subgroups, whereas D^j_- induces the group $C_{\infty v}$ and all its subgroups. As stressed in ref. 125, these results do not show any difference between the symmetry modifications induced by, let say, D^2_+ and D^4_+. This is understandable by considering that, although the nematic phase of symmetry $D_{\infty h}$ is characterized by an irreducible tensor of rank 2, of the form 4.2, any physical quantity represented by an irreducible tensor of rank 4 will also display non-zero components. Accordingly, to characterize an anisotropic phase arizing below the isotropic liquid, one has to specify the number of independent components of the irreducible tensors associated which each phase. As a consequence, the measurement of a sufficient number of tensorial physical quantities will define completely a given anisotropic phase [125].

General results are also established in ref. 125 for the verification of the Landau and Lifshitz conditions for the IR's D^j_+. For SO(3), the Landau condition is not fulfilled if the symmetrical third power of the IR D^j contains the identity representation of SO(3). If g is an element of SO(3), one can write :

$$\int_{SO(3)} [\chi^j(g)]^3 \, dg = \int_{SO(3)} \{\tfrac{1}{3}\chi^j(g^3) + \tfrac{1}{2}\chi^j(g^2)\chi^j(g) + \tfrac{1}{6}(\chi^j(g))^3\} dz \qquad (9.5)$$

using (9.3), (9.5) becomes :

$$\int_{SO(3)} [\chi^j(g)]^3 \, dg = \frac{1}{3} + \frac{(-1)^j}{2} + \frac{1}{6} \qquad (9.6)$$

which shows that if j is even the Landau condition is violated while for odd j it is fulfilled. The same result holds for the D^j_+ IR of O(3), while the IR's D^j_- always satisfy the Landau condition. In particular one can verify that at the transition to a nematic phase of symmetry $D_{\infty h}$, which corresponds to the IR D^2_+, the Landau condition is not fulfilled and the transition should be first-order (see §.5)

The Lifshitz condition, i.e. the existence of antisymmetric invariants linear in the first derivatives of the order-parameter, can be expressed here by the following group-theoretical condition: if there exists an integer n such that the representation $[D^j]^n \times D^1_- \times D^j_+$ contains the identity representation D^o_+, then the Lifshitz condition is not fulfilled. In ref. 125 it is shown that all the phases induced by the IR's D^j_- violate the Lifshitz condition, while the D^j_+ IR's satisfy this condition. For the D^j IR's of the SO(3) group there always exist an invariant of the Lifshitz type. In particular the D^1 and D^2 IR's allow respectively invariants of the type $\vec{P}$ curl $\vec{P}$ and $\varepsilon_{ijk}\eta_j\eta_l \frac{\partial}{\partial x_i}(\eta_k\eta_m)$ [7]

c) As an illustrative example let us brifly discuss the transitions associated with the IR's D^1_- that has been considered by Loginov [107].

As the IR D^1_- is a vector representation, one can take the electric polarization $\vec{P}$ as the order-parameter. The symmetry of D^1_-

allows existence for an invariant of the form P^2 div $\vec{P}$. Accordingly the thermodynamic potential can be written :

$$F = F_o + \int \{ \frac{\alpha}{2} P^2 + \frac{\beta}{4} P^4 + \gamma P^2 \operatorname{div} \vec{P} + \delta_1 \sum_i \left(\frac{\partial P_i}{\partial x_i}\right)^2 + \delta_2 (\operatorname{div} \vec{P})^2 \} d\vec{r} \quad (9.7)$$

where $\alpha = a_o (T - T_c)$ and except γ , all the other constants are assumed to be positive. The presence of the γ invariant leads to the result that a modulated structure arises below the isotropic phase. Writing the polarization under the form :

$$\vec{P} = P(\vec{r}).\vec{S}(\vec{r}) \qquad \text{with} \qquad S^2 = 1 \quad (9.8)$$

Assuming a one-dimensional periodic modulation, one can find an approximate solution for the minimization of (9.7) by expanding $P(\vec{r})$ and $\vec{S}(\vec{r})$ in Fourier series, and retaining only the zeroth and first harmonics (see chapter V). It comes :

$$P(\vec{r}) = \eta(\cos\theta + \sqrt{2}\sin\theta\cos\phi\cos\vec{k}\vec{r} + \sqrt{2}\sin\theta\sin\phi\sin\vec{k}\vec{r}) \quad (9.9)$$

and

$$S_x = \cos\psi , \quad S_y = \sin\psi\cos\vec{k}\vec{r}, \quad S_z = \sin\psi\sin\vec{k}\vec{r} \quad (9.10)$$

Introducing (9.9) and (9.10) in (9.7) and minimizing F with respect to $\vec{k}$ and ψ , one gets :

$$F = F_o + V \left[\frac{\alpha}{2}\eta^2 + \frac{\beta^*(\theta)}{4}\eta^4\right] \quad (9.11)$$

where V is the volume of the system, and :

$$\beta^*(\theta) = \beta\left[1 + \sin^2 2\theta + \frac{1}{2}\sin^4\theta - \frac{\gamma^2\sin^2\theta(1+\cos^2\theta)^2}{2\beta(\delta_1+2\delta_2)(1+\sin^2\theta)}\right. \quad (9.12)$$

which yields the equilibrium values for the wave vector :

$$k_x = 0, \quad k_y = k\sin\phi \quad , k_z = -k\cos\phi$$

with

$$k = \frac{\gamma\eta\sin\theta(1+\cos^2\theta)}{\sqrt{2}(\delta_1+\delta_2)(1+\sin^2\theta)} \quad (9.13)$$

Depending on the numerical values of the quantity:

$$\xi = \gamma^2/2\beta(\delta_1+\delta_2)$$

one finds three qualitatively different solutions for $\vec{P}(\vec{r})$. For $\xi<1$, one has $\theta^e = 0$, $k^e = 0$ and $\vec{P} = \vec{P}_o$, namely an homogeneous ferroelectric liquid. As shown by Khachaturyan [104] in finite size specimens, there will arise a domain structure in which the polarization will be helicoïdally twisted in space. For $1 < \xi < 16/7$, the equilibrium values are $\theta^e = \frac{\pi}{2}$, $P_x = 0$, $P_y = \eta\sqrt{2}\cos^2 kz$, $P_z = -\eta\sqrt{2}\sin kz \cos kz$, $k^e = \gamma\eta/2\sqrt{2}(\delta_1+\delta_2)$, i.e. the polarization undergoes a modulation in the yz plane, with different projections on the y and z axes. Finally for $\xi > \frac{16}{7}$, θ^e increases, and one has the following expressions for the polarization components : $P_x = 0$, $P_y = \eta \cos kz$ ($\cos\theta + \sqrt{2}\sin\theta\cos kz$), $P_z = -\eta\sin kz$ ($\cos\theta + \sqrt{2}\sin\theta\cos kz$). It can be noted that $\beta^*(\theta)$ becomes negative for $\xi > 2.79$ and thus for $\xi = 2.79$ one has a tricritical point separating a line of first order transitions from a line of continuous transitions.

REFERENCES

1. L.D. Landau, Zh. Eksper. Teor. Fis. 7, 627 (1937). Translated in "Collected papers of L.D.Landau" Edited by D. Ter Haar, Pergamon Press, London (1965).
2. P.G. de Gennes, Solid State Commun. 10, 753 (1972)
3. P.G. de Gennes, Phys. Lett. 30A, 454 (1969)
4. P.G. de Gennes, "The Physics of Liquid Crystals", Clarendon Press, Oxford (1974)
5. N.Boccara, "Symétries brisées", Hermann (1976)
6. S. Goshen, D.Mukamel, S.Shtrikman, Solid State Commun. 9, 649 (1971)
7. V.L.Indenbom, S.A.Pikin and E.B.Loginov, Sov.Phys.Crystallogr.21, 632 (1976)
8. S.A. Pikin, and V.L. Indenbom, Sov.Phys.Usp.21, 487 (1978)
9. V.L.Indenbom and E.B.Loginov, Sov.Phys.Crystallogr.26, 526 (1981)
10. V.L.Indenbom, E.B.Loginov, and M.A.Osipov, Sov. Phys.Crystallogr. 26, 656 (1981)
11. J.Prost, Advances in Physics 33, 1 (1984)
12. B.A.Tavger, Sov.Phys.Crystallogr. 5, 646 (1961)
13. F.Moussa,J.J.Benattar and C.Williams, Mol. Cryst.Liq.Cryst.99, 145 (1983)
14. C.W.Curtis, and I.Reiner, "Representation theory of finite groups and associative algebras, Interscience, New-York (1962)
15. S.L.Altmann, Induced representations in cyrstals and molecules", Academic Press, London (1977)
16. G.Ya.Lyubarski, "The application of the Group Theory in Physics", Pergamon Press, Oxford (1960)
17. H.Brand and H.Pleiner, J. de Physique 43, 853 (1982)
18. International Tables for X-ray Cristallography (Kynoch, Birmingham, 1952)
19. J.L.Martinaud, and G. Durand, Solid State Commun. 10, 815 (1972)
20. I.W.Smith, Y.Galerne, S.T.Largerwall,E.Dubois-Violette, G.Durand, J.de Physique 36, C1-237 (1975)
21. Nguyen Huu Tinh, C.Destrade and H.Gasparoux,Phys.Letters. 72A, 251 (1979)
22. D.H.Van Winkle and N.A.Clark, Phys.Rev.Letters 48, 1407 (1982)
23. L.Mamlok, J.Malthete, Nguyen Huu Tinh, C.Destrade and A.M.Levelut, J. de Physique Lettres 43, L 641 (1982)
24. L.A.Maksimov,I.Ya.Polishchuk and V.A.Somenkov,Solid State Commun. 44, 163 (1982)
25. S.Deonarine and J.L.Birman, Phys. Rev. B27, 2855 (1983)
26. W.Helfrich, J.de Physique 40, C3-105 (1979)
27. D.V.Kovalev, "Irreducible representations of the space groups" Gordon and Breach, New-York (1965)

28. J.Zak,A.Casher,H.Glück and Y.Gur :"The irreducible representations of space groups", Benjamin, New-York (1969)
29. S.C.Miller and W.F.Love :"Tables of irreducible representations of space groups and corepresentations of magnetic space groups".Pruett, Boulder, Colorado (1967)
30. A.Tardieu, and J.Billard, J. de Physique 37, C3-79 (1976)
31. M.M.M. Abdoh Srinivasa,N.C.Shivaprakash and J.Shashidara Prasad, Mol. Cryst.Liq.Cryst. 72, 225 (1982)
32. S.Goshen,D.Mukamel, and S.Shtrikman, Mol.Cryst.Liq.Cryst.31, 171 (1975)
33. H.Zimmermann, R.Poupko, Z.Luz and J.Billard,Z.Naturforsch,40a, 149 (1985)
34. F.D.Murnaghan, "The theory of group representations",Dover New-York (1963)
35. K.Rosciszewski, Acta Phys. Polonica A56, 891 (1979)
36. G.Sigaud,F.Hardouin,M.F.Achard,and H.Gasparoux,J.de Physique 40, C3-356 (1979)
37. F.Hardouin, Nguyen Huu Tinh, and A.M.Levelut, J.de Physique Lettres 43, L-779 (1982)
38. R.G.Priest, Mol. Cryst. Liq. Cryst. 60, 167 (1980)
39. P.E. Cladis, Phys. Rev. Letters 35, 48 (1975)
40. P.Tolédano and R.Tekaïa, Ferroelectrics 55, 167 (1984)
41. R.Tekaïa, Thèse de 3e cycle, University of Picardie (1984) unpublished
42. P.Tolédano, Jap.Jour.of Applied Physics 24 Suppl. 121 (1985)
43. P.Tolédano and R.Tekaïa, Jap.Jour. of Applied Physics, 24 suppl., 403 (1985)
44. R.Tekaïa and P.Tolédano, to be published in Phys. Rev. A
45. M.Boschmans, R.TekaIa and P.Tolédano, to be published in Ferroelectrics.
46. K.K.Kobayashi, Phys. Lett. A31, 125 (1970)
47. W.L. Mc Millan, Phys. Rev. A4, 1238 (1971)
48. R.Alben, Solid State Commun. 13, 1783 (1973)
49. J.W.Doane, R.S.Parker,B.Cvikl, D.L.Jonhson and D.L.Fishel, Phys. Rev. Letters 28, 1694 (1972)
50. M.Delaye, R.Ribotta and G.Durand, Phys.Rev.Letters 31, 443 (1973)
51. D.L.Jonhson, C.Maze,E.Oppenheim and R.Reynolds, Phys.Rev.Letters 34, 1143 (1975)
52. J.Thoen, H.Marynssen, and W.Van Dael, Phys.Rev.Letters 52,204(1984)
53. B.M.Ocko,R.J.Birgeneau,J.D.Litster,and M.E.Neubert,Phys.Rev.Letters 52, 208 (1984)
54. B.I.Halperin,T.C.Lubensky, and S.K.Ma, Phys.Rev.Letters 32, 292 (1974)
55. T.C.Lubensky,S.G.Dunn,and J.Isaacson, Phys.Rev.Letters 47, 1609 (1981)
56. T.C.Lubensky and A.J.Mc Kane,J. de Physique Lettres 43, L217 (1982)
57.J.M.Viner and C.C. Huang, Solid State Commun.39, 789 (1981)
58. A.de Vries, Mol. Cryst. Liq. Cryst. 10, 31 (1970)
59. G.Derfel, Mol. Cryst. Liq. Cryst. 82, 277 (1982)
60. J.Als-Nielsen, F.Christensen and P.S.Pershan, Phys.Rev.Letters 48, 1107 (1982)
61. J.H.Chen, and T.C.Lubensky, Phys.Rev. A14, 1202 (1976)
62. K.C.Chu, and W.L.Mc.Millan, Phys.Rev. A15, 1181 (1977)
63. D.Jonhson, D.Allender, R.de Hoff,G.Maze,E.Oppenheim and R.Reynolds Phys.Rev. B16, 470 (1977)
64. G.Sigaud,F.Hardouin and M.F.Achard, Solid State Commun.23, 35 (1977)

65. J.Swift, Phys. Rev. A14, 2274 (1976)
66. L.Benguigui, J.de Physique 40; C3-419 (1979)
67. C.C.Huang, and S.C. Lien, Phys.Rev.Letters 47, 1917 (1981)
68. R.De Hoff, R.Biggers,D.Brisbin, M.Mahmood,G.Grisden and D.L.Jonhson, Phys. Rev. Letters 47, 664 (1981)
69. C.R.Safinya, R.J.Birgeneau, J.D.Litster and M.E.Neubert, Phys. Rev Letters 47, 668 (1981)
70. R. De Hoff, R.Biggers, D.Brisbin, and D.L.Jonhson, Phys.Rev.A25, 472 (1982)
71. C.C.Huang, Solid state Commun. 43, 883 (1982)
72. S.Witanachi, J.Huang and J.T.Ho, Phys.Rev.Letters 50, 594 (1983)
73. C.R.Safinya,L.J.Martinez-Miranda, M.Kaplan, J.D.Litster and R.J. Birgeneau, Phys. Rev. Letters 50, 56 (1983)
74. D.Mukamel and R.M.Hornreich, J.Phys. C13, 161 (1980)
75. G.Grinstein and J.Toner, Phys.Rev.Letters 51, 2386 (1983)
76. P.E.Cladis, R.K.Bogardus, W.B.Daniels and G.N.Taylor, Phys.Rev. Letters 39, 720 (1977)
77. L.Liebert and W.B.Daniels, J.de Physique 38, L-333 (1977)
78. F.Hardouin, G.Sigaud, M.F.Achard and H.Gasparoux, Solid State Commun. 30, 265 (1979)
79. F.Hardouin and A.M.Levelut, J. de Physique 41, 41 (1980).
80. L.Benguigui and F.Hardouin, J.Physique Lettres, 42, L111 (1981)
81. W.H. de Jeu, Solid State Commun. 41, 529 (1982)
82. S.Takenaka, H.Nakai, and S.Kusabayashi, Mol.Cryst.Liq.Cryst. 100, 299 (1983)
83. S.Diele, G.Pelze, I.Latif, and D.Demus, Mol.Cryst.Liq.Cryst. 92, 27 (1983)
84. N.A.P. Vaz, Z.Yaniv and J.W. Doane, Mol.Cryst.Liq.Cryst.101, 47 (1983)
85. K.A.Smesh, R.Shashidar,G.Heppke and R.Hopf, Mol.Cryst.Liq.Cryst.99 249 (1983)
86. P.S.Pershan and J.Prost, J.de Physique Lettres 40, L27 (1979)
87. N.A.Clark, J.de Physique 40, C3-345 (1979)
88. K.Hida, J.Phys.Soc.Japan 50, 3869 (1981)
89. J.Billard. C.R.Acad.Sci.Paris 292, 881 51981)
90. Nguyen Huu Tinh, C.Destrade, J.Malthete and J.Jacques,Mol.Cryst. Liq.Cryst. 72, L 195 (1982)
91. N.A.P. Vaz, Z.Yaniv and J.W.Doane, Mol.Cryst.Liq.Cryst.92, L75 (1983)
92. M.J.Freiser, Phys.Rev.Letters 24, 104 (1970)
93. R. Alben ,Phys. Rev.Letters 30, 778 (1973)
94. L.J.Yu, and A.Saupe, Phys.Rev.Letters 45, 1000 (1980)
95. N.Isaert, private communication
96. H.Schröder, in "Liquid crystals of one-and-two-dimensional order" Ed.W.Helfrich and G.Heppke, Springer-Verlag (Berlin, 1980) p.196
97. A.Michelson, D.Cabib, and L.Benguigui, J.de Physique 38, 961 (1977)
98. F.Hardouin, A.M.Levelut,J.J.Benattar and G.Sigaud, Solid State Commun. 33, 337 (1980)
99. D.Guillon, and A.Skoulios, Mol.Cryst.Liq.Cryst. 91, 341 (1983)
100. C.Druon, J.M.Wacrenier, F.Hardouin, Nguyen Huu Tinh, and H.Gasparoux, J.de Physique 44, 1195 (1983)
101. A.M.Levelut,B.Zaghloul and F.Hardouin, J.de Physique Lettres 43, L83 (1982)
102.A.J.Leadbetter, J.C.Frost,J.P.Gaughan,G.W.Gray and A.Mosley,J.de Physique 40, 375 (1979)

103. G.Sigaud, F.Hardouin,M.F.Achard, and A.M.Levelut, J.de Physique 42, 107 (1981)
104. A.G. Khachaturyan, J.Phys. Chem.Solids 36, 1055 (1975)
105. S.A.Brazovskii, Sov.Phys. JETP 41, 85 (1975)
106. S.A.Brazovskii, Sov.Phys. JETP 42, 497 (1976)
107. E.B.Loginov, Sov. Phys. Crystallogr. 24, 637 (1979)
108. N.A. Clark and R.B.Meyer, Appl.Phys.Letters 22, 493 (1973)
109. N.A. Clark, and P.S. Pershan, Phys.Rev. Letters 30, 3 (1973)
110. R.Ribotta, G.Durand and J.D.Litster, Solid State Commun. 12, 27 (1973)
111. L.D.Landau and E.M.Lifhistz, Statistical Physics, Pergamon Press, London (1958)
112. R.B.Meyer, L.Liebert,L.Strzelecki, and P.Keller, J.de Physique 36 L 69 (1975)
113. P.Martinot-Lagarde, J.de Physique 37, C3-129 (1976)
114. P.Keller, L.Liebert and L.Strzelecki, J.de Physique 37,C3-27 (1976)
115. P.Pieranski, E.Guyon, and P.Keller, J.de Physique 36, 1005 (1975)
116. B.I.Ostrovski, A.Z.Rabinovich, A.S. Sonin,B.A.Strukov and N.I. Chernova, JETP Letters 25, 70 (1977)
117. Special issue of Ferroelectrics on ferroelectric Liquid Crystals Vol. 58 (1984)
118. N.A. Clark and S.T. Lagerwall, Appl.Phys.Letters 36, 899 (1980)
119. R.B.Meyer, Phys.Rev.Letters 22, 918 (1969)
120. M.A.Osipov and S.A.Pikin, Sov.Phys. JETP 53, 1246 (1981)
121. J.Prost, and J.P.Marcerou, J.de Physique 38, 315 (1977)
122. A.Derzhanski, A.G. Petrov and M.D. Mitov, J.de Physique 39,273 (1978)
123. W. Helfrich, Z.Naturforsch. Teil A26, 833 (1971)
124. J.Prost, and P.S. Pershan, J.Appl. Physics 47, 2298 (1976)
125. N.Boccara, Annals of Physics 76, 72 (1973)
126. M.Kléman and L.Michel, Phys.Rev.Letters 40, 1387 (1978)
127. I.M.Gel'fand, R.A.Minlos, and Z.Ya.Shapiro "Representation of the Rotation and Lorentz Groups, and their applications ", Pergamon Press, London (1963)
128. N.Ya.Vilenkin "Special functions and the theory of Group representations", American Math. Society, Providence, Rhode,Island (1968)
129. M.E. Rose, "Elementary theory of Angular Momentum" J.Wiley, New-York (1968)
130. E.P.Wigner, "Group theory", Academic Press, New York (1959)
131. B.R. Judd, "Operator Techniques in Atomic Spectroscopy", Mc Graw Hill, New-York (1963).

CHAPTER VIII

RECENT DEVELOPMENTS AND FIELD OF VALIDITY OF LANDAU'S THEORY

1. INTRODUCTION.

In this last chapter,we discuss two types of considerations.On the one hand,we present the ideas underlying attempts to use the Landau's theory in order to deal with recently encountered physical problems. In this presentation we restrict to an outline of the basic principles of the methods used because these attempts,though promising, have not yet reached a fully consistent explanation of the examined phenomena. We describe first,in §2, the theory which aims at accounting for the stability of the so-called "icosahedral-quasi-crystalline phases". This theory has been developped by a number of authors on the basis of a scheme initially sketched by Landau [1] ,and later extended and specified by Alexander and Mc Tague [2] ,for the investigation of the liquid-solid phase transition. In the second place,we summarize ,in §3, the main features of a theory of the influence of defects on phase transitions. The corresponding adptation of Landau's theory is due to Levanyuk and coworkers [3].

The second type of considerations discussed in the chapter concern the field of validity of Landau's theory. This is dealt with,in § 4. We first return on the basic assumptions of the theory,and sort out the specific point causing its general lack of validity from the standpoint of statistical physics. We then justify the preservation of the validity of the symmetry implications of the theory, and enumerate the situations in which the physical implications also remain valid.Finally,for the cases where the predictions of the Landau theory are incorrect,we briefly summarize the correct results supplied by the recent statistical theories. We also point out the role played in these theories by the symmetry concepts introduced by Landau's theory.

2. STABILITY OF ICOSAHEDRAL QUASI-CRYSTALLINE PHASES.

Icosahedral quasi-crystalline phases are ordered solid phases in which electron microscopic diffraction,and other types of diffraction experiments reveal unusual patterns of sharp diffraction peaks. Namely,in an appropriate orientation of the samples ,the pattern of diffraction spots displays a 5-fold rotational symmetry (fig. 1a), while for other appropriate sample-orientations the diffraction pattern shows a 3-fold rotational symmetry. The overall pattern is consistent with the symmetry of the regular icosahedron (Cf. fig 1b).

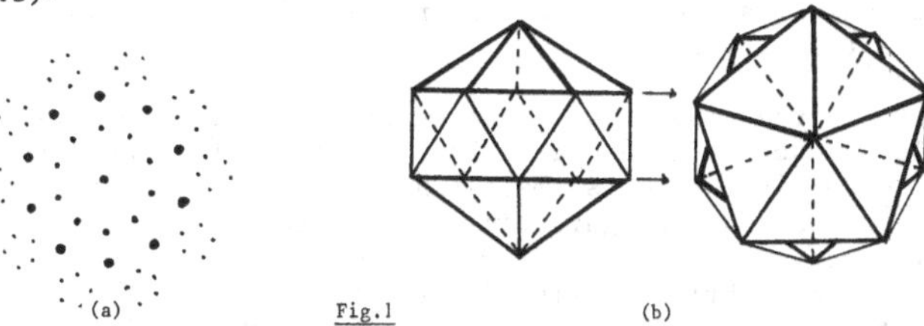

Fig.1

The occurence of the 5-fold rotation is incompatible with the existence of 3-dimensional crystallographic order (Cf. chap. III §2). This is in agreement with the fact that the observed diffraction spots do not generate one of the 14 Bravais lattices. The vectors joining these spots are not linear combinations with integral coefficients of 3 basic vectors only. They have been noted to be linear combinations,with integral coefficients,of the 12 vectors which join the center of a regular icosahedron to its vertices [5] (their set comprises in particular the vectors parallel to the icosahedron edges). It has also been shown [6] that the former 12 vectors are not independent on the field of integers ,and that 6 vectors are enough to generate by integral combination the entire pattern. This feature has been expressed as the existence in these systems of a "6-dimensional crystallographic order". In this respect,quasi-crystalline phases appear as a variety of incommensurate phases for which we have already pointed out the possibility to generate the diffraction pattern with help of (3+d) basic vectors (chap.V). However, an important distinction between the presently considered phases and the incommensurate phases analyzed in chap.V lies in the fact that the latter phases possess a 3-dimensional "basic" structure constituting a good approximation of the actual structure,while quasi-crystalline phases have no 3-dimensional periodic structure of reference.

The icosahedral phases are observed [4],in certain cooling conditions, below the range of stability of the liquid phase.This circumstance has led several authors to the belief that the existence of icosahedral phases could possibly be understood through an investigation of the liquid-solid phase-transition. In particular one should be able to understand the stability of these phases by means of the Landau theory of the liquid-solid phase transition. In such a theory the high-symmetry G_o is the symmetry of the liquid phase.

It is worth pointing out that ,up to now, the Landau theory of the transtion between a liquid,and an ordinary 3-dimensional crystalline solid has not yet been satisfactorily formulated, and that,therefore,the task of performing such a formulation for the liquid -icosahedral-solid phase transition,is not warranted success. The reasons underlying the difficulty of working out a theory are twofold. The first one pertains to the mathematical complexity of the group G_o. The liquid phase being homogeneous and isotropic, G_o is the centrosymmetric Euclidian group, i.e. the group generated by the translations $\vec{T} \in \mathfrak{T}$ of arbitrary vector, and rotations of SO(3) about any point, as well as by the inversion. We have already invoked in chapter VII the difficulties arising in deriving from this group the symmetries of liquid -crystal phases,on the basis of Landau's theory. The second difficulty pertains to the discontinuous character of the liquid -solid phase transition. Consequently, there is no universal symmetry scheme readily applicable (i.e. based on an irreducible order parameter). One has to elaborate a specific model defined by the number and nature of the irreducible representations (IR) involved in the transition.

In the case of the transitions towards liquid crystal phases, a simplified approach could be used because one could restrict to the investigation of the orientational order in the low symmetry phase. In the present case,one mainly wishes to understand the configuration of the $\vec{k}$-vectors observed in the diffraction experiments. These vectors correspond to the components of the Fourier transform of the equilibrium density $\rho_{eq.}(\vec{r})$,of the particles constituting the system in the considered phase. As emphasized in chap.III §8.1, and also in chap. V , these vectors comprise two sets.On the one hand there are the $\vec{k}$-vectors present in the Fourier transform of $\rho_o(\vec{r})$,which is the equilibrium density in the high symmetry phase. On the other hand, there are the k-vectors characterizing the translational symmetry properties of the order parameter components (primary and secondary ones) which possess a non-zero equilibrium value in the low-symmetry phase. In the present case the former set does not contain discrete vectors since it corresponds to the liquid phase. Hence the observed configuration of vectors should be entirely determined by the translational symmetry properties of the order-parameter components.

This justifies the common hypothesis of the simplified formulations of Landau's theory developped for icosahedral phases. These formulations do not follow the standard scheme summarized at the end of chapter II: consideration of the entire set of OP-components, construction of the corresponding free-energy invariant by G_0, derivation of the icosahedral phase as one of the low-symmetry phases, stable for a certain range of the expansion's coefficients.

Instead, the adopted procedures restrict to the consideration of a limited number of OP-components : the ones whose non-zero-value are expected to determine one of the phases of interest. The method compares the free-energy of the icosahedral phase to the free-energy of other phases selected on the basis of physical arguments. Besides, while the invariance of the free-energy by the translational symmetry is fully exploited, the rotational symmetry is only partly used.

2.1 Landau's model of the liquid-solid phase transition.

Consider the Fourier expansion of the density increment $\delta\rho(\vec{r})$:

$$\delta\rho(\vec{r}) = \int \eta(\vec{k}).e^{-i\vec{k}.\vec{r}}.d\vec{k} \tag{2.1}$$

We know, on the basis of eqs. (2.2") and (3.5) in chap. II, as well as of eq. (2.6) in chap. III, that eq. (2.1) hereabove is the standard Landau expansion of $\delta\rho(\vec{r})$ in terms of functions carrying IR's of an infinite group of translations. In the case of the continuous group of translations $\mathcal{T}_0$ of the liquid phase, the integral in (2.1) extends to all $\vec{k}$-vectors in the 3-dimensional reciprocal space, while for the discrete translation-group of a crystal, the sum can be restricted to the first Brillouin zone. With respect to $\mathcal{T}_0$, there is no equivalence up to a reciprocal lattice vector: any two distinct vectors will define unequivalent IR's of $\mathcal{T}_0$. In particular, a translationally invariant (totally-symmetric) function corresponds to $\vec{k}=0$. Also note that the reality of $\delta\rho(\vec{r})$ imposes $\bar{\eta}(\vec{k}) = \eta(-\vec{k})$.

We can classify the components $\eta(\vec{k})$ in (2.1) as a function of the modulus $k = |\vec{k}|$ of their $\vec{k}$-vectors. For each value of k, let us consider a finite set of components $\eta(\vec{k}_i)$ ($|\vec{k}_i| = k$). The quadratic contribution of these degrees of freedom to the Landau free-energy is (following eqs. (3.5) and (3.7) in chapter II) :

$$f_2 = \int \alpha(T,k)\{\sum_{|\vec{k}_i|=k} |\eta(\vec{k}_i)|^2\}.dk \tag{2.2}$$

This specific form, in which all the $|\eta(k_i)|^2$ terms have the same coefficient α stems from the fact that the isotropy of the liquid imposes the invariance of the $\alpha(\vec{k}_i)$ under the action of any rotation of the $\vec{k}_i$. Eq. (2.2) is similar to eq. (3.7) of chap. II, where all the components associated to the same IR had the same α coefficient. The $\eta(\vec{k}_i)$ hereabove do not necessarily carry an IR of G_o. Indeed we can note that the continuous set of $\eta(\vec{k}_i)$ (or, equivalently, of functions $\exp(-i\vec{k}.\vec{r})$),carries the regular representation of O(3) [7] ,which contains all the IR's of this group. Nevertheless, the formal identity between eq. (2.2) and eq. (3.7) in chap.II, leads to the conclusion that the primary components of interest are defined by the modulus k_o associated to the minimum of $\alpha(T,k)$ with respect to k.

Let us restrict to the set of components $\eta(\vec{k}_i)$ having $|\vec{k}_i|=k_o$.In the solid phase,one has $\alpha(T,k_o)< 0$,and, as usual, the stability of this phase is determined by the higher-degree terms in the Landau expansion,which can be constructed as invariant polynomials of the set $\eta(\vec{k}_i)$. In particular the invariance must hold with respect to the translation group $\mathcal{T}_o$.

Consider first the third-dgree terms. These terms do not have to be absent from the free energy since the transition is a discontinuous one. Their condition of translational invariance is expressed by eq. (4.4) of chap. III:

$$\vec{k}_{\ell_1} + \vec{k}_{\ell_2} + \vec{k}_{\ell_3} = 0 \qquad (2.3)$$

where the three vectors in (2.3) belongs to the set of vectors with moduli k_o appearing in (2.2). As already stressed above, eq.(2.3) unlike eq. (4.4) of chap. III holds strictly,and not up to a reciprocal lattice vector. Hence, the three vectors $\vec{k}_i$ in eq.(2.3) are necessarily different, and they form an equilateral triangle of edge k_o. Each equilateral triangle which can be formed by 3 such vectors,gives rise to 2 third-degree terms in the free - energy expansion,one for the set $\vec{k}_i$, and its complex conjugate ,associated to $(-\vec{k}_i)$.

The form of the third-degree contribution is:

$$f_3 = B(T,P,\ k_o)\Big[\sum_{|\vec{k}_{\ell_i}|=k_o} \eta(\vec{k}_{\ell_1}).\eta(\vec{k}_{\ell_2}).\eta(\vec{k}_{\ell_3})\Big] \qquad (2.4)$$

All the triangular terms are associated to the same coefficient B, due to the invariance of B with respect to the rotations (the triangles all have the same size ,and only differ in their orientations).

The form of the fourth-degree contribution to the free-energy derives from the same principles. The translational invariance gives rise to a condition similar to (2.3) with 4 vectors involved. The rotational invariance imposes, in addition,certain conditions to the different quartic terms. However,for the 4-th degree,the geometrical figure formed by 4 vectors is not entirely specified by the analog of condition (2.3): this figure involves 2 arbitrary angles. Accordingly,the expression,analog of (2.4), and relative to the 4-th degree-terms, is less simple (Cf. ref.8).

2.2. Alexander and Mc Tague's extension of the model.

In his work [1] ,Landau restricted the use of expression (2.4) to the investigation of the types of phase diagrams which can arise from the existence of the third degree term,and which involve isolated points of continuous transitions in the pressure-temperature plane. Alexander and Mc Tague [2] have shown that this expression could also be used to compare the stabilities of various solid phases with different symmetries. Their basic implicit assumption [9,10] is that the quartic contribution to the free-energy is "isotropic" in the space of the OP components $\eta(\vec{k}_i)$,i.e. it is of the form:

$$f_4 = C(T,P,k_o)(\Sigma \ |\eta(\vec{k}_i)|^2)^2 \qquad (2.5)$$

where the coefficient C is strictly positive. In that case,the only anisotropic term is the third-degree one. It is therefore this term which allows to select the stable directions in the OP-space, i.e. which determines the symmetry of the solid phase,stable below T_c (Cf. Chap.II, §3.6.2).Following a procedure already used in chapter II (§3.6.2), let us put $\eta(\vec{k}_i)=\rho\gamma(\vec{k}_i)$, with $\rho^2=(\Sigma|\eta(\vec{k}_i)|^2)$, and $(\Sigma\gamma(\vec{k}_i)^2)=1$. The stable phase corresponds to the minimum of the third degree-term (2.4):

$$B\rho^3 \sum_{|k_{\ell_i}|} \gamma(k_{\ell_1}).\gamma(k_{\ell_2}).\gamma(k_{\ell_3}) \quad , \qquad (2.4')$$

with respect to the $\gamma(\vec{k}_i)$ .If (2.4') is positive for a certain direction γ_i in the OP-space,it is negative for the opposite direction $(-\gamma_i)$. As both directions have the same invariance group, we can conclude that the symmetry below T_c does not depend on the sign of the B-coefficient: the stable phase corresponds to the set of $\gamma(\vec{k}_i)$ associated to the largest value for the expression:

$$f_3' = \sum_{|\vec{k}_{\ell_i}|=k_o} |\gamma(\vec{k}_{\ell_1}).\gamma(\vec{k}_{\ell_2}).\gamma(\vec{k}_{\ell_3})| \qquad (2.4'')$$

with,
$$\Sigma|\gamma(\vec{k}_{\ell i})|^2 = 1 \qquad (2.4''')$$

Note that for a given set of $\vec{k}_{\ell i}$ vectors,the extremum of (2.4') is realized by equal values of the $|\gamma(\vec{k}_{\ell i})|$. Hence, all the triangular terms in (2.4") are equal. If there are m such terms and p distinct $\gamma(\vec{k}_i)$ (not counting the $\gamma(-\vec{k}_i)$),we can write $f_3'=m.\gamma^3$,with $p.\gamma^2=1$ (eq. 2.4'''). Thus,

$$f_3' = \frac{m}{p^{3/2}} \qquad (2.6)$$

One can then use eq. (2.6) to compare the stabilities of various solid phases of physical interest, each of these phases being associated to a specific set of p vectors $\vec{k}_i$,forming differently oriented equilateral triangles. As pointed out by Alexander and Mc Tague,these vectors can be put in correspondance with the edges of 3-dimensional regular polyhedra which possess faces, consisting of identical equilateral triangles. Consider,for instance the regular octahedron (fig.2). Its edges define p=6 vector directions, while its faces correspond to 4 differently oriented triangles. Taking also into account the cubic terms in f_3' which are associated to the vectors $(-\vec{k}_i)$, we have m=8,and $f_3' = (4/3\sqrt{6})$.

Note that the edge-vectors of a regular octahedron generate a cubic face-centered (F) lattice in the 3-dimensional space (Cf. chap. III eq. (5.1)). In agreement with the remark made in the introduction of the present chapter, the translational symmetry of the solid phase corresponding to the "octahedron set of k-vectors", is described by a cubic F-Bravais reciprocal lattice. The direct lattice of the solid is therefore a body-centered cubic (I) lattice (BCC).

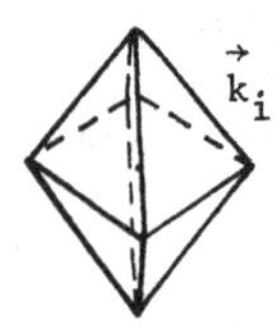

Fig. 2

2.3. Application of the model to the icosahedral phase.

Alexander and Mc Tague have noticed that the edges of a regular icosahedron could be put in correspondance with vectors $\vec{k}_i$ giving rise to a cubic term in the Landau expansion (2.2)-(2.4). The model in §2.2 is therefore relevant to discuss the stability of a solid phase whose diffraction spots are generated by the preceding $\vec{k}_i$ vectors. This phase is assumed to coincide with the icosahedral phase observed in experiments.

Figure 1 shows that a regular icosahedron has 10 differently oriented faces consttuted by identical triangles. Taking into account the contributions of the vectors $(-k_i)$, we can deduce that there will be 20 distinct terms (m=20) in the third degree invariants(2.4"). In addition, there are p=15 differently oriented edges. The value of f_3', which determines the free-energy of the icosahedral (ICOS) phase is:

$$f_3'(\mathrm{ICOS}) = \frac{4}{3\sqrt{15}} \tag{2.7}$$

This value being smaller than the one found for the BCC phase $(4/3\sqrt{6})$, we can conclude that, in the framework of the model of Alexander and Mc Tague, the ICOS phase is less stable than the BCC phase. This model does not explain the experimentally observed stability of this phase, on solidifying the appropriate liquid.

Several refinements of this model have been developped, in order to account for this stability. We briefly outline their basic features.

2.3.1. Influence of a fifth-degree term. "Edge Model".

Bak has noted [6,11], that the $\vec{k}_i$ vectors considered above, and parallel to the edges of the regular icosahedron, can be used to construct 6 differently oriented regular pentagons. Figure 1 shows the projection of one of these pentagons. The others are deduced by applying 3-fold rotations.

Forming pentagons with 5 $\vec{k}_i$-vectors implies the existence, in the Landau free-energy, of translationally invariant 5th-degree terms of the form:

$$D(T,P,k_o).\rho^5.\sum_{|\vec{k}_{\ell_i}|=k_o} \gamma(\vec{k}_{\ell_1}).\gamma(\vec{k}_{\ell_2}).\gamma(\vec{k}_{\ell_3}).\gamma(\vec{k}_{\ell_4}).\gamma(\vec{k}_{\ell_5}) \qquad (2.8)$$

with,

$$\sum_{i=1}^{5} \vec{k}_{\ell_i} = 0 \qquad (2.8')$$

There are 12 terms in (2.8), all with the same coefficient D (due to the rotational invariance), and with the same absolute value ($1/p^{5/2}$). In the same way as for the 3rd degree term, this contribution can be made negative by suitable orientation of the γ_i vector in the OP space, whatever the sign of D. If the 4th-degree term in the free energy is assumed to be isotropic [6,11] ,as in Alexander and Mc Tague's model, the total contribution of the anisotropic 3rd and 5th degrees terms to the icosahedral free-energy, is:

$$-\frac{4B\rho^3}{3\sqrt{15}} \quad -\frac{4D\rho^5}{75\sqrt{15}} \qquad (2.9)$$

The second term in (2.9) being determined by the specific configuration of the "icosahedral" $\vec{k}_i$ vectors, does not exist for the BCC phase. Hence, if its magnitude is large enough ,it can stabilize the ICOS phase. Note that the two terms in (2.9) do not necessarily have different orders of magnitude , since the considered transition is discontinuous, and that ρ is not infinitesimal.

2.3.2. Consideration of edge and vertex models.

Up to now, we have considered a set of vectors all having the same modulus k_o specified by the minimum of A(k), and assumed to correspond to the edges of a regular icosahedron. As mentioned in the introduction of the paragraph, the experimental data reveal that the diffraction pattern is rather generated by the 12 vectors $\vec{q}_j$ joining the center of the icosahedron to its 12 vertices. Their modulus q_o differs from k_o :$(k_o/q_o)=1.0515$.

Three models have been proposed, for the stabilization of the ICOS phase, whose essential common ingredient is the consideration of both the vertex $\vec{q}_j$ and the edge $\vec{k}_i$ vectors.

Geometrically, each $\vec{k}_i$ vector is the integral combination of two $\vec{q}_j$ vectors (e.g. $\vec{k}_1=\vec{q}_2 - \vec{q}_1$). Hence, the set of order-parameter components $\eta(\vec{k}_i)$ are harmonics of the set of components $\zeta(\vec{q}_j)$. Besides, the preceding geometrical relationship implies that one $\vec{k}_i$ vector and two $\vec{q}_j$ vectors form an isoceles triangle : there is a translationally invariant third- degree term in the Landau free-energy of the form $\eta(\vec{k}_i)\zeta(\vec{q}_j)\zeta(\vec{q}_k)$.

a) Stabilization by the third degree terms.

Kalugin, Kitaev and Levitov [12] consider that A(k) is minimum for $k=q_o$ (the vertex vectors), and that it has a quadratic dependence nearby the minimum :

$$A(k) = A(q_o) + a.\left(\frac{k - q_o}{q_o}\right)^2 \tag{2.10}$$

This dependence is similar to the one used in chapter V (§3.2) in order to estimate the influence of the harmonics of an incommensurate modulation.

In this model, the vertex-vectors $\vec{q}_i$ are associated to the primary OP $\zeta(\vec{q}_i)$. As the $\vec{q}_i$ vectors do not permit the construction of equilateral triangles, there are no cubic terms involving the sole $\zeta(\vec{q}_i)$ components. The harmonic components $\eta(\vec{k}_i)$ have non-zero values in the solid phase defined by the $\zeta(\vec{q}_i)$. These secondary OP give rise to the 20 third degree terms already discussed in § 2.2 and § 2.3.a) (actually these terms, if small, are of the same order of magnitude as sixth-degree terms in the $\zeta(\vec{q}_j)$).

Besides, there are "mixed" third degree terms of the form $\eta(\vec{k}_i)\zeta(\vec{q}_j)\zeta(\vec{q}_k)$ which are translationally invariant.

Consideration of the two preceding types of third degree terms determines a free-energy of the ICOS phase which is different from the one calculated hereabove (eq.2.7).

Clearly a similar scheme can be developped for the BCC phase. For this phase the ratio (k_o/q_o) of the moduli of the "fundamental" and "harmonic" vectors is 1.41. For given coefficients $A(q_o)$ and $\underline{a}$ in eq.(2.10), one will obtain a larger $A(k_o)$ coefficient in the BCC phase than in the ICOS phase. This has the consequence [12] that the negative contribution of the harmonics to the free-energy is larger in the ICOS phase than in the BCC phase.

Provided the $\underline{a}$ coefficient in eq. (2.10) is large enough (i.e. if $A(k)$ increases steeply enough in the vicinity of q_o), this effect can be shown [12] to be sufficient to stabilize the ICOS phase.

b) Stabilization by a fourth degree term.

The conclusions of Alexander and Mc Tague's model rely partly on the fact that the quartic term has the same value for the two phases which are compared (i.e. the BCC phase and the ICOS phase). This assumption is also part of the two models discussed in §2.3.1 and §2.3.2a). Mermin and Troian [13] have shown that consideration of the $\zeta(\vec{q}_i)$ as primary OP and of the $\eta(\vec{k}_i)$ as secondary OP gives rise to a specific "renormalized" quartic term in the primary free energy, which favours the ICOS phase with respect to the BCC one.

We had discussed this type of renormalization in the general theory of chap.II §5.3 : if a secondary OP η_i with faintness index f=2 exists, its elimination from the free energy by partial minimization with respect to the η_i components will yield a primary free energy with modified quartic terms, this modification being a negative contribution to the primary quartic terms (Cf. eq.(5.12) in chap. II).

In the present case, the coupling term between the $\eta(\vec{k}_i)$ and the $\zeta(\vec{q}_j)$ is of the form $\eta\zeta\zeta'$, thus showing that the value of the relevant faintness index is 2. Besides the $A(k_o)$ coefficient of $(\Sigma|\eta(\vec{k}_i)|^2)$ can be a small positive number, since it lies close to the minimum of $A(k)$ ($|(k_o-q_o)/q_o| \propto$

5%). In consequence ,the primary quartic contribution,fuction of the $\zeta(\vec{q}_j)$, which is generated by the elimination of the components $\eta(\vec{k}_i)$,can be a large negative contribution to the free energy of the ICOS phase,which will efficiently stabilize this phase [9, 13].

2.3.2. Triggering mechanisms.

Both the models of Kalugin et al [12] and of Mermin and Troian [13] make a distinction between a primary set of components (the vertex ones) and a secondary set (the edge components). This distinction is essentially of interest for continuous transitions. For discontinuous ones,when all the components can be of the same order of magnitude,the distinction is not necessarily relevant.In particular,we have examined a model of "triggered phase transitions" (chap.IV) in which the instabilities with respect to two sets of degrees of freedom could occur simultaneously.

Dvořak and Holakovsky [8] have considered a model of the stability of the ICOS phase,in which the "edge" set of OP triggers the "vertex" set of OP. This model completes the two former ones since it incorporates, in addition to the third degree terms, an anisotropic quartic term. These authors were able to show,with help of a numerical analysis of the complicated free energy thus constructed, that the ICOS- phase can be stable for some range of the expansion's coefficients.

The liquid-icosahedral phase transition is also described as a triggered phase transition in a very different model developped by Jaric [10] .In this model,the triggering OP is an orientational degree of freedom associated to a transition from the liquid phase to a liquid-crystal phase. We have seen in chap. VII that such transitions could be described by an OP carrying an irreducible representation $D^{(j)}$ of SO(3) (or of O(3)). We had considered in chap. VII the cases j=1,2 . Jaric considers the case of an OP denoted Q(j,m) corresponding to j=6 . This OP is shown to induce,among various low symmetry phases,a liquid-crystal phase with icosahedral orientational symmetry and continuous translational symmetry $\mathcal{G}_o$. The symmetry of this OP allows a third-degree coupling term of the form $Q(j,m)\zeta(\vec{q}_k)\zeta(-\vec{q}_k)$ between the orientational degree of freedom,and the vertex components defined in the preceding models. As shown in chap. IV, such a term is appropriate for triggering the onset of non-zero $\zeta(\vec{q}_k)$ components at a discontinuous transition provoked by the Q(j,m).

In the latter model, the ICOS phase arises as an indirect consequence of the latent onset of an icosahedral liquid-crystal phase.

3. INFLUENCE OF DEFECTS ON PHASE TRANSITIONS.

The influence of defects is often invoked to account qualitatively for the fact that certain experimental results relative to phase transitions depart from the behaviour expected ,either on the basis of Landau's theory, or, on the basis of statistical theories (Cf. §4 hereunder). This is especially true of the case of structural transitions [3,14] . The largest part of the theoretical works which have been undertaken to incorporate the influence of defects in the prediction of the anomalies which occur at a phase transition,pertains to statistical physical methods [14] .However some of the developped theoretical approaches can be considered as appropriate extensions of Landau's theory [3,5,16] .This is in particular the case of the approach of Levanyuk and coworkers [3] .In this paragraph we outline the principles of their method for studying the influence of point and extended defects on the characteristics of a phase transition.

3.1 Classification of defects.

Defects are physical objects defined by degrees of freedom involving a certain randomness, and which are coupled to the OP of the considered transition.

In the theory which deals with the influence of defects on phase transitions one implicitly assumes the existence of two types of systems. One is the pure system defined in the usual manner by the symmetries of the phases surrounding the transition,by the characteristics of an irreducible OP, and by a free-energy expansion of specified form. The other is the defective system defined by a free-energy density expressed as a function of the same OP as the pure system, as well as by the parameters representing the coupling to the degrees of freedom of the defects. The existence of this coupling modifies the local equilibrium value of the OP,at each temperature and pressure

One is lead to distinguish various types of defects on the basis of three essential features:

a) Spatial extension of the defects.

The extended character of a defect is a relative concept which implies the comparison of two characteristic lengths.The first one is the characteristic distance r_d over which the random values of the degrees of freedom defining the defects are correlated. The other one is the correlation length ξ associated to the OP of the pure system (Cf. §3.2 hereafter). One deals with an extended defect if $r_d >> \xi$. In the converse situation,the defect is localized. One can distinguish point-defects (localized in all directions),linear defects,and planar defects (extended in one or two dimensions).

b) Frozen-in,or mobile character of the defects.

A frozen-in defect is defined by values of degrees of freedom which are independent of time as well as independent of the state of the pure system in which the defect is enbedded. Several types of defects mobilities are of interest. For instance,one can consider the situation in which the parameters defining the state of the defects are additional variational degrees of freedom of the free-energy of the defective system whose equilibrium values are determined,in the same way as the values of the OP,by the minimum of the free-energ with respect to the entire set of the former physical variables [17] .One can also have defects which can "hop" between various states having frozen-in characteristics,the lifetime of a defect in each state being specified [3,15,16]

c) Coupling scheme between defect and OP.

Two main classes of defects differing by the mode of coupling to the OP have been studied [3] .In the first one the defect induces a local symmetry breaking ,by imposing a non-zero value to the OP of the system. The degenerate character of the symmetry breaking in the pure system implies the existence of several possible (degenerate) states for the defects. The preservation of the "average" symmetry of the system at macroscopic level,as well as the fluctuation of the defect concentration at this level,can be taken into account by assigning a random character to the non-zero value imposed by the defect to the OP.

Note that the symmetry-breaking defects have an action similar to the local application of a field conjugate to the OP.Such defects can therefore be represented as a set of local random fields acting on the pure system.

The second type of defects are the "symmetric" defects which preserve locally the symmetry of the pure system.In a phenomenological description,the change induced by such defects is expressed by a random shift of the local transition temperature with respect to the temperature T_c in the pure system.

Figure 3 shows illustrations of a symmetry-breaking ,and of a symmetric point defect in the case of a structural system. These defects are respectively realized by an intersticial impurity (having two degenerate locations)(figs.3a,b) and by a substitutional impurity (fig.3c).

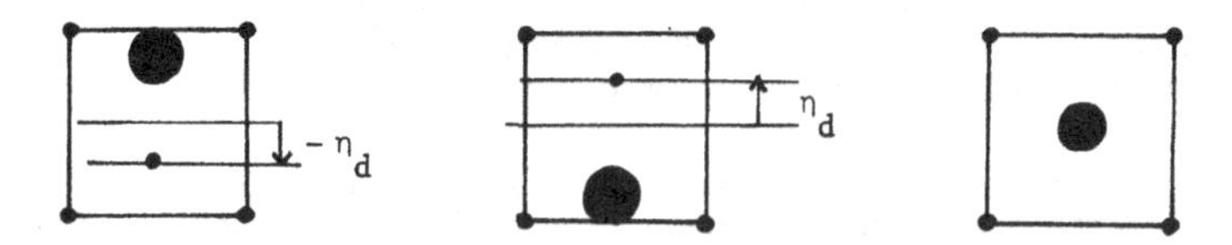

Fig. 3

3.2 Phenomenological theory for symmetry-breaking point defects.

In order to illustrate the Landau theoretical approach to the study of defective systems, let us examine the case of a set of symmetry-breaking point defects imposing a frozen-in local distortion (§3.1.b.,c). This approach closely follows the work of Levanyuk et al [3] .

3.2.1. Description of the model.

Each symmetry-breaking point defect can be represented as a sphere of volume v_d and of radius r_d ,on the boundary of which the value η_d of the OP-components is specified. For the sake of simplicity, a single OP component is considered. η_d can take two possible values $\pm|\eta_d|$ with probability (1/2).

At the macroscopic scale,the value of the OP-component in the defective system (outside the volumes of the defects) is a function $\eta(\vec{r})$ of the spatial coordinates (fig.4). Such a description by a "spatially modulated" OP is similar to the one adopted in the description of incommensurate phases (chap.V §2.3). Referring to the latter theory,we can derive the form of the free-energy density $f(\vec{r})$ of the system outside the volume occupied by the defects.

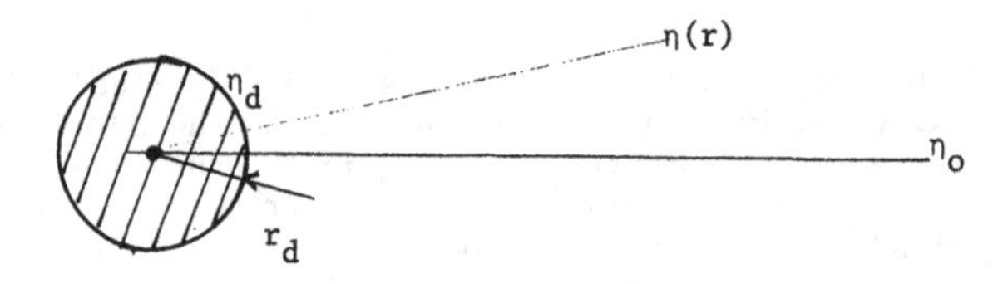

Fig.4

In the presence of a spatially uniform field ε conjugated to η ,we have:

$$f(\vec{r}) = f_o[\eta(\vec{r})] - \varepsilon.\eta(\vec{r}) \tag{3.1}$$

with,

$$f_o[\eta(\vec{r})] = \frac{\alpha}{2}\eta^2(\vec{r}) + \frac{\beta}{4}\eta^4(\vec{r}) + \frac{\kappa}{2}[\overrightarrow{\text{grad}}\eta(\vec{r})]^2 \tag{3.1'}$$

As compared to eq.(2.39) in chap.V , eq(3.1') hereabove differs by the absence of a Lifschitz invariant:such a term has no reason to exist in (3.1') since $\eta(\vec{r})$ is the OP of a structural transition corresponding to the minimum of $\alpha(\vec{k})$ with respect to $\vec{k}$. In eq. (3.1'), we take the signs of the coefficients compatible with the occurence of a continuous transition in the pure system: $\alpha=a(T-T_c)$,a> 0, β> 0,κ>0.

A basic assumption of Levanyuk's et al theory is the additivity of the influence of the defects . Each defect acts independently on the equilibrium of the defective system.The condition expressing the validity of this assumption will be worked out in §3.2.3 On the basis of this assumption, we are entitled to start the investigation of the system by examining the influence of a single defect.

3.2.2 . Calculation of $\eta_{eq}(r)$ for a single defect.

For a single defect imposing the value $\eta(r_d)=\eta_d$,the equilibrium of the defective system is determined by the absolute minimum of the free-energy:

$$F_d = f_d + \int_{|\vec{r}|>r_d} \{f_o[\eta(\vec{r})] - \varepsilon.\eta(\vec{r})\}\, d\vec{r} \tag{3.2}$$

This minimum is constrainted by the boundary conditions $\eta(r_d)=\eta_d$,and $\eta(\infty)=\eta_o$,where η_o is the equilibrium of the OP in the pure system. η_o corresponds to the minimum of the spatially uniform function $f(\eta)$ in eqs.(3.1) and (3.1'). In the absence of field,we have:

$$\eta_o(T\geqslant T_c) = 0 \qquad \text{and} \qquad \eta_o^2(T \leqslant T_c) = \frac{a(T_c-T)}{\beta} \tag{3.3}$$

In eq.(3.2) f_d is the free-energy of the defect. It is a mere constant in the case of a frozen-in defect. The integration is over the volume V of the system.

The extrema of F are determined by the variational equations associated to (3.2) (Cf. chap.V §4).

Taking the origin of the coordinates at the center of the defect,we note that the problem has the spherical symmetry. The single variational equation relative to the radial dependence of $\eta(\vec{r})$,is :

$$\kappa.\Delta\eta = \frac{\partial f}{\partial \eta} \qquad (3.4)$$

In order to solve this equation,let us assume that $|(\eta(\vec{r})-\eta_o)/\eta_o| \ll 1$,and let us put:

$$\eta(r) = \eta_o + \frac{\mu(r)}{r} \qquad (3.5)$$

Reporting (3.5) into (3.4) and retaining the relevant term of lowest degree in the expansion of $(\partial f/\partial \eta)$ in the vicinity of η_o,we obtain:

$$\kappa \frac{d^2\mu}{dr^2} = \mu \left[\frac{\partial^2 f}{\partial \eta^2}\Big|_{\eta=\eta_o}\right] \qquad (3.6)$$

Denoting,on the other hand,

$$\xi^2 = \frac{\kappa}{(\partial^2 f/\partial \eta^2|_{\eta=\eta_o})} \qquad (3.7)$$

we obtain the solution of (3.6) complying with the boundary conditions $\mu(\infty)=0$, and $\mu(r_d)=(\eta_d-\eta_o).r_d$:

$$\frac{\mu(r)}{r} = \eta(r)-\eta_o = (\eta_d - \eta_o).\frac{r_d}{r}.\ e^{-(r-r_d)/\xi} \qquad (3.8)$$

The spatial variation of $\eta(r)$ is represented on figure 5,for $T>T_c$ $(\eta_o=0)$, and for $T<T_c$ $(\eta_o\neq 0)$. The characteristic length ξ is a measure of the spatial extension of a local perturbation of the OP value. This length is termed the correlation length [18] . Referring to its expression (3.7), we note that ξ goes to infinity on either sides of T_c,in the absence of field ε ,since $\lim_{o} |\partial^2 f/\partial \eta^2|$ $(T\to T_c\pm)$ $=\alpha(T_c)=0$. The volume in the system which becomes influenced by the defect becomes infinite at T_c (fig.5).

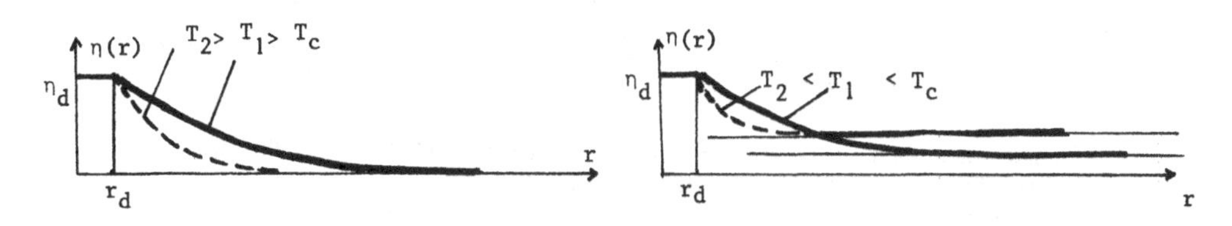

Fig.5

The fact that the considered defect is a point defect, as well as the assumed smallness of ($\eta(r)-\eta_o$) require $r_d \ll \xi$.

Using eq. (3.8),we can calculate the equilibrium free-energy of the system. Assuming ($r_d \ll \xi$), we obtain:

$$F_d^{eq.} = f_d + (V-v_d).f(\eta_o) + 2\pi\kappa r_d(\eta_d - \eta_o)^2 (1+\frac{r_d}{\xi}) \tag{3.9}$$

3.2.3. Physical properties of the defective system.

We now consider a system containing (N/V) defects per unit volume. In order to preserve the independence of their action,these defects must have a mean distance R_d satisfying the condition:

$$R_d \gg \xi \tag{3.10}$$

Since $\xi(T_c) = \infty$,the considered model can only be valid away from the immediate vicinity of the transition.

a) Spatial average of the OP and shift of T_c.

In certain experiments one measures the spatial average of the local macroscopic function $\eta(r)$ determined by the N-defects. For a defect (i) imposing locally the value $\eta_d(i)=\pm|\eta_d|$,spatial integration of (3.8) over the volume V of the system yields:

$$\bar{\eta}_i = \frac{1}{V}\int \eta(r).4\pi r^2.dr = \eta_o + \frac{\eta_d(i) - \eta_o}{V}.4\pi r_d.\xi^2 \tag{3.11}$$

Averaging $\bar{\eta}_i$ over the N defects, we obtain:

$$\bar{\eta} = \eta_o[1 - \frac{4\pi N r_d.\xi^2}{V}] \tag{3.11'}$$

Above T_c we have $\bar{\eta}=\eta_o=0$ in the absence of field ε. Below T_c we have, using (3.3) and $\xi^2=\kappa/2a(T_c-T)$:

$$\bar{\eta}^2 \quad \frac{a}{\beta}(T_c - \frac{\pi\kappa N.r_d}{V} - T) = \frac{a}{\beta}(T_c' - T) \tag{3.12}$$

Hence, below T_c, and in a range of temperatures compatible with (3.10), $\bar{\eta}$ has the same square-root temperature dependence as $\eta_o(T)$. However the effective transition temperature T_c' governing its variations is lower than T_c. The downward shift $(T_c - T_c')$ is proportional to the concentration (N/V) of the defects.

b) Susceptibility.

In the same manner as for the OP,the susceptibility $\chi=(\partial\eta/\partial\varepsilon)$ can be derived from eq.(3.8), by averaging the expression of $(\partial\eta/\partial\varepsilon)$ over the volume of the system ,and over the two distortions $\eta_d(i)=\pm|\eta_d|$ imposed by the N defects. One obtains:

$$\chi(T > T_c) = \frac{1}{a(T - T_c + \frac{4\pi N\kappa r_d}{aV})} \tag{3.13}$$

$$\chi(T < T_c) = \frac{1}{2a(T_c + \frac{4\pi N\kappa r_d}{aV} - T)} \tag{3.13'}$$

These equations show that χ has the same type of divergence as in the pure system, referred to the shifted temperature T_c'.

c) Specific heat.

The behaviour of the specific heat, on either sides of T_c can be derived from the expression (3.9) of the equilibrium free energy associated to one defect:

$$C_\varepsilon = - T\frac{\partial^2 F_d^{eq.}}{\partial T^2} \tag{3.14}$$

For N defects, a detailed calculation of (3.14) ,which we do not reproduce in details (Cf. ref.3) leads to:

$$\left(\frac{C_\varepsilon}{\Delta C}\right)_o = \left(\frac{N}{V}\right).\left(\frac{\pi\beta\sqrt{\kappa}}{a^{3/2}}\right).r_d^2 \cdot |\eta_d|^2.|T - T_c|^{-3/2} \qquad \text{for } T > T_c$$

$$\left(\frac{C_\varepsilon}{\Delta C}\right) = \sqrt{2} \cdot \left(\frac{C_\varepsilon}{\Delta C}\right)_o \qquad \text{for } T < T_c \tag{3.15}$$

In these equations ΔC is the step anomaly of the specific heat in the pure system ($\Delta C = a^2/2\beta$) (Cf. eq.(2.9) in chap.I).

Eq.(3.15) shows that unlike the cases of $\bar{\eta}$ and of χ , the behaviour of the specific heat differs qualitatively from that of the pure system. Instead of a step anomaly, one finds in the defective system a divergence with exponent (3/2) as a function of $|T-T_c|$. However the assumed constraints ($r_d \ll \xi$,and $R_d \gg \xi$) imply that the contribution of the defects to the specific heat is weak. Indeed, (3.15) can be put in the form, for $T < T_c$:

$$\frac{C_\varepsilon}{\Delta C} \sim \left(\frac{N}{V}\xi^3\right)\left(\frac{r_d^2}{\xi^2}\right).\left(\frac{\eta_d^2}{\eta_o^2}\right) \quad \ll 1 \tag{3.16}$$

3.2.4. <u>Extension of the theory</u>.

The validity of the above theory can be extended to a range of temperatures closer to T_c ,by releasing the assumption $R_d >> \xi$ .In this view,we can consider that whenever the correlation volumes relative to the different defects overlap, the same method as above can be used,provided that the boundary condition $\eta(\infty) = \eta_o$ is replaced by a modified condition $\eta(\infty) = \bar{\eta}$, where $\bar{\eta}$ is determined self-consistently from eq.(3.11').

On the other hand, the method described for the study of point defects can be applied to the cases of linear or planar defects (e.g. $\eta = \eta_d$ on a line).

These various extensions which have been worked out in refs.3 and 19 will not be examined here.

4.VALIDITY OF LANDAU'S THEORY.

In this paragraph we examine the Landau theory from the standpoint of statistical physics.

4.1 Validity of the theory and statistical fluctuations.

In the Landau theory the complete description of a state of a system is achieved by means of the density increment $\delta\rho(\vec{r})$. A given choice of $\delta\rho(\vec{r})$ specifies,with respect to a reference state,the overall configuration of the particles constituting the system.From the standpoint of the statistical theory [20,21] ,the function $\delta\rho(\vec{r})$ plays the role of a microstate. The Landau theory associates to a single particle configuration $\delta\rho(\vec{r})$ a free-energy $F(\delta\rho(\vec{r}))$. The minimum F_m of F with respect to the $\delta\rho(\vec{r})$ is identified to the equilibrium free-energy of the system. Actually,in the statistical theory, the equilibrium free-energy,$\mathcal{F}$,and the partition function Z are related to $F(\delta\rho)$ by:

$$Z = e^{-\mathcal{F}/kT} = \sum_{\{\delta\rho(r)\}} e^{-F(\delta\rho)/kT} \qquad (4.1)$$

where the sum is over all possible configurations $\delta\rho(\vec{r})$. The identification of $\mathcal{F}$ with F_m which is effected by the Landau theory consists in retaining,in the sum (4.1),the largest exponential term.In general such a procedure is fully justified. Indeed,in usual physical situations,the configuration $\delta\rho_m(\vec{r})$ which corresponds to F_m has a considerably larger statistical weight than any configuration differing appreciably from $\delta\rho_m(\vec{r})$. Large fluctuations are extremely improbable,and the statistical average of fluctuations is therefore small [20,21] .

However,it is possible to deduce from Landau's theory an expression of the statistical average of the fluctuations of the order-parameter which discloses that ,in the neighborhood of T_c ,one has an exceptional situation in which the statistical average of the fluctuations is not small. It is even infinite at T_c. This divergence of the fluctuations when $T \to T_c$ is clearly incompatible with the identification of $\mathcal{F}$ to F_m :there is a lack of consistency ,in this respect, of the Landau theory.

In the framework of Landau's theory,the existence of large fluctuations can be qualitatively understood by invoking the fact that the minimum of the Landau free-energy,as a function of the variational OP,becomes very shallow when $T \to T_c$ (Cf.chap I fig.3). Thus the probability $\exp(-F(\delta\rho)/kT)$ of a configuration $\delta\rho$ differing significantly from $\delta\rho_m$ becomes comparable to the probability $\exp(-F_m/kT)$ of $\delta\rho_m$.

This explanation can be made quantitative. It then constitutes the Levanyuk-Ginzburg criterion [22-24] which specifies the growth of the statistical average of the OP-fluctuations when $T \to T_c$. This criterion shows, in addition that the divergence of the fluctuations would not occur if the dimension of space, instead of being d=3, was d> 4. For the speculative dimensions d > 4, the Landau theory would be consistent.

Let us outline the main arguments of the Levanyuk-Ginzburg test of self-consistency of the Landau theory.

Assume that the considered transition has a one-dimensional OP η. The small fluctuations of $\delta\rho(\vec{r})=\eta.\phi(\vec{r})$, around the equilibrium value of η, can be conveniently described by means of smoothly varying functions $\eta(\vec{r})$, whose characteristic distance of variation is large as compared to the characteristic distance of variations of $\phi(\vec{r})$ (the latter distance is of the order of the interatomic distances in the system; Cf. chapV, §2.3). This smoothness can be expressed by stating that the Fourier transform of $\eta(\vec{r})$ only contains components with small vectors $\vec{q}$: $|\vec{q}| = q \ll \Lambda$. In this case, we can use a form of the Landau free-energy density $f(\vec{r})$ identical to the one in eqs.(3.1),(3.1'), provided the field in these equations is taken non-uniform: $\varepsilon = \varepsilon(\vec{r})$. If we denote $\delta\eta(\vec{r})$ a spatially non-uniform fluctuation of η, a standard statistical result can be deduced [20,25] from the expressions of the free-energy (3.1)-(3.1'), and from the partition function (4.1):

$$\langle\delta\eta(\vec{r}).\delta\eta(\vec{r}\,')\rangle = kT.\chi(\vec{r},\vec{r}\,') \quad , \tag{4.2}$$

where the brackets in the first member represent a statistical average [20], and where $\chi(r,r')$ is a generalized susceptibility defined by [25]:

$$\chi(\vec{r},\vec{r}\,') = \frac{\partial\langle\eta(\vec{r})\rangle}{\partial\varepsilon(\vec{r}\,')} \tag{4.2'}$$

Equation (4.2) holds both above and below T_c. Using the Euler-Lagrange equation (3.4), and operating a Fourier transformation, leads, as shown in ref.25 to:

$$\langle\delta\eta(\vec{r}).\delta\eta(\vec{r})\rangle = \chi(\vec{r},\vec{r}).kT \propto \int_0^{|\vec{q}|=\Lambda} \frac{d^{(d)}\vec{q}}{\vec{q}^{\,2}+\xi^{-2}} \tag{4.3}$$

where $d^{(d)}\vec{q}$ is the elementary volume in the d-dimensional space, and ξ the correlation length already encountered in §3 (eq.3.7). The temperature dependence of ξ, provided by the Landau theory, is $\xi^2 \propto (\kappa/|\alpha|) \propto |T-T_c|^{-1}$. After integration of (4.3) over the angular coordinates one obtains:

$$\langle\delta\eta(\vec{r}).\delta\eta(\vec{r})\rangle \propto \int_0^{\Lambda} \frac{q^{d-1}.dq}{q^2+\xi^{-2}} = \xi^{2-d}.\int_0^{\Lambda\xi} \frac{u^{d-1}.du}{1+u^2} \tag{4.4}$$

As the last integral in (4.4) is a non-infinitesimal number, we have, using the temperature dependences of $\xi(T)$ and of $\eta_{eq}(T)$ below T_c :

$$\frac{<\delta\eta(\vec{r}).\delta\eta(\vec{r})>}{\eta_{eq}^2} \propto \frac{\xi^{2-d}}{\eta_{eq}^2} \propto (T_c - T)^{(d-4)/2} \quad (4.5)$$

Hence ,for real systems (d=3) the statistical average of fluctuations becomes infinitely large ,as compared to the equilibrium value of the OP, when $T \to T_c$. As announced hereabove,this result is in contradiction with the basic assumption of Landau's theory $\mathcal{F}=F_m$. We can see that this contradiction precisely derives from the vanishing at T_c of the coefficient $\alpha(T)$ of the quadratic term in the Landau free-energy (since it is this vanishing which governs the divergence of ξ, and the vanishing of η_{eq}).

4.2. Preservation of the validity of the symmetry aspects of the theory.

The Landau theory has two prominent symmetry aspects.One is the irreducibility of the primary OP (chap.II §3.5). The other is the specification of the low symmetry group G as the invariance group of the equilibrium density increment $\delta\rho_{eq}(\vec{r})$ (chap.II §3.6). Let us show that despite the lack of correctness of Landau's theory in the range of divergence of the fluctuations,these two symmetry aspects have their validity preserved.

Returning on the definitions given in chap.II §3.1 we consider now the functions $\delta\rho_{eq}(\vec{r})$ and $\delta\rho_{eq}(\vec{r})$ as statistical averages over the configurations of particles composing the system. With this precaution,we need not change the definitions of the groups G_o and G. Accordingly the decomposition of any function $\delta\rho(\vec{r})$ characterizing a single configuration of the system into irreducible parts with respect to G_o can be performed without modification,as in eq.(3.4) of chap.II.Likewise the variational free-energy F associated to $\delta\rho(\vec{r})$ can be written,as in eq.(3.7), in the form of a sum of squares associated to irreducible degrees of freedom which are uncoupled up to this degree. On this basis,the argument illustrated by fig. 2 of chap.I remains valid : excluding an incidental identity of behaviour of the distinct coefficients $\alpha_i(T)$,we can conclude that these coefficients will not vanish simultaneously at T_c. As diverging fluctuations arise from the vanishing of the former coefficients,we deduce that a divergence will only concern the fluctuations of an irreducible set of degrees of freedom. This set can be termed the OP of the considered transition.

Hence the OP defined on the basis of the occurence of a singularity at the transition has the same irreducible character as the OP defined by Landau's theory. In the two approaches the mathematical ground is the same :it resides in the uncoupling of the quadratic invariants associated to distinct irreducible degrees of freedom.

On the same mathematical basis,Boccara [26] has provided a microscopic derivation of this property.

Let us now examine the nature of the low symmetry group G. As stressed above, G is the invariance group of the increment:

$$< \delta\rho_{eq} > = \Sigma < \eta_r^o > \phi_r(\vec{r}) \quad (4.6)$$

where the $<\eta_r^o>$ are the statistical averages of the OP components η_r^o ,and where the $\phi_r(\vec{r})$ are a set of normalized functions.Clearly, the $<\eta_r^o>$ carry the irreducible representation of the OP.Hence as already stated in §3.6,as well as in § 4.5 of chap.II, G is one of the isotropy groups in the carrier space of the OP-representation.The methods of determination of G,which were described in chap.II can therefore be used without modification.

4.3.Cases in which the thermodynamical results of the theory are preserved.

As shown by eq. (4.5),the lack of consistency of Landau's theory only concerns a range of temperatures adjacent to T_c,which is termed the critical range.Its extension will differ in different systems.Outside the critical range,the Landau theory is self-consistent,and its predictions regarding the temperature dependence of physical quantities are expected to be valid. It is worth pointing out,however,that the practical interest of these predictions will be restricted to systems in which the critical range is not too wide. Indeed,only below a narrow critical range will the mean equilibrium value of the OP be small enough to permit the adjustment of the experimental data to the simple laws of variations deduced from a Landau expansion truncated to the 4th or to the 6th degree. In the converse case, (i.e. if η_{eq} is large below the critical range of temperatures)one has to recur to a free-energy expanded up to a high degree. This free-energy will depend of many unknown coefficients (the coefficients of the independent polynomial invariants) whose fitting to the set of experimental data will be a less convincing test of the overall consistency of these data with the Landau theory.

It is generally the former situation which is found to prevail in real systems ,and in particular in the case of structural transitions [27,28] . In these systems,a 6th degree expansion is ,most of the time,sufficient to account satisfactorily for the experimental data in a wide range of temperatures below the critical range.

Aside from this general circumstance of validity of Landau's theory, there are also specific situations in which the thermodynamical results of Landau's theory keep their validity even in the neighborhood of T_c.These situations arise in categories of physical systems in which the microscopic interactions governing the transition anomalies decrease slowly with the distance (the interactions are long range). For these systems,the expression of the gradient term used in eq.(3.4) and controlling the form of the correlation volume is not correct.The actual term has a directional dependence. This has the consequence to confine the divergence of the fluctuations to certain isolated q-directions in the reciprocal space [29-33] . This confinement suppresses the divergence in (4.5) and preserves the self-consistency of the Landau theory [29,30] .

The category of systems pertaining to the former scheme is constituted by systems undergoing elastic phase transitions. In the framework of Landau's theory,these transitions are defined by the fact that their OP transforms according to the same one-dimensional IR as a symmetry-breaking strain-component. With respect to the ferroic classification of structural transitions(Cf. chap.III §6) they coincide with the "proper" or "pseudo-proper" ferroelastic transitions. For these transitions the validity of the thermodynamical results of Landau's theory has been confirmed by statistical theories [32] . Table 1 shows for some selected examples of real systems the excellent agreement between the theoretically expected temperature dependence of the OP (Cf.ref.14)($\eta \propto (T_c-T)^{\beta}$ with $\beta=(1/2)$),and of its associated susceptibility ($\chi \propto |T_c-T|^{-\gamma}$ with $\gamma=1$).

Table 1

SUBSTANCE	Landau exponent.	Experimental result.	Extension of the measurement range. $(T-T_c)/T_c$
LaP_5O_{14}	$\beta=0.5$ $\gamma=1$	0.5±0.007 1± 0.002	3.10^{-2} $\pm 2.10^{-2}$
Tanane	$\beta=0.5$ $\gamma=1$	0.5±0.01 1± 0.02	5.10^{-3} $+5.10^{-3}$
$KH_3(SeO_3)_2$	$\beta=0.5$ $\gamma=1$	0.5±0.01 1±0.02	10^{-1} $\pm 5.10^{-3}$

In addition to the preceding category of systems,there are systems for which the behaviour in the critical range,though differing in principle from the results of Landau's theory,coincide in practice with the latter results. This is the case of certain proper ferroelastics having multidimensional OP [32] ,and also of systems in which a uniaxial dipolar interaction governs the critical behaviour (e.g. uniaxial ferroelectrics [34]). For all these systems, the divergence of the fluctuations is very anisotropic in reciprocal space. The laws of variations of the physical quantities which are expected for these systems differ from the ones derived from Landau's theory by "logarithmic corrections" [34] . Such corrections have proven to be difficult to detect experimentally [35] . Hence in experiments,the observed behaviour will generally appear in agreement with the results of Landau's theory (Cf. for instance the results in ref.27).

4.4.Behaviour of physical quantities in the critical range.

In this paragraph we examine systems pertaining to the general situation pointed out in §4.1,i.e. in which the predictions of Landau's theory fail within the critical range of temperatures. The description of the statistical methods which permit the study of the critical range is out of the scope of this book. The basic ideas of these methods and the detailed procedures of their implementation can be found in a number of recent textbooks [36,37] . We will restrict here to an outline of two aspects of the problem.

In the first place,we will sort out the concepts defined by Landau's theory which are retained by Wilson's theory of the critical behaviour [36] .In the second place,we will summarize the physical predictions derived from the latter theory,which must be substituted to the predictions of Landau's theory.

4.4.1. Wilson's theory and the Landau free-energy.

Wilson's theory of the critical behaviour [36,37] aims at determining the physical properties of a system undergoing a continuous phase transition in the critical temperature range,i.e. in the range where the correlation length ξ diverges, and where relatedly,the physical behaviour is dominated by the fluctuations.

a) Hamiltonian density.

Consider the expression (4.1) of the partition function. Three remarks are of interest. In the first place,this expression shows that the Landau free-energy $F(\delta\rho)$ plays the same role as a Hamiltonian for the system. In the second place,among all the degrees of freedom which can be included in $\delta\rho$,the only ones which are concerned by a divergence of the fluctuations are a set of n-components carrying an irreducible order parameter (or,more generally a physically irreducible one) (Cf.§4.1-4.2). Finally, if one is only interested in the critical range ,the OP-components can be considered as almost uniform over large distances of the order of ξ (the scale is the interatomic distances in the system), and consequently one can restrict to smooth functions $\eta(\vec{r})$ in the free-energy density $f(\vec{r})$.

These remarks justify that in the framework of Wilson's theory,one adopts as an effective Hamiltonian for the study of the critical behaviour the Landau free-energy associated to the set of smooth functions $\eta_r(\vec{r})$ carrying an IR of G_o. It can be shown,in addition [37] that the critical behaviour is essentially determined by the fourth degree expansion in $f(\vec{r})$. Hence,in Wilson's theory,the critical behaviour is determined by an effective Hamiltonian density which is a polynomial expansion of the n components of the transition's order parameter and of their spatial derivatives. The homogeneous part of this density is a fourth degree expansion identical to the Landau free-energy of the transition. In accordance with chap.II §3, the Hamiltonian density can be written as:

$$\mathcal{H}(\vec{r}) = \frac{a(T-T_c)}{2}\left[\Sigma\ \eta_r^2(\vec{r})\right] + \sum_i \beta_i . Q_i\left[\eta_r(\vec{r})\right] + \Sigma\ (\vec{\nabla}\eta_r)^2 \qquad (4.7)$$

where the Q_i are the independent fourth degree invariants compatible with the irreducible symmetry of the OP. In (4.7), a simplified (isotropic) form of the gradient is adopted,and its coefficient is made equal to <u>one</u> by a suitable change of the length scale in the system.

b) Flow of Hamiltonians and fixed-point Hamiltonian.

A second step of Wilson's theory is to define an infinitesimal transformation of the Hamiltonian (4.7) termed the renormalization group transformation. This transformation generates a trajectory of Hamiltonians associated to an initial Hamiltonian of the form (4.7).The set of the various trajectories form a flow of Hamiltonians. The renormalization group transformation is characterized by a physical property and by a symmetry property.

Physically all the Hamiltonians belonging to a given trajectory of the flow are associated to the same critical behaviour [37] .

From the standpoint of symmetry,the transformation does not decrease the symmetry of the system considered. More precisely the invariance group of the initial Hamiltonian is preserved as a minimal symmetry along a trajectory. The latter property has the important consequence that all the Hamiltonians of a given trajectory can be expressed as functions of the same Q_i invariants as the initial Hamiltonian. These Hamiltonians will have the expression (4.7) with,however,different coefficients α and β_i (some of the β_i possibly being equal to zero).

It can happen that a trajectory will end at a fixed-point Hamiltonian, i.e. at a Hamiltonian which is transformed into itself by the applied transformation. This fixed point can be stable or unstable (Cf. ref. 37 for the definition of this concept).A fixed-point Hamiltonian $\mathcal{H}^*$,specified by coefficients (α^* ,β_i^*) determines the same critical behaviour as $\mathcal{H}$.It can be shown that the symmetry of the Hamiltonian along a trajectory is strictly preserved except possibly at the fixed point where it can increase. A fixed point Hamiltonian will often possess a higher symmetry than the other Hamiltonians of the trajectories ending at it[38]. Wilson's theory shows that if the flow contains a stable-fixed point Hamiltonian,a continuous transition is possible in the considered system. Besides,a method can be described,which allows to calculate on the basis of the characteristics of the flow in the vicinity of the fixed-point,the critical behaviour of the system,by means of an approximation method [36,39] .

In summary,in the implementation of Wilson's method,the procedure defined by Landau's theory provides the symmetry principles and results for constructing the relevant Hamiltonian density, and for specifying the parameter space (the α and the β_i) in which the transformation of Hamiltonians can be represented,and the fixed-point located. One can,in particular,use the classification of Landau free-energies performed in chapter II (§4.5),and enumerate in the same manner,the relevant Hamiltonian densities.

4.4.2. Behaviour in the critical range.

The theory of critical phenomena shows that the divergence of the fluctuations induces modifications with respect to the results of Landau's theory in several aspects of the behaviour of the system.

a) Critical exponents pertaining to primary quantities.

In the Landau theory,as well as in the statistical theory of the critical behaviour,it is found that the anomalous temperature dependence of the quantities related to the primary OP (equilibrium value of the OP, susceptibility,specific heat,...) can be described by power laws of the form $A_i \propto |T - T_c|^{x_i}$,where the x_i are termed critical exponents.A set of critical exponents relative to the various physical quantities (η, χ,C, ...) defines a critical regime .Within the range of temperatures where the fluctuations of the OP are large,a system can display several distinct critical regimes in adjacent temperature ranges. One of these regimes ,the asymptotic critical regime holds closest to the transition point.

As emphasized in chapters I and II,the Landau theory determines the same (incorrect) critical regime for all systems :in the framework of this theory, the critical behaviour has a universal validity. The results of Wilson's theory retain this characteristics of universality though in a less extensive way.As suggested in §4.4.1 ,the critical behaviour will be the same for all the systems possessing the same Hamiltonian density and the same spatial dimensionality. The statistical theory shows that the details of the Hamiltonian density are not all relevant,and that one can establish a classification of physical systems into "universality classes",each class being defined by the following 4 features [40] :

i) The spatial dimension of the system. In real systems one has,in general d=3. However in certain systems,a smaller "effective" dimension (d=2 or d=1) can sometimes be invoked.

ii)The number n of components of the order parameter of the transition. As already stressed above,this number is provided by the symmetry analysis of the considered transition according to the lines set in chapters I-VII.

iii) The number and form of the independent fourth degree invariants composing the homogeneous part of the Hamiltonian density. These characteristics are determined by the symmetry properties of the OP ,as shown in chap.II.

iv)The range and anisotropy of the interactions giving rise to the considered transition. This feature appears in the form of the spatially dispersive terms in the free-energy density. We have already discussed the case of a long range interaction (§4.3). Other interesting cases arise,for instance,whenever one must take into account spatial derivatives of degree higher than two in the coordinates (e.g. a Lifschitz point Cf. chap IV).We will not further discuss these cases here,and restrict to the general situation of short range interactions which gives rise to the square gradient reproduced in (4.7).In this situation,the application of Wilson's theory yields the following results for d=3:

- If the dimension n of the OP satisfies $n \leqslant 3$,the predictions of the theory are particularly simple since they show that the symmetry properties of the OP are irrelevant. The asymptotic critical behaviour expected only depends on the value of n [39] .We have recalled on table 2 the critical exponents respectively associated to the temperature dependences of the primary OP,to its corresponding susceptibility (above and below T_c), and to the specific heat (above and below T_c) [41] . The irrelevance of the detailed structure of the free-energy density results from the fact that for $n \leqslant 3$, the fixed point Hamiltonian has the isotropic symmetry O(n), whatever the symmetry of the initial Hamiltonian |39| .

- For $n \geqslant 4$, the value of n is not sufficient to specify the critical behaviour which also depends of the OP-symmetry. More precisely the behaviour depends on the number and form of the Q_i polynomials in (4.7). In some cases (Cf §d, and § 4.5 hereunder), the effect of fluctuations is to prevent the occurence of a continuous transition. In the other cases the possibility of a continuous transition is preserved,but the universality of the exponents is less extensive than for $n \leqslant 3$. The fourth line in table 2 show examples of predictions concerning the critical exponents in systems with n=4. [38,42] .

Table 2 . Critical exponents according to refs. 38,41,42.

n	β	γ,γ'	α,α'
1	0.325	1.241	+0.110
2	0.3455	1.316	-0.007
3	0.3645	1.386	-0.115
4	0.39	1.385	-0.166

b) Phase diagram below T_c,and thermodynamic order of the transition.

As illustrated by several examples in chapter I (§7) and chapter II (§4), the Landau theory shows that several phases with distinct symmetries can be stable below T_c for a given irreducible OP. The theory associates to each phase an interval of the expansion's coefficients (e.g. the β_i coefficients of the 4th degree terms). There are boundaries between the various low symmetry phases defined by relationships between the coefficients (Cf. fig.6 in chap.I).

We have seen in §4.2 hereabove,that the list of the possible symmetries of the phases stable below T_c is not modified by the fluctuations. However, the boundaries between these phases are modified [38,39] .The general trend of the modifications is that the interval of β_i coefficients in which a given phase is stable is reduced as compared to the result provided by Landau's theory[38] . Besides, some phases which are possibly stable below a continuous transition,in Landau's theory,have their stability destroyed by the fluctuations. In the framework of Wilson's theory, this circumstance will arise in two cases:

i) If the flow of Hamiltonian densities (4.7) does not contain a stable fixed point Hamiltonian.

ii)If the initial Hamiltonian density does not lye on a trajectory flowing towards the existing stable fixed point.

It has been pointed out that in certain cases, a converse effect could be induced by the fluctuations [43] . Thus ,if a cubic invariant exists in the free-energy,the Landau theory precludes the occurence of a continuous transition towards any low symmetry phase. If the initial coefficient of the cubic term is small, the fluctuations can cancel its effect and allow the occurence of a continuous transition [43] . Similarly, if the fourth degree term is not positive in every direction of the OP-space,the Landau theory predicts a discontinuous transition. The effect of the fluctuations, operating through the 6th degree term [44], can be to induce a continuous transition.

c) Secondary order-parameters.

When fluctuations are taken into account, the results relative to secondary OP differ in several respects from those worked out in Landau's theory (Cf. chap.II §5). Namely we have the following results:

i) In agreement with the situation found for the primary OP, the critical exponents relative to the temperature dependence of the secondary OP ζ, or to the corresponding susceptibility S are not those indicated by Landau's theory.

ii) The relationship between the primary and secondary exponents generally differs from that expected on the basis of Landau's theory. However, it remains that in both theories this relationship exclusively depends on the form of the coupling term between the two sets of quantities. As stressed in chap.II §5, the form of the coupling term derives from the respective symmetries of the primary and secondary OP, in a manner entirely described in chapters II (§5) and III (§6). For instance if we consider the example of a two-dimensional primary OP (n=2), and of secondary OP having a faintness index of 2 (e.g. the form of the coupling term is $\zeta(\eta_1^2 - \eta_2^2)$, the Landau theory states that $\zeta(T) \propto |T_c - T|$ for $T < T_c$, and that $\zeta \propto \eta^2(T)$. The statistical theory yields [40] $\zeta(T) \propto |T_c - T|^{0.82}$ and that $\eta(T) \propto |T_c - T|^{0.35}$ (hence $\zeta \neq \eta^2$). Likewise, the susceptibility, which has a step variation, in the framework of Landau's theory, has a weak divergence within the statistical theory ($S \propto |T - T_c|^{-0.007}$) which is identical to the divergence of the specific heat of the system at the transition point (Cf. table 2 for n=2).

iii) The effect of the coupling on the thermodynamic order can be much more drastic than in the Landau theory. We had stressed in chap.III §6 that any OP will be coupled to totally symmetric strain components, with a faintness index of 2. As shown in chap.II §5, the elimination of the strain components from the free-energy renormalizes negatively the coefficients of the 4th degree terms. This renormalization does not necessarily bring the coefficients of the 4th degree terms into a range precluding the possibility of a continuous transition. A stronger result holds in the presence of the fluctuations. For systems whose specific heat is expected to diverge, the renormalization induced by the strain components always bring the system away from the possibility of a continuous transition [40].

REFERENCES

|1| L.D. Landau Collected papers (Pergamon, N.Y. 1965) p.21

|2| S. Alexander and J. Mc Tague Phys. Rev. Letters 41, 702 (1978)

|3| A.P. Levanyuk, V.V. Osipov, A.S. Sigov, and A.A. Sobyanin Sov. Phys. JETP 49, 176 (1979)

|4| D. Schechtman, I. Blech, D. Gratias and J.W. Cahn Phys. Rev. Letters 53, 1951 (1984)

|5| D.R. Nelson ,and S. Sachdev Phys. Rev. B32,689 (1985)

|6| P. Bak Phys. Rev. B32,5764 (1985) ;M. Duneau and S. Katz Phys. Rev. Letters 54,2688 (1985)

|7| G.Ya. Lyubarskii. The applications of group theory in Physics (Pergamon press .London 1960).

|8| V. Dvorak and V. Holakovsky J. Physics C19 ,5289 (1986)

|9| S.M. Troian in International Workshop on aperiodic crystals. Journ. de Physique. 47,C3-271 (1986).

|10| M.V. Jaric Phys. Rev. Letters 55,607 (1985).

|11| P. Bak Phys. Rev. Letters 54,1517 (1985)

|12| P.A. Kalugin,A.Yu Kitaev, and L.S. Levitov JETP-letters 41,145 (1985).

|13| N.D. Mermin and S.M. Troian Phys. Rev. Letters 54,1524 (1985).

|14| J.C. Tolédano Ann. Télécommun. 39,278 (1984).

|15| B.I. Halperin,and C.M. Varma Phys. Rev. B14,4030 (1976).

|16| U.T. Höchli and A.D. Bruce J. Physics C13,1963 (1980).

|17| M.E. Fisher Phys. Rev. 176,257 (1968).

|18| E. Stanley Phase Transitions and Critical Phenomena (Clarendon,Oxford 1971).

|19| A.P. Levanyuk,A.S. Sigov,and A.A. Sobyanin Ferroelectrics 24,61 (1980)

|20| F. Reif Statistical and thermal Physics (Mc Graw Hill Tokyo 1965).

|21| R. Balian Physique Statistique.Cours de l'école Polytechnique (Paris 1986).

|22| A.P. Levanyuk Sov.Physics JETP 36,571 (1959)

|23| V.L. Ginzburg Sov. Physics Solid State 2, 1824 (1960)

|24| D.J. Amit J. Physics C7,3369 (1972).

|25| N. Boccara Symétries Brisées (Hermann Paris 1976).

|26| N. Boccara Solid State Commun. 11,39 (1972).

|27| F. Jona and G. Shirane Ferroelectric crystals (Pergamon N.Y. 1962).

|28| A. Lines and A.M. Glass Ferroelectric and related crystals (Oxford 1975).

|29| A.P. Levanyuk and A.A. Sobyanin Sov. Physics Solid State 16,2079 (1970).

|30| J. Villain Solid State Commun. 8,295 (1970).

|31| R.A. Cowley. Phys. Rev. B13 ,4877 (1976).

|32| R. Folk,H. Iro, and F. Schwabl Zeitschrift Physik B25,69 (1976).

|33| K. Parlinski Acta Physica Polonica A58,197 (1980).

|34| D. Stauffer Ferroelectrics 18,199 (1978)

|35| R. Frowein and J. Kötzler Phys. Rev. B25,3292 (1982)

|36| K.G. Wilson Phys. Rev. B4,3184 (1971).

|37| Phase Transitions and Critical Phenomena . Ed. C. Domb and M.S. Green. Vol. 6 (Academic Press .N.Y. 1976).

|38| J.C. Tolédano, L. Michel, P.Tolédano, and E. Brézin Phys. Rev. B31, 7171 (1985).

|39| E. Brézin, J.C. Le Guillou,and J. Zinn-Justin Phys. Rev. B10,892 (1974).

|40| A.D. Bruce. Advances in Physics 29,111 (1980)

|41| J.C. Le Guillou,and J. Zinn-Justin Phys. Rev. B16,1138 (1980).

|42| D. Mukamel and S. Krinsky Phys. Rev. B13, 5078 (1976).

|43| S. Alexander Solid State Commun.14,1069 (1974).

|44| D. Blankshtein ,and A. Aharony Phys. Rev. Letters 47,439 (1982).